XIX Latin American Symposium on Solid State Physics 2008

(SLAFES XIX)

Journal of Physics: Conference Series Volume 167

Puerto Iguazu, Argentina
5-10 October 2008

Editors:

Adriana Serquis
Pablo Bolcatto
Carlos Balseiro

ISBN: 978-1-61738-316-8

Printed from e-media with permission by:

Curran Associates, Inc.
57 Morehouse Lane
Red Hook, NY 12571
www.proceedings.com

Some format issues inherent in the e-media version may also appear in this print version.

Printed by Curran Associates, Inc. (2010)

Institute of Physics
Dirac House, Temple Back
Bristol BS1 6BE UK

Phone: 44 1 17 929 7481
Fax: 44 1 17 920 0979

techtracking@iop.org

TABLE OF CONTENTS

NANOPHYSICS, NANOMATERIALS AND NANOTECHNOLOGY

SUPERCONDUCTIVITY

THIN FILMS, SURFACES AND INTERFACES

LOW DIMENSIONAL SYSTEMS

MATERIALS: PREPARATION AND CHARACTERIZATION

MATERIALS: THEORY, COMPUTING AND SIMULATIONS

This volume contains selected papers which have been presented at the XIX Latin American Symposium on Solid State Physics (SLAFES XIX) held at Puerto Iguazú, Argentina, from 5--10 October 2008. The conference, covering all areas of Solid State Physics, is one of the most important and traditional meetings in Physics in our region. The Latin American Symposium on Solid State Physics is a forum where researchers and students from Latin America as well as leading scientists from other parts of the world get together to exchange information, strengthen collaborations and identify new challenges in Solid State Physics. This successful series of meetings has been organised in eight different countries, the last three held in Mérida, Venezuela (2002), La Habana, Cuba (2004) and Puebla, México (2006). Following the trends of previous events, SLAFES XIX included seven plenary talks, eighteen invited talks and contributions, and 28 oral and 255 poster presentations, covering mostly the latest experimental and theoretical advances in Nanophysics, Nanomaterials and Nanotechnology, Spintronics, Magnetism, New Materials, Superconductivity, Surfaces and Interfaces, Low-Dimensional Systems, Materials Preparation and Characterization, Theory and Computing Simulations of Materials among other topics.

The group of scientists participating had come from Argentina, Chile, Colombia, Cuba, Brazil, France, Spain, Switzerland and the USA We are indebted to all participants for their enthusiasm and contributions and to the members of the International Advisory Commitees. We also wish to thank to the rest of the Organizing Committee: Gustavo Lozano, Ana María Llois, Laura Steren and Edith Goldberg and very specially to Javier Schmidt, Gustavo Ruano, Marcelo Romero, Lucila Cristina and Juan Carlos Moreno for their invaluable assistance during the event. Finally we gratefully aknowledge the financial support the Consejo Nacional de Investigaciones Científicas y Técnicas (CONICET), Argentina, Agencia Nacional de Promoción Científica y Tecnológica (ANPCyT), Argentina, Secretaría de Estado de Ciencia, Tecnología e Innovación, Provincia de Santa Fe, Argentina, Centro Latinoamericano de Física (CLAF), Universidad Nacional del Litoral, Santa Fe, Argentina, Institute of Physics (IOP) Publishing, The European Physical Journal (EPJ), EDP Science, Societa Italiana di Fisica and Springer.

Adriana Serquis, Carlos Balseiro and Pablo Bolcatto

Intraband Auger effect in InAs/InGaAlAs/InP Quantum Dot Structures

T Gebhard[1], D Alvarenga[2], P L Souza[3], P S S Guimarães[4], K Unterrainer[5], M P Pires[6], G S Vieira[7] and J M Villas Boas[8]

[1,5]Center for Micro&Nanostructures, TU, Vienna, Austria
[2,4]Departamento de Fisica, UFMG, Belo Horizonte, Brazil
[3]LabSem/CETUC, PUC, Rio de Janeiro, Brazil
[6]Instituto de Fisica, UFRJ, Rio de Janeiro, Brazil
[7]Divisão de Fisica Aplicada, IEA, São José dos Campos, Brazil
[8]Walter Schottky Institue, TU, Munich, Germany

Email: Thomas.gebhard@tuwien.ac.at

Abstract. Intraband photocurrent and absorption measurements were performed on InAs/InGaAlAs/InP quantum dot structures. A full three-dimensional theoretical model has been employed to identify the observed photocurrent as a bound to bound transition, where the final state is about 200 meV deep below the conduction band continuum. The reported results strongly suggest that an Auger process plays a fundamental role in generating the observed intraband photocurrent.

1. Introduction

Semiconductor quantum dot (QD) structures have raised much interest in the past years for application in mid-infrared photodetection due to their potential for efficiently coupling normal incident light and for higher operation temperature as compared to quantum well structures [1-6]. In most QD structures for this purpose, the intraband optical transitions occur in the 5-20 μm range between two bound states or between a bound and a quasibound state. Although they produce more selective detectors, they have the disadvantage of requiring an extra carrier transport mechanism to generate the observed photocurrent (PC). This mechanism has been often attributed to thermal excitations even for deep final bound states [1-4]. Another possibility to extract the carriers from the final bound energy state to the continuum, where they can easily participate in the PC, is via an Auger effect.

Theoretical studies in the 90's showed that intraband Auger scattering in quantum dots should be more effective than the interband mechanism[7]. The activation of photoexcited electrons by Auger scattering in interband transitions has already been shown experimentally for InAs/InP QD structures [2] and Auger scattering in undoped and *n*- and *p*-type doped InGaAs quantum dots has also been theoretically investigated more recently [9]. In this communication we present results that demonstrate that Auger processes can indeed play a fundamental role in generating an intraband PC, broadening the possibilities for designing QD structures for highly selective mid-infrared photodetectors.

2. Experiment and Discussion

The samples under investigation are grown on a 150 nm thick InP buffer layer deposited at 630°C on a semi-insulating InP substrate followed by a 500 nm thick *n*-doped InGaAs, which is the bottom contact layer. Afterwards a 109 nm thick layer of InGaAlAs is grown with 16% Al content.

[1] Thomas.gebhard@tuwien.ac.at

The InAs QDs are then deposited for 5.5 s at 520°C and they are annealed in an arsine atmosphere for 12 s. They are covered by a 16 nm thick InP layer while the temperature is ramped up to 600°C. This sequence is repeated 10 times. A 109 nm thick layer of the quaternary material is then grown and finally a 250 nm *n*-doped InGaAs contact layer is deposited. The doping level at the contact layers is 1.0 x 10^{18} cm^{-3}. All ternary and quaternary layers are lattice matched to the InP substrate and grown at 600°C. The QD samples were grown with three different doping levels: one nominally undoped sample and 2 samples with 2 and 4 electrons per dot. The composition of the quaternary material was chosen so as to maximize the dot density [11]. Using atomic force microscopy (AFM) an average dot height and density of (9 ± 1.5) nm and 1.5 x 10^{10} cm^{-2}, respectively, have been measured on uncapped QD control samples. Transmission electron microscopy (TEM) images showed lens shaped QDs with a base diameter of approximately 50 nm and confirmed the QD height average of 9 nm.

Photocurrent (PC) and absorption measurements were performed using Fourier Transform Infrared Spectroscopy (FTIR). Figure 1 shows the photocurrent spectra at 5 K for the investigated samples measured with normal incident light and no external bias. A narrow PC peak, with a full width at half maximum of 13 meV, is observed around 190 meV for the undoped sample. The sample is nominally undoped, but the presence of residual doping and carrier diffusion from the contacts lead to a population of the QD so that intraband absorption can occur. A ten times stronger and three times broader signal was observed for the same structure with a doping of about 2 and 4 electrons per dot. The broadening of the PC-peak as a function of doping can be explained by the inhomogeneous broadening of the quantum dot sizes. If the doping is low only a small fraction of the QDs is populated. The QDs which will be populated first are the larger ones because their ground state energy is lower. As the doping increases, more QDs will be populated. The dispersion in QDs sizes and, consequently in ground state energies, lead to the broadening of the PC peak. Additionally, higher energy peaks can also be populated in some QDs, further contributing to the PC peak broadening. Absorption measurements were performed in order to verify whether absorption occurs at other wavelengths, as well. They were performed in waveguide geometry using the same structures as for the PC measurement but with 20 instead of 10 layers. A peak centred around 185 meV–195 meV has been observed in full agreement with the PC spectra (inset fig.1).

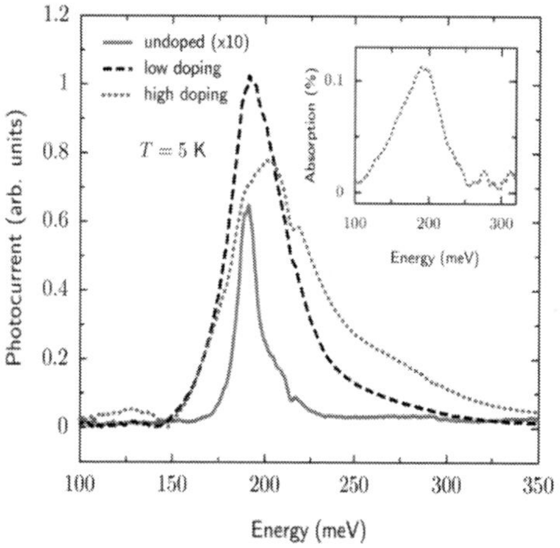

Figure 1: PC response for different doping; the inset shows the absorption with respect to the InP substrate at 5K.

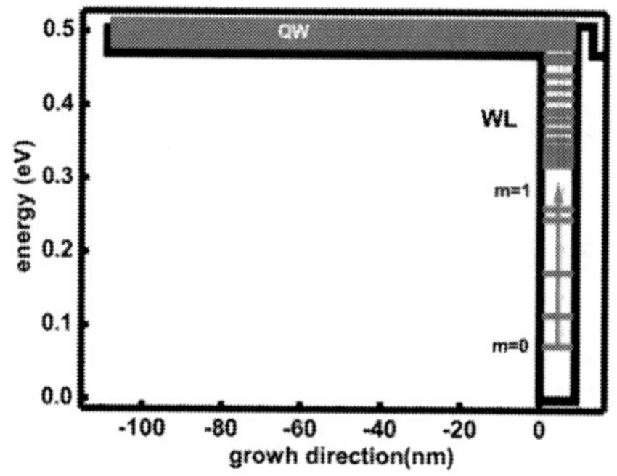

Figure 2: Conduction band structure profile and calculated energy levels for the investigated samples for quantum numbers *m*=0 (solid lines) and *m*=1 (dashed lines). The arrows show the optical transitions for which the oscillator strength is largest. The wetting layer (WL) states are also shown. QW states for quantum well.

A 3 dimensional effective mass model was applied in the theoretical calculations assuming lens shaped QDs (as observed in the TEM images). The physical parameters were taken from [12]. A more sophisticated treatment of the theoretical model can be found in [11]. The results of the calculations can be seen in fig. 2, which is drawn to scale. The offsets used can be seen on the figure.

The horizontal axis refers to the position in the growth direction. The solutions for the quantum numbers $m=0$ (solid line) and $m=1$ (dashed line) are superimposed in the QD structure profile. The transition between the ground state and the fourth excited state, corresponding to an energy of 197 meV, has the largest oscillator strength. Now the important point to note is that the final state of this transition is still 200 meV from the continuum which implies that the excited electrons cannot contribute directly to the PC.

Figure 3 shows the photoluminescence (PL) at 80 K produced by the quantum dot (dashed line). One observes a broad peak around 680 meV which agrees well with the optical transition theoretically predicted between the electron and hole fundamental energy levels. For comparison, the InAs, InP and InGaAlAs energy gaps at room temperature are 0.35, 1.35 and 1.0 eV. In the same figure the solid line represents the interband PC at the same temperature. One clearly sees that both have a peak around 700 meV, meaning that PC is produced by electrons transferred from the valence band to the electronic ground state of the QD, which is deep in the conduction band. This gives further support to the fact that the intraband PC can be due to an absorption transition which has its final state deep inside the QD. As has been already claimed by Landin *et al* [8], the interband PC from the QDs is produced by an interband Auger effect, where the photo-excited electron receives additional energy from the recombination of another electron with the photogenerated hole. A scheme of such a process is depicted in the inset of Fig. 3.

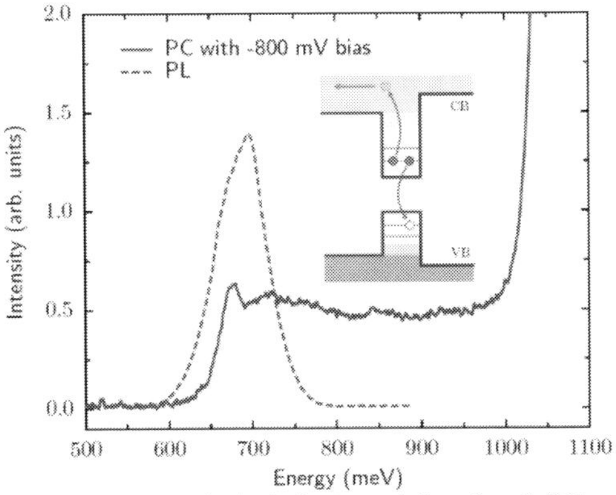

Figure. 3: PL (dashed line) and interband PC (solid line) spectra for the undoped sample at 80 K. The inset illustrates the interband Auger effect. The offsets and energy gap are not to scale.

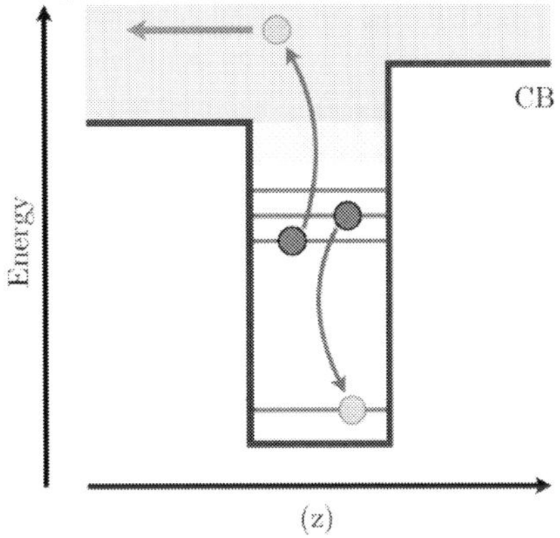

Figure 4: Scheme of the intraband Auger effect

The final state of the intraband transition that produces the PC is 200 meV below the continuum. The thermal energy is very small at low temperature so that thermal assisted extraction of such photo-excited carriers is rather inefficient. We attribute the generation of the observed intraband photocurrent to the absorption of a photon from the QD ground state to the fourth excited state followed by the promotion of the photo-generated electron to the continuum via an intraband Auger effect. A photo-excited electron can be promoted to the continuum by an energy transfer from an electron that relaxes from a higher energy state down to the ground state. The suggested intraband Auger scattering can be seen schematically in fig 4. The theoretical calculations showed that there are many possible excited states available to participate in this process. One should also consider that the FTIR measurement is carried out with a broadband light source, increasing the probability of having two electrons simultaneously excited on the same QD.

The temperature dependence of the observed intraband PC gives further support to this attribution. The intensity of the PC shows nearly no temperature dependence between 5 and 60 K (fig. 5), as expected for an Auger scattering process. Moreover, dynamical processes in the device could cause a non-vanishing population of higher states. The dark current in the devices is a result of thermal emission and re-capture of electrons. Time resolved measurements showed that electron capture occurs very fast into high lying states and is then slowed down for lower lying states [13]. As a consequence we can expect that higher lying states are populated.

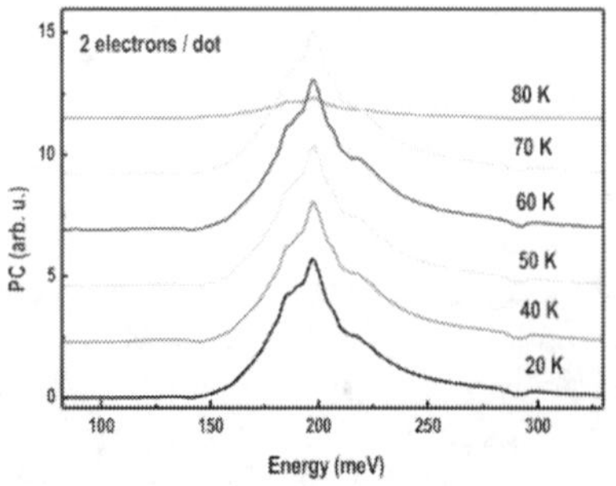

Figure 5: Temperature dependence of the medium doped sample. The amplitude stays constant up to 70K

In summary, we have carried out a detailed study of the intraband optical response of quantum dot structures for mid-infrared photodetection in order to verify the possibility of carriers in a rather deep bound state generating a photocurrent. The obtained photocurrent and absorption results, together with a realistic theoretical calculation, strongly suggest that an Auger process can produce a significant current. Photoluminescence together with interband photocurrent further supports this attribution. The reported results indicate that quantum dot photodetector structures could be designed to operate based on the more selective bound-to-bound transitions of quantum dots and still produce an efficient device.

Acknowledgements We thank I. G. Solórzano and H. F. Filho for the TEM and AFM measurements. This work has benefited from fruitful discussions with G. Bastard. Financial support from the WPI, FWF SFB IR-ON, EC NoE SANDIE, CAPES, CNPq, FINEP, FAPERJ and FAPEMIG is acknowledged. JVB is supported by the Alexander von Humboldt Foundation.

References

[1] Lim H, Zhang W, Tsao S, Sills T, Szafraniec J, Mi K, Movaghar B, Razeghi M, Phys. Rev. B **72**, 085332 (2005).
[2] Liu H C, Opto-electronics Rev. **11**, 1 (2003).
[3] Boucaud P, Sauvage S, C. R. Physique **4**, 1133 (2003).
[4] Towe E, Pan D, IEEE J. Sel. Topics Quantum Electronics **6**, 408 (2000).
[5] Bhattacharya P, Su X H, Chakrabarti S, Ariyawansa G, Perera A G U, Appl. Phys. Lett. **86**, 191106 (2005).
[6] Chakrabarti S, Stiff-Roberts A D, Su X H, Bhattacharya P, Ariyawansa G, Perera A G U, J. Phys. D: Appl. Phys. **38**, 2135 (2005).
[7] Pan J L, Phys Rev. B **49**, 11272 (1994).
[8] Landin L, Pettersson H, Kleverman M, Borgström M, Zhang X, Seifert W and Samuelson L, J. Appl. Phys. **95**, 8007 (2004).
[9] Nilsson H H, Zhang J Z, Galbraith I, Appl. Phys. Lett. **91**, 161113 (2007).
[10] Borgström M, Pires M P, Bryllert T, Landi S M, Seifert W, Souza P L, Journal of Crystal Growth **252**, 481 (2003).
[11] Souza P L, Gebhard T, Lopes A J, Pires M P, Villas-Boas J M, Unterrainer K, Guimarães P S S, Vieira G S and Studart N, Appl. Phys. Lett. **90**, 173510 (2007).
[12] Vurgaftman I, Meyer J R and Ram-Mohan L R, J. Appl. Phys. **89**, 5816 (2001).
[13] Müller T, Schrey F F, Strasser G, and Unterrainer K, Appl. Phys. Lett. **83**, 3572 2003

GaInAs Quantum Dots (QD) grown by Liquid Phase Epitaxy (LPE)

Ortiz Vázquez F. E.[1], **Mishurnyi V. A.**[1], **Gorbatchev A. Yu.**[1], **De Anda F.**[1], **Elyukhin V. A.**[2]

[1]Universidad Autónoma de San Luis Potosí, Instituto de Investigación en Comunicación Optica, Av. Karakorum 1470, Col. Lomas 4a Sec., San Luis Potosí, SLP, CP 78210, México
[2]CINVESTAV-IPN, Av. IPN 2508, Col. San Pedro Zacatenco, México D.F., CP 07360, México

E-mail: fcoe_ov@prodigy.net.mx, andre@cactus.iico.uaslp.mx

Abstract. The majority of the semiconductor structures with QD today are grown by MBE and MOCVD. It is known that the best material quality can be achieved by LPE because, in contrast to MBE and MOCVD, this method is realized at near-equilibrium conditions. To develop QD LPE technology first of all it is necessary to find out a growth technique allowing the crystallization of epitaxial materials with very small volume. This can be done by means of different techniques. In this work we apply a low temperature short-time growth method, which allows the production not only of single, but also of multilayer heterostructures. We have grown $Ga_xIn_{1-x}As$ QD on GaAs (100) substrates at 450 C. The details of the QD formation, depending on composition of the $Ga_xIn_{1-x}As$ solid solutions, have been studied by atom-force microscopy. The photoluminescence spectra of investigated samples show, in addition to a short-wave GaAs related peak, a longer wavelength line, which disappears after removal of the grown GaInAs material using an etching solution. This fact, together with atom-force microscopy results can be interpreted as a proof that QD heterostructures were grown successfully by LPE.

1. Introduction

Starting from the 1970's, liquid phase epitaxy (LPE) became the main method of the production of different multilayer structures. In nowadays semiconductor's technology the overwhelming majority of devices based on multilayer structures, including QD's, are grown by means of molecular beam epitaxy (MBE) and gas phase epitaxy, especially metal-organic chemical vapor deposition (MOCVD). This happened because the LPE has several shortcomings in comparison with MBE and MOCVD. But in spite of the advantages of MBE and MOCVD, the LPE technique is still an important technological method for growing various epitaxial films and multilayer structures. First of all the quality of materials grown by this method is excellent. For example, the internal quantum efficiency of radiative recombination for lightly doped direct-bandgap III-V binary compound and solid solutions grown by LPE is close to 100% [1,2]. This phenomenon can be explained because the crystallization process in LPE is carried out at conditions not far from equilibrium. So the purpose of the work was to develop the LPE growth technology and to investigate the QD structures on the base of GaInAs/GaAs system. To solve this problem first of all it is necessary to find a LPE growth method of epitaxial layers with very small thickness (in another words a very small volume of the growing materials). In the present work we have used low growth temperature and a short time of contact between substrate and liquid phase. It is known, that in the MBE and MOCVD technologies, when there is a lattice mismatch between the substrate and epitaxial layer, QD can be grown due to a self organizing mechanism [3]. In the present work we have used LPE and studied the influence of the lattice mismatch between GaAs substrate and $Ga_xIn_{1-x}As$ epitaxial layers on the process of QD crystallization. In other words we have investigated the growth peculiarities of very thin $Ga_xIn_{1-x}As$ epitaxial films as a function of composition (x). By "epitaxial films" we mean not only complete epitaxial layers but also separated islands (including the QDs), growing on the substrate's surface during an epitaxial process due to the lattice mismatch between substrate and growing material. Previous reports on the growth of QD's by LPE include the growth of InAsSbP [4] and InAs [5] QD's on GaAs substrates, InSb QD's on InAs [6,7] and GaAs [6] substrates and PbSe [8] on GaSb substrates. In works [4,6,8] the growth experiments were done using very short times of contact, of the order of ms, between the liquid solution and the substrate. To achieve such short times the contact was done trough a narrow slit (1 mm wide) and using electric motors to obtain high slider speeds. In Ref. [5] a special pulsed cooling method was applied to a supersaturated liquid phase at 400 C. There are several reviews [9-11] about the growth by LPE of quantum structures, such as quantum wells and QD's.

2. Experimental

The usual routine employed to prepare the growth experiments is described in Ref [12]. All samples were grown at a temperature of 444 C (note that the usual and typical growth temperatures in the LPE of Ga - In - As system are in the interval 550-650 C). Growth time in the most cases was 15 seconds, but for some samples it was near 1 second. Prior to the crystallization of the GaInAs solid solutions or pure InAs, a buffer GaAs layer with thickness of a few microns were grown on the surfaces of the GaAs substrates. The morphology of the grown samples was studied by means of atom force microscope (AFM). The photoluminescence spectra of the structures were recorded using a 100 mW Ar laser with a wavelength of 514 nm and a 0.5 m spectrometer. An InGaAs photodetector was used for detection in conjunction with conventional lock-in techniques.

3. Results and discussion

First of all, to estimate the composition of GaInAs (and accordingly the lattice mismatch between substrates and epitaxial layers), the phase diagram of the Ga-In-As system at the used growth temperature has been calculated by means of usual procedure, applied for the different III-V systems [13].

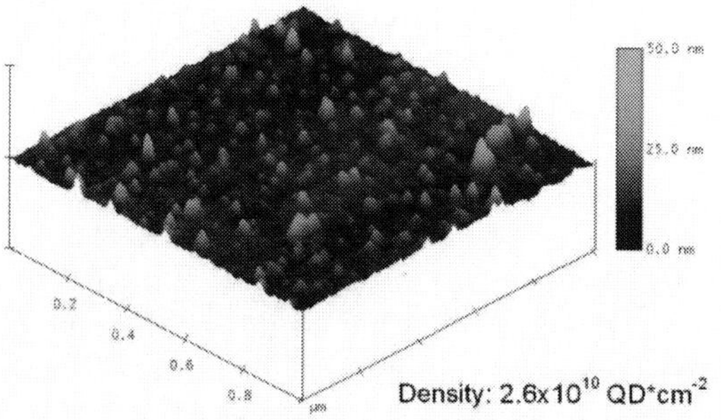

Figure 1. AFM Image of the structure grown from the In-As liquid phase at 444 ^oC during 15 s. The density of nano-islands is about of $2.6{\cdot}10^{10}$ cm^{-2}. The average diameter and height are $\sim$ 16 nm and 4 nm, accordingly.

When the growth experiments were done at 444 C, complete epitaxial $Ga_xIn_{1-x}As$ layers could be obtained in the composition range $1 \leq x \leq 0.75$. In contrast, at a growth temperature of 550 C, this composition range was reduced to $1 \leq x \leq 0.93$ as usual for this system. This can be explained by the big difference between the thicknesses of the epitaxial layers grown at high and low temperatures due to the solubility differences of As in a solution of Ga and In as can predicted from the phase diagram.

Table 1. The data of the $Ga_xIn_{1-x}As$ epitaxial layers obtained at 444 C.

Growth Time	X_{In}/X_{Ga}	*Kind of Structure*	*Density* QD/cm^2	*Calculated Composition* $Ga_xIn_{1-x}As$
15 s	15	O		0.95
15 s	20	O		0.87
15 s	25	O		0.8
15 s	30	X	$\approx 8.8 \cdot 10^8$	0.7
15 s	40	X	$\approx 2.3 \cdot 10^9$	0.35
15 s	50	X	$\approx 3.0 \cdot 10^9$	0.2
15 s	Pure In	X	$\approx 2.6 \cdot 10^{10}$	0
$\sim$1 s	Pure In	X	$\approx 1.3 \cdot 10^{10}$	0

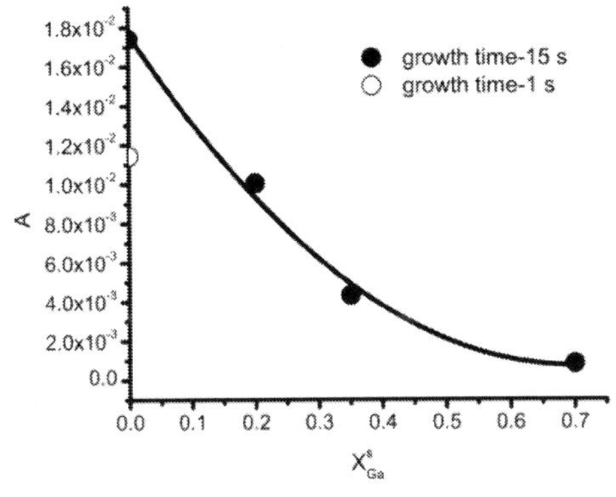

Figure 2. The dependence of the magnitude of A as function of the composition of the solid phase.

If the composition x of the $Ga_xIn_{1-x}As$ solid solutions decreases from x = 1 to $x \leq 0.75$, the quality of epitaxial films becomes worse as evidenced by an increase of the full widths at half maximum and a decrease of the intensity of the photoluminescence spectra. With our crystallization regime, complete epitaxial $Ga_xIn_{1-x}As$ films cannot be grown on the GaAs surface if their composition is $x < 0.75$. In these cases the ternary solid solutions are crystallized as separated islands. To study the morphology of such islands, AFM images of the structures with different compositions of the $Ga_xIn_{1-x}As$ solid solutions have been investigated. These images were processed to estimate the density of grown QD's. As an example, the image corresponding to pure InAs is shown in Fig. 1. All these data are assembled in Table 1. The column titled "Kind of Structure" indicates with an "X" the cases when separated Islands were observed and with an "O" the growth of complete layers.

It is possible to see from this table that the QDs density increases with the In concentration in the solid phase. This is likely related to the increase of lattice mismatch. The highest density of the QD is achieved in the case of InAs/GaAs structures.

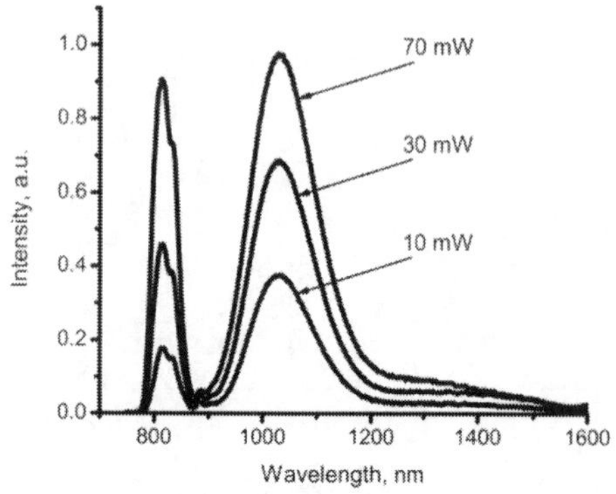

Figure 3. The PL spectra at 22 K and different levels of excitations (Ar laser power: 10, 30 and 70 mW) of the InAs/GaAs nano-islands structure.

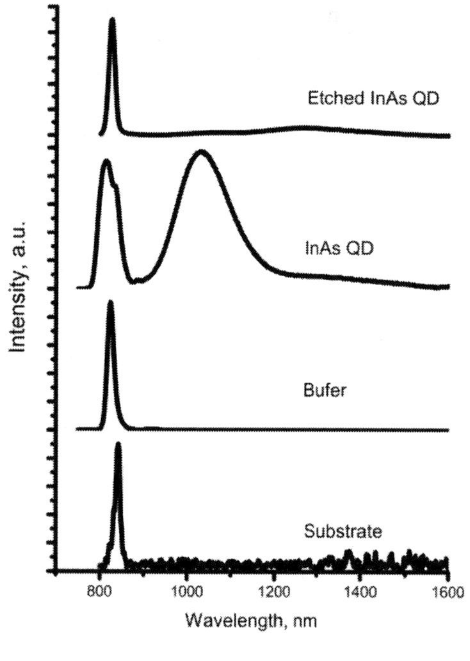

Figure 4. The PL spectra at 22 K of the GaAs substrate, the GaAs substrate with buffer layer, the structure with InAs QDs and the structure with removed InAs QDs.

As an indication of the total volume or weight of material forming the QDs, a quantity A was defined as the product of the average height, diameter, and density. The dependence of the magnitude of A, as function of the composition of the solid phase is shown in Fig. 2. It is possible to see that its dependence looks like the corresponding liquidus isotherm in the Ga-In-As system.

The photoluminescence spectra (PL) at 22 K and different levels of excitations of the InAs/GaAs nano-islands structure are shown on the Fig. 3. The first line on these spectra with a peak at 814 nm is associated with the GaAs buffer layer and the second one with a peak at 1030 nm could be related to the emission from the InAs QD's since its position does not change with excitation level. To further test this assumption we have used a chemical etching where the InAs nano-islands have been removed from the GaAs surface. The PL spectra of structure before and after etching are shown on Fig. 4. It is possible to see from this figure that the second line disappears after the InAs nano-islands have been removed, which proves that QDs were grown successfully in our LPE experiments.

4. Conclusions

Using a short-time, low-temperature LPE growth technique it is possible to obtain InAs/GaAs nano-islands or QDs. The existence of these nano-islands has been proved by means of atom force microscope images and photoluminescence spectra. The highest density of such nano-islands (in the case of InAs/GaAs structures) was $2.6 \cdot 10^{10}$ cm^{-2} .

5. Acknowledgments

The work was partially supported by CONACYT and FAI at UASLP.

6. References

[1] Alferov Zh I, Andreev V M, Garbuzov D Z and Rumiantsev V D 1975 *Sov. Phys. Semicond.* **9** 305
[2] Garbuzov D Z and Khalfin V B 1993 *Quantum well lasers*, ed. P S Zory Jr (Academic Press, Inc.) p. 277
[3] Pchelyakov O P, Bolkhovityanov Yu B, Dvurechenskii A V, Sokolov L V, Nikiforov A I, Yakimov A I and Voigtländer 2000 *Semiconductors* **34** 1229
[4] Krier A, Labadi Z and Hammiche A 1999 *J. Phys. D: Appl. Phys.* **32** 2587
[5] Maronchuk I E, Maronchuk A I, Kulyutkina T F, Naydyonkova M V and Chorny I V 2005 *Poverkhnost* No. 12, 97
[6] Krier A, Huang X L and Hammiche A 2001 *J. Phys. D: Appl. Phys.* **34** 874
[7] Moiseev K D, Parkhomenko Ya A, Ankudinov A V, Guschina E V, Mikhailova M P, Titkov A N and Yakovlev Yu P 2007 *Pisjma v ZhTF (Rus.)* **33** 50
[8] Huang X L, Labadi Z, Hammiche A and Krier A 2002 *J. Phys. D: Appl. Phys.* **35** 3091
[9] Krier A, Huang X L and Labadi Z 2007 *Liquid Phase Epitaxy of Electronic, Optical and Optoelectronic Materials* Ed Peter Capper and Michael Mauk (West Sussex: John Wiley & Sons, Ltd) chapter 8 pp 227-257
[10] Mishurnyi V A 2007 *Liquid Phase Epitaxy of Electronic, Optical and Optoelectronic Materials* Ed Peter Capper and Michael Mauk (West Sussex: John Wiley & Sons, Ltd) chapter 2 pp 21-43
[11] Mishurnyi V A, de Anda F, Elyukhin V A, Hernandez I C 2006 *Critical Reviews in Solid State and Material Sciences* **31** (Taylor Francis) pp 1-13
[12] Mishurnyu V A, De Anda F, Gorbatchev A Yu, Kudriavtsev Yu, Elyukhin V A, Pritskij T, Pelosi C, Bocchi C, Bre B Y and Ortiz Vazquez F E 2008 *Thin Solid Films* **516** 8092
[13] Charykov N A, Litvak A M, Mikhailova M P, Moiseev K D and Yakovlev Yu P 1997 *Semiconductors* **31** 344

Tetrahedral 1B4Sb nanoclusters in GaP:(B, Sb)

V. A. Elyukhin[1]

[1] Departamento de Ingeniería Eléctrica-SEES, CINVESTAV-IPN, Avenida IPN 2508, Col. San Pedro Zacatenco, C. P. 07360, México, D. F., México

E-mail: elyukhin@cinvestav.mx

Abstract. Self-assembling conditions of 1B4Sb tetrahedral nanoclusters in GaP doped with boron and Sb isoelectronic impurities are represented in the ultradilute and dilute limits of the boron and Sb contents, respectively. The fulfilled estimates demonstrated the preferential complete or almost complete allocation of boron atoms in 1B4Sb nanoclusters at temperatures of 500 °C and 900 °C, respectively. The significant decrease of the sum of the free energies of the constituent compounds is the main origin of self-assembling. The reduction of the strain energy is the additional cause of this phenomenon.

1. Introduction

Self-assembling (SA) of 1N4Al and 1As4Ga tetrahedral nanoclusters in GaAs:(Al, N) and zinc blende AlN:(Ga, As), respectively, was predicted in Ref. [1]. The co-doping of a binary compound *BD* with cation *A* and anion *C* isoelectronic impurities transforms it into $A_xB_{1-x}C_yD_{1-y}$ quaternary alloy. In these alloys an exchange of lattice sites between cations or anions leads to the reaction between bonds $nA\text{-}C + nB\text{-}D \rightarrow nA\text{-}D + nB\text{-}C$, ($1 \leq n \leq z$, where z is a coordination number). The larger sum of the cohesive energies of the bonds participating in the reaction provides their preferential formation. A decrease of the strain energy after SA is an additional cause of the impurity complex formation in such alloys. The main origin of SA in GaAs:(Al, N) and AlN:(Ga, As) is a preferential AlN and GaAs bonding over AlAs and GaN one. The equality conditions of the free energies of the random alloys and alloys in which all minority impurities are in the nanoclusters were represented [1]. Later the phase diagram of AlN:(Ga, As) with the partial location of minority impurities in the nanoclusters was developed [2].

Size dispersion is the main problem of the semiconductor structures with the zero-dimensional nanoobjects such as quantum dots. A technology of the monodisperse quantum dots is not developed yet. SA of the boron and Sb isoelectronic impurity nanoclusters is proposed as a line of attack on this problem that is the formation of monodisperse semiconductor nanoobjects embedded in the wide gap matrix. GaP:(B, Sb) is proposed as a semiconductor with 1B4Sb nanoclusters. One of the reason of SA in GaP:(B, Sb) can be a decrease of the lattice strains after the formation of 1B4Sb nanoclusters from isolated atoms. The tetrahedral covalent radii of boron and Sb are smaller and larger than those of Ga and phosphor, respectively. Isolated boron and Sb atoms dilate and shrink, respectively, a crystal lattice of GaP:(B, Sb). The dilation and shrinking compensate each other after SA of 1B4Sb nanoclusters. Thus, SA is preferential from the strain energy standpoint. GaP:(B, Sb) is a GaP-rich $B_xGa_{1-x}Sb_yP_{1-y}$ alloy. Moreover, in such alloy an exchange of lattice sites between

Journal of Physics: Conference Series **167** (2009) 012003 doi:10.1088/1742-6596/167/1/012003

cations or anions alters the quantities of bonds according to the reaction *n*Ga-Sb + *n*B-P → *n*Ga-P + *n*B-Sb (*n* = 1,..,4) or vise versa. The larger sum of the cohesive energies of bonds participating in the reaction provides their preferential formation. It can be an additional reason of the SA. The nanocluster may be represented as a quantum dot with a size of the order of 1 nm and with the composition $Ga_{0.75}B_{0.25}Sb$ that is determined by Ga-Sb and B-Sb bonds in the wide band gap GaP-rich alloy. The SA conditions in the binary semiconductor compounds co-doped with the cationic and anionic isoelectronic impurities are represented by the phase diagrams [2]. The aim of the paper is to consider a phase diagram of GaP:(B, Sb).

2. The Model

The free energy of a GaP-rich $B_xGa_{1-x}Sb_yP_{1-y}$ cluster ordered alloy $f = h^0 + u_{SE} - Ts_C$ is as a sum of the enthalpies of the constituent compounds h^0, strain energy u_{SE} and configurational entropy term $-Ts_C$. The clustering degree is represented by the cluster order parameter α that is a fraction of boron atoms situated in 1B4Sb nanoclusters. This parameter can vary from zero to unity. The alloys with $\alpha = 0$, $0 < \alpha < 1$ and $\alpha = 1$ are homogeneous, partially cluster ordered and completely cluster ordered ones, respectively [2]. The ultradilute and dilute limits for the boron and Sb contents, respectively, are considered. The arrangement of boron atoms should be random and this fact is taken into account when describing the items of the sum. The sum of the enthalpies of the constituent compounds is given as

$$h^0 = xyh^0_{BSb} + x(1-y)h^0_{BP} + (1-x)yh^0_{GaSb} + (1-x)(1-y)h^0_{GaP} + \\ + \alpha(1-x)y\left(h^0_{BSb} - h^0_{BP} - h^0_{GaSb} + h^0_{GaP}\right), \quad (1)$$

where, x and y are the concentrations of boron and Sb in the cation and anion sublattices, respectively. The variation of the sum (1) as a result of SA depends on the relation $h^0_{BSb} - h^0_{BP} - h^0_{GaSb} + h^0_{GaP}$. Therefore, only last item of the sum (1) should be taken into account. The value of this relation is equal to the similar relation between the enthalpies of formation. The enthalpies of formation of BP, GaSb and GaP are available [3]. The enthalpy of formation of BSb was obtained from the cohesive energy of BSb and atomization energies of boron and Sb as $h^{0f}_{BSb} = u^{Coh}_{BSb} - u^{At}_B - u^{At}_{Sb}$. The atomization energies were taken from [4]. The cohesive energy of BSb was estimated by the Sanderson's theory [4] as a sum of the ionic and covalent components that are given as

$$u^C_{BSb} = \frac{1.39 M_{BSb}(R_B + R_{Sb})k}{R_{BSb}} t_i + \frac{4.19\times10^{-3}(R_B + R_{Sb})\sqrt{u_B u_{Sb}}}{R_{BSb}}(1-t_i) = 0.185\,\text{MJ/mole},$$

where M_{BSb} is the Madelung constant of BSb, R_B and R_{Sb} are the covalent radii of boron and Sb, respectively, k is the repulsion coefficient, t_i is the average of the partial charges on two atoms forming B-Sb bond, R_{BSb} = 0.516 nm [5] is the distance between the nearest atoms in BSb, u_B and u_{Sb} are the atomization energies of boron and Sb, respectively. The obtained value of the relation between the enthalpies of formation $h^{0f}_{BSb} - h^{0f}_{BP} - h^{0f}_{GaSb} + h^{0f}_{GaP} = -1.502\times10^5\, J/mole$ demonstrates the significant preferential B-Sb and Ga-P bond formation over the B-P and Ga-Sb one.

The strain energy of the GaP-rich $B_xGa_{1-x}Sb_yP_{1-y}$ alloy is given as

$$u_{SE} = xu_B + yu_{Sb} + \alpha y\left(u_{1B4Sb} - u_B - 4u_{Sb}\right) \quad (2)$$

where u_{1B4Sb}, u_B and u_{Sb} are the strain energies caused by the impurity nanoclusters and isolated impurities, respectively. As it is seen from (2) the strain energy does not depend on the clustering degree since the relation $y >> x$ is fulfilled. The strain energies caused by nanoclusters and isolated atoms were estimated as a sum of two items. The first item is the strain energy of quadruples of tetrahedral cells situated around the central atoms of nanoclusters or around isolated boron and Sb impurities. This part was described using the valence force field model by the approach developed for the strain energy estimation of III-V ternary alloys [6]. The averages of the bond bending elastic constants of the compounds were used as elastic constants of the angles between the unlike bonds. The bond stretching and bond bending elastic constants of GaSb and GaP were taken from [7]. The bond stretching and bond bending elastic constants of BSb and BP were estimated by the valence force field model [7] using the stiffness coefficients taken from [8] and [9], respectively. The other contribution is the deformation energy of the alloy outside the quadruples. This part was represented as the energy of the elastic media with radial displacements inversely proportional to the square of a distance from the central atoms of the quadruples. The strain energies caused by isolated boron and Sb atoms and 1b4Sb nanoclusters are equal, respectively, to $u_B = 1.017\times10^5$ J/mole, $u_{Sb} = 1.215\times10^5$ J/mole and $u_{1B4Sb} = 1.189\times10^5$ J/mole. Thus, SA of 1B4Sb nanoclusters reduces almost five times the strain energy and this fact also demonstrates the significant B-Sb and Ga-P bond preferential formation over the B-P and Ga-Sb one.

The third item is given as

$$-Ts_C = RT\left[x\ln x + (1-x)\ln(1-x) + (y-4\alpha x)\ln\frac{y-4\alpha x}{1-4\alpha x} + (1-y)\ln\frac{1-y}{1-4\alpha x}\right]. \quad (3)$$

The number of configurations was considered as a product of two factors. The first one is a number of arrangements of the cations. There is a random distribution of the cations because of the chosen ultradiliute limit for the boron content. The second one is a number of the arrangements of anions situated outside 1B4Sb nanoclusters at the fixed allocation of isolated boron atoms and nanoclusters. The mean distances between isolated boron atoms and 1B4Sb nanoclusters are significantly larger then those between the nearest atoms. In such a case, spatial correlations among boron atoms that are outside nanoclusters are very small. In fact, impurity nanoclusters and isolated boron atoms should be distributed randomly in the cluster ordered alloy.

3. Results and Discussion

The boron and Sb contents $10^{-8} \le x \le 10^{-4}$ and $4x \le y \le 10^{-2}$, respectively, were considered. The ultradilute limit for the boron concentration was chosen in order to avoid the interactions between nanoclusters. In such a case nanoclusters should create bound states. In the case of the dilute limits the formation of nanoclusters will change the band gap. GaP and GaP-rich alloys are mainly grown by the metalorganic chemical vapor deposition at temperatures of 500 °C and 900 °C if tertiarybutylphosphine and phosphine are used, respectively, as precursors of phosphor. Therefore, the temperature range between these temperatures was chosen for the estimates. The clustering degree is determined by the condition $\frac{df}{d\alpha} = 0$ [2]. The fulfilled estimates demonstrated that at temperature of 500 °C the preferential thermodynamically phase state should be the completely cluster ordered ($\alpha = 1$) in the entire boron content range and at the Sb content $y \ge 4x$. At higher temperature of 900 °C only negligibly small fraction of boron atoms ($1-\alpha < 0.001$) are isolated in the entire boron content range and at the Sb content $y \ge 4x$. Thus, at both considered temperatures the alloys should be the completely cluster ordered or almost completely cluster ordered phase ones. In comparison with the alloys such as zinc blende AlN-rich $Ga_xAl_{1-x}As_yN_{1-y}$, ZnTe-rich $Mg_xZn_{1-x}O_yTe_{1-y}$ and ZnTe-rich $Sr_xZn_{1-x}O_yTe_{1-y}$ [2; 10-11] GaP-rich $B_xGa_{1-x}Sb_yP_{1-y}$ alloys demonstrate the larger cluster ordering.

The bond composition of 1B4Sb nanoclusters corresponds to $B_{0.25}Ga_{0.75}Sb$ alloy. The bandgap of such alloy should be smaller than that of GaP since the bandgap of BSb is smaller than that of GaSb [12]. Thus, 1B4Sb nanoclusters in GaP-rich $B_xGa_{1-x}Sb_yP_{1-y}$ alloy are zero-dimensional uniform narrow bandgap objects in wide bandgap semiconductor.

4. Conclusions

The self-assembling conditions of 1B4Sb nanoclusters in GaP:(B, Sb) were obtained in the ultradilute boron and dilute Sb limits. The decrease of the free energy contribution of the constituent compounds

Journal of Physics: Conference Series **167** (2009) 012003 doi:10.1088/1742-6596/167/1/012003

and reduction of the strain energy after the formation of 1B4Sb nanoclusters are origins of self-assembling. The transformation of B-P and Ga-Sb bonds into more preferential B-Sb and Ga-P ones and decrease of the strain energy are the origins of self-assembling. In the considered concentration ranges all or almost all boron atoms should be in 1B4Sb nanoclusters at temperatures of 500 °C and 900 °C, respectively.

References

[1] Elyukhin V A, Sánchez-R V M and Elyukhina O V 2004 *Appl. Phys. Lett.* **85** 1704
[2] Elyukhin V A, Sánchez-R V M and Escobosa A 2006 *Appl. Phys. Lett.* **89** 173110
[3] *Landolt-Börnstein* (vol 17d) 1984 ed O Madelung (Berlin: Springer)
[4] Sanderson R T 1971 *Chemical Bonds and Bond Energy* (New York: Academic) p 15
[5] Bouhafs B, Aourag H, Fernat M and Certier M 2000 *J. Phys.: Condens. Matter* **12** 5655
[6] Elyukhin V A and Sorokina L P 1986 *Sov. Phys. Dokl.* **31** 342
[7] Martin R M 1970 *Phys. Rev. B* **1** 4005
[8] Azuhata T, Soka T and Suzuki K 1996 *J. Phys.: Condens. Matter* **8** 3111 (1996)
[9] Ozolin'sh G V, Averkieva G K, Ievin'sh A. F and Goryunova N A 1963 *Sov. Phys. Crystallogr.* **7** 691
[10] Elyukhin V A, Sorokina L P and Rodriguez de Santiago M 2007 *Physica Status Solidi (b)* **244** 1639
[11] Elyukhin V A 2008 *Appl. Phys. Lett.* **92** 103115
[12] Dalui S, Das S N, Hussain S, Paramanik D, Verma S and Pal A K 2007 *J. Cryst. Growth* **305** 149

NANOSTRUCTURES BASED IN BORO NITRIDE THIN FILMS DEPOSITED BY PLD ONTO $Si/Si_3N_4/DLC$ SUBSTRATE

W. S. Román[1], H. Riascos[1], J. C. Caicedo[2], R. Ospina[3], L. Tirado-Mejía [4]

[1]Grupo Plasma, Láser y Aplicaciones, Universidad Tecnológica de Pereira –Colombia.
[2]Grupo de Películas Delgadas, Universidad del Valle, Cali –Colombia.
[3]Laboratorio de Plasma, Universidad Nacional de Colombia, sede Manizales.
[4]Laboratorio de Optoelectrónica, Universidad del Quindío, Armenia – Colombia

E-mail: hriascos@utp.edu.co

Abstract. Diamond-like carbon and boron nitride were deposited like nanostructered bilayer on Si/Si_3N_4 substrate, both with (100) crystallographic orientation, these films were deposited through pulsed laser technique (Nd: YAG: 8 Jcm^{-2}, 9ns). Graphite (99.99%) and boron nitride (99.99%) targets used to growth the films in argon atmosphere. The thicknesses of bilayer were determined with a perfilometer, active vibration modes were analyzed using infrared spectroscopy (FTIR), finding bands associated around 1400 cm^{-1} for B – N bonding and bands around 1700 cm^{-1} associated with C=C stretching vibrations of non-conjugated alkenes and azometinic groups, respectively. The crystallites of thin films were analyzed using X-ray diffraction (XRD) and determinated the h-BN (0002), α-Si_3N_4 (101) phases. The aim of this study is to relate the dependence on physical and chemical characteristics of the system $Si/Si_3N_4/DLC/BN$ with gas pressure adjusted at the 1.33, 2.67 and 5.33 Pa values.

1. Introduction

Pulsed laser deposition (PLD) is a conceptually simple technique that generates reactive species under highly non-equilibrium conditions, which cannot be produced by any other technique. This leads the way to the growth of materials having unique properties, and indeed, PLD is proving to be one of the most attractive techniques to synthesize amorphous carbon-based thin films such as diamond-like-carbon (DLC) [1] and nitride thin films. The physical properties of carbon-based materials, such as hardness and electrical conductivity, depend to a large extent on the deposition conditions such as gas pressure and substrate temperature as well as on the structural arrangement of the atoms in the films.

Boron nitride (BN) is a synthetic ceramic compound, which crystallizes in crystallographic structures similar to the allotropic forms of carbon. Two most important crystalline forms are hexagonal (h-BN) and cubic (c-BN) one, which are similar to graphite and diamond respectively. Graphite-like phase (h-BN) is a soft chemically inert and stable in air atmosphere insulator, what makes it suitable compound for insulating protective and self lubricating coatings, while cubic phase (c-BN) is the second hardness compound after diamond. Cubic form has also special advantage with

[1] hriascos@utp.edu.co

respect to diamond like its inertness with respect to iron even at high temperatures. Therefore c-BN thin films can be applied as super-hard wear-resistant corrosion resistant coatings for cutting tools, optical and electronic devices etc. [1-2]. The most important advantages of PLD are high deposition rate in laser pulsed, stoichiometry transfer, possibility to deposition without substrate heating and easy preparation of multilayer coatings [3].

In this paper, we report on low-temperature preparation of BN thin films, using pulsed laser deposition method. We prepared BN films on top of the DLC layers and DLC thin films were grown on silicon and silicon nitride as Si/Si_3N_4 substrates, through pulsed laser ablation of a sintered BN target, in the environment of Argon. The composition and chemical structure of the prepared films were studied using X-ray spectroscopy (XPS) and Fourier transform infrared (FTIR) spectroscopy.

2. Experimental Details

The experiments have made in usual PLD configuration consisting of laser system, a multi-port stainless steel vacuum chamber equipped with a gas inlet, a rotating target and a heated substrate holder. We used a Nd:YAG laser that provides pulses at the wavelength of 1064 nm with 9 ns pulse duration and repetition rate 10 Hz, the laser fluence kept to 8 J/cm^2. The laser beam was focused with an f=23 cm glass lens on the target at the angle of 45°, with respect the normal. The target was rotating 2.2 rpm to avoid fast drilling. The distance between the target and the substrate was 6.5 cm. Before deposition the vacuum chamber was evacuated down to 2 x 10^{-5} Pa by using rotary pump and turbo-molecular pump, then a continuous flow of argon, 99.99% purity, was introduce and the argon pressure was fixed from 1 to 20 Pa by changing the argon flux. Fig. 1 shows a schematic diagram of the PLD system used in this study [4]. The DLC films were grown at room temperature and under different argon pressures on the Si_3N_4 substrate, whereas the BN films were grown on top of the DLC layers with the same experimental parameters of growth. For all thin films the deposition time was 10 min. IR absorption was investigated using the Fourier transform infrared (FTIR) spectrophotometer (Shimatzu model prestige 21). The crystallinity of the films was evaluated by using a X-ray diffraction (XRD) apparatus (Bruker, D8 Advance) with a Cu Kα radiation. The nano-structural analysis of film morphology and crystal nucleation was performed on a Cambridge 360 SEMyEDS.

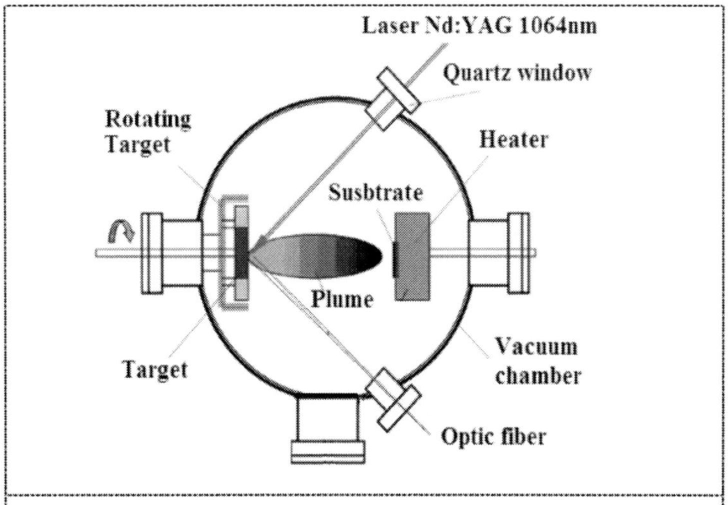

Figure 1. Schematics of the laser ablation setup.

3. Results and discussion

When prepared PLD for 10 min, the film has a thickness of approximately 350 nm. The BN and DLC films prepared on Si substrates show a very smooth appearance, and are stable and adherent to the substrates under ambient conditions.

3.1. X-ray diffraction analysis

X-ray diffraction patterns of DLC/BN for different gas pressure are shown in Fig. 2. All spectra reveal crystalline structures. XRD patterns showed that the major phase in all the three samples was β-Si_3N_4, which indicated that adding h-BN did not hinder the $\alpha \rightarrow \beta$ phase transformation of Si_3N_4. The c-BN phase was not detected in XRD patterns, which may due to either the amounts being too small to be detected or forming the intergranular amorphous phase. In samples grown at 2.67 and 1.33 Pa, additional peak that is close to (0002) peak of h-BN was observed, suggesting the presence of a strong local boron nitride structure [8]. The (0002) diffraction peak in the XRD spectra of the B_x-N_y samples suggests the presence of an h-BN-like phase. The amount of this hexagonal B_x-N_y phase in the samples can be expected to be proportional to the intensity of the (0002) diffraction peak.

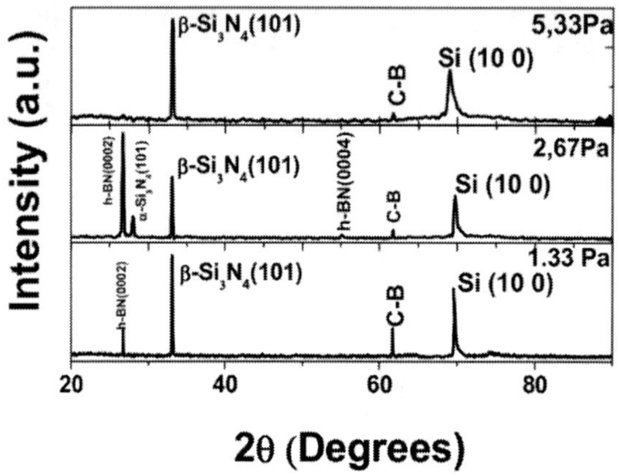

Figure 2. X-ray diffraction patterns of DLC/BN films grown by PLD at three different gas pressure on Si3N4 substrates.

3.2. FTIR analysis

We show in Fig. 3, typical IR absorption spectra obtained in the range 500–3600 cm^{-1}, at three values of nitrogen partial pressure. In spectrum, two strong absorption bands with dips at approximately 1400 cm^{-1} and 768 cm^{-1}, are clearly identified. These two bands can be interpreted as the in-plane B–N, stretching vibration and the out-of-plane B–N–B bending vibration [2,3,5], respectively. This indicates a characteristic of the h-BN phase, with a sp^2 bonding structure. In addition, there is also a small but distinct absorption near 1080 cm^{-1} corresponding to the transverse optical mode (TO) of c-BN. However, this absorption might also be due to B–C bonds [6]. The film prepared at 2.7 Pa exhibits a strong absorption characteristic of the in-plane B–N stretching vibration at 1400 cm^{-1} and its out-of-plane B–N–B bending vibration gets weaker. The maximum of peak of the tensile vibrations of B-N bond is observed to be shifted towards lower frequencies (the peak appears at 1371cm^{-1} in the reference h-BN crystal). This suggests that there is tensile stress in the BN layer [7]. Such stress could improve the overall layer adhesion by compensating compressive stress, usually encountered in the areas with the sp^3-type phase obtained by PLD method and frequently responsible for layer delamination. The spectrum obtained from the layer produced at 5.3 Pa is characterized by a reduced intensity of the signals received from the sp^2-BN form and contains only a small intensity peak with the maximum at about 1388 cm^{-1}. This could have been caused by the decrease of amount of crystalline h-BN and the tendency of amorphous BN to increase its share in the entire layer volume. The broadening in the absorption band suggests that the film is highly disordered.

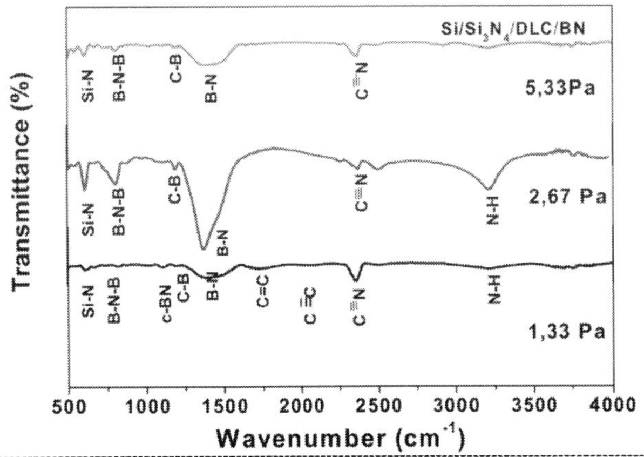

Figure 3. Typical infrared spectra of DLC/BN thin films for different values of argon partial pressure grown by PLD.

3.3. SEM Characterization

SEM investigations were carried out to observe the surface morphology of the deposited films. Fig. 4 shows the SEM micrographs of the surface of the BN films deposited at different pressures. The deposited films have small droplets and crystallites. It is thought that the droplets consist of elemental boron, ejected from the melt, due to the inhomogeneity of the power density distribution of laser beam on the target. Fragment consists of h-BN, which are typically ejected in ablation process of the soft targets such as h-BN. For films deposited at 5.3 Pa the surfaces contain mostly small size (few tens of nanometer) boron droplets. At 1.33 Pa, the size of the melted boron droplet increases on the surface and the films deposited at 2.67 Pa have a rougher surface. In summary the surface morphology, however, becomes smoother and grain size smaller with increasing the gas pressure.

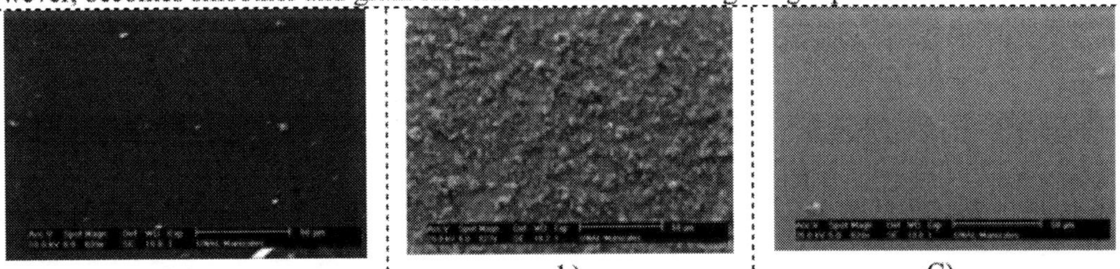

a) b) C)

Figure 4. Typical SEM images of the surface of the BN films deposited at the a). 1.33, b). 2.66 and c). 5.3 Pa of Ar gas pressure by Nd:YAG laser ablation at 8 J/cm^2.

4. Conclusion

DLC/BN bilayer films were deposited PLD using simultaneously graphite and h-BN as targets in Ar atmosphere. The gas pressure in the vacuum chamber was varied from 1.33 Pa to 5.33 Pa changing the argon flux. The pressure changes induce important variations on formation of crystallographic structures of h-BN(0002) and h-BN(0004) phases.

The Films grown at 2.67 Pa has shown a shift of the absorption band of B-N and B-N-B phases around 1400 cm^{-1} and 768 cm^{-1} respectively, indicating a high influence of gas pressure on the molecular formation of DLC/BN films. The SEM analysis of the film grown at 5.33 Pa showed more regularity and continuity on its surface than the other bilayer films grown.

Acknowledgements
This work was supported by the Universidad Tecnológica de Pereira –Colombia.

References

[1] A. A. Voevodin, M.S.Donley, *Surf. Coat.Technol.* **82** (1996) 199.
[2] C.W. Ong, X.A. Zhao, K.F. Chan, et al., *Thin Solid Films* **307** (1997) 152
[3] T. Sugino, H. Hieda, *Diamond Relat. Mater.* **9** (2000) 1233.
[4] H. Riascos, G. Zambrano, P. Prieto. *Braz. J. Phys.* **34** (4b) (2004) 1583.
[5] A. Lousa, J. Esteve, S. Muhl, E. Martinez, *Diamond Relat. Mater.* **9** (2000) 502.
[6] M. Dinescu, A. Perrone, A.P. Caricato. *Appl. Surf. Sci.* **127-129** (1998) 692.
[7] Y. Yap, T. Aoyama, Y. Wada. *Diamonds and Related Materials* **9**, (2000), 592–595
[8] G. C. Chen, S.B. Lee and J.H. Boo, *surf Rev Lett.* **10** (2003) 629.

Journal of Physics: Conference Series **167** (2009) 012005 doi:10.1088/1742-6596/167/1/012005

Enhanced photoconductivity and fine response tuning in nanostructured porous silicon microcavities

R Urteaga[1], O Marín[1], LN Acquaroli[1], D Comedi[2] JA Schmidt[1,3], RR Koropecki[1,3]

[1] INTEC-UNL-CONICET, Guemes 3450 - 3000 Santa Fe.
[2] CONICET y LAFISO, Departamento de Física, FACET, Universidad Nacional de Tucumán.
[3] Departamento de Ingeniería de Materiales, FIQ – UNL

E-mail: rkoro@intec.ceride.gov.ar

Abstract. We used light confinement in optical microcavities to achieve a strong enhancement and a precise wavelength tunability of the electrical photoconductance of nanostructured porous silicon (PS). The devices consist of a periodic array of alternating PS layers, electrochemically etched to have high and low porosities - and therefore distinct dielectric functions. A central layer having a doubled thickness breaks up the symmetry of the one-dimensional photonic structure, producing a resonance in the photonic band gap that is clearly observed in the reflectance spectrum. The devices were transferred to a glass coated with a transparent SnO_2 electrode, while an Al contact was evaporated on its back side. The electrical conductance was measured as a function of the photon energy. A strong enhancement of the conductance is obtained in a narrow (17nm FWHM) band peaking at the resonance. We present experimental results of the angular dependence of this photoconductance peak energy, and propose an explanation of the conductivity behaviour supported by calculations of the internal electromagnetic field. These devices are promising candidates for finely tuned photoresistors with potential application as chemical sensors and biosensors.

1. Introduction

Photonic structures made of nanostructured semiconductors have generated great interest in the last ten years mainly due to the availability of simple preparation methods and potential applications in control of the emission, propagation, and detection of light. In particular, photonic structures made of porous materials, such as porous silicon (PS), allow the sensing of chemical or biological species through changes in the photonic response of the structure [1-4].

PS can be prepared by electrochemical anodization of crystalline silicon in a fluorine-containing electrolyte [1]. The porosity of PS (i.e. the ratio between the void space volume and the total volume of the material), depends on preparation conditions such as current density and electrolyte composition. Provided the size of the pore structure is smaller than the light wavelength, PS behaves as an effective medium with a dielectric constant averaged among silicon, air and other components eventually adsorbed on pore surfaces.

An important feature of the PS preparation process is that a porosity (and also a dielectric function) in-depth profile can be built by changing the current density in time. In particular, a periodic multilayer with specific photonic response, such as one-dimensional photonic crystals like Bragg mirrors, or optical microcavities, can be made by using periodic current patterns in time. Since all the layers of the device are porous, any analyte existent in the surrounding atmosphere penetrates to all the layers. This eventually leads to changes in the refractive index at the pore surfaces, leading to a change in the optical response that can be used for sensing purposes.[1,2]

In this work we show that the energy resolved photocurrent of PS devices may be managed using multilayers with an appropriate design. We show that the light confinement in that multilayers leads to finely tuned photocurrent spectra.

2. Experimental details

The complex refractive index and rate of increase of the layers thicknesses were determined in correlation with the current densities by preparing two series of single layers for two different current densities, (6.4 and 124 mA/cm^2), with different anodization times. A Teflon cell was employed, with a p-type (1-4 mΩcm) silicon wafer used as the anode in which the PS layer was etched, and a platinum wire as the cathode. The electrolyte was a 2:1 solution of ethyl alcohol and hydrofluoric acid (50%).

The reflectance spectra of the layers were acquired using a reflectance accessory at 15° and fitted using a two stages computer code based in the Looyenga-Landau effective medium model [5]. Initial estimations of the two parameters (porosity and thickness of the film) are obtained in a first stage by using a genetic algorithm. These estimations are then used as input for a second conventional algorithm in the final fit stage.

The resulting refractive indexes and extinction coefficients spectra for each current density were the same for all the layers prepared during different anodization times, within an error of 5-7%. Furthermore, the thicknesses were proportional to the etching times for both current densities. The resulting porosity values for the two layers were p_1=52% (for 6.4mA/cm^2) and p_2=88% (for 124mA/cm^2). The refractive indexes at 700nm were 2.24+0.0053i and 1.27+0.001i respectively. Therefore, a rate of increase of the optical thickness was computed for each current density, and used to compute –as a function of the wavelength, and for each current density– the etching time necessary to obtain a layer with an optical thickness of (1/4)λ.

A computer controlled current source was used to follow a current density time pattern that had been designed to create an optical microcavity centred at a wavelength of around 700 nm. The cavity was a layer with an optical thickness of λ/2, having a low porosity p_1 which was surrounded by two 4 period Bragg mirrors. The Bragg mirrors were made of alternating layers of low (p_1) and high (p_2) porosities, all of them with optical thickness of (1/4)λ. The total thicknes of the sample is about 1.88μm. After the etching of the multilayer, a short pulse (4 sec) of 300-350 mA/cm^2 was applied in order to separate the porous multilayer from the monocrystalline substrate. This freestanding sample was transferred to a transparent, conducting SnO_2 film-coated glass slide. An aluminium layer was evaporated as a back contact.

The sample was illuminated through the transparent contact with a monochromatized light beam chopped at 17 Hz from a tungsten-halogen lamp. The linewidth of the resulted beam was lower than 1nm, which is negligible compared with the cavity bandwidth. A 60 mV DC bias was applied between both contacts and the photocurrent was measured as a function of the wavelength, by using a lock-in amplifier. The angular tuning was studied by acquiring the whole spectra for different incidence angles with the same experimental setup as described above, and also with an s-polarized He-Ne laser beam.

3. Results and Discussion

Figure 1 shows photocurrent raw data in the 550-900 nm range. A sharp photocurrent peak (FWHM of about 17 nm) can be observed. This feature can be attributed to the microcavity resonance, as demonstrated by the measured and computed reflectance spectrum shown in the insets. The spectral response of the conductance is a result of the intensification of the electromagnetic field within the cavity at the resonance energies. The field intensity in each layer can be computed using the transfer matrix formalism as follows: consider the light propagation through a single layer characterized by a complex dielectric function limited by media with refractive indexes η_o on the input side and η_s on the output side, respectively.

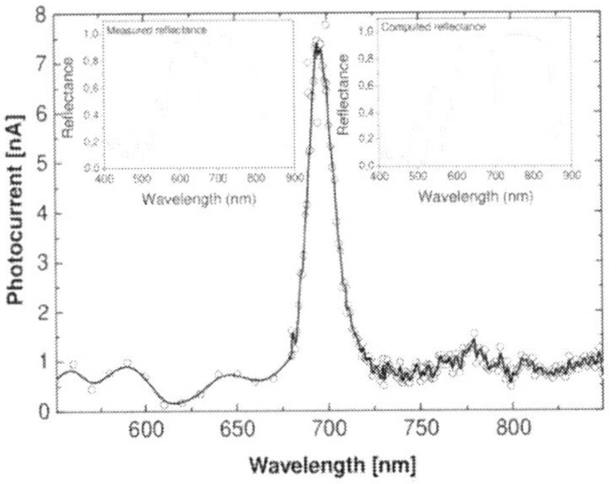

Figure 1: Photocurrent spectra for an optical microcavity with a λ/2 cavity centered in 700 nm. The insets show the measured (left) and computed (right) reflectance spectra for the freestanding PS multilayer. Both the FWHM of the photocurrent response, and the measured reflectance of the resonance are 17 nm, larger to that of the 9 nm of the computed one,

The fields E_o, B_o and E_i, B_i in the output and the input sides are related, for normal incidence, by the transfer matrix [6,7]:

$$\begin{bmatrix} E_i \\ B_i \end{bmatrix} = M . \begin{bmatrix} E_o \\ B_o \end{bmatrix} \qquad (1)$$

where M is the transfer matrix [6,7].

Equation (1) can be generalized for a multilayer of an arbitrary number of N layers. The relationship between the electric field in the layer j of the stack E_j and electric field in the input side can be obtained using the transfer matrix:

$$\tilde{M}_j = \begin{bmatrix} m_{11}^j & m_{12}^j \\ m_{21}^j & m_{22}^j \end{bmatrix} = \prod_{k<j} M_k \qquad (2)$$

Taking into account the boundary conditions, the enhancement factor that relates the field in the layer j with the input intensity for the wavelength λ can be calculated as [7],

$$\left(\frac{E_j(\lambda)}{E_o(\lambda)} \right) = (1 + r(\lambda)).m_{22}^j - \eta_0(\lambda).(1 - r(\lambda)).m_{12}^j \qquad (3)$$

where $r(\lambda)$ is the reflectance of the whole multilayer (which can be computed using the entire transfer matrix), $\eta_o(\lambda)$ is the complex refractive index of the input medium, and $E_o(\lambda)$, $E_j(\lambda)$ are the incident electric fields (in the input side) and the field in the layer j respectively, all for the wavelength λ.

The field intensity enhancement in the layer j is therefore:

$$I_j(\lambda) = \left(\frac{E_j(\lambda)}{E_o(\lambda)} \right)^2 . \qquad (4)$$

The field intensity in different points within each layer can be computed in the same way, substituting each layer by a number of thinner layers. Figure 2 shows, in a 3-D graph, the field intensity enhancement within the multilayer as a function of the depth and the photon energy. As can be observed, there is a large enhancement (about a factor of 14) for a narrow energy band in the cavity layer and in its neighbourhood.

We estimate the behaviour of the photoconductance of the microcavity by using a very simple model considering that the product μτ between carrier mobility and recombination time is independent on light intensity.

Journal of Physics: Conference Series **167** (2009) 012005 doi:10.1088/1742-6596/167/1/012005

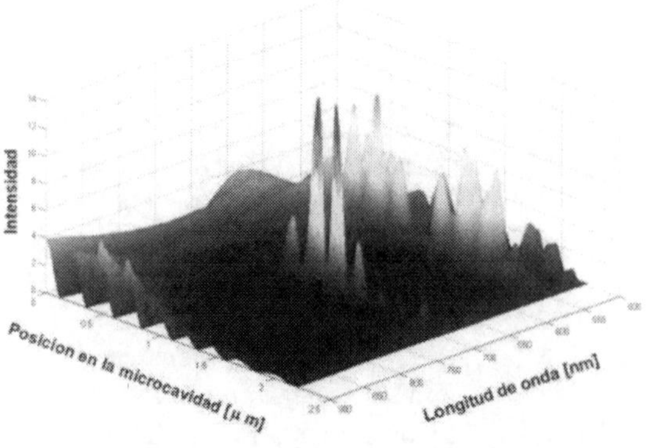

Figure 2: Intensity of the electromagnetic field inside the microcavity as a function of the position and the wavelenght.

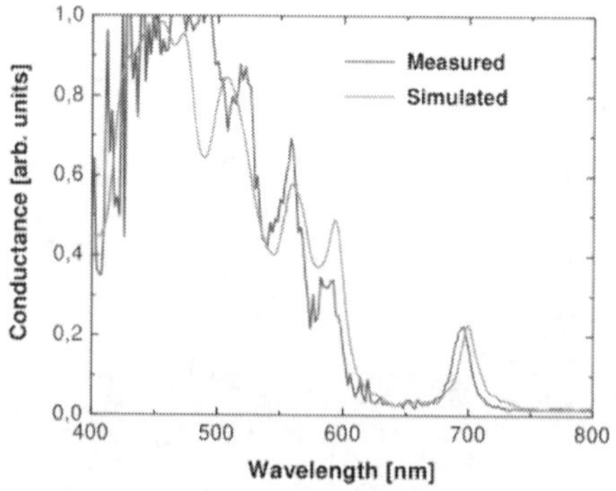

Figure 3: Comparison of the measured and simulated (Eq. 2) photocurrent spectra for a microcavity centered at 700 nm. The value of the Rose factor was 0.75.

Within this simple model the photoconductance of each layer for a given light intensity can be considered proportional to the absorption coefficient α. The dependence on the light intensity can be taken into account through the Rose parameter [8]. Hence, the conductance $G_j(\lambda)$ of the j layer is

$$G_j(\lambda) \propto \frac{\left[I_j(\lambda) \cdot \dfrac{k_j(\lambda)}{\lambda} \right]^{\gamma}}{t_j}, \qquad (5)$$

where k_j and t_j are the extinction coefficient and the thickness of the j layer for the wavelength λ.
The photocurrent I_p is proportional to the conductance of the whole set of N layers:

$$I_p \propto \left[\sum_{j=1}^{N} G_j^{-1} \right]^{-1} = \left[\sum_{j=1}^{N} \frac{t_j}{\left(I_j(\lambda) . \dfrac{k_j(\lambda)}{\lambda} \right)^{\gamma}} \right]^{-1} . \qquad (6)$$

Figure 3 shows the I_p values computed with this expression, using a Rose factor $\gamma = 0.75$ and the enhancement factors $I_j(\lambda)$ obtained by equation 4, for the present microcavity configuration. The experimental photocurrent spectrum (corrected for the light source spectrum and the monochromator spectral response) is also shown. The scales were adjusted to match the spectra. As it can be observed, the simple model reproduces quite well the general behaviour of the experimental photocurrent spectrum. A device having the same width, entirely made using a single porosity p_1 or p_2 will have a conductance lower by a factor of 16 or 3 respectively compared with the resonant device. Nevertheless, in case of a single layer the tunability is missed.

Since the light propagation characteristics through the multilayer depend on the incidence angle, a tuning of the mode wavelength can be performed by rotating the sample. The angular dependence of the photocurrent spectrum is shown in Figure 4, where the photocurrent spectra for different angles were plotted. The data for these spectra were also corrected to take into account the spectral response of the measurement system. As it can be observed, the tuning of the photocurrent peak can be performed finely, without major changes of the shape (the FWHM at 50° is about 20nm).

All the mentioned features can be used for sensing purposes. As an example, a microcavity can be tuned to obtain maximum photocurrent when a laser is used. The tunability of such a system is shown in Figure 5.

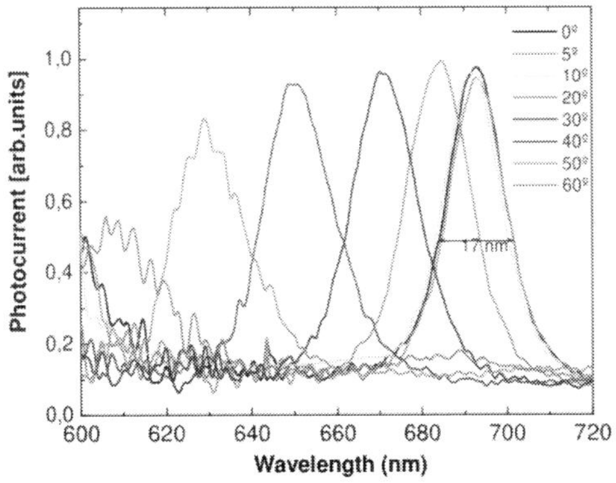

Figure 4: Photocurrent spectra of a microcavity for different incidence angles.

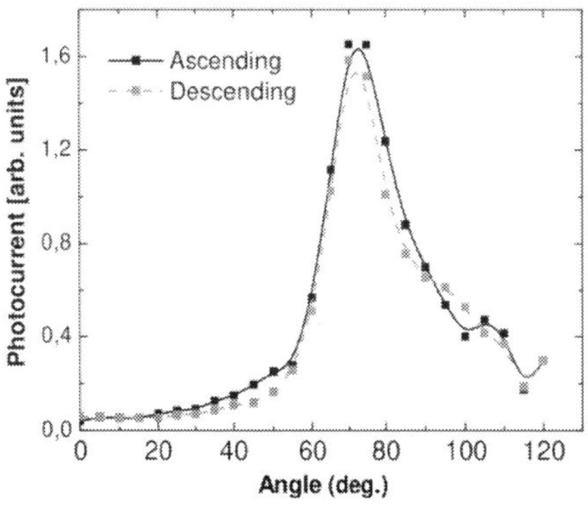

Figure 5: Changes in the photocurrent of a microcavity produced by change of the incidence angle of an He-Ne laser beam

As the whole structure of the microcavity is porous all the component layers are connected to the surrounding atmosphere. Therefore any change of the refractive indexes of the layers, induced by the presence of analytes in the atmosphere, produces a change in the tuning of the microcavity. Since the spectral peak is very sharp, large changes are expected to be produced in the photocurrent due to these tuning changes, allowing the sensing of an analyte. The refractive index of PS changes when exposed to non specific flammable solvents like ethylic alcohol or isopropyl alcohol. A change of 2% in the refractive index produces a variation of up to 5% in the conductance of this device. The pore network surface has to be functionalized for specific analyte sensing.

4. Conclusions

Nanostructured porous silicon has been used to make an optical microcavity in order to enhance the photoconductance by photon confinement in the cavity. The device is a multilayer consisting in a half-wavelength optical thickness layer, limited by two periodic arrays alternating layers with high and low refractive indexes, all of them with a quarter-wavelength optical path. The reflectance spectrum and the internal field intensity were computed, showing a large field enhancement at the cavity for wavelengths corresponding to the resonance. The measured FWHM of the resonance of the reflectance spectrum was 17 nm with Q = 40. A finely tuned spectral photo-response has been experimentally obtained, with the same Q value, in accord with the results predicted using a simple model. These features result as a consequence of an enhancement of the electric field intensity by a factor up to 14 in the mode of the cavity.

5. References

[1] Bisi O, Ossicini S, L. Pavesi L 2000 *Surface Science Reports*, **38**, 1
[2] Pavesi L, Mazzoleni, C. Tredicucci, A. Pellegrini, V. 1995 *Appl. Phys. Lett.*, **67**, 3280.
[3] Palestino G, Légros R, Agarwal V, Pérez E, Gergely C, 2008 *Sensors and Actuators B, Chem.* doi: 10.1016/j.snb. 2008.07.013.
[4] Ouyang H, Striemer C C, and Fauchet P M 2006 *Appl. Phys. Letters* **88** 163108
[5] Theiss W. 1997 *Surface Science Reports*, **29**, 91-192
[6] Cisneros, J. I. 2000 Ondas Electromagnéticas Fundamentos e aplicacoes. Livro Texto
[7] Born M , Wolf E, 1980 *Principles of optics*, 6th Ed. Pergamon Press, Oxford.
[8] Rose A 1978 *Concepts in Photoconductivity and Allied Problems*, Robert Krieger Publishing Co., Huntington, New York.

Secondary electron emission in nanostructured porous silicon

GD Ruano[1], **J Ferrón**[1,2] **,RR Koropecki**[1,2]

[1] INTEC-UNL-CONICET, Güemes 3450 - 3000 Santa Fe.
[2] Departamento de Ingeniería de Materiales, FIQ – UNL

E-mail: gdruano@ceride.gov.ar

Abstract. We studied the reversible reduction induced by ion bombardment of the secondary electron emission (SEE) yield. This effect has been modelled as due to changes in dynamically sustained dipoles related with ions and electrons penetration ranges. Such charge configuration precludes the escape of electrons from the nanoporous silicon, making the SEE dependent on the flux of impinging ions. Since this dipolar momentum depends on the electric conduction of the porous medium, by controlled oxidation of the nanoporous structure we change the conduction features of the sample, studying the impact on the SEE reduction effect. Li ion bombardment was also used with the intention of changing the parameters determining the effect. FT-IR and Auger electron spectroscopy were used to characterize the oxidation degree of the samples at different depth scales

1. Introduction

Since efficient photoluminescence of porous silicon (PS) was discovered by Canham [1], this material has been extensively studied. Pores in PS can be tailored in different size scales ranging from microns to nanometers. The most interesting features appear in the nanoscopic range, where quantum effects are involved. The expected utility of PS covers a wide range of applications. An incomplete list includes its possible use as electron cold emissor [2], electroluminescent devices [3], plasma and field emission flat displays [4, 5]. To fully accomplish any of these goals, a full understanding of secondary electron emission (SEE) processes is needed. Although the physics of SEE in thin dielectric films is a matter that has been widely studied [6-8], size effects in electronic properties of nanoporous silicon introduce new features opening an interesting field of basic and applied physics.
In this work we present results for ion bombardment effects on secondary electron emission of nano, meso and macroporous, pure silicon and oxidized films.

2. Experimental details

Nanoporous silicon (NPSi) samples were prepared by anodization of monocrystalline n-type (1-10Ωcm) and p-type (7-17Ωcm) silicon substrates using a solution 1:2 HF (50%): ETOH, as electrolyte and current density of 20mA/cm2. P-type samples were prepared in the dark and n-type under illumination having a luminic intensity of 140 mW/cm2 provided by a dichronic lamp. In order to prepare mesoporous silicon, p-type (1-4 mΩcm) wafers were anodized using the same electrolyte and current density as for NPSi. Macroporous silicon was obtained from p-type silicon (7-17 Ωcm) using a 1:9 HF (50%): Dimethilformamide. The current density used was 3mA/cm2 in this case. In all the cases the samples were rinsed with ethyl alcohol after preparation and then dried under N_2 flow. Thicknesses of the samples were estimated at around 1 μm for thin samples and 10-30 μm for the thicker ones. The thicker nanoporous samples showed crashes but they had a uniform aspect. The porosity of the mesoporous samples were estimated from the fit of the reflectance spectra taken from thin layers, using the Looyenga approximation. The porosity of NPSi was estimated using the gravimetric method. The values of porosities were around 0.9 for the nanoporous sample and 0.55 for the mesoporous samples.

The SEE spectra were measured in a PHI 590A Auger electron spectrometer, equipped with a single cylindrical analyzer. The base pressure was 2 x 10-10. To evaluate oxidation role in the studied effect in NPSi, separed samples pieces were annealed by exposure to air in an oven at 200 °C during periods ranging from 0 to 480 minutes. In order to compare different kind of PS samples; macro, meso and nano porous were annealed during 30 min just to stabilize them. To acquire the full spectra, the sample was polarized (-6 V). The spectra were acquired in the integrated mode, under a 400 eV - 750 nA/ mm^2 primary electron beam bombardment.

The sample surface was ion bombarded during SEE measurement in order to observe the reducing effect. Ion beam energy and current density was always 4 keV and 90 nA/mm^2. These conditions were chosen to make negligible the ion induced SEE as compared to the electron one. The oxygen amount was determined by means of Auger electron spectroscopy (AES) monitoring the O_{KLL} transition, using the same electrons spectrometer in the differentiated way, and IR transmission spectra were obtained in an FTIR commercial spectrometer (model Spectrum RX-FTIR).

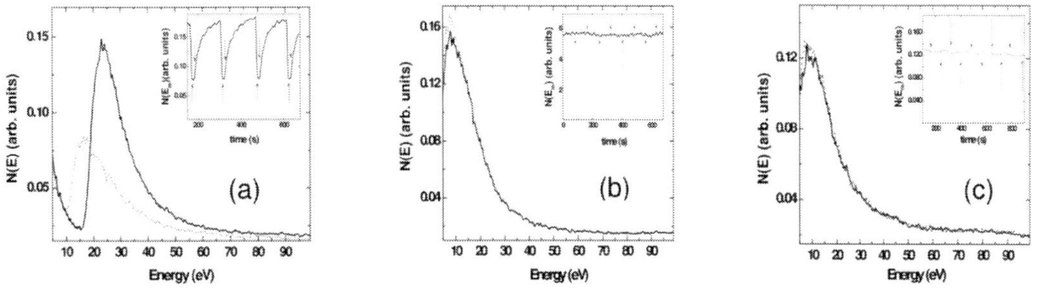

Figure 1. SEE spectra from a thick samples of NPSi (a), mesoporous silicon (b), and macroporous silicon (c). Full lines: only electron produced SEE, dotted line: SEE in the case of ion and electron simultaneous bombardment. Insets: Evolution of the maximum in SEE signal. Arrows up (down) indicates when ion gun is on (off).

3. Results and Discussion

The SEE spectra for samples exposed to the e-beam, and for the same samples exposed to both electrons and to ions are shown in Figure 1. Fig. 1 (a) shows the spectra of a ~ 30 μm thick oxidized n-type sample. When the ion beam is turned on the effect we are reporting can be fully appreciated, i.e. a strong signal decrease and an energy shift of the SEE spectrum is observed. The temporal evolution of the SEE signal, at fixed energy corresponding to the maximum of the electron induced SEE spectrum, is also shown (Insets in fig.1). The SEE reduction effect is clearly reversible. The effect was qualitatively similar in the case of unintentionally oxidized n-type nanoporous silicon. We found this effect negligible for the other tested (oxidized) samples, i.e. mesoporous (Figure 1b), macroporous (Figure 1c), and non porous crystalline silicon (not shown).

In order to reduce the ion induced energy shift effect (Fig. 1a) during the study of the signal evolution in the transient, we performed the experiments using a thinner (~1 μm) sample. The evolution of the maximum of the SEE spectrum is shown in Fig. 2 (a) for different electron beam currents. When the ion beam is turned off, the SEE signal evolves until the initial e-beam induced SEE condition is reached. Both, the initial and the final value of the SEE yields in these experiments depend on the electron beam current. The curves show an exponential behavior and can be fitted (dot lines in Fig. 2(a)) with the function:

$$I_S = I_o\left[1 - A\cdot\exp(-\frac{t}{\tau})\right] \quad (1)$$

3.1 Model.

The whole behaviour may be interpreted in terms of a capacitive effect, were τ is a time constant. The depth of electron penetration (range) is mainly determined by its energy and the solid density. The smaller the PS density, the greater the electrons range. Assuming a linear dependence we can estimate the range to be around 200 nm for 0.5 keV electrons. Being the scale size of the nanostructure in the few nm, the continuous medium approach model is valid. Therefore a negative layer is implanted deep in the PS layer.

Conversely, the range of Ar^+ ions with energies around 4 keV is around 5 nm for bulk silicon, which results of the same order or even smaller than the nanostructure dimensions, so that the range for ions in PS is similar to the silicon bulk case. Since secondary electrons emission originates close the surface, no matter the type of bombardment particle, i.e. electrons, ions, or electron plus ions, the surface layer became depleted of electrons. If only the primary e-beam is operating an internal dipole is created due to the electron depleted surface layer and negative cumulated charge at a depth defined by the impinging electron range. This process occurs for all dielectrics exposed to energetic primary e-beams [9] therefore a surface positive layer and a depth negative layer are created as shown in the scheme of Fig 2(b). On the other hand, the whole energy of the primary e-beam is deposited in the region in between these layers creating electron-hole pairs, and enlarging the conductivity. Thus, a steady state in the charge distribution will be reached as a result of the electron and hole drifts induced by the internal field of the dipole. An additional charge relaxation effect will occur between the implanted negative region and the crystalline silicon substrate. This relaxation takes place slowly due to the large PS resistivity originated in the quantum confinement enlarging of the gap. Therefore, the dipolar momentum results from a dynamic balance between charging and relaxation processes.

When the ion beam strikes the sample, having the electron beam previously turned on, the positive surface charge increases due to the smaller ion range, enhancing the value of the internal dipolar momentum but without electron-hole generation in the bulk. Since the electric field near the positive surface layer preclude the electron emission, the SEE signal should decay, as it actually occurs (see Fig.1 (a)). When the ion beam is turned off, the implanted charge exponentially decay [9] so the SEE yield should recover the only e-beam irradiation value, following the same rule, shown in Fig.2. Finally, considering the surface dipole as a capacitor, and taking into account that the greater the primary e-beam current the greater the e-h pairs generation and the greater the conductivity of the

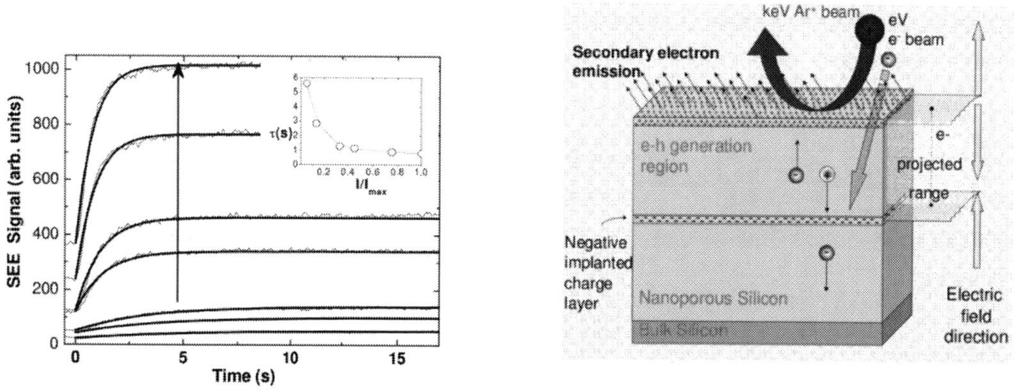

Figure 2. (a) Full lines: SEE evolution for different primary electron beam current after ion gun shutdown. Dotted lines: curve fit with equation (1).The arrow indicates the sense of increasing of the electron beam current. In the inset, the evolution of τ with the electron beam current refered to the maximum one is shown. (b) Schematic picture of the charging process during e-beam (and also ion beam) exposure.

irradiated layer, the relaxation of the charge should follow an exponential decaying with a time constant proportional to the inverse of the conductivity. Therefore the model predicts a decrease of τ with e beam intensity, as it is shown in Fig.2.

Li^+ ions were employed to induce the SEE reduction effect. The intention was to vary, based on the deeper penetration of Li ions as compared to Ar ones, the dipole distance between the charged regions. The effect was initially observable, but it rapidly vanished because of the doping induced by Li bombardment into the NPSi structure. This effect strongly supports the idea that the conductivity of the surface layer plays a fundamental role in the existence of the effect.

3.2 Structure effect.

Due to the fractal nature of the PS nanostructure, the characteristics of the surface remain essentially the same after ion bombardment and thus effect remains after long term ion bombardment.

The mesoporous and the macroporous samples have larger conductivities than nanoporous silicon one, allowing a faster charges rearrangement, turning negligible the surface charge and the induced dipole. Thus, this model predicts that the ion induced SEE reduction should operate for nano but not for meso and macroporous samples as we actually found in our experiments (Figs.1 a and b).

3.3 Oxidation effect

NPSi oxidation affects its conductance in two ways; the macroscopic conductance decreases due to the growth, at the expense of silicon, of an insulator phase of silicon oxide. On the other way, since dimensions of nanostructures are in close relation to the degree of oxidation of NPSi samples [10] the confinement gap increases lowering the conductivity.

Oxidation effects were studied in a detailed way through electron induced SEE spectroscopy with and without bombardment from p-type NPSi samples thermally oxidized in air at 200ºC. These samples were preferred due to its larger stability as compared to n-type ones.

For oxidation experiments, different fragments of the same p-type NPSi were isothermal annealed during different time intervals (up to 8 hours) at atmospheric pressure, being one of them loaded into the spectrometer immediately after fabrication, i.e. with no annealing. The oxygen amount was checked by AES and FT-IR, showing both techniques the same evolution. Since FT-IR monitors the whole sample whereas AES sees only a very narrow region near the surface we infer that oxygen is incorporated uniformly inside the porous layer

Fig. 3 shows the reversible reduction coefficient -defined as the relative change of the SEE produced by ion bombardment- versus the oxygen content -evaluated by integration of the stretching band of Si-O bonds in siloxane bridges. As it can be observed, more oxidized samples show larger reduction effects. In the inset of Fig. 3 we can see higher SEE for stronger oxidized samples. The behavior shown in this figure supports our idea that larger dipole means larger reducing effect.

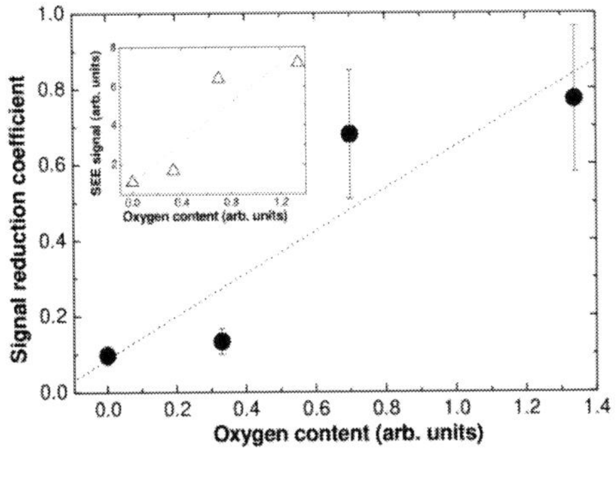

Figure 3. Signal reduction coefficient versus oxygen content. In the inset: SEE signal as a function of the oxygen content.

Journal of Physics: Conference Series **167** (2009) 012006 doi:10.1088/1742-6596/167/1/012006

3. Conclusions

In NPSi samples a reversible reduction in electron induced SEE exists when simultaneous bombardment with ions occurs. This phenomenon can be understood as the result of stationary dipoles modification coming from the balance between generation and recombination of carriers by charged particle bombardment. The difference between ion and electrons ranges has an important role in the kinetics of SEE variation. The reduction in secondary emission is more important for oxidized samples, and turns to be reversible due to nanoporous structure fractal feature.

5. References

[1] Canham LT 1990 *Appl. Phys. Lett.* **57** 1046

[2] Takai M, Yamashita M, Wille H, Yura S, Horibata S, and Ototake M 1995 *Appl.Phys.Lett.* **66** 4

[3] Koshida N, Sheng X, and Komoda T 1999 *Applied Surface Science*, **146**, 371

[4] Tam HL, Kung EW, Zhangb XX, Gongb ML, Cheah KW 2003 *Optoelectronics Proceedings of the Sixth Chinese Symposium* 120

[5] Kleps I, Nicolaescu D, Lungu C, Musa G, Bostan C , Caccavale F 1997 *Applied Surface Science* **111** 228

[6] Fitting HJ, Cornet N, Touzin M, Goeuriot D, Guerret-Piécourt C, Tréheux D 2007 *J. of the Eur. Cer. Soc.* **27**

[7] Xue-Chun Li, You-Nian Wang 2006 *Thin Solid Films* **506– 507** 307

[8] Baragiola RA 1982 *Radiation Effects* **61** 47

[9] Thiel BL, Toth M 2005 *J. Appl. Phys* **97** 051101

[10] Koropecki RR, Arce RD, Schmidt JA 2004 *Phys. Rev. B* **69** 205317

XIX Latin American Symposium on Solid State Physics (SLAFES XIX) IOP Publishing
Journal of Physics: Conference Series **167** (2009) 012007 doi:10.1088/1742-6596/167/1/012007

Au nanoparticles films used in biological sensing

M Rosales Pérez[1], R Delgado Macuil[1], M Rojas López[1], V L Gayou[1], J F Sánchez Ramírez[2]

[1]Centro de Investigación en Biotecnología Aplicada del IPN, Tepetitla Tlaxcala México C.P. 90700

[2]CICATA Legaria Instituto Politecnico Nacional, México Distrito Federal.

E-mail: mrosalespe@ipn.mx

Abstract. *Lactobacillus para paracasei* are used commonly as functional food and probiotic substances. In this work Au nanoparticles self-assembled films were used for *Lactobacillus para paracasei* determination at five different concentrations. Functionalized substrates were immersed in a colloidal solution for one and a half hour at room temperature and dried at room temperature during four hours. After that, drops of *Lactobacillus para paracasei* in aqueous solution were put into the Au nanoparticles film and let dry at room temperature for another two hours. Infrared spectroscopy in attenuated total reflectance sampling mode was used to observe generation peaks due to substrate silanization, enhancement of Si-O band intensity due to the Au colloids added to silanized substrate and also to observe the enhancement of *Lactobacillus para paracasei* infrared intensity of the characteristic frequencies at 1650, 1534 and 1450 cm^{-1} due to surface enhancement infrared absorption.

1. Introduction

The development and consumption of functional probiotic foods has been increasing alongside awareness of their beneficial effects in promoting gut health as well as in disease prevention and therapy, and this has raised interest in health-promoting foods [1]. The contribution of probiotic bacteria, mainly lactobacilli and bifidobacteria, is to maintaining or improving microbial balance in the gut [1], and investigations are currently under way into their role in reducing the risk of cancer, influencing immunomodulatory features and preventing food allergies [2], counteracting hypercholesterolemia [1], and alleviating the symptoms of lactose intolerance [3]. The benefits derived from a regular intake of probiotic foods are also correlated to their ability to inhibit pathogens [4] and protect humans from gastrointestinal diseases. Nowadays, foods fortified with health-promoting probiotic bacteria are mainly produced with fresh milk or milk derivatives such as yogurt, cheese, ice cream, and desserts [5]. Functional food industries are now focusing on new non-dairy foods that can contribute to a regular assumption of probiotics in individuals with lactose intolerance or with a diet lacking milk-derived products. The suitability of cereals as a substrate for growing probiotic bacterial strains has been evaluated [6] but at present only a few functional cereal-based products are available on the market, including oat meal gruel mixed with a fruit drink containing *Lactobacillus plantarum* 299v and a rose hip drink with oats fermented by *L. plantarum* 299v [7]. To colonize the gastrointestinal tract, probiotic strains need to be ingested as large populations and on a

daily basis. Therefore, food manufacturers are trying to include probiotic strains in foods and beverages that are part of a normal diet to provide health defences while enjoying meals and to differentiate such functional products from concentrated probiotic preparations available as capsules, powders, or liquids. Research is currently under way to obtain a variety of probiotic products such as dry sausage fermented by *Lactobacillus rhamnosus* strains [8] fruit pieces containing *Bifidobacterium* spp [9], dried fruits vacuum-impregnated with *Saccharomyces cerevisiae* and *Lactobacillus casei* subsp. *rhamnosus* [10], and others products with different bacterial contents [11, 12].
On the other side, nanoparticles are widely used in the field of electronics, magnetism, catalysis, and others because of their different properties [13,14,15]. The ability of gold nanoparticles to provide a stable immobilization of biomolecules retaining their bioactivity is a major advantage for the preparation of biosensors [16], whereas in spectroscopy analysis the Surface Plasmon Resonance (SPR) effect could be used to enhance the biological signal via the SEIRA (Surface Enhancement Infrared Absorption) or SERS (Surface Enhancement Raman Scattering) effect.
In this work, the presence of *Lactobacillus para paracasei* was determined via the SEIRA effect associated to the SPR effect of Au nanoparticles films. The infrared response was observed using five different *Lactobacillus para paracasei* concentrations.

2. Materials and Methods

2.1 Bacterial strains and culture conditions

Lactobacillus para paracasei was obtained from Pulque, Zacatlán, Puebla. The probiotic potential of *L. para paracasei* was confirmed by adherence and antagonism tests (figure 1). Working cultures were prepared by inoculation of *L. para paracasei* strain CIBA22 in MRS broth (Man Rogosa Sharpe, Oxoid) at 37°C during 24 h in anaerobic conditions. Working cultures were sub cultured twice before used in experiments. Bacterial suspensions were prepared, the bacterial concentration of *L. para paracasei* was determinate by optical density at 600 nm (OD600) using a MacFarland Nephelometer. For long-term storage, stock cultures were prepared mixing fresh medium (8 ml) with glycerol (2 ml, Merck) and then freezing 1-ml aliquots of this mixture at -70°C in sterile cryovials (Nalgene).
The concentrations of *L. para paracasei* used in this work were: 3×10^8 UFC/ml, 9×10^8 UFC/ml, 1.5×10^9 UFC/ml, 2.1×10^9 UFC/ml and 2.7×10^9 UFC/ml, **UFC** means Units Formed Colonies.

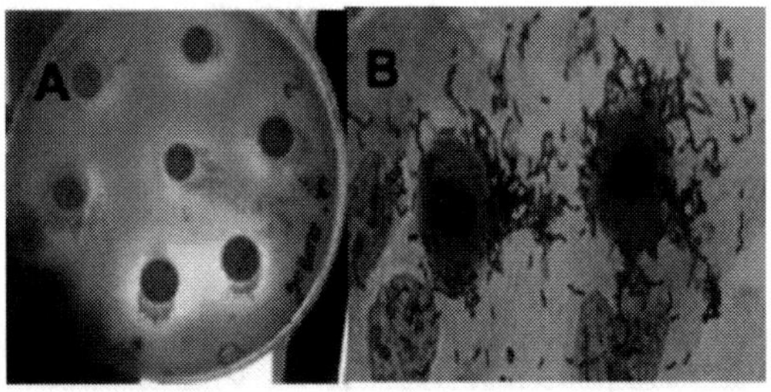

Figure 1. *Lactobacillus para paracasei. A.* Antagonism against pathogenic bacterial EHEC. B. Adherence on HEp-2 cells.

2.2 Colloids Solutions

Colloidal dispersions of gold (Au) nanoparticles with different sizes were prepared in a typical synthesis; solutions of $HAuCl_4$ (0.096 mmol in 25 ml of water) and poly (N-vinyl-2-pyrrolidone) (PVP, 100 mg in 20 ml of water) were prepared by dissolving the $HAuCl_4$ crystals and PVP in water. Both solutions were mixed to produce an Au (III) ion solution containing PVP as a protective polymer. Then an aqueous solution of ascorbic acid (AA, 0.096 mmol in 5 ml of water) was added to the resulting solution at room temperature. Colloidal dispersions of Au nanoparticles were formed after addition of AA solution in the mixture solution. The amount of metal ions was altered from 0.096 to 0.186 mmol in order to control the size of Au particles. Three colloids solutions were obtained with Au nanoparticle sizes: 21, 27 and 31 nm; labelled as Au-1, Au-2 and Au-3 respectively.

Journal of Physics: Conference Series **167** (2009) 012007 doi:10.1088/1742-6596/167/1/012007

In figure 2 are shown an Electron Micrograph of Au nanoparticles as well as an histogram of their distribution with particle size obtained by Transmission Electron Microscopy.

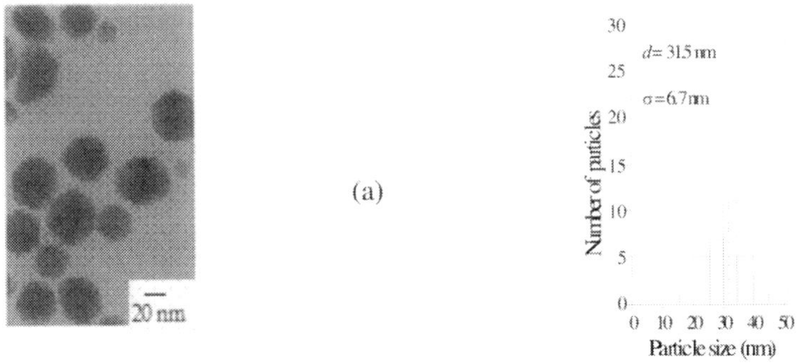

Figure 2. Electron micrograph and histogram of Au nanoparticles solution.

2.3 SEIRA Substrates

Functionalized substrates were done using the aminopropil trimetoxysilane at 1% in an acetic acid water solution. Corning glass substrates were submerged in this solution by one hour and then were cleaned using water. After that, silanized substrates were immersed in each colloid solution at two different times: a half (M) and one hour (H), at room temperature and dried by four hours at room temperature.

Deposition of Au nanoparticles was analyzed by infrared vibrational spectroscopy. The SEIRA effect on the silane molecules can be observed in figure 3, the peaks associated to silane bonds (dash line) increased their intensity due to the Au nanoparticles present in the functionalized layer (dark and grey continuous line). The peaks associated to silane have an amplification of one magnitude order; the value goes from 0.025 to 0.30 in Y axis.

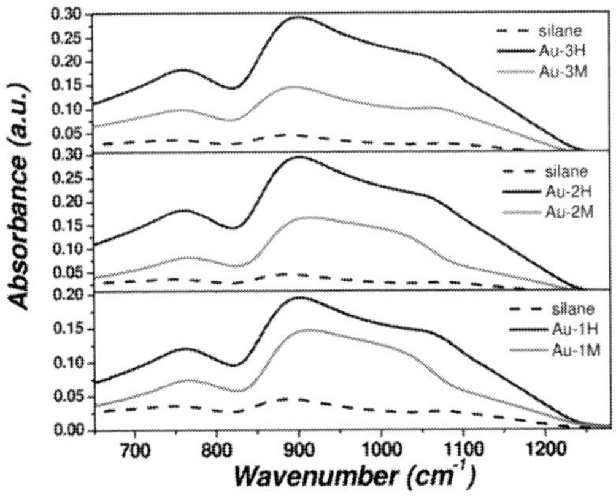

Figure 3. Infrared absorption spectra of SEIRA substrates.

Attenuated Total Reflexion sampling mode (ATR) of an Infrared spectrometer Bruker Vertex 70 was used to obtain the spectra of the figure 3. The spectrum of a glass substrate was used as background in all the measurements; the ATR crystal used was a ZnSe which has one reflection. The exposure time was 30 seconds for each sample.

Journal of Physics: Conference Series 167 (2009) 012007 doi:10.1088/1742-6596/167/1/012007

3. Results and Discussion

Figure 4, shows the infrared absorption spectrum of *L. para paracasei* in aqueous solution form. The water baseline correction has been applied to this spectrum. The characteristic peaks associated to the bacteria are present at 2100, 1635, 1481, 1430, 1345, 1288 and 1186 cm^{-1}.

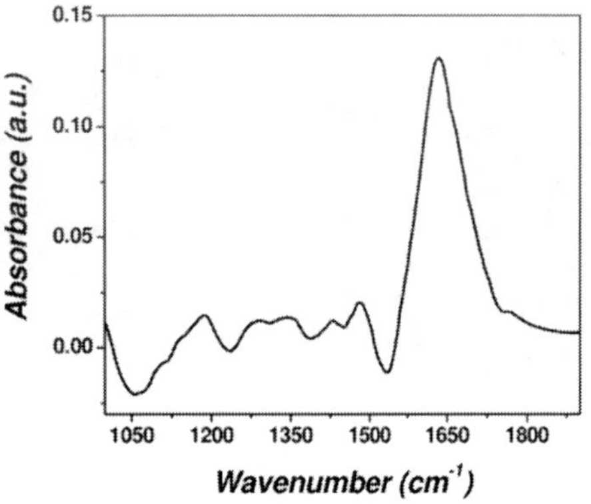

Figure 4. Infrared absorption of *L. para paracasei* infrared vibrational response in aqueous solution.

A drop of *L. para paracasei* in aqueous solution was added on the silanized substrate and on the Au nanoparticles functionalized films (glass with silane and Au). The samples were dried during two hours and then the ATR-FTIR spectra were taken.

Figures 5a and 5b show the behaviour obtained for the 9 x 10^8 UFC/ml and 2.1 x 10^9 UFC/ml, respectively. The spectra of corning (grey light continuous line), silane film (dark continuous line) and Au nanoparticles film (gray continuous line) were also added; these three spectra were translated vertically. Finally in the figure 5, dark dash line is for *L. para paracasei* on silane film and grey dash line is for *L. para paracasei* on Au nanoparticles film. The *L. para paracasei* spectra were not altered in anyway to put in evidence the SEIRA effect on the *L. para paracasei* molecules.

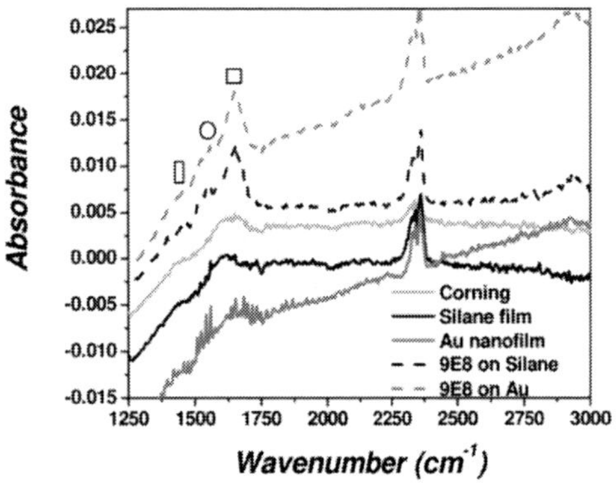

(a)

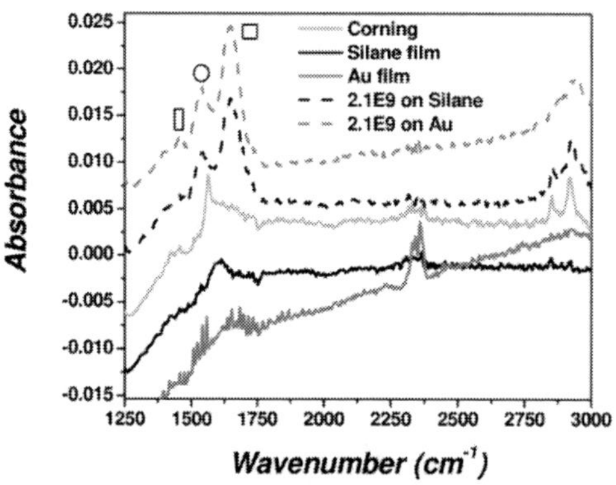

(b)

Figure 5. Infrared spectra for the *L. para paracasei* in SEIRA substrate.

The peak at 2300 cm^{-1} is associated to the environmental ozone. In both figures, the Au nanoparticles generate undefined spectra (dark gray continuous line). The spectra of *L. para paracasei* on silane and Au (dashed lines) show peaks at 1650, 1534 and 1450 cm^{-1} (square, circle and rectangle respectively), which could be associated only to *L. para paracasei* (see figure 4). Also in the CH region (region between 2700 to 3000 cm^{-1}) is possible observe the new peaks generation at 2940 and 2852 cm^{-1}.

Table 1. Values of peak intensity for the figures 5a and 5b spectra, associated to the bacterial concentration, in silane and Au nanoparticles substrates.

Wavenumber cm^{-1}	*L. para paracasei* concentration			
	9 x 10^8 UFC/ml		2.1 x 10^9 UFC/ml	
	Silane	Au nanoparticles	Silane	Au nanoparticles
1650	0.012	0.018	0.016	0.025
1534	0.0068	0.011	0.011	0.018
1450	0.0034	0.0071	0.0064	0.013

The amplification factor observed for *L. para paracasei* is two when the SEIRA effect is present, for the 1450 cm^{-1} peak. This amplification factor is lower, when compared with a one magnitude order observed for the silane films in the figure 3; the experimental values are shown on table 1. This behaviour could be due to the lower concentration of the bacterial on the film.

4. Conclusions

We have proved the use of silane and Au nanoparticles functionalized films, deposited on glass substrates to observe the enhancement of the infrared absorption intensity of the biological element *Lactobacillus para paracasei*. The mean peaks observed for this material, located at 1650, 1534 and 1450 cm^{-1} where enhancement in intensity which could be due to the SEIRA effect.

5. References

[1] Saarela M, La¨hteenma¨ki L, Crittenden R, Salminen S and T Mattila-Sandholm 2002 *Int. J. Food Microbiol* **78** 99–117.
[2] Murch S 2005 *Arch. Dis. Child.* **90** 881–88.
[3] Salminen S, Ouwehand A C and Isolauri E 1998 *Int. Dairy J.* **8** 563–572.
[4] Cross M L 2002 *FEMS Immunol. Med. Microbiol.* **34** 245–253.
[5] Lourens-Hattingh A and Viljoen B C 2001 *Int. Dairy J.* **11** 1–17.
[6] Charalampopoulos D, Wang R, Pandiella S S and Webb C 2002 *Int. J. Food Microbiol.* **79** 131–141.
[7] Molin G 2001 *Am. J. Clin. Nutr.* **73** 380s–385s.
[8] Erkkila¨ S, Suihko M L, Eerola S, Peta¨ja¨ E and Mattila-Sandholm T 2001 *Int. J. Food Microbiol.* **64** 205–210.
[9] Maguina G et al 2002 *Abstr. Annu. Meet. Food Expo., abstr.* 15E-25.
[10] Betoret N, Puente L, Díaz M J, García M J, Gras M L, Martínez- Monzó J and Fito P 2003 *J. Food Eng.* **56** 273–277.
[11] Shimakawa Y, Matsubara S, Yuki N, Ikeda M and Ishikawa F 2003 *Int. J. Food Microbiol.* **81** 131–136.
[12] Ouwehand A C, Kurvinen T and Rissanen P 2004 *Int. J. Food Microbiol.* **95** 103–106.
[13] Jia J, Wang B, Wu A, Cheng G, Li Z, Dong S. 2002 Anal Chem **74**:2217
[14] Tang DY, Xia BY, Zhang YQ 2008. Microchim Acta **160**: 367–374
[15] Liu Y, Geng T, Gao J 2008 Microchim Acta **161**: 241–248
[16] José M. Pingarrón, Paloma Yañez-Sedeño, Araceli González-Cortés 2008 Electrochimica Acta **53** 5848–5866.

Journal of Physics: Conference Series **167** (2009) 012008 doi:10.1088/1742-6596/167/1/012008

Effects of phonon dimensionality in the specific heat of multiwall carbon nanotubes at low temperatures

Guillermo A Jorge[1], V Bekeris[1], C Acha[1], M M Escobar[2,3], S Goyanes[2], D Zilli[4], A L Cukierman[4] and R J Candal[5]

[1] Laboratorio de Bajas Temperaturas, Departamento de Física, FCEyN-UBA, Pab. 1, Ciudad Universitaria (1428), Buenos Aires, Argentina.
[2] Laboratorio de Polímeros y Materiales Compuestos, Departamento de Física, FCEyN-UBA, Pab. 1, Ciudad Universitaria (1428), Buenos Aires, Argentina.
[3] Departamento de Química Inorgínica, Analítica y Química Física, FCEyN-UBA, Pab. 2, Ciudad Universitaria (1428), Buenos Aires, Argentina.
[4] PINMATE, Departamento de Industrias, FCEyN-UBA, Pab. Industrias, Ciudad Universitaria (1428), Buenos Aires, Argentina.
[5] Instituto de Fisicoquímica de Materiales, Ambiente y Energía, CONICET-UBA, Ciudad Universitaria (1428) Buenos Aires, Argentina

E-mail: gjorge@df.uba.ar

Abstract.
We have measured the specific heat at constant pressure, C_p, of three different samples of multiwall carbon nanotubes (MWNT). For all samples, C_p departs from a graphitic behavior at $T < 120$ K. C_p measurements show a temperature threshold from a linear regime for intermediate temperature to a higher-order power law for low temperatures. Moreover, it was found that this crossover only depends on the internal structure of the individual MWNT and not on the spatial order of the MWNT within a bundle.

1. Introduction

Carbon nanotubes have attracted great attention due to their potential technological applications. They present a variety of interesting physical properties that have been studied extensively over the past years. Particularly, the knowledge of the vibrational properties of these systems is of technological interest since the electron-phonon interaction is one of the mechanisms that determine their electronic transport properties. In spite of the intense experimental work done on single-wall carbon nanotubes (SWNT) and multiwall carbon nanotubes (MWNT), their specific heat at constant pressure (C_p) at low temperatures has not been comprehensively studied, specially if we refer to MWNT.

Within the few experimental works on the specific heat of MWNT, Yi *et al* [1] measured C_p in a bundle of MWNT, and found that it varied linearly with temperature in the range 10-300 K. It was attributed to a prevalence of the out-of-plane modes in the essentially bidimensional vibrational excitation spectrum within each tube, with a weak intertube coupling. Later on, Mizel *et al* measured bundles of SWNT and MWNT [2], finding that for the latter the specific heat was similar to that of graphite, meaning that the intertube coupling was comparable to interplane coupling in graphite. Finally, Masarapu *et al* worked with a bundle of aligned

MWNT [3] and obtained a linear specific heat for $T>40$ K, and a dimensional crossover to a higher power law at lower temperatures. This was assigned to a dimensional change in the excitation spectrum from 3D at low temperatures to a reduced dimensionality at higher temperatures.

In this work, we present a systematic survey on the specific heat at constant pressure in three samples of MWNT with different geometries as a function of temperature. We observed a crossover whose characteristic temperature depends on the internal structure of the individual MWNT that constitute the sample.

To our knowledge, a systematic study of this feature in different kinds of MWNT has not been done previously. Due to the scarce and disperse specific heat data on MWNT and the lack of a theoretical or numerical framework, we expect the work presented in this paper to help understand the basic physical phenomena involved.

2. Experimental

Three different classes of MWNT were were studied: sample A (90 nm external diameter) sample B (30 nm) and sample C (25 nm). Samples A and B were synthesized by CVD of analytical grade iron(II) phthalocyanine, under Ar/H_2 atmosphere in a flow reactor consisting of a quartz glass tube. It was heated by a dual furnace fitted with independent temperature controllers, following the procedure reported elsewhere [4]. Growing conditions involved temperatures of around 880°C, a total gas flow rate of 30 ml/min and Ar/H_2 ratio of 1:1. Carbon nanotubes of sample C were synthesized using a catalyst of Fe nanoparticles diluted in a SiO_2 matrix. The nanoparticles were reduced in a 9% H_2/N_2 flux (110 cm^3/min, 180 Torr) at 600°C for 5 hs. Carbon was deposited using acetylene gas diluted in H_2 at 9% maintaining the same flow and pressure, at 700°C for 3 hs. The nanotubes of sample C were purified using an excess of HF/H_2SO_4 (1:1 v/v) in order to eliminate the silica support and rinsed with HCl to solubilize the iron nanoparticles. Then, the sample was filtered, washed with distilled water, and dried at 80°C for 24 hs.

As a result of the different growing conditions of samples $A-B$ and C, they have different structural properties. SEM microscopy shows that samples A and B growth in aligned bundles, whereas sample C has a forest distribution. TEM images also reveal in the former an internal bamboo-like structure within each MW, not present in the latter.

The specific heat measurements were made with a thin silicon nitride membrane microcalorimeter. It consists in a silicon frame of 11x11 mm^2 with a 5x5 mm^2 SiN membrane, 1 μm thick [5]. On the front side of the membrane, a platinum resistor heater and a NbSi composite thermometer were made by sputtering and photolitography. On the back side of the membrane, a gold conduction layer is deposited in order to thermalize the thermometer and the heater, and the sample is attached on it with a thermally conductive compound. The film thermometers on the membrane were calibrated in temperature against a factory-calibrated Cernox thermometer. The method used to measure specific heat is the thermal relaxation time technique [6].

3. Results and discussions

It is known that the electronic specific heat of carbon nanotubes is more than two orders of magnitude lower than the phonon counterpart [7], even when the nanotubes are metallic. Therefore, the specific heat measured at constant pressure can be regarded as coming entirely from the vibrational degrees of freedom of the system, and has a strong relation with the phonon density of states.

The measured C_p for the three samples is displayed in Fig. 1. We include also the curve of C_p of graphite as a reference [8]. The curves depart from the graphite curve for $T < 120$ K. It is clear that samples A and B follow almost the same curve, whereas sample C presents

Journal of Physics: Conference Series **167** (2009) 012008 doi:10.1088/1742-6596/167/1/012008

differences at low temperatures with respect to the others. First of all, samples A and B have a marked peak at T_P= 60 K, not present in sample C. This peak will be extensively studied in a forthcoming paper, so we will not discuss its origin here. Secondly, we observe a shoulder that is produced at T_{DC}= 33 K in samples A and B, and T_{DC}= 52 K in sample C. All these characteristic temperatures are marked in Fig. 1.

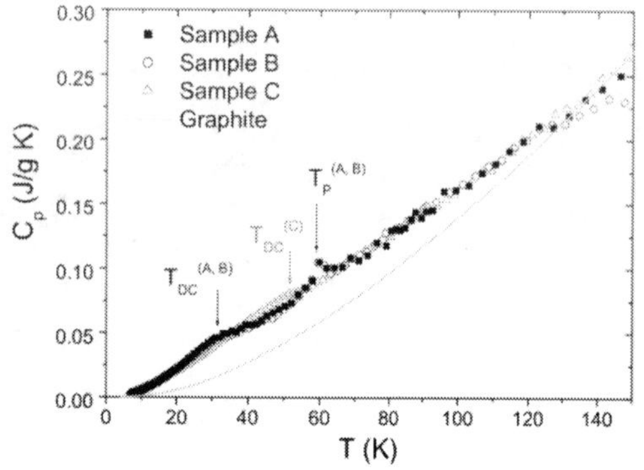

Figure 1. Specific heat of sample A, B and C as a function of temperature. The C_p of graphite is shown for comparison [8]. The characteristic temperatures T_{DC} and T_P are marqued.

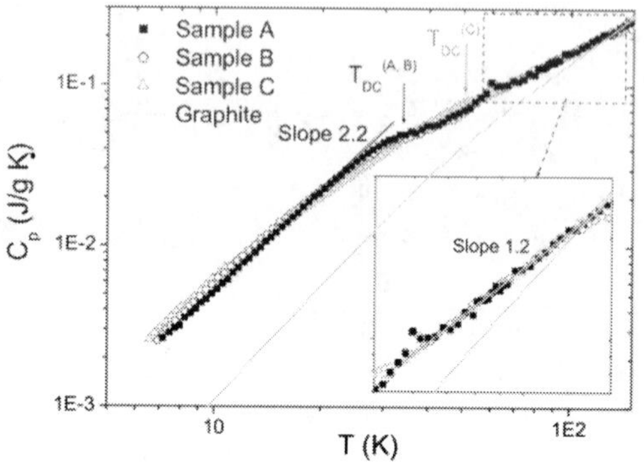

Figure 2. The same data on Fig. 1 with logarithmic axes. The different slopes and characteristic temperatures are marqued. The inset shows an expansion of the rectangle.

In Fig. 2 the same curves are displayed in a bilogarithmic graph. The curves present an approximate slope of 2.2 below T_{DC}. As the temperature is increased, the slope changes to 1.1 above T_{DC} (see the expanded region in the inset of Fig. 2). This crossover is seen clearly in the plot of C_p/T vs. T showed in Fig. 3. In this figure, the characteristic temperature manifests as a maximum in C_p/T. Above T_{DC}, the curves become almost flat indicating a pseudo-linear regime in all samples. Bellow T_{DC}, the vibrational behavior is different among them, as seen in the figure. The crossover at T_{DC} is similar to that observed by Masarapu et al [3], but we obtained a sample dependent shifting of the temperature T_{DC}, that has not been observed previously.

A similar crossover occurs in graphite, in which C_p has a cubic behavior at very low temperatures. Since the graphene layers are weakly coupled, phonon modes with displacements perpendicular to the layers have a very low Debye energy, causing a saturation of the specific heat associated with those modes when the temperature is raised. Hence, the specific heat at intermediate temperatures has a quadratic variation [9]. In our case, we observed a similar behavior, but the crossover is more pronounced, so that we can assign a well-defined crossover temperature T_{DC} for each sample.

For an isolated perfect nanotube, considered as a graphene layer rolled up, it is known that at very low temperatures it can be considered a perfect 1D system [9]. As the temperature is raised (or the radius is increased) radial modes can be excited and the system undergoes a dimensional crossover to a 2D regime. The order of magnitude for the temperature at which this change takes place is $T_{CR} \approx \hbar v / k_b R$, with v being the typical sound velocity and R de radius of the tube [7]. For our samples, this temperature is less than 5 K for all diameters, so we effectively have MWNT formed by essentially 2D individual tubes.

A multiwall nanotube can be regarded as a set of concentric tubes with increasing diameter. Then, there will be a coupling between the internal tubes. If the nanotubes are ordered in bundles, as in the case of samples A and B, there is also coupling between neighbor tubes.

Both couplings introduce phonon dispersion perpendicular to the tube axis. Hence, at very low temperature the system will behave as three-dimensional. As the temperature increases, the transversal modes will saturate their contribution to the specific heat, and the internal phonon structure of the constituents will show up. The mode relevant in a graphene layer is the out-of-plane acoustic mode that has a quadratic dispersion relation, giving a nearly constant phonon density of states and, consequently, a linear specific heat. In the case of tubes large enough this behavior will show up at intermediate temperatures. That is what we observed in Figs. 1, 2 and 3. At $T < T_{DC}$, the system tends to show a $T^{2.2}$ behavior, but with an increasing slope at lower temperatures. Above T_{DC}, the linear behavior in all samples is a consequence of the dominant phonon modes of the individual tubes. The reason why the crossover is not smooth, like in graphite, is unknown to us.

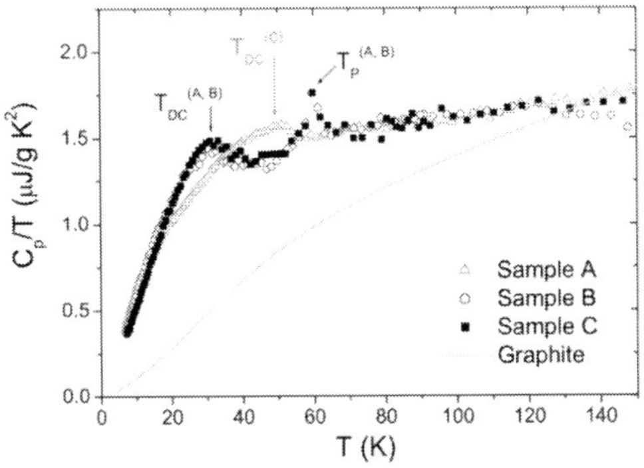

Figure 3. C_p/T vs. T for sample A, B and C. The temperature T_{DC} and T_P are marqued.

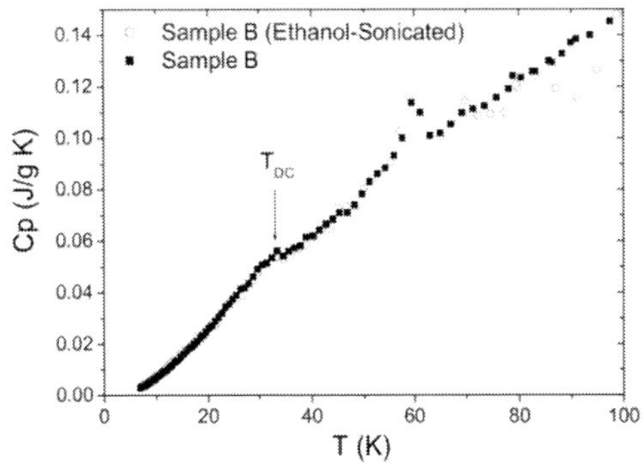

Figure 4. Comparison between C_p of sample B and ethanol-sonicated sample B.

From the measurements we can realize that the characteristic crossover temperature is different for samples $A - B$ (T_{DC}= 33 K) than for sample C (T_{DC}= 52 K). The different temperatures could be due to different internal structure of the MWNT or different arrangement of the MWNT within a bundle. We already mentioned that samples $A - B$ were synthesized by the same process, having an ordered bundle structure. In order to determine whether this structural order is relevant in this problem, we dispersed and sonicated sample B in ethanol and measured again. The spatial order of the MWNT was eliminated by this process, as seen by SEM micrography. The results are shown in Fig. 4, where a plot of the original sample B (full symbols) is displayed alongside the ethanol-sonicated sample (open symbols). The fact that both curves almost coincide in the whole range measured confirms that the spatial order of the individual MWNT within a bundle is not a relevant parameter for the observed crossover.

4. Conclusion

We present specific heat measurements at low temperatures in three different MWNT samples. For all cases a crossover at a well defined temperature, T_{DC}, was observed, separating a low temperature quadratic from a high temperature linear dependence. The main factor that influences T_{DC} is the internal structure of the individual MWNT. This could explain the differences between T_{DC} of samples $A - B$ and T_{DC} of sample C. The latter undergoes the crossover at a higher temperature, possibly due to the enhanced coupling between internal tubes caused by the absence of bamboo-like defects. Further work is needed to prove that it is the change in phonon dispersion relation rather than a change in dimensionality what describes our results.

Acknowledgments
This work was supported by CONICET (PIP5609) and ANPCyT (PICT2006-01201).

References
[1] Yi W, Lu L, Dian-lin Z, Pan Z W and Xie S S 1999 *Phys. Rev.* B **59** R9015–R9018
[2] Mizel A, Benedict L X, Cohen M L, Louie S G, Zettl A, Budraa N K and Beyermann W P 1999 *Phys. Rev.* B **60** 3264–3270
[3] Masarapu C, Henry L L and Wei B 2005 *Nanotechnology* **16** 1490–1494
[4] Huang S, Dai L and Mau A 1999 *Journal of Physical Chemistry* B **103** 4223–4227
[5] Denlinger D W, Abarra E N, Allen K, Rooney P W, Messer M T, Watson S K and Hellman F 1994 *Review of Scientific Instruments* **65** 946–959
[6] Bachmann R, F J DiSalvo J, Geballe T H, Greene R L, Howard R E, King C N, Kirsch H C, Lee K N, Schwall R E, Thomas H U and Zubeck R B 1972 *Review of Scientific Instruments* **43** 205–214
[7] Benedict L X, Louie S G and Cohen M L 1996 *Solid State Communications* **100** 177–180
[8] Nihira T and Iwata T 2003 *Phys. Rev.* B **68** 134305
[9] Dresselhaus M S and Eklund P C 2000 *Advances in Physics* **49** 705–814

Experimental and numerical study of transverse flux shaking in MgB_2 superconductors

J Luzuriaga[1], A Badía- Majós[2], G Nieva[1], J L Giordano[3], C López[4], A Serquis[1] and G Serrano[1]

[1] Centro Atómico Bariloche, CNEA, Inst. Balseiro, UNC, Argentina
[2] Dep. de Física de la Materia Condensada–ICMA, Universidad de Zaragoza–CSIC, Spain
[3] Departamento de Ciencias de la Ingeniería, Universidad de Talca, Chile
[4] Dep. Matemáticas, Universidad de Alcalá de Henares, Spain

E-mail: `luzuriag@cab.cnea.gov.ar`

Abstract. Magnetization measurements in the mixed state of thick strips of carbon nanotube doped MgB_2 in crossed fields configurations are reported, together with numerical simulations performed with a geometry equivalent to the sample shape. The samples were subjected to magnetic field components along mutually perpendicular directions, an oscillatory field in one direction and a remanent magnetization in the perpendicular direction. The magnetic response along the oscillatory field and the magnetic relaxation perpendicular to it are observed and simulated using the critical state theory. A remarkable quantitative agreement between the experiment and the theory was obtained.

1. Introduction

The application of a transverse ac field has been found to induce relaxation of the non equilibrium perpendicular magnetization as has been recently reported in high-T_c superconducting samples.[1, 2]

From the theoretical point of view, the critical state theory seems to capture the essential physics behind the experimental facts. Of special note are the so-called shaking effects on the flux lines. They have been predicted, either for thin strips within a purely analytical approach,[3, 4, 5] or for arbitrary section strips in numerical studies.[6] Remarkably, the irreversible component $\mathbf{M}_{irr}$ perpendicular to the oscillating magnetic field relaxes in the complete absence of thermal activation effects. However, it was also shown[6] that relaxation may stop in some metastable state or eventually lead to a null remanent magnetization depending on the actual experimental conditions.

We study the shaking effect on both components of $\mathbf{M}$ for MgB_2 samples. In our experiment a magnetic field is ramped with a typical period of around an hour, and the changes are large enough to put the sample in the critical state. Therefore this shaking effect is different from that reported in experiments of the type performed by Pasquini *et al.*[7] which use smaller and faster flux shaking that can change pinning of the flux lines. The practical absence of anisotropy and the high pinning forces, make MgB_2 a good candidate for measuring the connection between the phenomenological critical state approach and flux shaking effects. We have matched the geometry of the measured sample and the simulations, and also paid particular attention to the magnetic field dependence of the critical current. In this way, all the parameters which enter into the calculation are fixed, and we are therefore able to test the validity of the approximations. We obtain full agreement between the theoretical prediction and experiments at a quantitative level. A single phenomenological relation $J_{c\perp}(B)$, i.e. the perpendicular (field dependent) critical current density has been used. Such quantity is obtained from a detailed analysis of the one-dimensional $M(H)$ response [8]. This gives confidence in the applicability of the critical state model[6], and proves that the essential physics is captured by it.

2. Experimental details

For the experiments a polycrystalline MgB_2 sample doped with a 10at-% concentration of single walled carbon nanotubes (SWCNT) was used. The $T_c \sim 36$ K [9]and its dimensions were $0.31 \times 0.66 \times 3.0$ mm. The longest dimension (3mm) was set parallel to the rotation axis used for establishing the crossed field configuration. The smallest dimension, (0.31 mm), is matched with the y-axis, and the remaining (0.66 mm) with the x-axis.

We have used a custom built probe[10] with a rotating sample holder, inside a Quantum Design SQUID magnetometer which allows rotation around an axis perpendicular to the magnetic field direction. The crossed field configuration was achieved experimentally by placing the sample so that the y-axis was parallel to the magnetic field direction. Then, it was cooled in this position down to 10 K in zero field. Subsequently, a field H_y was applied and then removed, producing a remanent magnetic moment along the y-axis. With no magnetic field present so as to avoid magnetic induction effects, the sample was rotated 90° placing the x-axis parallel to the field direction. Following this, H_x was cycled several times.

Taking advantage of the SQUID's capability for simultaneously recording the parallel and perpendicular components of magnetization with respect to the magnet's field, we measured the vector $(M_x(t), M_y(t))$ while oscillating the applied field H_x. All the experiments reported were performed at 10 K. We have checked that at this temperatures flux creep effects are negligible, i.e. $\Delta M/M < 0.01$ in a typical measurement period of five hours.

3. Experimental and Numerical results

In Fig. 1 we show measurements of M_y as a function of time compared with the results of the simulations, both in linear and logarithmic scales.

Experimental time scales are small with regard to flux creep and long compared with transients associated with flux changes. The initial field amplitude for the trapped magnetic field was $\mu_0 H_{y,a} = 0.4$ T while $H_{x,a}$ has been swept up to maximum values $\mu_0 H_{x,a} = 0.2$; 0.4 ; 0.6; 0.8 and 1.2 T.

Two regimes are visible, as predicted in Ref. [6], depending on the amplitude of the cycling field $H_{x,a}$. From the graphs, it is clear that the measured M_y tends to a finite saturation value for low values of the amplitude $H_{x,a}$ and to zero for values of $H_{x,a}$ above the penetration field H_{px}. The excellent agreement between the simulations and the measurements can be clearly seen. It should be remarked that the measurement of low values of M_y is subject to larger experimental errors due to some unshielded induction of the oscillating field along the x-axis. This is apparent in particular in the logarithmic plots, where the small discrepancies are enlarged by the scale.

In Fig. 2 we show the measurements of M_x which follows the cycle imposed by the field, together with the corresponding numerical results, showing also quantitative agreement. The magnetization along x is cyclic after the first complete cycle in H_x. It should be noticed that M_x is always in a metastable state during the cycle so that the sample as whole is not in thermodynamic equilibrium, even when the value of M_y is close to zero (or to the reversible equilibrium value). In experiments such as those of Beidenkopf *et al*[11] which use perpendicular fields and aim at measuring a thermodynamic state this fact could perhaps be relevant, especially if the dimension d cannot be neglected.

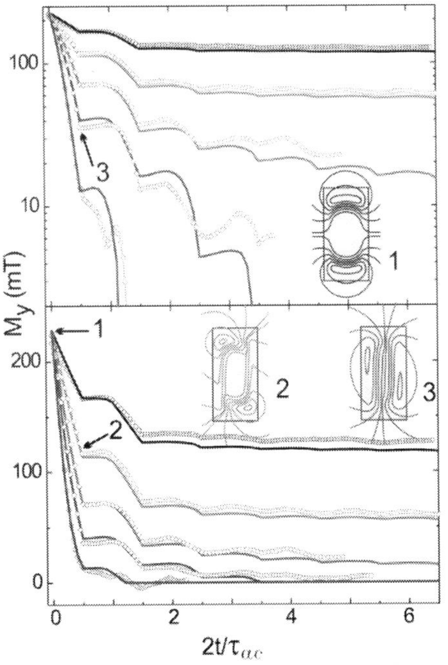

Figure 1. Plots of $M_y(t)$ against time both in logarithmic and linear scales. Symbols: measurements, full lines: model. Notice the two regimes, for $H < H_{px}$ a finite magnetization remains, for $H > H_{px}$ the magnetization collapses to zero. The curves, from top to bottom correspond to $\mu_0 H_{x,a} = 0.2$; 0.4; 0.6; 0.8 and 1.2 T. The insets show the magnetic flux lines penetrating the sample: (1) just after inducing the remanent state for $\mu_0 H_{y,a} = 0.4$ T and rotating the sample, (2) subsequent to the first half-cycle of H_x for an amplitude $\mu_0 H_{x,a} = 0.4$ T and (3) the same for $\mu_0 H_{x,a} = 0.8$ T

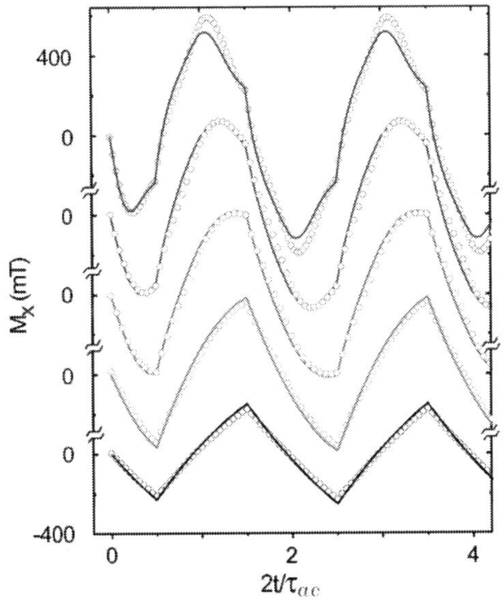

Figure 2. Magnetization along the x-axis $M_x(t)$ as a function of time. Symbols: measurements, full lines: model. The curves correspond to $\mu_0 H_x = 0.2$; 0.4 ; 0.6; 0.8 and 1.2 T. For higher values of H_x the response of M_x also shows a greater amplitude.

As a final experimental issue, we stress that the quantitative analysis of the sample's response $(M_x(t), M_y(t))$ has required a careful determination of the critical current density dependence $J_{c\perp}(B)$.In the low field region problems may arise when extracting $J_{c\perp}(B)$ from the width of the hysteresis loops $\Delta M(H)$[12] so we used a thin disk from the same batch of material. From ΔM obtained by measuring with the magnetic field along the axis of the disk, we have estimated a zero field critical current $J_{c0} = 6.0 \times 10^5 A/cm^2$ which is in agreement with other measurements in this compound.[9] From the same magnetization loop, we have found that for fields below 3 T, one can use the following fit $J_{c\perp}(B) = \frac{J_{c0}}{(1+B/B_0)^2}$ with $B_0 = 1.25$ T.

We have also done some preliminary measurements in which the shielding currents run mainly

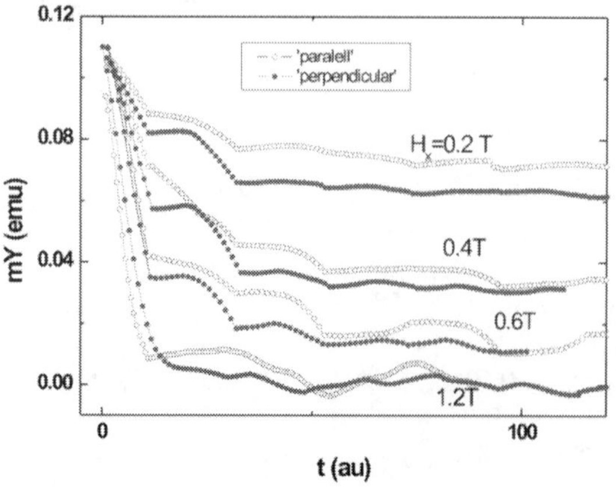

Figure 3. Comparison of the remanent magnetization in two different configurations of the oscillating field. The remanent magnetization m_Y of 0.4 T is applied in the same way, but the oscillating field is along the sample's long axis (black open circles)or along the shorter side of the parallelepiped (blue closed circles,the same data as in Fig. 2)

parallel to the oscillating field in order to test the model in this different condition. We used the same sample, but placed in the holder so that the rotation axis is along the x direction of the sample. The remanent magnetization is established in the same way as before, in the y direction of the sample, but now (after rotation) the oscillating field is parallel to z, the longer dimension of the sample. Consideration of the geometry shows that most of the shielding current associated with M_y is now parallel to the oscillating field. Results are shown in Fig. 3 (open circles) compared to the earlier measurements (closed circles). The relaxation is slightly slower when the shielding currents run parallel to the oscillating field, as may be expected from the lower Lorentz force in this configuration. Numerical simulations are being performed to compare quantitatively with the measurements.

3.1. Theoretical modeling

With a high accuracy, predictions of the macroscopic response of type-II superconducting samples, i.e. average magnetic moment vs. applied magnetic field, are typically done in terms of the usual critical state theory.[8] In this approach, flux quanta are treated by the phenomenological relation $|\mathbf{J} \times \mathbf{B}| \leq F_p$ between the volume pinning force and the average values of the current density and magnetic flux density. Thus, a maximum value for the pinning force is equivalent to a critical value in the component of the current density, perpendicular to the local magnetic induction ($J_\perp \leq J_{c,\perp}$). Finally, corresponding to a large dissipation when flux is eventually unpinned, one may treat the response of the sample by the Maxwell equations for the average fields, plus the condition $J_\perp = J_{c,\perp}$ in a magnetoquasistationary approach. In physical terms, the dissipation related time constant may be neglected, and evolution is linked to the external excitation process by Ampère's and Faraday's laws.

The application of the above ideas to flux shaking experiments for non-idealized geometry is not simple, and has been done either under *ad hoc* simplifying hypotheses for extreme geometries[3, 4] or by using numerical minimization techniques.[6] In our case, variational solutions of Faraday's law when the sample undergoes a magnetic process as shown in Fig. 3 have been obtained under the mutual inductance approach (eq.(2) in Ref. [6]).

In the experiments we measure the decay of a remanent magnetization, while in Ref. [6] the field penetration was calculated so new calculations have been performed relying on the same physical principles and similar numerical methods. Because the field-independence of $J_{c\perp}$ may not be assumed here, a local field dependence of the critical current $I_{c,i}(B_i)$ has been used. Such refinement was needed for a more realistic approach to the experimental facts. As explained above, $J_{c,\perp}(B)$ was derived from a specifically designed experiment.

4. Discussion

The results of our calculations are shown in Figs. 1 and 2, in which the comparison between theory and experiment is displayed. The general trends are those described in Ref. [6], i.e. the remanent magnetization along the y-axis relaxes toward a constant $M_{y,\infty}$, by means of the *ac* oscillation of $H_x(t)$. The actual value of $M_{y,\infty}$ depends on the amplitude of the applied *ac* field $H_{x,a}$ (going to zero as $H_{x,a}$ increases). Finally, the relaxation process takes place in a step-like descent, modulated by the *ac* cycle. However, some peculiarities have been observed in the experiment that may be clearly ascribed to the field dependence of the critical current. To be specific, small bumps are visible in the step-descent of M_y that cannot be obtained from the field independent $J_{c\perp}$ approximation. Additionally, in the present case, the $M_x(t)$ response does not reach a saturation as predicted in the conditions of Ref. [6].

In conclusion, transverse flux experiments in MgB_2 samples have revealed a number of features that provide new insights in the problem of applying mutually perpendicular field components to type-II superconductors.

Acknowledgments

This work was partially supported by Chilean **FONDECyT** Project 1040668, Argentina **ANPCyT** PICT00-03-08937 and **UNCuyo** 06/C252, and Spanish **CICyT** Projects MTM2006-10531 and MAT2005-06279-C03-01. JLG was partially supported by University of Talca (Chile), and RAICES Program (SECyT Argentina), while on sabbatical leave at CAB.

References

[1] Willemin M, Rossel C, Hofer J, Keller H, Erb A, and Walker E 1998, *Phys. Rev.* B **58**, R5940.
M. Willemin, A. Schilling, H. Keller, C. Rossel, J. Hofer, U. Welp, W. K. Kwok, R. J. Olson, and G. W. Crabtree 1998, *Phys. Rev. Lett.* **81**, *4236*.
[2] Vanderbemden Ph, Hong Z, Coombs T A, Denis S, Ausloos M, Schwartz J, Rutel I B, Babu N H, Cardwell D A, and Campbell A M 2007, *Phys. Rev.* B **75**, 174515.
[3] Brandt E H and Mikitik G P 2002, *Phys. Rev. Lett.* **89**, 027002.
[4] Mikitik G P and Brandt E H 2003, *Phys. Rev.* B **67**, 104511.
[5] Brandt E H 1996, *Phys. Rev.* B **54**, 4246.
[6] Badía A and López C 2007, *Phys. Rev.* B **76**, 054504.
[7] Pasquini G, Perez Daroca D, Chiliotte C, Lozano G S and Bekeris V 2008 ,*Phys. Rev. Lett.* **100**, 247003.
[8] Bean C P 1964, *Rev. Mod. Phys.*, 31-39.
[9] Serquis A, Serrano G, Moreno S, Civale L, Maiorov B, Balakirev F, and Jaime M 2007, *Supercond. Sci. Technol.* **20**, L12.
[10] D. Casa,1995 Thesis for *'Licenciado'* degree (Unpublished).
[11] Beidenkopf H, Avraham A, Myasoedov Y, Shtrikman H, Zeldov E, Rosenstein B, Brandt E H and Tamegai T 2005, *Phys. Rev. Lett.* **95**, 257004.
[12] Sanchez A and Navau C 2001, *Supercond. Sci. Technol.* **14**, 444.

Giant magnetostructural coupling in $Gd_{2/3}Ca_{1/3}MnO_3$

Correa V. F.[1], **Sänger N.**[2], **Jorge G.**[3], **Nieva G.**[1], **Haberkorn N.**[1]

[1] Comisión Nacional de Energía Atómica, Centro Atómico Bariloche, 8400 S. C. de Bariloche, Argentina

[2] Fachbereich Physik, Universität Konstanz, D-78457 Konstanz, Germany

[3] Departamento de Física, FCEyN, Universidad de Buenos Aires, Argentina

E-mail: victor.correa@cab.cnea.gov.ar

Abstract. We report high magnetic field magnetostructural studies on $Gd_{2/3}Ca_{1/3}MnO_3$ single crystals. A giant linear magnetostrictive effect is observed in a wide temperature range ($T < 120$ K). Above 25 K a large hysteresis is seen reflecting the Mn magnetic moments ordering. At lower temperature ($T < 15$ K), a rather complicated field dependence arising from the competition between the Mn and Gd magnetic sublattices is observed. The relevance of the Gd ions in the low temperature behavior is further corroborated by specific heat experiments.

1. Introduction

Manganite compounds are mostly known for their spectacular colossal magnetoresistance (CMR). But CMR is not the only superb effect observed in these systems. Less known, but as formidable as CMR, they show what is called giant linear magnetostriction (MS): any linear dimension of the sample is strongly affected by an applied magnetic field. The effect is comparable in magnitude ($\lambda = \Delta L/L \geq 10^{-3}$ at several Tesla) to the highest MS values ever reported. In some members of the family, even giant volume MS ($\Delta V/V$) is seen. In this sense, they are not only excellent candidates for applications in magnetic memories making use of their CMR properties but also good candidates for applications as actuators, sonar systems or ultrasound devices due to their giant MS properties.

For some doping levels, manganites show a metal-insulator (MI) transition associated with the onset of Mn ions ferromagnetic ordering at the Curie temperature T_c. As it happens with magnetoresistance, the magnetostrictive effect is particularly large around T_c, implying that CMR and giant MS could be related phenomena [1]. In fact, their overall field dependence is quite similar displaying, both resistivity ρ and λ, a rapid increase at low fields followed by saturation above the same field that magnetization saturates.

In zeroeth order, metal-insulator transition as well as magnetoresistance can be easily understood in terms of a double-exchange mechanism: ferromagnetic ordering (spontaneous or field induced) favors metallic character. However, as pointed out by Millis et al. [2], many aspects of manganites can not be explained by such a simple mechanism and the need of a more sophisticated model was suggested. Particularly, the need to include the Jahn Teller (JT) effect to better describe the MI transition.

In this way, by introducing not only double-exchange hopping but also JT distortion, Zeeman and elastic energies a simple correlation between metallic character, magnetoresistance and magnetostriction can be obtained [3]: at certain temperature, the JT effect in Mn ions favors the lattice distortion at the cost of elastic energy. When lowering the temperature, the ions order ferromagnetically inducing a MI transition. The system can further lower the energy by killing the JT splitting, thus placing more electrons at the bottom of the e_g bands and gaining elastic energy. If around T_c a magnetic field is applied, magnetic ordering is even more favorable. Then, CMR and giant MS naturally follow.

So far so good with metallic manganites. But, what happens with doping levels for which no MI transition is observed? Well, it turns out to be that CMR as well as giant MS are detected down to very low temperature [4]. Naively, it can be thought that the exchange is not strong enough as to induce spontaneous magnetic ordering but an external magnetic field does it, giving CMR and giant MS. And, what happens if the system is further complicated introducing a second magnetic ion?

In this work we present magnetostriction experiments on single crystals of the insulating ferrimagnet $Gd_{2/3}Ca_{1/3}MnO_3$ [5, 6, 7]. The Gd and Mn magnetic sublattices compete each other giving rise to a non monotonic field dependence at low temperature. At higher temperature ($T \geq 25$ K) a large hysteresis associated with the ferrimagnetic order is seen. Even though a negligible magnetoresistance was reported [5], the magnetostructural effect is giant in a wide temperature range suggesting that an extra mechanism besides double-exchange and Jahn Teller distortion should be considered for a proper description of this system.

2. Results

$Gd_{2/3}Ca_{1/3}MnO_3$ single crystals were grown by the floating zone technique [7]. Magnetostriction experiments were carried out using a capacitive dilatometer with sub-Angstrom resolution [8] while specific heat measurements were performed in a SiN membrane micro-calorimeter with the thermal relaxation time technique [9].

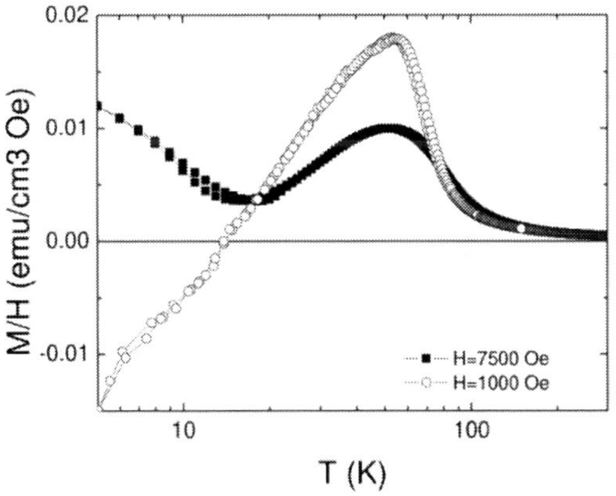

Figure 1. Temperature dependence of the magnetization/magnetic field ratio at two different applied fields (for FCC and FCW procedures). Small differences between FCC and FCW around T_{comp} are discussed in Ref. 7.

Fig. 1 displays the temperature dependence of the magnetization/magnetic field ratio (M/H) for field cooling cooling (FCC) and field cooling warming (FCW) procedures at two different applied fields: 1000 and 7500 Oe. It is the typical behavior of ferrimagnetic systems. Around $T_c \sim 80$ K the Mn magnetic moments start to order ferromagnetically. The Gd moments behaves basically as paramagnetic free ions reacting to the internal field created by the ferromagnetic Mn sublattice. Hence, they start to align in the opposite direction giving rise to the peak seen around 50 K. At $T_{comp} \sim 15$ K, the two sublattices magnetism compensate and the magnetization M vanishes. Below T_{comp} the Gd moment is larger and M is negative. Applying a field higher than the coercive field the net magnetization can be flipped over switching to positive as the 7500 Oe curve shows.

Journal of Physics: Conference Series **167** (2009) 012010 doi:10.1088/1742-6596/167/1/012010

Linear magnetostriction $\Delta L_c/L_c$ along the c-axis (direction [002] of the orthorrombic *Pnma* phase) in a parallel field can be seen in Figs. 2 and 3. Large hysteresis is observed above 25 K (Fig. 2). Also, and not shown here, a strong lattice relaxation effect (with a characteristic time of the order of 15 min.) is found between 40 and 70 K. Both phenomena have already been reported [10] in other manganite systems and they were adscribed to metastable states arising from a phase coexistence of ferromagnetic and paramagnetic domains.

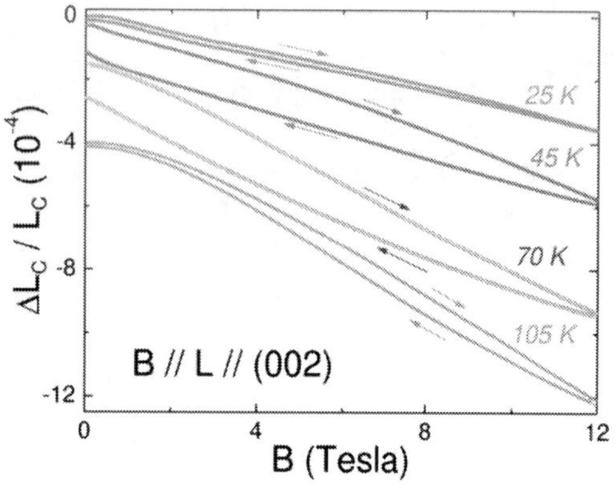

Figure 2. High temperature ($T > T_{comp}$) field dependence of c-axis magnetostriction. Curves are vertically shifted for clarity.

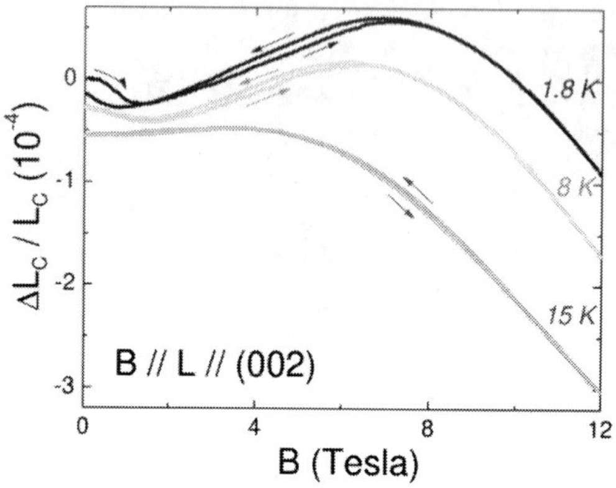

Figure 3. Low temperature ($T < T_{comp}$) field dependence of c-axis magnetostriction. Curves are vertically shifted for clarity.

Below T_{comp} the behavior changes dramatically: a non monotonic field dependence where λ changes sign twice is observed (see Fig. 3). The effect is more remarkable as T is lowered and it may be a consequence of the interplay between Mn and the now more relevant *Gd-Mn* interaction as can be inferred from magnetization curves.

Further evidence of the role played by *Gd* moments at low temperature can extracted from zero field specifc heat experiments shown in Fig. 4. Even though no particular feature is associated with T_c, C_p flattens below T_{comp} showing even a small bump. It is interesting to note as well that, down to 5 K, C_p is still quite large. This means an appreciable amount of missing entropy at low T that could be possible related to a short-range magnetic order of *Gd*.

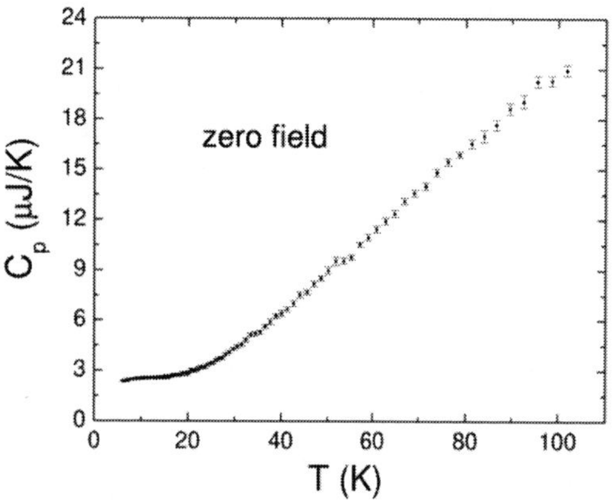

Figure 4. Zero field specific heat vs. temperature.

3. Conclusions

Magnetostriction experiments were carried out in $Gd_{2/3}Ca_{1/3}MnO_3$ single crystals. Even though no magnetoresistance is found [5], a giant magnetostructural effect is seen in the whole temperature range studied (1 K $\leq T \leq$ 120 K). The overall behavior can be separated in two well differentiated regimes. The high T regime ($T \geq T_{comp} \sim$ 15 K) is dominated by the Mn interactions showing large hysteresis associated with the magnetic order. On the other hand, λ offers a non monotonic field dependence in the low T regime ($T \leq T_{comp}$) reflecting the increasing strength of *Gd*-Mn and *Gd*-*Gd* interactions and their competition with Mn magnetism. These results suggest that extra microscopic mechanisms (other than double-exchange and Jahn Teller effect), like superexchange, Mn-*Gd* magnetic interactions and/or a strain dependence of the exchange/hopping parameters should be considered for a suitable description of the observed phenomena.

Finally, we would like to stress the following issue: a extrapolation of the linear field dependence of magnetization curves [5] up to the expected full polarization of both *Gd* and Mn sublattices (8.34 mu_B/unit cell) gives a saturation field of about 30 Tesla. Doing the same linear extrapolation of the magnetostriction curves, we get λ(30 Tesla) $\sim$ 3-4 $\cdot 10^{-3}$, the largest value ever observed in manganites.

Acknowledgments

The authors would like to acknowledge B. Alascio and D. García for helpful discussions and R. Fuentes for technical assistance. V.F.C., G.J., G.N. and N.H. are members of CONICET, Argentina. Work partially supported by ANPCyT PICT05-32900.

References

[1] Ibarra M R, Algarabel P A, Marquina C, Blasco J and García 1995 *Phys. Rev. Lett.* **75** 3541
[2] Millis A J, Littlewood P B and Shraiman B I 1995 *Phys. Rev. Lett.* **74** 5144
[3] Ghatak S K and Chaudhuri I 2003 *J. Magn. Magn. Mater.* **261** 442
[4] Kimura T, Tomioka Y, Asamitsu A and Tokura Y 1998 *Phys. Rev. Lett.* **81** 5920
[5] Snyder G J, Booth C H, Bridges F, Hiskes R, DiCarolis S, Beasley M R and Geballe T H 1997 *Phys. Rev.* B **55** 6453
[6] Peña O, Bahout M, Ghanimi K, Duran P, Gutierrez D and Moure C 2002 Spin reversal and ferrimagnetism in (Gd,Ca)MnO_3 *J. Mater. Chem.* **12** 2480
[7] Haberkorn N, Larregola S, Franco D and Nieva G 2008 Inhomogeneous ferrimagnetic-like behavior in $Gd_{2/3}Ca_{1/3}MnO_3$ single crystals *J. Magn. Magn. Mater.* (in press) *Preprint* arXiv:0808.3922
[8] Schmiedeshoff G M, Lounsbury A W, Luna D J, Tracy S J, Schramm A J, Tozer S W, Correa V F, Hannahs S T, Murphy T P, Palm E C, Lacerda A H, Bud'ko S L, Canfield P C, Smith J L, Lashley J C and Cooley J C 2006 *Rev. Sci. Instrum.* **77** 123907
[9] Denlinger D W, Abarra E N, Allen K, Rooney P W, Messer M T, Watson S K and Hellman F 1994 *Rev. Sci. Instrum.* **65** 946
[10] Matsukawa M, Akasaka K, Noto H, Suryanarayanan R, Nimori S, Apostu M, Revcolevschi A and Kobayashi N 2005 *Phys. Rev.* B **72** 064412

Titania Doping Effect on Superconducting Properties of MgB_2 Bulk Samples

G. Serrano[1], G. Bridoux[2], A. Serquis[1]

1- Instituto Balseiro - Centro Atómico Bariloche and CONICET, (8400) S. C. de Bariloche, Río Negro, Argentina

2- Instituto Balseiro - Centro Atómico Bariloche, (8400) S. C. de Bariloche, Río Negro, Argentina

serranog@ib.cnea.gov.ar

Abstract. In this work we study the microstructural and superconducting properties of doped and undoped bulk MgB_2 samples prepared by solid-state reaction, with 0 and 2.5 %at. nominal TiO_2 nanotubes contents, annealed at different temperatures in the 750-900°C range. We discuss the T_c, J_c and H_{c2} performance and their correlation with the different synthesis parameters.

Index Terms — MgB_2, Titania Nanotubes doping, material synthesis effects.

1. Introduction

Since the discovery of superconductivity of MgB_2 countless studies have been conducted with the aim of improving their properties for potential technological applications. [1]-[5] Many groups have worked to improve the critical current density by the creation of defects through dopantes or additions that may serve as vortex-pinning centers, such as SiC, [6][9] CNT [10][14] and nano-TiO_2 [15]. These additions also improve the critical field through the chemistry substitution at the boron sites. At the same time, Gurevich et al. proposed that this chemical substitution affect the electron scattering between σ and π bands, and the diffusivity into the bands. These variations have a strong correlation with temperature dependence and values of critical fields.[17]-[19]

In this work we studied the influence of titania oxide (TiO_2) nanotubes (NT) incorporation into MgB_2 bulk samples over the microestructural (*a*-lattice parameter, MgO amount) and superconducting properties (critical current density, critical fields and critical temperature), and the influence of synthesis parameters.

2. Experiment Details

2.1. Sample preparation

The samples were prepared by solid-state reaction using as starting materials magnesium powder (-325 mesh, 99.8% Assay), amorphous boron powder (-325 mesh, 99.99% Alfa Aesar) and titania oxide nanotubes (outer and inner diameter ~10nm and 5-7nm respectively).[20] The powders were mixed using ball milling inside a N_2 atmosphere globe box, during 1 hour. The undoped sample was prepared only with Mg and B with 1.2:2 ratio and sintered at 900°C during 30 min and then naturally cooled down until room temperature. On the other hand, doped samples were made with addition of 2.5%at TiO_2NT, and sintered at temperatures of 750, 850 and 900°C during 150, 80 and 30 min respectively, the table 1 shows these synthesis parameters.

2.2. *Microstructural characterization*
X-ray diffraction patterns were taken in a Phillips PW 1700 with radiation Cu K_α (1.5417 Å). The powders were mixed with Si standard in order to correct the peaks position of MgB_2 and MgO phases. The amount of MgO phase was calculated making a semi-quantitative analysis between MgB_2 and MgO amounts. The *a*-lattice parameter was determined taking into account the shifts in all the peaks.

2.3. *Superconducting characterization*
D.c. magnetization, *M*, of ~ 6 x 0.2 x 0.3 mm^3 long pieces of the samples, has been measured with the applied magnetic field H parallel to the long axis of parallelepiped, using a Quantum Design SQUID magnetometer up to 5T. We determined T_c from *M* vs *T* curves with the onset transition criteria. J_c was calculated from magnetization loops, *M*(*H*), using the Bean model assuming uniform current density.

Resistivity measurements up to 12T using the four probe method was used to determine the upper critical field (H_{c2}) and irreversibility field (H_{irr}) for all samples listed in table 1. The residual resistance ratio (*RRR*) was determined as the ratio $\rho(290)/\rho(40)$. The T_c from resistivity measurements was determined using the 50% transition criteria and the bar error as the difference between 10 and 90% of the transition.

Table 1. Characteristics of MgB_2 undoped and TiO_2NT doped samples

Sample	Composition: MgB_2 +...	Sintering Temperature	Sintering Time
NT00-900	undoped	900°C	30 min.
NT25-900	2.5%TiO_2NT	900°C	30 min.
NT25-850	2.5%TiO_2NT	850°C	80 min.
NT25-750	2.5%TiO_2NT	750°C	150 min.

3. Results and discussion

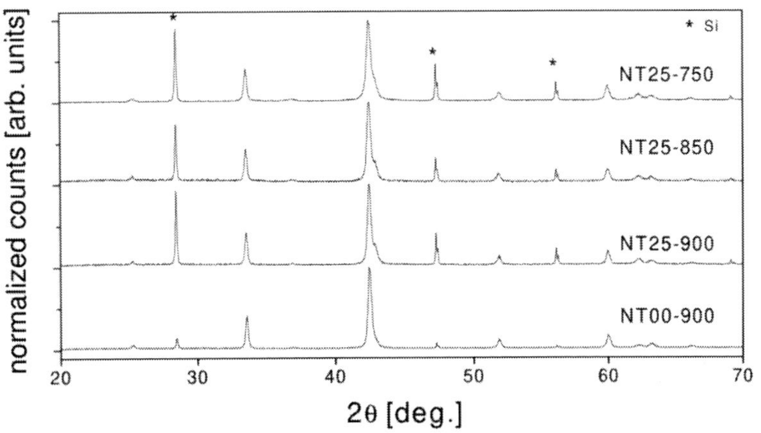

Figure 1. Normalized x-ray diffraction pattern for all samples listed in Table 1. The symbol * indicate the Si peaks used as internal standard.

3.1. X-rays analysis

The x-rays diffraction patterns for all samples listed in Table 1 are shown in Figure 1. All samples have MgO phase, observed as a peak at $2\theta \sim 42°$. The *a*-lattice parameters were calculated from the shifts of MgB_2 peaks and these are shown in Figure 2a. NT25-900 sample sintered at 900°C displays lower *a*-lattice parameter than NT-900. Doped samples sintered at lower temperatures exhibit higher *a*-lattice parameters than NT25-900; this may indicate that part of oxygen from titania nanotubes is introduced in the MgB_2 structure. Figure 2b displays the amount of MgO phase calculated from semi-quantitative analysis that compare the intensity and positions of the patterns. It can be observed that doped samples have more MgO content than NT00-900 due to the presence of TiO_2NT as oxygen source, furthermore, the MgO amount increases at lower sintering temperature.

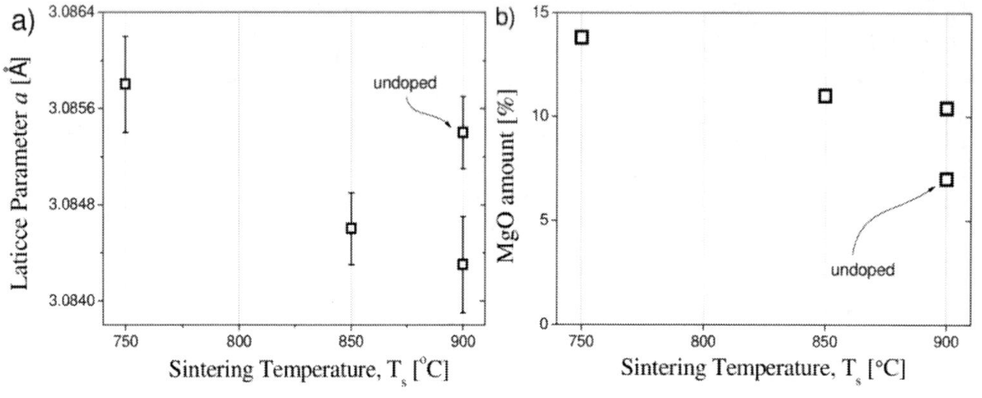

Figure 2. a) Lattice parameter *a* (calculated from peaks shifts), and **b)** MgO amount (estimated from semi-quantitative analysis) as a function of sintering temperature for samples listed in table 1.

3.2. Magnetization measurements

Figure 3 shows the J_c field dependence up to 5T for all samples investigated, calculated from magnetization loops at 5, 20 and 26.5K. At 5K samples NT00-900 and NT25-850 have the same J_c values and higher than the others samples, but at 20 and 26.5K all doped samples have lower J_c values than sample NT00-900. This result can be explained taking into account that doped samples have more amount of MgO, and hence the grains connectivity is reduced, this effect is of first order in comparison to vortex-pinning defects such as nanotubes. Figure 3 also shows the Kishan et al. [15] data for undoped and 2%nano-TiO_2, at 5 and 20K, in order to compare with our samples. Their undoped and doped samples have lower J_c values than our NT00-900 sample, indicating that sintering parameters are more important than the kind of addition.

Table 2. Microstructural and superconducting properties of MgB_2 undoped/doped TiO_2NT samples

Sample	a-lattice parameter [Å]	MgO amount [%]	T_c^{Mag}[K]	T_c^{Trans} [K]	RRR	Active area fraction, A_f
NT00-900	3.0854 ± 0.0003	7.0	38.2 ± 0.1	38.62 ± 0.08	2.66	0.26
NT25-900	3.0843 ± 0.0004	10.4	38.4 ± 0.1	38.82 ± 0.08	2.82	0.25
NT25-850	3.0846 ± 0.0003	11.0	38.3 ± 0.1	38.67 ± 0.08	2.54	0.25
NT25-750	3.0858 ± 0.0004	13.8	38.1 ± 0.1	38.51 ± 0.08	2.41	0.13

The T_c's obtained by magnetization are shown in Table 2. T_c's decreases with higher a-lattice parameters. All width transitions are similar and lower than 0.4K, indicating that all samples are very homogeneous.

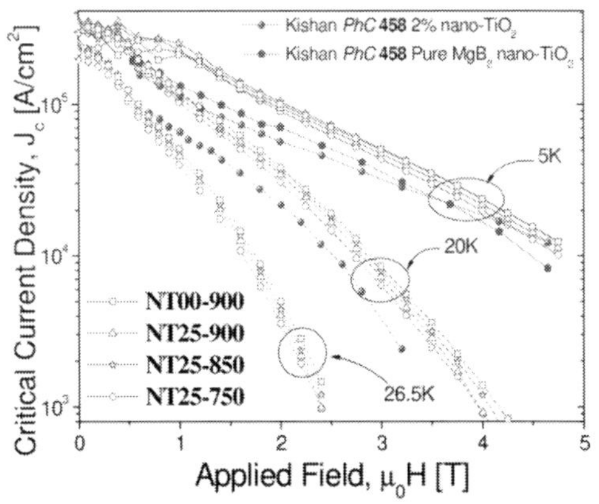

Figure 3. Critical current density obtained by magnetization as a function of applied field for all samples listed in table 1. The lines are only a guide to eye.

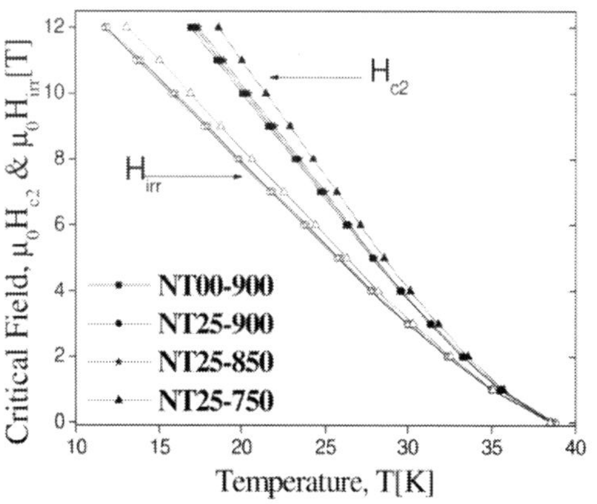

Figure 4. Temperature dependence of H_{c2} and H_{irr} obtained from resistivity measurements until 12T for all samples investigated. The lines are only a guide to eye.

3.3. H_{c2} and T_c obtained by transport measurements

Figure 4 shows the H_{c2} and H_{irr} as function of temperature, for all samples listed in table 1. We observed that doped samples sintered at higher temperatures have a small improvement in the H_{c2} and H_{irr} values; in contrast sample NT25-750 has an important H_{c2} enhancement of 1T at 20K.

The *RRR* and T_c obtained by transport measurements are shown in Table 2, both are correlated with a-lattice parameter, and increase with doping level but decrease at low sintering temperature. This is caused by a shift to dirty limit.

The active area fraction A_f was calculated according to the formula proposed by Rowell et al. [21] (see Table 2). These values estimate the degree of grain connectivity. We observed that A_f increases with the sintering temperature, indicating that samples with higher sintering temperature have a better grain connectivity.

4. Summary

A study on the effect of 2.5%TiO_2NT doping and sintering temperatures in the 750-900°C range is presented.

Variations in the a-lattice parameter with doping and sintering temperature were observed. At T_s=900°C, the doped sample has a lower a-lattice parameter but when the T_s decreases the a-lattice parameter increase, probably as a consequence of oxygen substitution into boron sites. The semi-quantitative analysis showed that all doped samples have more MgO amount, due to titania nanotubes addition work like an oxygen source.

The $J_c(H)$ of undoped sample is higher than the values of all doped samples for all the studied temperatures. This is correlated with the MgO amount present in the samples, which strongly suggest that the MgO amount is the main responsibly of J_c behavior compared with any kind of vortex-pinning defects.

A strong correlation between H_{c2} and H_{irr} with a-lattice parameter was observed. This is in agreement with a theoretical model of Gurevich, in which any kind of substitution produces variation in the temperature dependence of H_{c2}.[17] An H_{c2} improvement of 1T at 20K was measured for NT25-750 sample.

The titania nanotubes additions does not show a correlated enhancement between H_{c2} and J_c in contrast with carbon nanotubes addition, when both H_{c2} and J_c improve with additions.

Both $T_c^{Trans.}$ and RRR present a good correlation with a-lattice parameter, indicating that the oxygen plays an important role in the transport mechanism, and that doped samples tend to the dirty limit.

5. Acknowledgments

The present work was supported by CONICET, Universidad Nacional de Cuyo Proyecto 06/C269, Agencia Nacional de Promoción Científica y Tecnológica PICT 1250.

References

[1] Nagamatsu J; Nakagawa N; Muranaka T; Zenitani Y and Akimitsu J (2001) *Nature (London)* 410, 63-64
[2] Buzea C and Yamashita T (2001) *Supercond. Sci. Technol.* 14, R115-R146
[3] Larbalestier D C *et al* (2001) *Nature (London)* **410**, 186-189
[4] Flükiger R; Suo HL; Musolino N; Beneduce C; Toulemonde P and Lezza P (2003) *Physica C* **385**, 286-305, and references therein.
[5] Serquis A; Liao X Z; Zhu Y T; Coulter J Y; Huang J Y; Willis J O; Peterson D E; Mueller F M; Moreno N O; Thompson J D; Nesterenko V F and Indrakanti S S (2002) *Journal of Applied Physics* **92**, 351
[6] Dou S X, Soltanian S, Horvat J, Wang X L, Munroe P, Zhou S H, Ionescu M, Liu H K and Tomsic M (2002) *Appl. Phys. Lett.* **81**, 3419–21
[7] Buzea C and Yamashita T (2001) *Supercond. Sci. Technol.* **14**, R115–46
[8] Liao X Z, Serquis A, Zhu Y T, Huang J Y, Civale L, Peterson D E, Xu H F and Mueller F M (2003) *J. Appl. Phys.* **93,** 6208–15
[9] Zhou S H, Pan A V, Qin M J, Liu H K and Dou S X (2003) *Physica C* **387**, 321–7
[10] Dou S X, Yeoh W K, Horvat J and Ionescu M (2003) *Appl. Phys. Lett.* **83**, 4996–8
[11] Yeoh W K, Horvat J, Dou S X and Munroe P (2005) *IEEE Trans. Appl. Supercond.* **15**, 3284–7
[12] Yeoh W K, Kim J H, Horvat J, Dou S X and Munroe P (2006) *Supercond. Sci. Technol.* **19**, L5–8
[13] Serquis A, Serrano G, Moreno M S, Civale L, Maiorov B, Balakirev F and Jaime M (2007) *Supercond. Sci. Technol.* **20**, L12–L15
[14] Serrano G, Serquis A, Dou S X, Soltanian S, Civale L, Maiorov B, Balakirev F and Jaime M (2008) *Journal of Applied Physics* **103**, 023907
[15] Kishan H, Awana V P S, de Oliveira T M, Sher Alam, Saito M, de Lima O F (2007) *Physica C* **458**, 1–5
[16] Fu B Q, Feng Y, Zhao Y, Pradhan A K, Cheng C H, Ji P, Liu X H, Liu C F, Yan G, Zhou L (2003) *Physica C* **386**, 659–662
[17] Gurevich A (2003) *Phys. Rev. B* 67, 184515-184528
[18] Gurevich A *et al* (2004) *Supercond. Sci. Technol.* **17**, 278–286
[19] Braccini V *et al* (2005) *Phys. Rev. B* 71, 012504-012508
[20] Santillan M J, Moreno M S, Boccaccini A R, submitted to Nanotechnology (2008)
[21] Rowell J M (2003) *Supercond. Sci. Technol.* **16**, R17

Influence of the introduction and formation of artificial pinning centers on the transport properties of nanostructured Nb_3Sn superconducting wires

L B S Da Silva, C A Rodrigues, C Bormio-Nunes, N F Oliveira Jr, and D Rodrigues Jr

Superconductivity Group, Department of Materials Engineering (DEMAR)
Escola de Engenharia de Lorena (EEL), Universidade de São Paulo (USP)
Polo Urbo-Industrial, Gleba AI-6 - P.O. Box 116 - Lorena, SP, Brazil

E-mail: lucas_sarno@ppgem.eel.usp.br, durval@demar.eel.usp.br

Abstract. The formation of nanostructures projected to act as pinning centers is presented as a highly promising technique for the transport properties optimization of superconductors. However, due to the necessity of nanometric dimensions of these pinning centers, the heat treatment (HT) profiles must be carefully analyzed. The present work describes a methodology to optimize the HT profiles in respect to diffusion, reaction and formation of the superconducting phases. After the HT, samples were removed for microstructural characterization. Measurements of transport properties were performed to analyze the influence of the introduction of artificial pinning centers (APC) on the superconducting phase and to find the flux pinning mechanism acting in these wires. Fitting the volumetric pinning force vs. applied magnetic field (F_p vs. $\mu_0 H$) curves of transport properties, we could determine the type and influence of flux pinning mechanism acting in the global behavior of the samples. It was concluded that the maximum current densities were obtained when normal phases (due to the introduction of the APCs) are the most efficient pinning centers in the global behavior of the samples. The use of HT with profile 220°C/100h+575°C/50h+650°C/100h was found as the best treatment for these nanostructured superconducting wires.

1. Introduction

Since the discovery of the Nb_3Sn superconductors many efforts have been expended to improve the transport properties in these materials. One example for Nb_3Sn application is the ITER Project, that uses high current densities to create magnetic fields close to 12 Tesla [1-2].

In previous works [3-4] our group projected and manufactured a Nb_3Sn superconductor wire with APC of Cu(Sn), using the solid-liquid diffusion method. The produced wire has 1.0 mm of diameter, 1,451,610 Nb filaments, 36 elements of Sn, internal stabilization of Cu and Ta used as diffusion barrier. These wires showed significant improvement in their transport properties, in comparison to other conventional superconductors produced by the same approach [5]. On the other hand, the superconducting properties of these nanostructured wires can be improved by the optimization of the heat treatment (HT) profiles.

The optimization of the HT profiles is a stage of extreme importance due to the small dimensions of the APCs and Nb filaments. These profiles must be carefully analyzed. The temperatures and times of treatments must permit the diffusion, reaction and formation of the superconducting phases in the matrix with nanometric distribution of the initial material preventing, however, the extreme increase of the average grain size. On the other hand, the position of the APCs in the superconducting matrix also depends essentially on the HT profile [3]. This will lead to the optimization of the magnetic flux pinning and, consequently, to the increase of the critical current densities [6].

In this paper, the HT profiles for Nb_3Sn superconductor wires Cu(Sn) APCs in nanometric-scale, produced by RODRIGUES, 2007 [3], were analyzed in an attempt to improve the transport properties by changes in HT. Microstructural characterization and transport measurements were performed in an attempt to find the superconducting properties and the pinning mechanisms acting in the samples.

2. Experimental procedure

For the experiments in the present work, samples of RODRIGUES, 2007 [3] were used. These wires were deformed until 1.06 mm in diameter, which had, before HT, 1,064,514 Nb filaments with average diameter of 410 nm, 36 cores of Sn and internal stabilization of Cu surrounded by Ta diffusion barrier. The APCs had an average diameter of 40 nm. These dimensions were determined by electron microscopy using a Field Emission Gun SEM (FEG-SEM).

Samples were removed for the heat treatment optimization. Cu was electrodeposited on both ends of the samples to avoid leak of Sn during the heat treatments at high temperatures. After that, the samples were encapsulated in quartz tubes under an argon atmosphere of 20 mTorr. The annealing HT already started with the required temperature and finished with quenching to room temperature. The HT profiles were used following the steps showed in Table 1.

Table 1. Heat treatment (HT) profiles.

Samples	Temperature/Time
1	575°C/200h
2	650°C/200h
3	700°C/100h
4	220°C/100h+650°C/50h
5	480°C/50h+650°C/50h
6	575°C/50h+650°C/50h
7	220°C/100h+480°C/50h+650°C/100h
8	220°C/100h+575°C/50h+650°C/100h

After HT, samples were removed for microstructural characterization using a JSM6330F Field Emission Gun Scanning Electron Microscope (FEG-SEM). These samples were polished and etched with nitric acid.

Superconducting characterization of the samples was performed in order to determine the existence and quality of the superconducting phases, critical temperatures T_c, and critical current densities J_c, consequently determining the influence of the heat treatments on the superconducting phase and on the transport properties.

The critical temperature measurements were performed using the four probe method at the Laboratory of Superconductivity of the DEMAR-EEL-USP. These measurements used the excitation current of 500 mA.

The transport properties were measured also using the four probe method in a magnet capable to generate a continuous magnetic field up to 17 T installed at the Solid State and Low Temperature Laboratory of the Department of Material Physics and Mechanics, Institute of Physics, USP, SP. The criterion of 10 µV/cm was used to determine the critical currents [7]. All measurements were performed at 4.2 K (liquid helium bath). The volumetric pinning force $F_p(H,T{=}4.2K){=}J_cB$ versus μ_0H curves were found using the transport J_c.

3. Heat Treatment Analysis

Figure 1 are micrographs using a FEG-SEM in the secondary electrons mode, comparing the microstructures of the Nb filaments formed before HT (a) and after HT at: (b) 575°C/200h, (c) 650°C/200h and (d) 700°C/100h, samples 1, 2, and 3, respectively. It can be observed the coalescence of the filaments from the first bundling (127 monofilaments) in just one filament with APCs into the superconducting phase. It can be seen a large number of Cu regions used as APCs in the large Nb filament, especially for the sample heat treated at low temperature. For high temperature, the Cu APCs were smaller in number, whereas the average cross sections increased due to the increase of the temperature gradient and, consequently, faster diffusion of Sn into the Nb filaments.

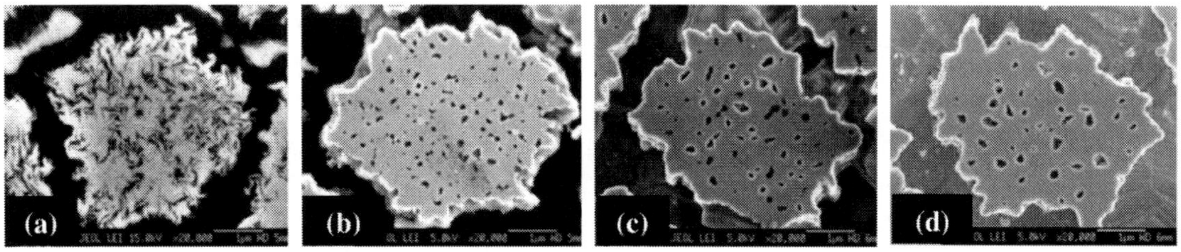

Figure 1. Microstructure of the Nb filaments (a) before HT, and after HT at (b) 575°C/200h, (c) 650°C/200h and (d) 700°C/100h. Scale bars with 1 μm.

4. Superconducting Analysis

Table 2 shows the measurements of critical temperature T_c and the dispersion ΔT_c. The values of T_c were obtained as the middle of the transition, calculated between the 10% and 90% of the transition height. The half-width of the superconductor-to-normal transition defined ΔT_c [8]. All samples showed almost the same critical temperatures T_c around 17 K. The values of ΔT_c give the quality of the Nb_3Sn superconducting phase formed after HT. These values of T_c around 17 K were expected for Nb_3Sn wires due to the differential thermal contraction between Cu, Sn, Nb, and Nb_3Sn, increasing the stress on the superconducting phase and decreasing T_c [6].

Table 2. Measurement of critical temperature.

Samples	T_c (K)	ΔT_c (K)
1	16.6	0.4
2	17.3	0.5
3	16.4	0.3
4	17.1	0.3
5	17.3	0.2
6	16.6	0.3
7	16.5	0.5
8	16.7	0.3

Figure 2a shows the curve of critical current densities J_c as a function of the applied magnetic field $\mu_0 H$ for all samples after heat treatment. These transport measurements describe the global behavior of all samples. It can be observed that the sample 8 (Table 1) had the best J_c behavior for high magnetic fields, while the sample 7 presented the worst result for all magnetic field analyzed.

Figure 2b shows the curve of pinning force $F_p=J_cB$ as a function of the applied magnetic field $\mu_0 H$ for all samples after heat treatment. Table 3 is a compilation of the results from the mathematical fitting of the F_p curves presented in the Figure 2b. These fitting were calculate using the expression

$$F_p = K\left[x\, b^{1/2}(1-b)^2 + (1-x)\,b(1-b)\right] \tag{1}$$

where K is a constant of proportionality, b ($=B/B_{C2}$) is the reduced magnetic field, and x is the contribution of the flux pinning mechanism due to the grain boundaries. These expressions are adapted from Dew-Hughes original works [9-10], where the first term, $b^{1/2}(1-b)^2$, corresponds to the mechanism of magnetic flux pinning due to the grain boundaries expected for Nb_3Sn superconductors and the second term, $b(1-b)$, is the contribution of the normal phase (or with different superconducting properties) due to the introduction of the APCs into the superconducting phase.

Analyzing the behavior of the curves in Figure 2b and in Table 3, it can be observed that the sample 8, which showed the best values of J_c, had flux pinning centers only due to the normal phase. This behavior can also be seen by the values of x and $b(F_{pmax})$, which showed 0% of grain boundaries contribution on the flux pinning and maximum of F_p at reduced magnetic field next to 0.5, expected for pinning centers due to the normal phases. Moreover, this sample showed the highest value of upper critical magnetic field B_{C2}. These results can be explained in terms of the HT, where intermediate steps are required to form a homogeneous superconducting phase with APCs in the Nb_3Sn phase.

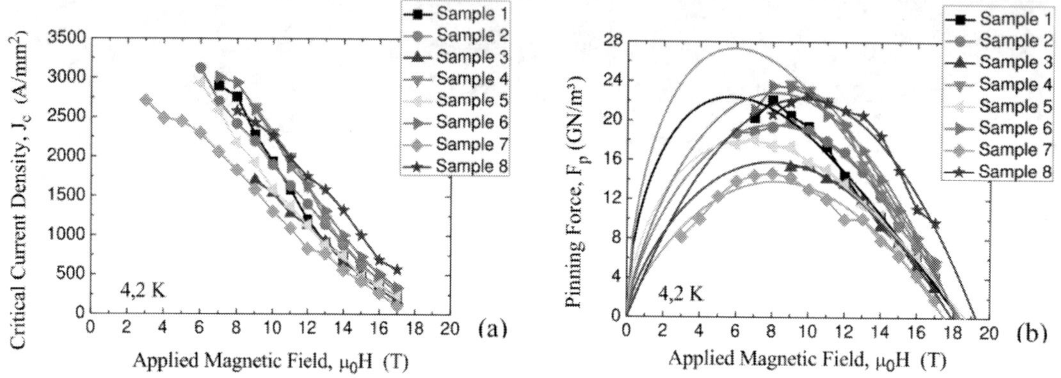

Figure 2. Transport Measurements (a) J_c vs μ_0H and (b) F_p vs μ_0H.

Observing the values of x in Table 3 and the behavior of the curves in Figure 2, it can be supposed that the high values of J_c for samples 4, 6 and 8 have distinct explanations. Sample 4 had large area of homogeneous Nb_3Sn phases, but without excessive growth of the grains. For this sample the flux pinning had contributions of 50% due to the grain boundaries and 50% due to normal phase. Due to the low value of x it can be supposed that sample 6 presented an excessive increased of the average Nb_3Sn grain size. On the other hand, the introduction of the APCs acted effectively on the flux pinning. Finally, sample 8 showed flux pinning essentially due to the normal phases.

Table 3. Results from the mathematical fitting of the experimental F_p using equation (1). F_p obtained from the transport properties.

Samples	K (GN/m³)	x	B_{C2} (T)	$B(F_{pmax})$ (T)	F_{pmax} (GN/m³)	$b(F_{pmax})$
1	93.02	0.53	18.42	5.80	22.50	0.31
2	81.97	0.20	18.32	8.08	19.51	0.44
3	6.09	0.18	17.93	7.89	15.78	0.44
4	114.09	0.50	18.40	5.95	27.28	0.32
5	75.82	0.42	18.67	6.47	17.93	0.35
6	95.24	0.17	18.17	8.03	22.73	0.44
7	56.92	0.12	17.51	7.99	13.78	0.45
8	89.03	0.00	18.88	10.36	22.20	0.55

The worst behavior of J_c happened for the samples heat treated with intermediate step of HT at 480°C and with only one step of HT at 700°C. These results can be explained by distinct processes: for samples 5 and 7, the formation of the δ-phase in the Cu-Sn diffusion layer consumes large amount of Sn atoms, changing the availability of Sn atoms and prejudicing the formation of a more homogeneous superconducting Nb_3Sn phase [11]; for sample 3, the excessive increase of the superconducting average grain size decreases the grain boundary densities and pinning. The excessive increase of the APCs cross section, like those in Figure 1, decreases the flux pinning due to the difficult for the Cooper pairs to surpass the potential barrier of the normal phases [6].

It could be seen that the HT at 575°C/200h (sample 1) already had enough condition to promote the formation of the superconducting phases capable of transporting high current densities, with high upper critical magnetic field values.

5. Conclusion

During this work it could be observed that conventional heat treatments (HT) [12-13] could not be applied for the nanostructured superconducting wires and news profiles of HTs were required.

All samples showed almost the same critical temperatures next to 17 K, values obtained from transport measurement of the superconducting transition.

It could be seen that the HT at 575°C already showed the formation of the superconducting phase capable of transporting high current densities and with high upper critical magnetic field values. On the other hand, samples heat treated with intermediate step at 480°C or with only 700°C one-step treatment showed the worst results for J_c, F_p and B_{c2}. These results can be explained by the extreme increase of the average grain size, showed by the microstructures. The sample treated at 220°C/100h+575°C/50h+650°C/100h showed the best results, due to the major contribution of the normal phases in the superconducting phase and, consequently, in the flux pinning.

Furthermore, the 650°C/200h showed to be a good option to heat treat the samples without intermediate steps of diffusion, if a faster HT profile is needed for some application.

Acknowledgements

The authors wish to thank the Institute of Physics, USP, São Paulo, Brazil, for the use of the 17T magnet during the transport measurements; LME-LNLS, Campinas, Brazil, for the use of FEG-SEM; and to CNPq and CAPES, Brazil, for the financial support. DRJ and NFOJ are CNPq researchers.

References

[1] Chaowu Z, Sulpice A, Lian Z, Soubeyroux J L, Verwaerde C and Hoang G K 2007 *IEEE Trans. Appl. Superc.* **17** 7-12.
[2] Zhang P X, Zhou L, Tang X D, Li C G, Wu Y, Li K, Yan G, Yang M, Feng Y, Liu X H, Weng P D and Lu Y F 2006 *Physica C* **445-448** 819-822,.
[3] Rodrigues C A and Rodrigues Jr D 2007 *IEEE Trans. Appl. Superc.* **17** 2627-2630.
[4] Rodrigues C A and Rodrigues Jr D 2006 *Journal of Physics: Conference Series* **43** 43-46.
[5] Rodrigues Jr D, Rodrigues C A, Silveira E B and Romão E G M 2005 *IEEE Trans. Appl. Superc.* **15** 3389-3392.
[6] Suenaga M 1981 *In:* Schwartz B B and Foner S *Eds. Superconductor material science-Metallurgy, fabrication and applications.* (New York: Plenum Press) chapter 4 p. 201.
[7] Rodrigues C A and Rodrigues Jr D 2004 *Physica C* **408-410** 921-922.
[8] Murase S, Itoh K, Wada H, Noto K, Kimura Y, Tanaka Y and Osamura K 2001 *Physica C* **357-360** 1197-1200.
[9] Dew-Hughes D 1974 *Phil. Mag. B* **30** 293.
[10] Dew-Hughes D 1987 *Phil. Mag. B* **55** 459.
[11] Yamasaki H and Kimura Y 1982 *Cryogenics* **22** 89-93.
[12] Müller M, Schulz H and Kirchmayr H 2005 *Physica C* **419** 115-120.
[13] Müller H and Schneider Th 2004 *Physica C* **401** 325-329.

XIX Latin American Symposium on Solid State Physics (SLAFES XIX) IOP Publishing
Journal of Physics: Conference Series **167** (2009) 012013 doi:10.1088/1742-6596/167/1/012013

Vortex transport entropy in the $H-T$ diagram of high T_c superconductors

G. Bridoux, G. Nieva, and F. de la Cruz

Centro Atómico Bariloche and Instituto Balseiro, Comisión Nacional de Energía Atómica, Av. E. Bustillo 9500, R84002AGP S. C. de Bariloche, Argentina

E-mail: `gbridoux@yahoo.com.ar`

Abstract.

The combination of Nernst effect and electrical resistivity measurements allows to extract the transport entropy carried by moving vortices. In high T_c superconductors, the vortex-like fluctuations close and above T_c can be detected with these tools if local phase coherence is still present. In this work we study the vortex transport entropy in the two milestone high T_c, $YBa_2Cu_3O_{7-\delta}$ (YBCO) and $Bi_2Sr_2CaCu_2O_{8+\delta}$ (BSCCO). While below T_c the YBCO entropy displays a mean field like behavior, close and above T_c the entropy reveals typical features of strong superconducting fluctuations. The lower dimensionality in BSCCO enhances the strength of superconducting fluctuations in a wider region of the $H-T$ diagram and a mean field treatment can not be applied. In this region the vortex transport entropy remains unaffected by the presence of correlated defects.

1. Introduction

The displacement of vortices in the mixed state of type II superconductors under the influence of a thermal gradient, ∇T, perpendicular to the internal magnetic field, $\mathbf{B}$, produces a Josephson electrical voltage perpendicular to both, $\mathbf{B}$ and ∇T. This is the Nernst signal, which is essentially zero in the normal state [1], [2] and consequently it is an outstanding tool to detect thermal induced vortex displacements, in particular at fields or temperatures approaching the normal state.

In the linear response regime [3], if $\nabla T \| \hat{\mathbf{x}}$ and $\mathbf{B} \| \hat{\mathbf{z}}$, the Nernst electric field is proportional to $(\nabla T)_x$. Within this regime, the thermal force per unit vortex length is [4]

$$F_T(H) = S_\phi(H,T)(\nabla T)_x \tag{1}$$

where $S_\phi(H,T)$ is the transport entropy per unit vortex length. In this limit and if vortex-pinning is absent, the impedance to vortex displacement is the flux flow vortex viscosity, associated with the flux flow electrical resistivity, ρ_f. Thus, the Nernst signal is found [2]

$$e_N(H,T) = \frac{E_y}{(\nabla T)_x} = \frac{\rho_f S_\phi(H,T)}{\phi_0} \tag{2}$$

where E_y is the Nernst electrical field and ϕ_0 the flux quantum. We see that from the measurement of the Nernst signal and flux flow resistivity the vortex transport entropy can be determined. Theoretical calculations [5] have shown that within the limits of the Ginzburg-Landau theory the transport energy, $U_\phi(T,H) = TS_\phi(H,T)$ is related to the thermodynamic equilibrium magnetization $M(T,H)$ by $U_\phi(T,H) = -\phi_0 M(T,H) L_D(T)$ and consequently [3]:

$$U_\phi(T,H) = \frac{\phi_0}{4\pi} \frac{(H_{c2}(T) - H)}{\beta_A(2\kappa_2^2 - 1) + 1} L_D(T) \tag{3}$$

where $\beta_A = 1.16$ in an hexagonal vortex lattice, κ_2 is the Ginzburg-Landau parameter and L_D is a numerical function, both weakly T dependent close to T_c [5],[3]. Expression (3) is valid for applied magnetic fields, H, and temperatures close to $H_{c2}(T)$ or $T_{c2}(H)$, respectively. It shows that the vortex transport energy tends linearly to zero with either $(H_{c2}(T) - H)$ or $(T_{c2}(H) - T)$ depending on which variable, H or T, is controlled in the experiments.

From (3) and since $S_\phi(H,T) = 0$ at either $T = 0$ or for $H < H_{c1}(T)$ we see that a maximum in the Nernst signal as a function of H and T should appear. It is known that vortex displacement is reduced by the presence of structural defects acting as pinning centers [3], and consequently the maximum of e_N is usually determined by the field or temperature where pinning reduces vortex mobility to zero. In low T_c materials this fact restricts the $H - T$ window in which the Nernst signal can be used to extract $S_\phi(T,H)$. Despite of this, pioneering work [6] allowed to show a well defined maximum of e_N in the $H - T$ phase diagram, in qualitative agreement with theoretical expectations.

The discovery of an anomalous Nernst signal above T_c in high-T_c superconductors by Ong and collaborators [7] has triggered intensive research along this decade. While these authors have suggested an scenario dominated by strong phase fluctuations of the superconducting order parameter or vortex-like fluctuations, in agreement with [8] and [9], another explanations based in amplitude fluctuations within the G-L theory context [10],[11] have also been proposed. In high T_c superconductors, the wide extension of the vortex liquid in the $H - T$ diagram [12] gives us an unique opportunity to study the vortex transport entropy in these materials. This work presents Nernst effect and electrical transport measurements in the two milestone high-T_c superconductors, YBCO and BSCCO with and without columnar defects, CD. In BSCCO-OPT, its extremely high anisotropy ($\gamma \sim 150$) should enhance fluctuations effects [13] in contrast to YBCO-OPT, in which $\gamma \sim 7$.

2. Experimental Details

The YBCO and BSCCO single crystals were grown using a self flux technique and annealed for optimal doping as described in [14] and [15]. The columnar defects, nearly parallel to the c crystal axis, with a density of $B_\phi = 3\,\mathrm{T}$ were created by irradiation at TANDAR irradiation facility, Argentina [16],[17]. The Nernst and electrical transport measurements techniques were described elsewhere [14],[17],[18] with $\mathbf{H} \| \mathbf{c}$.

3. Results and discussion

Fig.1 shows the field and temperature dependence of the Nernst signal in the YBCO-OPT crystal without CD. It is clearly seen that sweeping T at fixed H or sweeping H at fixed T a maximum in e_N is reached (T_{max} and H_{max} respectively). Below T_c, both maxima seem to coincide in the case of YBCO-OPT, as seen in Fig.1 and in the $H - T$ diagram of Fig.2 (b).

When U_ϕ is extracted in the YBCO crystals with and without CD using the Nernst and electrical transport measurements across equation (2), a corresponding maximum below T_c does not appear in U_ϕ, as shown in Fig.2 (a). This evidence indicates that the e_N maximum in YBCO-OPT below T_c is controlled by vortex-pinning and mobility and it has not influence in the thermodynamic limit. More important, the U_ϕ for samples with and without CD coincide in a wide range of temperatures close and above T_c for all fields investigated, from 3 to 10 T within our experimental resolution [19], an example is shown in Fig.2 (a) at $H = 8\,\mathrm{T}$. This

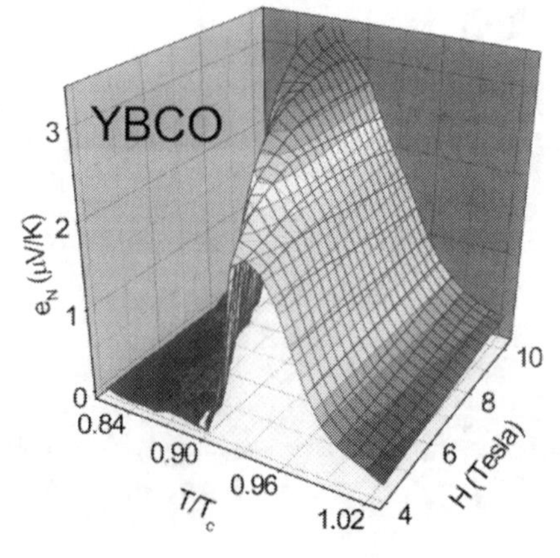

Figure 1. The Nernst signal as a function of T/T_c and H in YBCO-OPT.

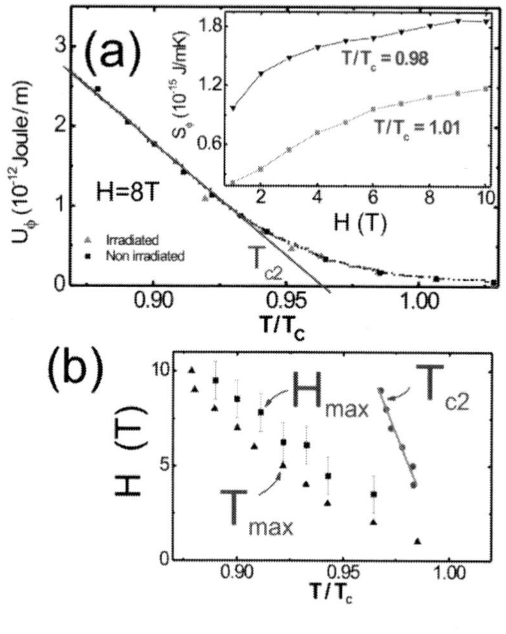

Figure 2. (a) U_ϕ vs. T/T_c in YBCO at $H = 8\,\mathrm{T}$ for the irradiated (red triangles) and the non-irradiated sample (black squares, interpolation with small symbols). The mean field like linear behavior is also shown (blue line). Inset: S_ϕ vs. H in non-irradiated YBCO at $T/T_c = 1.01$ and 0.98. (b) $H-T$ diagram. The line T_{max} (up triangles) and H_{max} (squares) for the non-irradiated YBCO sample and T_{c2} (circles and linear fit) are shown.

result strongly suggests that the vortex liquid and the superconducting fluctuations behave like an ideal system in this regime, unaffected by the presence of structural defects.

From equation (3) is possible to extrapolate the U_ϕ mean field like linear behavior to $U_\phi = 0$ to obtain T_{c2}, as shown in Fig.2 (a). The results are shown in the $H - T$ diagram of Fig.2 (b) and at first glance, they agree with a mean field behavior of YBCO-OPT. Close and above T_c, there is a U_ϕ depart from the mean field like behavior, as seen in Fig.2 (a). When the field is swept at fixed T in this region, a maximum in e_N is reached, H_{max}, see Fig.1, and it also appears when S_ϕ is extracted, as shown in the inset of Fig.2 (a). In linear order in H, these results are in concordance with the predicted 3D-anisotropic behavior of S_ϕ by Ussishkin et al. [10]. In this theoretical context, amplitude fluctuations of the order parameter are the most relevant contribution to S_ϕ.

Fig.3 shows the field and temperature dependence of the Nernst signal in the BSCCO-OPT crystal without CD. In accordance with the Ginzburg criteria [13], the higher anisotropy of BSCCO-OPT compared with that of YBCO-OPT amplifies fluctuations effects, extending the Nernst signal to a wider region of the $H-T$ diagram, see Fig.3. Since mobility has a smooth variation close and above T_c [19], S_ϕ and e_N present the same shape in the $H-T$ diagram, see equation (2). Hence, from Fig.3 it follows that the low dimensionality of BSCCO-OPT prevents that a mean field treatment can be applied to this compound.

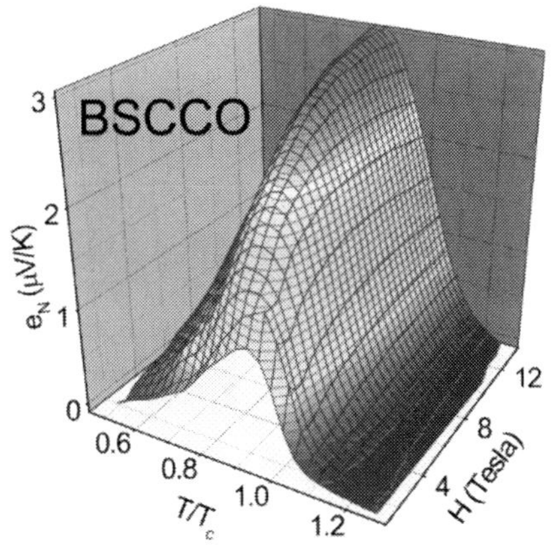

Figure 3. The Nernst signal as a function of T/T_c and H in BSCCO-OPT.

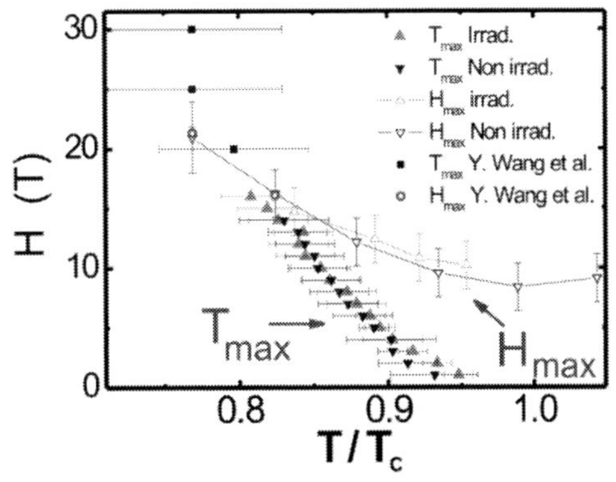

Figure 4. $H-T$ diagram of BSCCO. The lines T_{max} and H_{max} for the irradiated (filled and open up triangles respectively) and non-irradiated samples (filled and open down triangles respectively) are shown. T_{max} (squares) and H_{max} (circles) at high magnetic fields correspond to [20].

In BSCCO-OPT, T_{max} does not coincide with H_{max} in the whole range of fields investigated, see Fig.3. If the high magnetic field data of Ong and collaborators are taken into account

[20], both maxima in BSCCO seem to merge in one single maximum, as shown in the $H-T$ diagram of Fig.4. The exactly same behavior occurs in the BSCCO sample with CD, see Fig.4. Consequently, these BSCCO features are mobility and pinning independent, and they also appear in the thermodynamic properties S_ϕ and U_ϕ.

In this laminar material, the S_ϕ behavior can be governed by phase [9] or amplitude fluctuations [10] of the order parameter. In both cases [9],[10],[21], the 2D vortex transport energy follows the magnetization, $U_\phi \propto M$. Hence, the magnetization in BSCCO-OPT should also display these maxima features, as previous measurements anticipate [22].

In BSCCO-OPT, the loss of metallicity along the c axis direction, which is reflected in the c axis resistivity [19],[23] it is the responsible of the fluctuations enhancement and the depart from the mean field like behavior. In YBCO-OPT, the c axis resistivity is metallic [24], fluctuations effects are less pronounced and a mean field treatment is still applicable. In this sense it would be of great interest a study of the vortex transport entropy and the c axis resistivity when the anisotropy is changed gradually, like in oxygen deficient YBCO samples. Previous measurements in these crystals [25],[26],[27] confirm their similarity with BSCCO-OPT.

4. Conclusion

In conclusion, the study of the vortex transport entropy in the two paramount high-T_c has revealed new features of the vortex liquid and of the vortex-like fluctuations. While in YBCO-OPT a mean field like linear behavior of U_ϕ is still present, the higher anisotropy of BSCCO-OPT enhances fluctuations effects and it prevents the application of a mean field like treatment. Below T_c, there are no maxima in the S_ϕ of YBCO-OPT, but there is a splitting of maxima in the corresponding S_ϕ of BSCCO-OPT. Close and above T_c a maximum sweeping H at fixed T, H_{max}, appears in the S_ϕ of both compounds. This a typical fingerprint of superconducting fluctuations [9],[10]. In this region the vortex liquid remains unaffected by the presence of structural defects.

References

[1] Sondheimer E. H. 1948 *Proc. R. Soc. London, Ser. A* **193**, 484
[2] Wang Y et al 2001 *Phys. Rev. B* **64**, 224519
[3] Campbell A M and Evetts J E 2001 *Advances in Physics* **50**, 1249, and references there in.
[4] Stephen M J 1966 *Phys. Rev. Lett.* **16**, 801
[5] Houghton A and Maki K 1971 *Phys. Rev. B* **3**, 1625, and references there in.
[6] Solomon P R and Otter F A 1967 *Phys. Rev.* **164**, 608
[7] Wang Y et al 2006 *Phys. Rev. B* **73**, 024510, and references there in.
[8] Emery V J and Kivelson S A 1995 *Nature* **374**, 434
[9] Podolsky D, Raghu S and Vishwanath A 2007 *Phys. Rev. Lett.* **99**, 117004
[10] Ussishkin I, Sondhi S L and Huse D A 2002 *Phys. Rev. Lett.* **89**, 287001
[11] Mukerjee S and Huse D A 2004 *Phys. Rev. B* **70**, 014506
[12] Blatter G et al 1994 *Rev. Mod. Phys.* **66**, 1125, and references there in.
[13] Ginzburg V L 1960 *Sov. Phys., Solid State* **2**, 1824
[14] de la Cruz F, López D, and Nieva G 1994 *Philos. Mag. B* **70**, 773
[15] Kaul E E and Nieva G 2000 *Physica C* **341-348**, 1343
[16] Silhanek A et al 1999 *Phys. Rev. B* **60**, 13189
[17] Bridoux G 2008 Ph.D. thesis Instituto Balseiro (unpublished).
[18] Bridoux G, Nieva G and de la Cruz F 2007 *Physica C* **460-462**, 841
[19] Bridoux G, Nieva G and de la Cruz F 2008 *Phys. Rev. Lett.* **101**, 117002
[20] Wang Y et al 2003 *Science* **299**, 86
[21] Skocpol W J and Tinkham M 1975 *Rep. Prog. Phys.* **38**, 1049
[22] Wang Y 2005 *Phys. Rev. Lett.* **95**, 247002
[23] Briceño G, Crommie M F and Zettl A 1991 *Phys. Rev. Lett.* **66**, 2164
[24] Corvalán M S 1996 Master thesis Tucumán National University (unpublished).
[25] Yan Y F et al 1995 *Phys. Rev. B* **52**, R751
[26] Wang Y et al 2002 *Phys. Rev. Lett.* **88**, 257003
[27] López D et al 1996 *Physica C* **260**, 211

Dissipative dynamics of superconducting hybrid qubit systems

Enrique Montes, Jesús M Calero and John H Reina
Departamento de Física, Universidad del Valle, A.A. 25360, Cali, Colombia
E-mail: enriquem@univalle.edu.co and j.reina-estupinan@physics.ox.ac.uk

Abstract. We perform a theoretical study of composed superconducting qubit systems for the case of a coupled qubit configuration based on a hybrid qubit circuit made of both charge and phase qubits, which are coupled via a $\sigma_x \otimes \sigma_z$ interaction. We compute the system's eigen-energies in terms of the qubit transition frequencies and the strength of the inter-qubit coupling, and describe the sensitivity of the energy crossing/anti-crossing features to such coupling. We compute the hybrid system's dissipative dynamics for the cases of i) collective and ii) independent decoherence, whereby the system interacts with one common and two different baths of harmonic oscillators, respectively. The calculations have been performed within the Bloch-Redfield formalism and we report the solutions for the populations and the coherences of the system's reduced density matrix. The dephasing and relaxation rates are explicitly calculated as a function of the heat bath temperature.

1. Introduction

A quantum computer uses quantum two-level systems or quantum bits (qubits) which must be essentially interacting systems weakly or, even more desirable, completely decoupled from dissipative environments [1]. The low temperature superconductors, like aluminum, are good candidates for this purpose. For this reason, in the last few years, interest in the study of superconducting-qubit systems has grown [2, 3]. In these kind of systems, the circuit must be cooled to a temperature at which the thermal energy $\beta^{-1} \equiv k_B T$ is less than the transition energy $\hbar\omega_{0,1}$ between the qubit basis states, say $|0\rangle$ and $|1\rangle$. Another important requirement is that $\hbar\omega_{0,1} << \Delta$, where Δ is the energy gap of the superconducting material [4]. The basic elements used in the preparation of superconducting qubits are the Josephson junctions, which are one of the few electronic systems which, besides being nonlinear, also exhibit the property of a very weak dissipation [5]. Quantum information processors require a qubit coherent dynamics in order to allow the construction of quantum logic gates and circuits [1]. The hybrid systems have had an impact in the development of these circuits because they can bind the scales of atomic motion to macroscopic degrees of freedom [6]. Some solid-state systems, such as the so-called superconducting qubits, allow the implementation of such hybrid systems in their respective circuits which can be connected in different ways, in the form of simple circuit elements [7].

2. Model

Hybrid quantum circuit.— A generic superconducting qubit can be modeled by the Hamiltonian

$$\widehat{H}_{qb} = -\frac{1}{2}\epsilon\widehat{\sigma}_z - \frac{1}{2}\Delta\widehat{\sigma}_x, \qquad (1)$$

where ϵ and Δ denote the longitudinal and transversal parameters of the corresponding qubit, and σ_i are the usual Pauli matrices. Charge and flux qubits are two different types of superconducting qubits for quantum computing. In short, the charge qubit has the advantage of a more flexible controllability via external parameters: it can be conveniently controlled by a voltage gate or an applied magnetic flux. These external control parameters appear in the longitudinal (σ_z) and transverse (σ_x) terms of the circuit's reduced Hamiltonian. For the flux qubit, the longitudinal term can be controlled by the applied magnetic flux, but it is hard to control the transversal term via an external parameter.

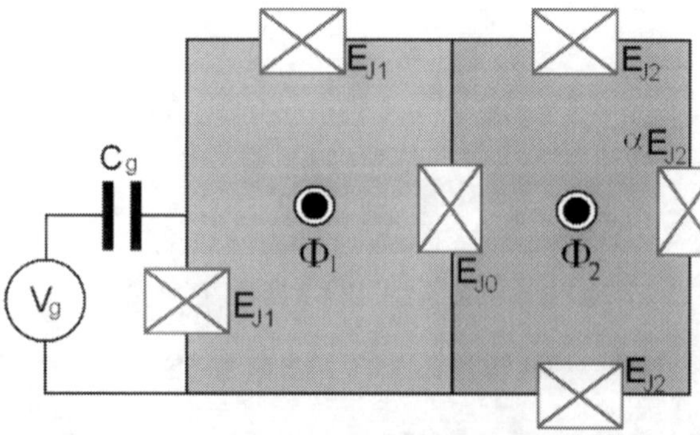

Figure 1. Hybrid quantum circuit: charge (left) and flux (right) qubits are effectively coupled due to the Josephson junction E_{J0}. E_{Ji} denotes the Josephson energy of each junction (crossed boxes). The charge ($i = 1$) and the flux ($i = 2$) qubits are crossed by externally controlled magnetic fluxs Φ_i.

The hybrid quantum circuit we study in this work is a system that couples a charge and a flux qubit, as schematically shown in Figure 1, with an effective interaction due to a Josephson junction that binds them. This system has been proposed in order to control the transversal term of the flux qubit with the charge qubit [7]. A calculation of the explicit Hamiltonian for the hybrid qubit system reads

$$\widehat{H}_{2q} = \epsilon_1(V_g)\widehat{\sigma}_z^{(1)} - \Delta_1\widehat{\sigma}_x^{(1)} + \epsilon_2(\phi_2)\widehat{\sigma}_z^{(2)} - \Delta_2\widehat{\sigma}_x^{(2)} + \chi\widehat{\sigma}_x^{(1)}\widehat{\sigma}_z^{(2)}, \tag{2}$$

where ϵ_1 and Δ_1 are the longitudinal and transversal parameters of the charge qubit, and likewise ϵ_2 and Δ_2 for the flux qubit. The interaction between the qubits gives rise to an effective $\sigma_x \otimes \sigma_z$ geometric term [7] with strength χ.

3. Coupling to the bath

In order to study the system's dissipative dynamics, we modeled the dissipative environment as either one or two baths of harmonic oscillators which are coupled to the σ_z components of each qubit. The first case considers each qubit coupled to its own bath of oscillators in the form $\widehat{\sigma}_z^{(i)}\widehat{X}^i$, where $\widehat{X}^i$ $(i = 1,2)$ denotes the bath coordinates. The full interacting Hamiltonian reads

$$\widehat{H}_{2B} = \sum_{i=1,2}\left(-\frac{1}{2}\epsilon_i\widehat{\sigma}_z^{(i)} - \frac{1}{2}\Delta_i\widehat{\sigma}_x^{(i)} + \frac{1}{2}\widehat{\sigma}_z^{(i)}\widehat{X}^i\right) + \chi\sigma_x^{(1)}\sigma_z^{(2)} + \widehat{H}_{B1} + \widehat{H}_{B2}, \tag{3}$$

where $\widehat{H}_{Bi}$ $(i = 1,2)$ is the Hamiltonian of the i-th bath. In the second case, we consider that the qubits are coupled to a common bath. Hence, the full system's Hamiltonian reads

$$\widehat{H}_{1B} = \sum_{i=1,2}\left(-\frac{1}{2}\epsilon_i\widehat{\sigma}_z^{(i)} - \frac{1}{2}\Delta_i\widehat{\sigma}_x^{(i)}\right) + \frac{1}{2}\left(\widehat{\sigma}_z^{(1)} + \widehat{\sigma}_z^{(2)}\right)\widehat{X} + \chi\widehat{\sigma}_x^{(1)}\widehat{\sigma}_z^{(2)} + \widehat{H}_B, \tag{4}$$

where $\widehat{X}$ is the bath coordinate and $\widehat{H}_B$ represents the Hamiltonian of the common bath.

Bloch-Redfield formalism.— This formalism is derived from a projector operator approach and provides an important tool for finding a set of coupled master equations which describe the dynamics of the reduced density matrix for a given system in contact with a dissipative environment. The Liouville equation for the total density operator ρ_T (of the whole system) is given by

$i\hbar\frac{d}{dt}\rho_T(t) = [\widehat{H}(t), \rho_T(t)]$. The corresponding reduced density matrix obeys, in the Born-Markov limit, the so-called Redfield equation [8]:

$$\dot{\rho} = -i\omega_{nm}\rho_{nm}(t) - \sum_{kl} R_{nmkl}\rho_{kl}(t), \tag{5}$$

where the first term represents the reversible motion in terms of the transition energies $\hbar\omega_{nm} \equiv E_n - E_m$, and the second term describes the system's relaxation. The Redfield relaxation tensor R_{nmkl} includes the dissipative efects of the system-environment coupling:

$$R_{nmkl} = \delta_{lm}\sum_r \Gamma^{(+)}_{nrrk} + \delta_{nk}\sum_r \Gamma^{(-)}_{lrrm} - \Gamma^{(+)}_{lmnk} - \Gamma^{(-)}_{lmnk}, \tag{6}$$

and its components are given by the 'golden rule' expressions

$$\begin{aligned} \Gamma^{(+)}_{lmnk} &= \hbar^{-2}\int_0^\infty dt e^{-i\omega_{nk}t}\left\langle \tilde{H}_{I,lm}(t)\tilde{H}_{I,nk}(0)\right\rangle, \\ \Gamma^{(-)}_{lmnk} &= \hbar^{-2}\int_0^\infty dt e^{-i\omega_{lm}t}\left\langle \tilde{H}_{I,lm}(0)\tilde{H}_{I,nk}(t)\right\rangle. \end{aligned} \tag{7}$$

Here, $\tilde{H}_I(t) = \exp(\frac{i\widehat{H}_B t}{\hbar})H_I\exp(\frac{-i\widehat{H}_B t}{\hbar})$ is the interaction term in the interaction picture, and the bracket denotes the thermal average over the bath's degrees of freedom.

4. Results and Discussion

System's eigen-energies.— These are worked out in the singlet/triplet basis states. In this basis, the Hamiltonian H_{2q} of the two-qubit system assumes the matrix form:

$$\widehat{H}_{2q} = -\frac{1}{2}\begin{pmatrix} \epsilon & \eta+\chi & 0 & -\chi \\ \eta+\chi & 0 & \eta-\chi & 0 \\ 0 & \eta-\chi & -\epsilon & -\chi \\ -\chi & 0 & -\chi & 0 \end{pmatrix}, \tag{8}$$

where we have introduced $\epsilon = \epsilon_1 + \epsilon_2$, $\Delta\epsilon = \epsilon_1 - \epsilon_2$, $\eta = \frac{\Delta_1+\Delta_2}{\sqrt{2}}$, $\Delta\eta = \frac{\Delta_1-\Delta_2}{\sqrt{2}}$, and considered qubits with the same transition energies $\epsilon_1 = \epsilon_2$ and transversal coupling $\Delta_1 = \Delta_2$. Thus, $\Delta\epsilon = \Delta\eta = 0$.

In Figure 2, we present the system's energies for three different scenarios, in which two of the parameters ϵ, η, χ have been fixed and the third one is varied. In Figure 2(a) we have varied the longitudinal parameter ϵ, showing a crossing between the energy of the singlet (E_1) and one of the energies of the triplet (E_3). In Figure 2(b) the influence of the transversal parameter η has been plotted. We notice the appearance of the same crossing observed in Figure 2(a), between the singlet and the E_3-triplet state, which indicates that these states do not interact. Figure 2(c) shows the dependence on the coupling term χ. Now, a crossing between the energy of the singlet state and that of the triplet state E_4 takes place. Note that for large values of χ, there is a degeneracy between the singlet state E_1 and the triplet state E_3. This feature means that for large positive values of the parameter χ, a large ferromagnetic coupling is present between the qubits.

Dissipative dynamics and temperature effects.— The solution of the Redfield equation $\dot{\rho}_{nm} = -\sum_{kl} R_{nmkl}\rho_{kl}(t)$ allows to analyse the system's reduced dynamics, for which explicit expressions have been found for the components of Equation (7), for both types of coupling to the environment. In the case of independent coupling we get

$$\Gamma^{(+)}_{lmnk} = \Gamma^{(-)}_{lmnk} = \frac{1}{4\beta\hbar}\left[\sigma^{(1)}_{z,lm}\sigma^{(1)}_{z,nk}\alpha_1 + \sigma^{(2)}_{z,lm}\sigma^{(2)}_{z,nk}\alpha_2\right], \tag{9}$$

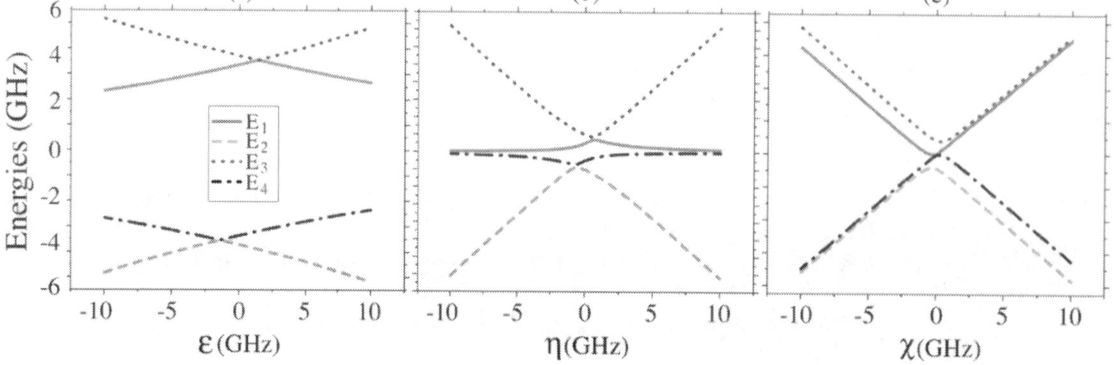

Figure 2. Energy spectrum of the coupled hybrid two-qubit system. (a) $\chi = 10$ GHz, $\eta = 1$ GHz, and ϵ is varied, (b) $\chi = 1$ GHz, $\epsilon = 1$ GHz (η is varied), and (c) $\epsilon = \eta = 1$ GHz (χ is varied). In all the graphs, E_1 is the singlet state and E_2, E_3, and E_4 are the triplet states.

where the α_i's correspond to the bath coupling parameters for the specific cases of Ohmic spectral densities $J_i(\omega) = \alpha_i \hbar\omega/\left(1 + \frac{\omega^2}{\omega_c^2}\right)$, and $\sigma_{z,ij} \equiv \langle E_i | \sigma_z | E_j \rangle$. For the case of collective coupling, $J(\omega) = \alpha\hbar\omega/\left(1 + \frac{\omega^2}{\omega_c^2}\right)$, and the rates become

$$\Gamma^{(+)}_{lmnk} = \Gamma^{(-)}_{lmnk} = \frac{\alpha}{4\beta\hbar}\left[\sigma^{(1)}_{z,lm}\sigma^{(1)}_{z,nk} + \sigma^{(1)}_{z,lm}\sigma^{(2)}_{z,nk} + \sigma^{(2)}_{z,lm}\sigma^{(1)}_{z,nk} + \sigma^{(2)}_{z,lm}\sigma^{(2)}_{z,nk}\right]. \tag{10}$$

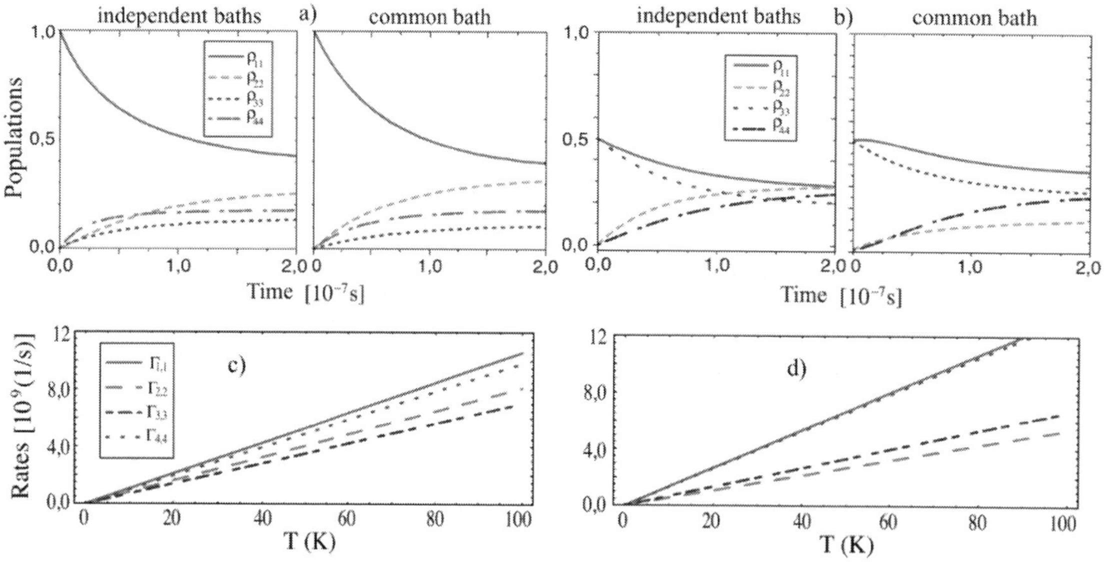

Figure 3. Populations dynamics, $\rho_{ii}(t)$. The density matrix calculation has been performed for the initial conditions: a) pure state $|E_1\rangle$, and b) maximally entangled Bell state $|\Phi^+\rangle$. Lower graphs show the relaxation rates (see main text) as a function of the bath temperature for the cases of c) independent and d) collective coupling to the environment. $\alpha_i = \alpha = 10^{-3}$.

The reduced density matrix has been computed from the system of differential equations (5). Figure 3 shows the time dependence of the system's populations ρ_{ii} for a coupling to two

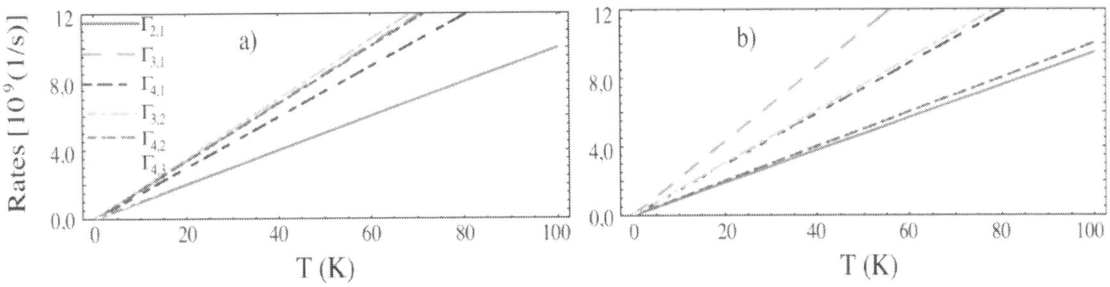

Figure 4. Decoherence rates as a function of the reservoir temperature for: a) independent and b) collective bath coupling. $\alpha_i = \alpha = 10^{-3}$.

different kinds of environments: i) independent and ii) collective or common baths. Figures 3(a) and 3(b) correspond to the initial conditions $|E_1\rangle$, and $|\Phi^+\rangle = \frac{1}{\sqrt{2}}(|00\rangle + |11\rangle)$, respectively. The population ρ_{11} has a slightly faster decay in the case of collective coupling for the initial condition $|E_1\rangle$, whereas this behaviour is reversed for the case of the entangled initial condition $|\Phi^+\rangle$. Thus, the dissipative population dynamics is very sensitive to initial state preparations.

The behaviour of the relaxation rates $\Gamma^{i/c}_{n,n} = Re(R^{i/c}_{nnnn})$ as a function of the bath temperature is shown in Figures 3(c) and (d), for the initial conditions $|E_i\rangle$. The superscripts i, and c stand for independent (Fig. 3(c)) and collective (Fig. 3(d)) couplings respectively. In agreement with graph 3(a), this shows that the rate Γ_{11} tends to a higher value for the case of collective coupling. In Figure 4, we have plotted the dephasing (decoherence) rates $\Gamma^{i/c}_{n,m} = Re(R^{i/c}_{nmnm})$, as a function of the bath temperature, for the cases of a) independent and b) collective coupling. We note that, in both cases, the decoherence rates are of the same order of magnitude. Furthermore, it is observed that Γ_{21} is the minimum dephasing rate, while Γ_{31} displays the higher rate.

By comparing the decoherence and the relaxation rates obtained, we conclude that both types of dissipation mechanisms occur on the same time scale. This implies a long decoherence time, which is a key requirement in the construction of quantum devices that are expected to work as quantum information processors.

Acknowledgments

This work was partially supported by Colciencias grant 1106-452-21296, the Excellence Centre for Novel Materials (CENM), and the exchange program PROCOL (DAAD-Colciencias).

References

[1] Nielsen M A and Chuang I L 2000 *Quantum Computation and Quantum information* (Cambridge)
[2] Makhlin Y, Schön G and Shnirman A 1999 *Nature* **398** 305
[3] Nakamura Y, Pashkin Y A and Tsai J S 1999 *Nature* **398** 786
[4] de Gennes P G 1966 *Superconductivity of Metals and Alloys* (New York: W H Benjamin)
[5] Ibach H and Luth H 1995 *Solid State Physics* (Springer)
[6] Neeley M *et al.* 2008 *Nature Phys.* **4** 523
[7] Xiao-Ling H *et al.* 2007 *Phys. Rev.* B **76** 024517
[8] Weiss U 1999 *Quantum Dissipative Systems* (World Scientific)
[9] Makhlin Y, Schön G and Shnirman A 2001 *Rev. Mod. Phy* **73** 357

Magnetic stripes in the UCoGe superconductor

Pablo de la Mora[1] and O Navarro[2]

[1]Departemento de Física, Facultad de Ciencias, Universidad Nacional Autónoma de México, Apartado Postal 70-542, 04510 México DF, México
and Instituto de Investigación en Materiales, Universidad Nacional Autónoma de México, Campus Morelia.

[2]Instituto de Investigación en Materiales, Universidad Nacional Autónoma de México, Apartado Postal 70-360, 04510 México DF, México

E-mail: delamora@unam.mx

Abstract. The magnetic superconductor UCoGe is analyzed with an electronic structure package, spin-orbit coupling and intra-atomic repulsion (via Hubbard U_H) were included. The possibility of an antiferromagnetic configuration is studied, but it is found to be unstable, also the non-collinear magnetization seems to be ruled out. The magnetization is given mainly by the Co-atoms with $M = 0.6$ μ_B/f.u. The U-atoms have two magnetic moments; $M = 0.07$ and 0.32 μ_B/f.u. arranged in an alternated planes along the *b*-direction forming magnetic stripes.

1. Introduction

Recently UCoGe was found to be both superconducting, $T_c = 0.8$ K, and ferromagnetic, $T_C = 3$ K, with a small magnetic moment, $M = 0.03$ μ_B/f.u. (Huy *et al.* 2007). Electronic quantum mechanical calculations have shown that both the U-atoms and the Co-atoms have a magnetic moment much larger than the experimental one, all the atoms are aligned ferromagnetically giving a total value of $M \approx 0.78$ μ_B/f.u. (de la Mora and Navarro 2008, Diviš 2008). In an attempt to conciliate these two different values various ferro- and antiferromagnetic calculations were performed.

2. Computational details

Electronic quantum mechanical calculations were done with the WIEN2k package (Blaha *et al.* 2001), which is a Linearized Augmented Plane Wave (FP-LAPW) method based on Density Functional Theory (DFT). Spin-orbit coupling is included in a second-variational way, and the strong correlations in uranium were included via LDA+U_H (SIC) (contributed by Pavel Novák). The Generalized Gradient Approximation of Perdew, Burke and Ernzerhof (1996) was used for the treatment of the exchange-correlation interactions. The energy threshold to separate localized and non-localized electronic states was -6 Ry. The muffin-tin radii were: 2.5a_0 for uranium, 2.43a_0 for cobalt, 2.48a_0 for rhodium, and 2.15a_0 for germanium (a_0 is the Bohr radius). The criterion for the number of plane waves was $R_{MT}^{min} \times K^{max} = 9$ and the number of k-points was 462 (7×11×6), for the case with all the atoms independent the number of k-points was 225 (5×9×5).

3. Results and discussion

The UCoGe compound has a deformed MgB_2 structure; it has undulated Co-Ge graphene-like sheets intercalated by U atoms (figure 1). In contrast to MgB_2, the bonds are not covalent but metallic.

The electronic structure calculations were performed using the cell parameters of Huy *et al.* (2007) and the atomic positions of Canepa *et al.* (1996). Spin-orbit interaction was included for uranium and cobalt, the intra-Coulombic repulsion U_H was also included for uranium 5d, with a $U_H = 0.362$ eV (de la Mora and Navarro 2008).

In our previous work (de la Mora and Navarro 2008), in order to have a better description of the system, all the atoms in the unit cell (which contains four formula units) were treated as independent atoms, that is, they could have different values in their electronic properties.

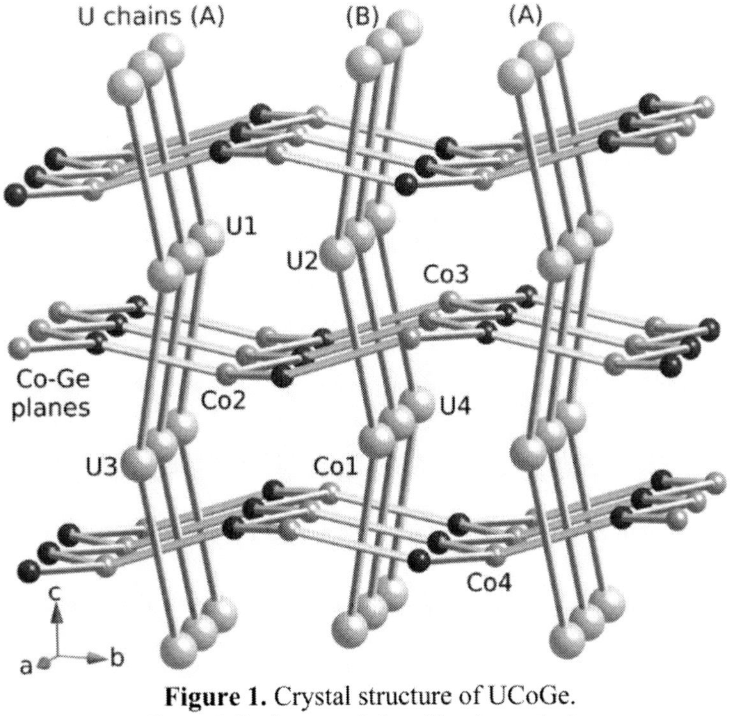

Figure 1. Crystal structure of UCoGe.
Co and Ge form undulated horizontal planes,
U form vertical zigzag chains.

Table 1. Energies, in meV/f.u., of the different magnetic configurations.
The most stable, the 'all spins up' *c*-direction configuration, is taken as reference.

Magnetization direction	Magnetic moment ordering	Energy	ΔE relative to 'all spins up'
	All spins up	10.8	0
b	U1 and U2 up	35.3	24.5
	U1 and U3 up	23.4	12.6
	U1 and U4 up	85.2	74.4
	All spins up	0	0
c	U1 and U2 up	27.6	27.6
	U1 and U3 up	39.8	39.8
	U1 and U4 up	61.2	61.2

In the present paper the magnetic moments of the unit cell were arranged in all the possible antiferromagnetic orders; a) all spins up (ferromagnetic), b) U1 and U2 spins up (U3 and U4 spins down), c) U1 and U3 spins up and d) U1 and U4 spins up. These calculations were done for the *b*- and *c*- magnetic moment directions which are the most stable orientations. The magnetic moments are given in table 1, the antiferromagnetic configurations have higher energy than the corresponding ferromagnetic one. The ferromagnetic configuration in the *c*-direction is the most stable, the experimental results on single crystals (Huy *et al.* 2008) find that the magnetization is in this direction, but the experimental magnetic moment is smaller.

Table 2. Magnetic moments (μ_B), with the magnetization in the *b*- and *c*-direction for the ferromagnetic case.

Atom	Direction		
	b	*c*	canting *a*-direction
U	0.094	0.073	0.070
	0.116	0.319	-0.124
	0.108	0.073	-0.069
	0.103	0.322	0.122
Tot/f.u.	0.105	0.179	
Co	0.625	0.610	0.033
	0.608	0.589	-0.041
	0.613	0.609	-0.033
	0.621	0.589	0.042
Tot/f.u.	0.615	0.586	
Tot/f.u.	0.722	0.764	

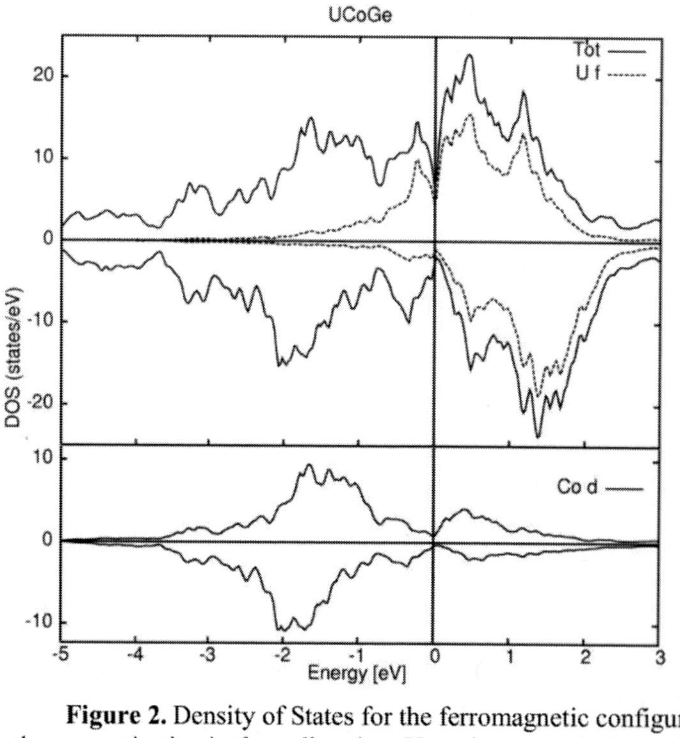

Figure 2. Density of States for the ferromagnetic configuration with the magnetization in the *c*-direction. Up-spin: upward, down-spin: downward.

In our previous work (de la Mora and Navarro 2008) the most stable configuration for the magnetic moment was in the *b*-direction. In the present work we found that the system is quite unstable and the convergence conditions should be quite strict, giving as a result that the *c*-direction is the most stable configuration.

The total magnetic moments (spin plus orbital) of the atoms for the *b*- and *c*-direction are given in table 2. U has large spin and orbital moments but in opposite directions which almost cancel. Cobalt has moderate spin and small orbital moments, both in the same direction. Therefore the main contribution comes from cobalt.

For the *c*-direction there is a deviation (canting) of the magnetic moments in the *a*-direction, while for the *b*-direction there is no canting. This canting is large for the total moments, the spin and orbital moments have opposite signs in their main component and the individual canting is relatively small (< 6°). This small value probably rules out a fully non-collinear magnetic order.

An interesting result, for the ferromagnetic configuration with the magnetization in the *c*-direction, is that there is a clear differentiation of the magnetic moment values into two groups, and these are grouped around the U-chains, see figure 1. The U-atoms in the A-chain have a small magnetic moment, ≈0.07 μ_B, while in the B-chain the value is considerably larger, ≈0.32 μ_B. The Co-atoms contiguous to these chains follow the same trend, although the magnetic moment difference is much smaller; the Co-atoms contiguous to the A-chains have ≈0.59 μ_B, while for the B-chain it is ≈0.61 μ_B. Therefore there is a modulation of the magnetic moment, or stripes, along the *b*-direction.

For the *b*-direction there is no magnetic moment differentiation, probably due to the fact that there is no canting.

The Density of States (DOS) is shown in figure 2. At the Fermi energy there is a narrow and deep drop, for the up-spin. For the down-spin the drop is larger by a factor of 3.5, and it is not so narrow, as if the material wanted to have a half-metal character.

4. Conclusions

In an attempt to conciliate with the small experimental magnetic moment, electronic structure calculations were done on UCoGe with different ferro- and antiferromagnetic configurations. The latter ones were found to have higher energy than the corresponding ferromagnetic one, being the ferromagnetic configuration in the *c*-direction the most stable, as a consequence it was not possible to find the small magnetic moment. The small deviations of the different components of the magnetic moments from the principal direction seem to rule out a non-collinear magnetic arrangement. There is a narrow partial gap at the Fermi energy which is deeper for the down-spin-DOS. The Co-atoms have a magnetic moment due mainly to the spin M = 0.6 μ_B/f.u. The U-atoms have large spin and orbital contributions in opposite directions giving a small total moment. The value of this total moment has two values, M = 0.07 and 0.32 μ_B/f.u., alternating in the *b*-direction forming magnetic stripes.

Acknowledgements

This work was done with support from DGAPA-UNAM under projects PAPIIT IN105207 and IN108907.

References

Blaha P, Schwarz K, Madsen G K H, Kvasnicka D and Luitz J 2001 WIEN2k, An Augmented Plane Wave + Local Orbitals Program for Calculating Crystal Properties Karlheinz Schwarz, Techn. Universität Wien, Austria (ISBN 3-9501031-1-2)

Canepa F, Manfrinetti P, Pani M and Palenzona A 1996 *J. Alloys Compd.* **234** 225-230

de la Mora P and Navarro O 2008 *J. Phys.: Cond. Matter* **20** 285221

Diviš M 2008 *Physica* B **403** 2505

Huy N T, Gasparini A, de Nijs D E, Huang Y, Klaasse J C P, Gortenmulder T, de Visser A, Hamann A, Görlach T and Löhneysen H v 2007 *Phys. Rev. Lett.* **99** 067006

Journal of Physics: Conference Series **167** (2009) 012015 doi:10.1088/1742-6596/167/1/012015

Huy N T, de Nijs D E, Huang Y and de Visser A 2008 *Phys. Rev. Lett.* **100** 077002
Perdew J P, Burke K and Ernzerhof M 1996 *Phys. Rev. Lett.* **77** 3865

EFFECT OF SUBSTRATE TEMPERATURE ON THE OPTICAL, STRUCTURAL AND MORPHOLOGICAL PROPERTIES OF In_2Se_3 THIN FILMS GROWN BY A TWO-STEP PROCESS

Clavijo J.[1], Romero E.[2] and Gordillo G.[3]
[1, 2] Departamento de Química, Universidad Nacional de Colombia, Bogotá.
[3] Departamento de Física, Universidad Nacional de Colombia, Bogotá.

E-mail: ggordillog@unal.edu.co

Abstract. Polycrystalline γ - In_2Se_3 thin films with adequate properties to use them as buffer layer in solar cells, were grown on corning glass substrates using a novel procedure which includes the formation of the α - In_2Se_3 phase in a first step followed by thermal annealing in Se ambient to activate the formation of the γ - In_2Se_3 phase. X-ray diffraction (XRD) measurements revealed that the substrate temperature strongly affects the phase in which the indium selenide films grow; at substrate temperatures of around 300ºC the indium selenide grow in the α-In_2Se_3 phase, whereas the samples deposited at temperatures between 300 and 550ºC grow with a mixture of the α-In_2Se_3 and γ-In_2Se_3 phases. The α-In_2Se_3 samples change into the γ-In_2Se_3 phase when subjected to heat treatment around 550ºC in Se ambient. Spectrophotometric measurements also revealed that the phase in which the indium selenide films grow, significantly affects the optical gap Eg. Eg values of 1.47 eV and 2.11 eV were determined for the α-In_2Se_3 and γ-In_2Se_3 films respectively, indicating that this γ-In_2Se_3 compound has better properties to perform as buffer layer in thin film solar cells. The effect of substrate temperature on the structural, optical and morphological properties was investigated using XRD, spectral transmittance and atomic force microscope (AFM) measurements. Theoretical simulation of the XRD pattern carried out with the help of the PowderCell package, allowed us to identify the phases associated to the X-Ray reflections, with a good degree of confidence.

1. Introduction

Indium Selenide is a promising semiconductor material because it has suitable properties for solar cells applications [1], and for other applications such as polymer Schottky diodes, capacitors and batteries [2]. Although In_2Se_3 can grow in the α, β, δ or γ phases [3], only the γ phase has suitable energy bandgap to use it as buffer layer in heterojunction solar cells [4]. In_2Se_3 is a III-VI semiconductor compound with n-type conductivity and energy bandgap ranging from 1.4 to 2.19 eV for the α and γ phases, [1, 2].

[1] Tel./fax:571-3165000 Ext. 13017 E-mail address: *jiclavijop@unal.edu.co*
[2] Tel./fax:571-3165000 Ext. 13017 E-mail address: *erromerom@unal.edu.co*
[3] Tel./fax:571-3165000 Ext. 13017 E-mail address: *ggordillog@unal.edu.co*

Several techniques have been used to grow In_2Se_3 thin films, such as: electrochemical synthesis [5], selenization [6] and co-evaporation of precursors [7]. In this work the γ - In_2Se_3 thin films were grown using a novel procedure which includes two stages: in the first stage, the α - In_2Se_3 phase was grown by co-evaporation of In and selenium and in the second stage the γ - In_2Se_3 phase is achieved through thermal annealing of the α - In_2Se_3 compound in selenium ambient. The influence of the deposition conditions on the optical constants, crystallographic phase and morphological properties were investigated by spectral transmittance, XRD and AFM measurements.

2. Experimental

The In_2Se_3 films were prepared using a deposition system constituted by the evaporation chamber connected to a vacuum system which allows working at pressures of about 10^{-5} mbar; a tungsten boat is used to evaporate In, a tantalum effusion cell to evaporate selenium and a thickness monitor (Maxtec TM-400) with a quartz crystal as sensor is used to measure the flux of the evaporated elements. The substrate temperature was controlled with a programmable PID controller (Eurotherm 900C).

It is well known from the In-Se binary phase diagram that the γ - In_2Se_3 phase can be achieved at temperatures greater than 500°C [8]. However, this was not possible in this work because at high temperatures the selenium re-evaporates. For this reason, a novel procedure was developed to get the γ - In_2Se_3 phase: this was accomplished in a two stage process. In the first stage, a layer of α - In_2Se_3 is grown by simultaneous evaporation of In and Se, keeping the substrate temperature at 300° C. During this stage the flux of In is kept around 3 Å/s and the flux of Se in 15 Å/s by controlling the evaporation temperature of Se around 260°C. In the second stage, the α - In_2Se_3 phase is converted into the γ - In_2Se_3 phase through thermal annealing at 550°C in Se atmosphere. In table 1 are listed the parameters that allowed us to grow In_2Se_3 in the γ phase, through a process in two steps.

Table 1. List of deposition parameters used to grow In_2Se_3 thin films in the γ phase with different thicknesses.

Deposition parameters	1st stage	2nd stage
In – Flux (Å/s)	4	-
Se-evaporation temp. (° C)	260	260
Substrate temperature (° C)	300	550
Evaporated mass of In (g)	0.02 - 0.3	-
Total Evaporated mass of Se (g)	1.0	

The transmittance measurements were carried out with a VIS – IR Oriel spectrophotometer. The XRD measurements were performed using the Cu Kα radiation of a Shimadzu-6000 diffractometer, and the AFM images were obtained with a PSI AP-0100 Scanning Probe Microscope. The film thickness was measured with a Veeco Dektak 150 surface profiler.

3. Results and discussion

3.1 Optical characterization

The In_2Se_3 samples were characterized through spectral transmittance measurements, in order to study the effect of the substrate temperature on the optical constants (refractive index n, absorption coefficient α and optical gap E_g). Fig.1 shows typical transmission spectra of In_2Se_3 films grown in the α and γ phases, respectively.

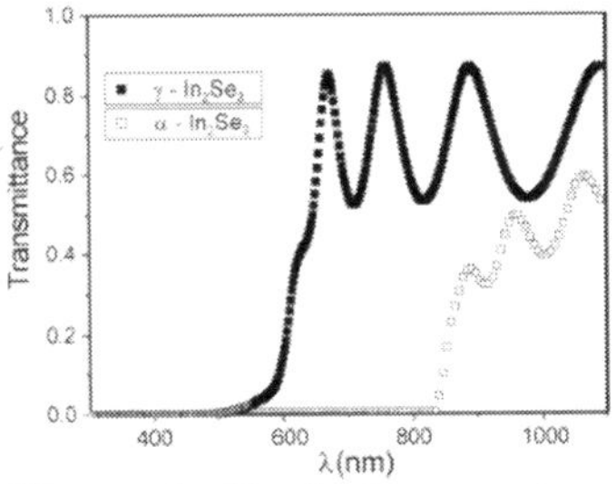

Figure 1. Typical transmittance curves of In_2Se_3 thin films grown in the α and γ phases.

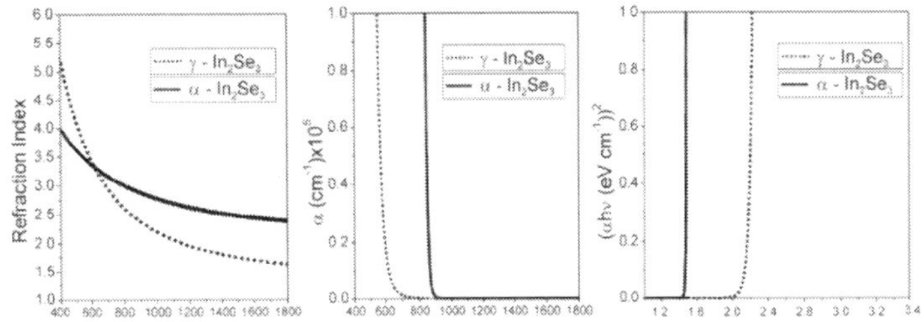

Figure 2. Typical curves of n vs.λ, α vs. λ and $(\alpha h\nu)^2$ vs. $h\nu$, corresponding to both, α - In_2Se_3 and γ - In_2Se_3 thin films. n is the refraction index, α is the absorption coefficient and $(\alpha h\nu)^2$ is a quantity proportional to (Eg - hν), used to get the energy band gap by extrapolating the $(\alpha h\nu)^2$ vs. $h\nu$ curve.

It is observed in fig. 1 that the transmittance of the α - In_2Se_3 films is smaller than that of the films grown in the γ- In_2Se_3 phase; this α - In_2Se_3 behavior is seemingly caused by photon absorption via states within the gap, which contributes to the absorption process simultaneously with the fundamental absorption. On the contrary, in In_2Se_3 samples grown in the γ phase the photon absorption is mainly achieved through the fundamental absorption mechanism. On the other hand, it is observed that the cut off wavelength of the γ- In_2Se_3 films is significantly smaller than that of the α - In_2Se_3 films, indicating that the optical gap Eg of the α - In_2Se_3 films is smaller than that of the γ - In_2Se_3 films. The difference in the Eg value seems to be caused by the fact that the lattice constant a, of the α -In_2Se_3 films is significantly smaller than that of the γ - In_2Se_3 films (see section 3.2).

In Figure 2 are depicted typical curves of n vs. λ, α vs. λ and $(\alpha h\nu)^2$ vs. hν corresponding to both, α - In_2Se_3 and γ - In_2Se_3 films which were obtained from the transmission spectra showed in figure 1 by using calculations based on a procedure detailed in references [9, 10]. The optical gap Eg for both types of In_2Se_3 samples was determined by extrapolating the $(\alpha h\nu)^2$ vs. $h\nu$ curve; E_g values of 1.45 and 2.15 eV were obtained for the α - In_2Se_3 and γ - In_2Se_3 films, respectively.

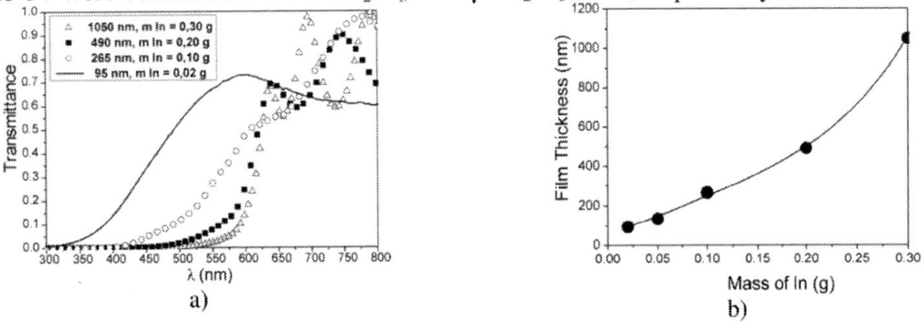

Figure 3: a) Transmittance curves corresponding to γ - In_2Se_3 thin films of different thickness and b) variation of the thickness as a function of the In mass evaporated.

Figure 3 shows transmittance curves corresponding to γ - In_2Se_3 films with thicknesses ranging from 95 to 1050 nm and curves of thickness as a function of the mass of In evaporated; the thickness was varied by varying the mass of In evaporated, as indicated in table 1. It is observed that the cut off wavelength is shifted toward lower values of λ when the film thickness decreases. We have not found a definitive explanation of this behavior. However, we consider that the shift of the wavelength cutoff observed when the film thickness increases from values less than 100 nm in which the films are in the coalescence stage, up to values of the order of μms in which a continuous film is formed, could be associated to changes in the band structure, induced by changes in the different interaction processes which take place during the different growth stages of the thin films. In particular, the superposition degree of electron clouds of neighboring atoms, which affect the band width, is feasible to increase by increasing the film thickness. The widening of the energy gap observed in the In_2Se_3 films when its thickness decreases, constitutes an advantage when this layer is used as optical window in solar cells.

3.2 Structural characterization

The influence of the phase in which the In_2Se_3 films grow on their structural properties was also studied through XRD measurements. Figure 4 compares experimental XRD spectra of In_2Se_3 films grown in the α and γ phases, with one obtained theoretically. The results were analyzed using the data reported in the JCPDS data base for these types of compounds (JCPDS cards 00-034-1279 and 00-040-1407) and theoretical simulation of the experimental XRD pattern, using the PowderCell package. This procedure allowed us to identify the phases present in the samples with a good degree of confidence. The results revealed that the samples deposited at low temperatures (T=300°C) contain only the α - In_2Se_3 phase with hexagonal type structure. The results also indicate that samples in the α-In_2Se_3 phase change into the γ-In_2Se_3 phase, also with hexagonal structure, when subjected to thermal annealing in Se atmosphere at temperatures around 550°C; however, the lattice constant a of the γ - In_2Se_3 films is greater than that of the α - In_2Se_3 samples. The values of the corresponding lattice constants were, a= 4.025±0.003 Å, c= 19.23±0.01 Å for the α - In_2Se_3 films and a= 7.128±0.001 Å, c= 19.384±0.002 Å for the γ - In_2Se_3 films. It was also found that the samples deposited at temperatures ranging from 300 to 550°C, grow with a mixture of the α-In_2Se_3 and γ-In_2Se_3 phases.

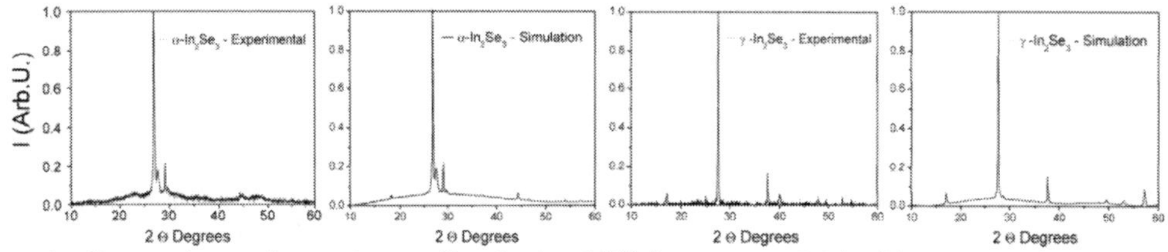

Figure 4 : Comparison of experimentally obtained XRD spectra of thin films grown in the α-In_2Se_3 and γ - In_2Se_3 phases, with those simulated using the Powder Cell software

3.1 Morphological characterization

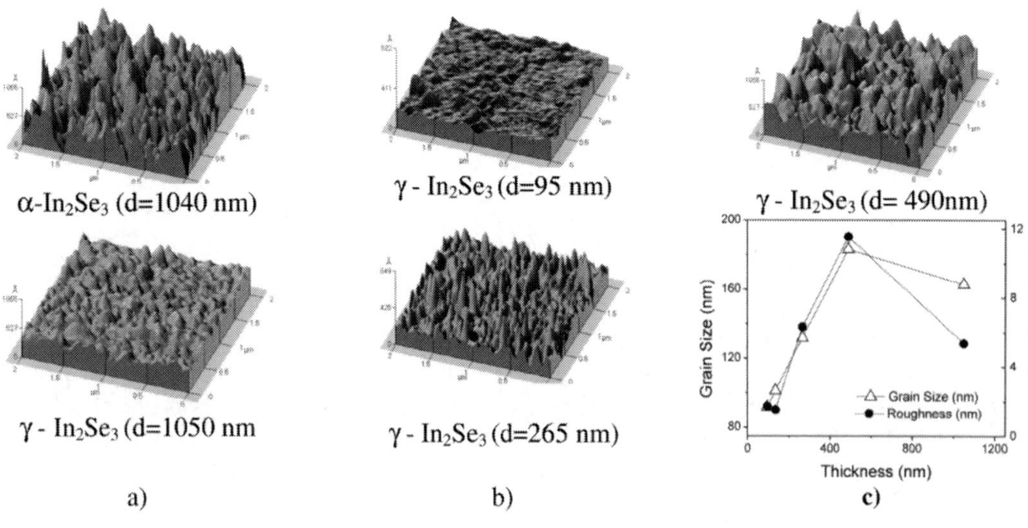

Figure 5. (a) Comparison of the AFM image of a typical α-In_2Se_3 thin film with one of a In_2Se_3 thin film of similar thickness, grown in the γ phase b) AFM images corresponding to γ - In_2Se_3 thin films with thicknesses varying between 95 and 1050 nm and c) Variation of the grain size and roughness of γ - In_2Se_3 thin films as a function of the thickness.

In Fig. 5a are compared the AFM images taken to typical In_2Se_3 thin films grown in the α and γ phases, with the same thickness. AFM images, showing the effect of the film thickness on the morphology are also depicted in Fig. 5b.
It is clear from the results of fig.5 that the phase in which the In_2Se_3 thin films grow, significantly affects the films' morphology. In general, both types of In_2Se_3 films present similar values of grain size; however, the roughness of the samples grown in the α phase is greater than that of the films grown in the γ phase. On the other hand it was found (see Fig. 5b) that the increase of the film thickness affects both, the average roughness and the average grain size which were determined analysing data from the AFM images, with the help of tools provided by the ProScan image analysis software. The curves depicted in Fig. 5c show the effect of the film thickness on the grain size and roughness of γ - In_2Se_3 thin films, determined from a statistical analysis carried out with the ProScan Data Acquisition software.

Conclusions

γ- In_2Se_3 thin films with adequate optical and structural properties to be used as buffer layers in solar cells were grown using a novel procedure which includes the formation of the α - In_2Se_3 phase in a first step followed by thermal annealing in Se ambient to activate the formation of the γ - In_2Se_3 phase. Also, it was found also that the substrate temperature affects the phase in which the In_2Se_3 films grow; in general, at substrate temperatures around 300ºC the indium selenide films grow in the α phase, between 300 and 550ºC there is a mixture of the α-In_2Se_3 and γ-In_2Se_3 phases; γ -In_2Se_3 films were achieved submitting the α-In_2Se_3 samples to a heat treatment around 550ºC in Se ambient. Finally, the α - In_2Se_3 thin films has found to have an optical gap significantly smaller than the γ - In_2Se_3 ones, indicating that the latter have better properties to perform as buffer layers in thin film solar cells.

AFM measurements showed that the phase in which the In_2Se_3 thin films grow and their thicknesses, significantly affect the films' morphology. In general, both type of In_2Se_3 films present similar values of grain size; however, the roughness of the samples grown in the α phase is greater than that of the films grown in the γ phase. On the other hand, it was found that the increase of the film thickness affects both, the average roughness and the average grain size.

Acknowledgements

This work was supported by the Universidad Nacional de Colombia and Colciencias.

References

[1] M. Emziane, S. Marsillac, J. C. Bernede, Materials Chemistry and Physics 62 (2000) **84**
[2] A. Ateş, M. Kundakçi, A. Astam, Physica E 40 (2008) **2709**
[3] D. Y. Lyu, T.Y. Lin , J. H. Lin, Solar Energy Materials & Solar Cells 91 (2007) **888**
[4] G. Gordillo, C. Calderón, Sol. Eng. Mater. Sol. Cell. 77 (2003) **163**
[5] R. Vaidyanathan et al, Journal of Electroanalytical Chemistry 559 (2003) **55**
[6] F. Hergert, S. Jost, R. Hock, M. Purwins, Journal of Solid State Chemistry 179 (2006) **2394**
[7] C. Amory, J.C. Bernede, E. Halgand, S. Marsillac, Thin Solid Films 431-432 (2003) **22**
[8] Y.-C. Huang et al, Journal of Crystal Growth 310 (2008) **1679**
[9] M. Gracia, F. Rojas, G. Gordillo, 20th European photovoltaic Solar Energy Conference, Barcelona, Spain, (2005)
[10] R. Swanepoel, J. Phys. E: sci. Instrum., **16** (1983) **1214**

Photoconduction in silicon rich oxide films

J. A. Luna-López[a], M. Aceves-Mijares[b], J. Carrillo- López[a], A. Morales-Sanchez[b], F. J. Flores-Gracia[a], G. Garcia-Salgado[a]

[a]CIDS-ICUAP, Benemérita Universidad Autónoma de Puebla. Ed. 103 D and C, col. San Manuel, Puebla, Pue. México 72570.

[b]INAOE, Luis Enrique Erro No. 1, Apdo. 51, Tonantzintla, Puebla, México 72000.

E-mail : jluna@buap.siu.mx, jluna@inaoep.mx

Abstract. Photoconduction of silicon rich oxide (SRO) thin films were studied by current-voltage (I-V) measurements, where ultraviolet (UV) and white (Vis) light illumination were applied. SRO thin films were deposited by low pressure chemical vapour deposition (LPCVD) technique, using SiH_4 (silane) and N_2O (nitrous oxide) as reactive gases at 700 °C. The gas flow ratio, Ro = $[N_2O]/[SiH_4]$ was used to control the silicon excess. The thickness and refractive index of the SRO films were 72.0 nm, 75.5 nm, 59.1 nm, 73.4 nm and 1.7, 1.5, 1.46, 1.45, corresponding to Ro = 10, 20, 30 and 50, respectively. These results were obtained by null ellipsometry. Si nanoparticles (Si-nps) and defects within SRO films permit to obtain interesting photoelectric properties as a high photocurrent and photoconduction. These effects strongly depend on the silicon excess, thickness and structure type. Two different structures (Al/SRO/Si and Al/SRO/SRO/Si metal-oxide-semiconductor (MOS)-like structures) were fabricated and used as devices. The photocurrent in these structures is dominated by the generation of carriers due to the incident photon energies (~3.0-1.6 eV and 5 eV). These structures showed large photoconductive response at room temperature. Therefore, these structures have potential applications in optoelectronics devices.

1. Introduction

Currently, nanoparticles (as silicon nanocrystals, silicon nanoclusters, amorphous silicon and compounds) in different materials have been subjected to an intense study due to their excellent optoelectronics properties, where the absorption-emission mechanism represents one of the most interesting problems in modern solid state physics. Another important problem has been the compatibility of these materials with the silicon technology in order to integrate optoelectronic functions in silicon. One of these materials with nanoparticles, compatible with silicon technology, is the silicon rich oxide (SRO), which is a material with silicon excess formed by multiple phases (SiO_2, SiO_x and Si-np crystalline or amorphous) [1]. SRO films can be obtained by different techniques, such as chemical vapour deposition (CVD), LPCVD (low pressure), PECVD (plasma enhanced), silicon implantation into thermal silicon dioxide, sol–gel, sputtering, etc [2, 3, 4]. Silicon excess can be controlled with the flow ratio (Ro) between N_2O and SiH_4 (Ro = $[N_2O]$ / $[SiH_4]$) as the reactant gasses. This material has been studied due to its interesting structural, electrical and optical properties [5, 6], which have been applied to different kinds of applications, such as waveguides, non-volatile memory and light detection devices [3-7].

In this work, SRO films were deposited by LPCVD technique. Different photoconductive responses were observed when the silicon excess was varied. An important parameter in this structure is the thickness of SRO films. Thinner films have a better photoconductive interaction with the light on the surface of the film and with the SRO/Si interface. Therefore, due to radiative recombination between the Si-nps and defects [1] within the SRO films when illuminated, the carrier generation is provoked. Electron-hole (e-h) pairs are generated by the fast interaction between incident photons with Si-nps in the SRO films, improving the transport of the photo-excited carriers in SRO thin films towards the contacts. This is clearly observed by the high and fast response of the photocurrent.

On the other hands, we describe the fabrication of Metal-Oxide-Semiconductor-like structures with a single SRO layer (as Al/SRO/Si) and with double SRO layer (as Al/SRO/SRO/Si). In the Al/SRO/SRO/Si structure, the second SRO layer has the purpose of to isolate (as a dielectric) the effects of the silicon bulk.

SRO films with thickness smaller than 75 nm and a large silicon excess (SRO_{10}) contain Si-nps (as Si-nc) which interact with light within a wavelength range. White or visible light generates a high photocurrent when the size of Si-nps is larger than 5 nm [7], but for nanoparticles (as Si-nanocluster or compounds) smaller than 1 nm UV light generates the highest photocurrent. In both cases, the measurement of photoconduction is carried out between two contacts of the Al/SRO/Si structures. Using a double layer of SRO it is possible to obtain a larger photoconduction.

2. Experiment

SRO films were deposited on N type Silicon (100) substrates with resistivity of 2-5 Ω-cm. SRO layers were obtained in a horizontal LPCVD hot wall reactor using SiH_4 (silane) and N_2O (nitrous oxide) as reactive gases at 700 °C. The gas flow ratio, Ro = $[N_2O]/[SiH_4]$, was used to control the amount of silicon excess in the SRO films. Ro = 10, 20, 30 and 50 were used for this experiment. The pressure was varied for each Ro from 1.64 - 2 Torr. After deposition, the samples were thermally annealed at 1000° C in N_2 atmosphere for 30 minutes. Aluminum (Al) contacts were made onto the SRO surface by evaporation and standard photo-lithography. The area of the gate electrode was A= 0.089 cm^2. For structures with double layer, which consisted of SRO_{10}/SRO_{30} and SRO_{10}/SRO_{50}, contacts were made with circular gate electrodes of area A= 0.01 cm^2. Ellipsometric measurements were made with a Gaertner L117 ellipsometer to obtain the thickness and refractive index of the SRO films, whose values are shown in Table I.

Table 1. Refractive index and thickness results to samples utilized.

Ro	Refractive index	Thickness (Å)
10	1.775 ± 0.005	720 ± 28
20	1.55 ± 0.03	755 ± 25
30	1.46 ± 0.01	591 ± 3
50	1.45 ± 0.01	734 ± 15

Current versus voltage (I-V) measurements were performed at room temperature in dark and under illumination conditions by using a computer controlled Keithley 6517A Electrometer in a screening box. The voltage sweep was done at a rate of 0.1 V s^{-1}. Illumination was performed with an UV lamp (UVG-54) with output power of 6.12 mW/cm^2, and white light (1.7 - 4 eV approximated range) with output power of 2.19 $\mu W/cm^2$. Both powers were measured by means of a radiometer (IL1 400A). Photocurrent was measured between two devices in samples with one and two SRO layers, as depicted in Fig. 1 a) and b).

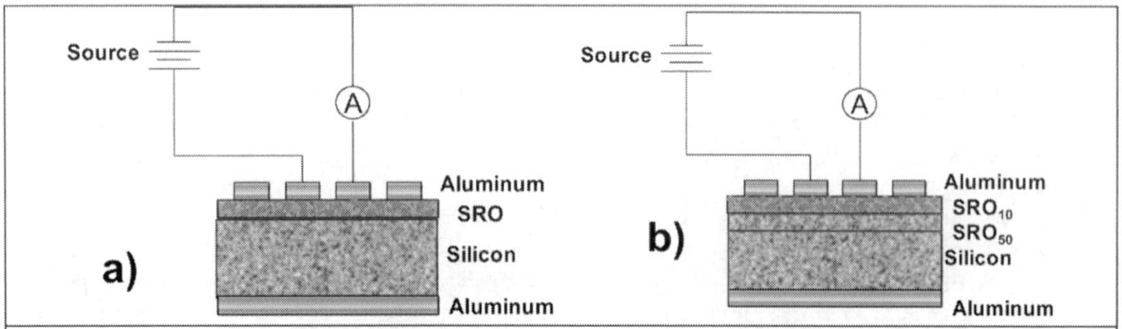

Figure 1. The circuit used for current-voltage measurements for a) one and b) two layers SRO thin films.

3. Results

Figure 1 shows the experimental circuit used for current-voltage measurements under dark and illuminated conditions of the Al/SRO/Si structures with one and two layers of SRO.

Figure 2 a), b) and c) show the dark current density (J_{dark}) of the Al/SRO/Si structure with one layer of SRO under forward and reverse bias. The photocurrent for structures with Ro = 10, 20 and 30, shows an increase due to the photoconduction in the films. Photocurrent density ($J_{pc} = J_{light} - J_{dark}$) is a function of the wavelength and the applied bias on the SRO films with different flow ratios Ro = 10, 20 and 30, and thickness values of 72 nm, 75 nm and 59 nm, respectively. The thickness and Si-nps contributes with an efficient generation of e-h pairs that allows a better photoconduction when the films are illuminated at different wavelengths. For Ro = 10 and 20, the measured photocurrent has a higher photo-response than the corresponding to Ro = 30. We can see a very important result: the photocurrent for SRO with Ro = 10, 20 and 30 has a symmetric behavior and it can be generated with positive or negative bias.

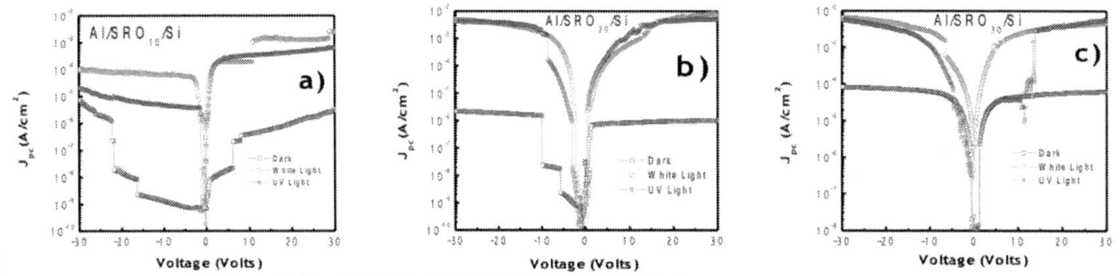

Figure 2. IV characteristics were taken under dark conditions and with different illumination on Al/SRO/Si MOS like structures. a) SRO_{10}, b) SRO_{20}, c) SRO_{30}.

The current in the Al/SRO_{10}/Si structure increases rapidly to about 4 μA/cm^2 at 0.5V when it is illuminated with white light and under forward bias, For voltages above 2 V, the current is about 0.19 mA/cm^2. Beyond 10 V, the current saturates to 1 mA/cm^2. On the other hand, when the Al/SRO_{10}/Si structure is reversed bias, the current increases rapidly. Beyond -10 V, the current density is 0.1 mA/cm^2 and it stays at this value. The forward and reverse current ratios with respect to dark current have about 4 magnitude orders. This behaviour is only for SRO_{10} with white light. When the structure is illuminated with UV light, the photocurrent ratio decreases near to 1 magnitude order respect to the value with white light excitation. For structures with SRO_{20} and SRO_{30}, the current have a similar behaviour when they are illuminated with white and UV-light.

Figures 3 a) and b) show the structures with double SRO layer (Al/SRO_{10}/SRO_{30}/Si and Al/SRO_{10}/SRO_{50}/Si). For the Al/SRO_{10}/SRO_{30}/Si structure, the dark current is approximately 0.1 mA/cm^2 in forward and reverse bias. When white light is applied the photocurrent increased near two orders of magnitude, while the photocurrent corresponding to UV light increased only 1 order of magnitude. In all ways the behaviour is symmetrical for forward and inverse bias. When white light is applied to the Al/SRO_{10}/SRO_{50}/Si structure, the forward current decreased to 1.8 pA/cm^2 at 1 volts as a discharge. When the SRO film is charging newly, the current rise at 1.2 V, then it increases nonlinearly to 0.074 mA/cm^2 at 3.2 V. Beyond 17 V, the current saturates to 1.3 mA/cm^2. On the other hand, the inverse current with white light applied remains constant at 5 nA/cm^2 from 0 to 5 V. The current starts to rise at -7 V, increases nonlinearly around 0.721-0.818 mA/cm^2 from -8 to -30 V. With UV illumination, the current decreased approximately 1 order of magnitude in relation to current measured under white light illumination. But a high photoconduction is obtained between the two contacts with white light and UV illumination. The photocurrent increases from 4 to 5 orders of magnitude. In all experiment conditions, the SRO films are operating as a photoconductor between the two Aluminium contacts.

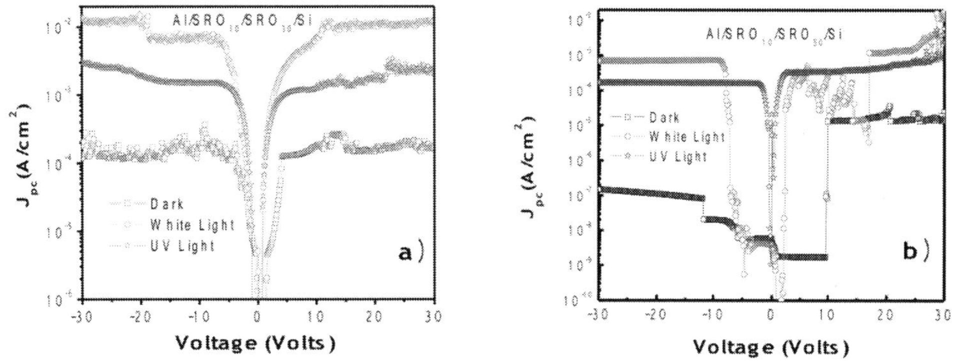

Figure 3. I-V characteristics under dark conditions, white and UV-light on Al/SRO/SRO/Si MOS like structures. a) SRO_{10}/SRO_{30}, b) SRO_{10}/SRO_{50}.

4. Analysis and discussion

We now discuss some possible mechanisms related to the electronic and physical properties of the Al/SRO/Si structures. The dark current is small respect to the photocurrent obtained in each one of the MOS-like structures with SRO films. Furthermore, the dark current is even very small in the two SRO layers structure; it is a direct result of the fact that $SRO_{30\text{ and }50}$ films have nanoparticles as silicon nanocluster smaller than 1 nm embedded in a SiO_2 matrix, they have smaller conduction that $SRO_{30\text{ and}}$ SRO_{50} layers. In principle there is a very low conduction between the SRO_{50} layers with the silicon bulk. Therefore, the film acts as a good insulator in the dark, but the SRO_{10} film begins to be a excellent conductor of electric current when it is exposed to the light.

This photoconduction is due to the following phenomena. In the case of SRO_{10}, with white light exposure, the photons reach the SRO film due to their penetration wavelength on the Al/SRO/Si structure. They produce photo excitation by the interaction between Si-nps (as Si-nc), and e-h pairs are generated and becomes to conductive paths towards the Al contacts. On the other hand, SRO_{20} or SRO_{30} layer devices have the possibility of photoconduction based upon the interaction of defects with Si-nps (nanocluster) with sizes smaller than 2 nm. If e-h pairs are generated in each nanoparticles or defect of the SRO film, this would correspond to a very high electron/hole pairs' density.

The possible photoconduction mechanisms depend on the silicon excess in the SRO films. Further detailed conduction mechanism study is on course and will be published elsewhere. It's clear that the enhancement in the photoconduction under irradiation is related with silicon excess and thickness in the structure MOS-like. Therefore, we have the following statements: First, the larger size of Si nanoparticles due to the silicon excess provides a conductive path across the film, allowing high photocurrent with white light illumination, such as in the SRO_{10} case. With low silicon excess, we have nanoclusters smaller than 1 nm, they interact with the defects and produce high photocurrent with UV light. Second, the light interacts in better way on the thin films with thickness smaller than 75 nm. The UV light produces an inversion layer, and consecuently produces more efficient charge separation and collection surface, which is the case of SRO_{20} and SRO_{30}. A very simple MOS-like structure that shows high photoconduction in the UV-Vis wavelength range is reported, it is possible the creation of e-h pairs in the SRO layer and in the inversion region generated when voltage is applied. On the others hand, SRO films have a different absorption edge that depends on silicon excess. This absorption edge allows the photons' absorption at different wavelengths [7]. This result is very important to understand why the SRO films can be excited at different wavelengths and have an excellent photoconduction.

5. Conclusions

We demonstrated that SRO thin films have photoconduction based on Si nanoparticles (silicon excess). They provide high photocurrent between path conductive towards the two contacts under forward and reverse bias when the samples are exposed to UV and White light. It's clear that there are several key factors as silicon excess, Si nanoparticles size, thickness and MOS-like structure, which allow that SRO films to act as a photoconductor. First, the larger size of Si nanoparticles due at silicon excess provide a conductive path across the film, permitting to obtain high photocurrent with white light illumination, such as in the case of SRO_{10}. With low silicon excess, we have nanocluster smaller than 1 nm that interact with the defects permitting to obtain high photocurrent with UV light. Second, the thin films with thickness smaller than 75 nm have better interaction with the light and provoked a inversion layer, this condition allows more efficient charge separation and collection with UV light, which is the case of SRO_{20} and SRO_{30}. A very simple structure that shows high photoconduction in the UV-Vis wavelength range is reported. The same architecture may be used as a photodetector and light emitter.

Acknowledgement

J. Alberto Luna thanks CONACYT for support received through a postdoctoral scholarship. This work has been partially supported by project CONACyT. The authors also thank Pablo Alarcón, Mauro Landa, Ignacio Juárez and Netzhualcoyotl Carlos for helping in the preparation of the samples.

References

[1] J. A. Luna-Lopez, A. Morales-Sanchez, M. Aceves-Mijares, Z. Yu, and C. Dominguez, J. Vac. Sci. Technol. A 27 (2009), 57-62.
[2] D. J. DiMaria, J. R. Kirtley, E. J. Pakulis, D. W. Dong, T. S. Kuan, F. L. Pesavento, T. N. Theis, J. A. Cutro, and S. D. Brorson, J. Appl. Phys. **56** (1984), 401.
[3] D. J. DiMaria, D. W. Dong, and F. L. Pesavento, J. Appl. Phys. **55** (1984), 3000.
[4] F. Ay and A. Aydinly, Optical Materials **26** (2004), 33.
[5] A. Luna-Lopez, M. Aceves-Mijares and O. Malik, Sensors and actuators A, V 132 (2006), 278-282.
[6] T. A. Burr, A. A. Seraphin, E. Werwa, and K. D. Kolenbrander, Phys. Rev. B, Condens. Matter, V 56 (1997), 4818–4824.
[7] M. Aceves, A. Malik, and R. Murphy, *Sensors and Chemometrics* edited by M. T. Ramirez-Silva *et al.* (Research Signpost, India, 2001).

Electrical and optical properties of thin films with a SnS_2 – Bi_2S_3 alloy grown by sulphurization

A Dussan[1], F Mesa[1], M Botero[2] and G Gordillo [1]
[1]Departamento de Física, Universidad Nacional de Colombia, Bogotá Cr.30 N° 45-03, Colombia
[2]Departamento de Física, Universidad Central, Bogotá Cr.5 N° 21A-03, Colombia

E-mail: ggordillog@unal.edu.co

Abstract. In this work, thin films of SnS_2 with increased Bi content were grown by sulphurization of a thin film of Sn:Bi alloy, at temperatures around 300°C. The effect of the Bi concentration on the optical, electrical and structural properties was determined through measurements of spectral transmittance, conductivity and x-ray diffraction XRD respectively. It was found that the optical constants (refractive index n, absorption coefficient α and energy gap Eg) and the electrical conductivity are significantly affected by the Bi concentration. In particular, a variation of the energy gap between 1.44 and 1.63 eV and a change of the conductivity greater than three orders of magnitude were observed when the content of Bi in the Sn:Bi alloy varied between 0 and 100 %. The analysis of the XRD measurements allowed us to find that the SnS: Bi films grow with a mixture of the SnS_2 and Bi_2S_3 phases, independently of the Bi content.

1.Introduction

Compounds based on the Sn - S system have been widely studied due to their potential application in optoelectronic and photovoltaic devices [1, 2]. Among them, SnS and SnS_2 are considered promising materials for these types of applications, especially due to the fact that the elements Sn and S are abundant in nature and do not contribute to pollution during the deposition process. Since the SnS is p-type and the SnS_2 is n-type, these compounds have demonstrated to be suitable to fabricate n-SnS_2/p-SnS hetero junctions [3]

Sn –S based compounds have been deposited by different techniques such as: vacuum evaporation of the compound [4], two-stage process [5] and electrochemical deposition [6]. In this work, SnS_2 films with increased Bi content have been grown by sulphurization of the metallic precursor species. Results will be reported concerning the effect of the deposition conditions on the phase in which the samples grow and on the optical, structural and electrical properties.

It is well known that tin sulphide films can grow in different phases such as SnS, SnS_2, Sn_2S_3, etc. [7]; therefore, it was necessary to perform an exhaustive study of the synthesis parameters to achieve samples containing only the SnS_2 phase. In this study, it was found that the parameters that affect critically the phase in which the samples grows and their properties are the sulphurization temperature and the flux of S evaporated.

[1] Tel./fax:571-3165000 Ext. 13080 E-mail address: adussanc@unal.edu.co

2. Experimental

The Bi doped tin sulphide thin films SnS_2 with increased Bi content were grown by sulfurization of the metallic precursors, in a two stage process. These processes are described as follows. In the first stage, a thin film is deposited on a glass substrate by evaporating a mixture of Sn and Bi from the same boat, at ambient temperature. In the second stage, the film of the Sn:Bi alloy is annealed for 40 minutes at temperatures around 300ºC, in a sulphur environment produced by evaporation of elemental sulphur from a tantalum effusion cell. SnS_2 films with increased Bi content were deposited using the same parameters that allowing to get tin sulphide films in the SnS_2 phase; the concentration X of Bi in the film, defined as $X = \frac{mols\ of\ Bi}{mols\ [Bi + Sn]} x100$ was obtained varying the Bi mass to Sn mass ratio. During the deposition process, pressure remained constant at $2x10^{-5}$ mbar and the total flux of the metallic species (Sn and Bi) was 2.5 Å/s.

A thickness monitor (Maxtec TM-400) with a quartz crystal as sensor, was used to measure the flux of the metallic species and a PID temperature controller was used to control the temperature of both, substrate and sulphur evaporation. The deposition temperature was varied from 200 to 400 ºC during sulphurization process. It was found that the samples deposited by sulphurization of the metallic precursor at temperatures lower than 300ºC, grows predominantly in the SnS_2 phase. The thickness of the films was ~ 600 nm.

Conductivity as a function of temperature was measured using Van der Pauw technique. Gold contacts were deposited by evaporation and current was measured with a Keithley 617 electrometer. The conductivity type of the SnS_2 with increased Bi content films was determined through thermoelectric power measurements and the phase in which the samples grow was determined by XRD measurements carried out using the Cu Kα radiation of a Shimadzu-6000 diffractometer.

3. Results and discussion

3.1 Structural characterization

The SnS_2 with increased Bi content films were characterized through XRD measurements in order to get information regarding the phase, crystalline structure and lattice parameters. Figure 1 shows typical XRD spectra of SnS_2 films deposited varying the atomic content of Bi in the Sn:Bi alloy, according to the relation: x= [mols of Bi/mols of (Bi+Sn)].

The experimental XRD spectra were analysed by comparison with those reported in the JCPDS data base. Theoretical simulations of the XRD spectra of figure 1 were also performed, in order to improve the confidence degree of the analysis on these spectra. It was found that the SnS_2 films grow with tetragonal structure; however, the phase in which the SnS_2 films grow is affected by the content of Bi in the Sn:Bi alloy (see table 1).

It was found that the films obtained by sulphurization of Sn grow in the SnS_2 phase, whereas those obtained by sulphurization of Bi grow in the Bi_2S_3 phase. On the other hand, the results revealed that the samples deposited by sulphurization of a Sn:Bi alloy grow with a mixture of the Bi_2S_3 and SnS_2 phases. No reflections corresponding to ternary compounds were identified.

In table 1 are listed the structure and phases identified in the studied SnS_2 samples varying Bi concentration between 0 and 100% as well as values of the corresponding lattice constants.

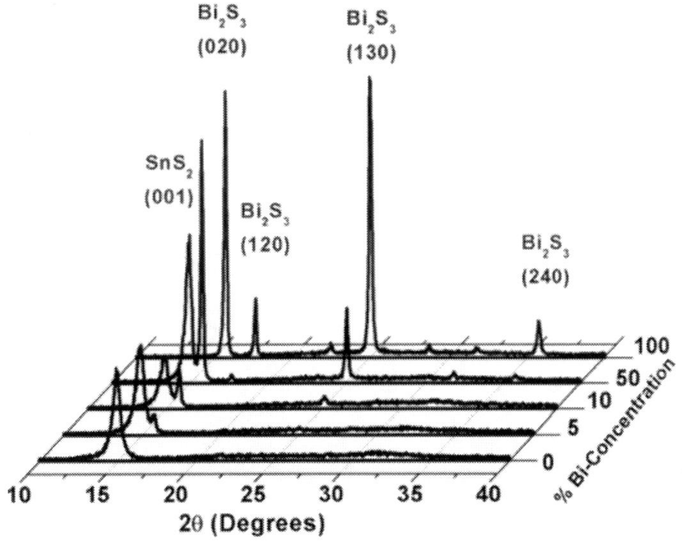

Figure 1. Typical XRD spectra of SnS_2 thin films, with Bi concentrations varying between x=0 (SnS_2) and x=1 (Bi_2S_3).

Table 1. Structure and phases identified in the studied SnS_2 with increased Bi content samples and values of the corresponding lattice constants.

Bi-Concentration (%)	**Phase**	**Structure**	**Lattice constants (Å)**		
			a	b	c
0	SnS_2	Tetragonal	3.625		5.877
0 <x ≤100%	SnS_2	Tetragonal	3.625		5.877
	Bi_2S_3	Orthorhombic	11.215	11.350	4.070
100 %	Bi_2S_3	Orthorhombic	11.215	11.35	4.070

3.2 Optical characterization

Spectral transmittance measurements were done to a representative number of samples of SnS_2 with increased Bi content. Figure 2 shows typical transmittance spectra of SnS_2 thin films synthesized with Bi concentrations varying between 0 and 100%; in Figure 3 are shown curves of n vs. λ and $(\alpha h\nu)^2$ vs. hν. The refractive index (n), absorption coefficient (α) and optical gap (Eg) were calculated using experimental data of the spectral transmittance measurements and theoretical calculations done with the help of a program developed using the procedure based on Swanepoel method [8].

These results show that the increase of Bi concentration leads to a shift of the cut-off wavelength λ_c toward greater values of wavelength, indicating that the incorporation of Bi affects the band structure of the SnS_2 films and therefore their optical gap. This behaviour is seemingly caused by the formation of the secondary Bi_2S_3 phase when Bi is incorporated to the SnS_2 lattice, because this compound has an Eg value significantly lower than that of the SnS_2. It is observed that the optical gap of the SnS_2 with increased Bi content films varies between 1.45 and 1.63 eV which are considered adequate for the fabrication of p-SnS/n-SnS_2: Bi heterojuntion solar cells.

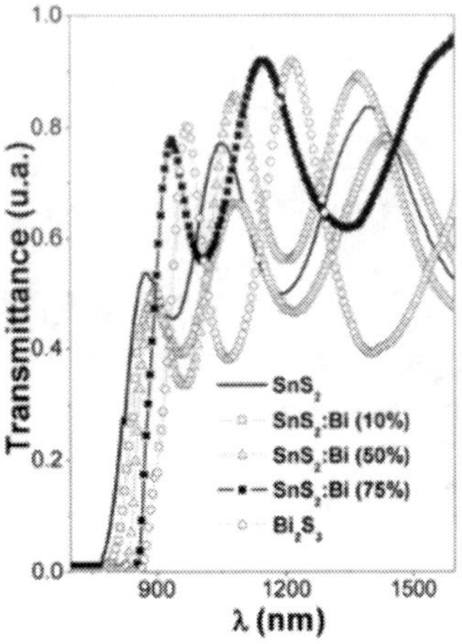

Figure 2. Influence of the Bi concentration on the transmittance of SnS_2 thin films with increased Bi content.

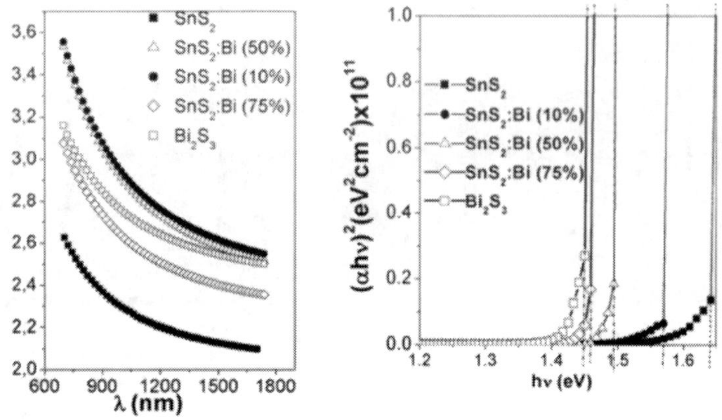

Figure 3. Curves of n vs λ y $(\alpha h\nu)^2$ vs. hν corresponding to SnS_2 thin films prepared with Bi concentrations varying between 0 and 100%.

3.3 Electrical characterization

In figure 4 are shown curves of ln σ vs. 1000/T corresponding to SnS_2 thin films deposited varying the concentration of Bi in the evaporated material between 0 and 100%.

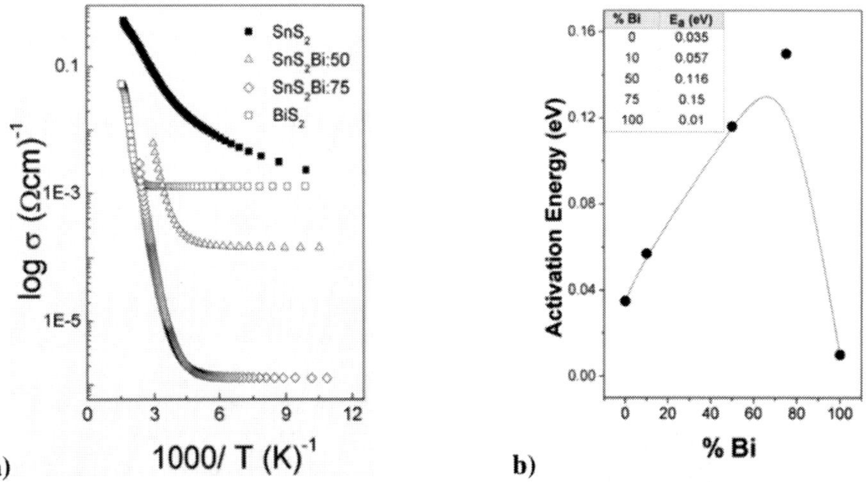

a) b)

Figure 4. a) Curves of ln σ vs. 1000/T corresponding to SnS_2: Bi films deposited varying the Bi concentration between 0 (SnS) and 100% (Bi_2S_3) and b) Variation of the activation energy as a function of de Bi concentration.

It was observed, that the increase of Bi concentration in SnS_2 samples, induces a mixture of phases (see figure 1) affecting the electrical conductivity of the compound. The results of figure 4 shows that the ln σ vs T curves present two different slopes, indicating that the conductivity is affected by two different transport mechanisms, each one predominating in a different temperature range; on the other hand, it is observed that the electric conductivity of the SnS_2 films (at room temperature) is significantly reduced when the Bi concentration is increased. This behavior is caused by the increase of the amount of Bi_2S_3 in the SnS_2 samples when the Bi concentration increases, because, in general, the Bi_2S_3 phase has lower conductivity than the SnS_2 phase.

The linear behavior of the ln σ vs 1000/T curves, demonstrates that the conductivity of the SnS_2 with increased Bi content films in the high temperature range is dominated by transport of free carriers in extended states of the conduction band. The results of figure 4b) show, additionally, that the Bi concentrations significantly affect the activation energy (E_a) of the SnS_2 with increased Bi content samples. The values of E_a obtained in the range 300–423 K (in which Arrhenius behavior is observed) are in the inset of figure 4b).

It was also found from thermoelectric power measurements that the SnS_2 with increased Bi content films present n type conductivity, independently of the Bi content.

Conclusions

SnS_2 with increased Bi content thin films were grown through a novel procedure based on the sulphurization of the metallic precursors with elemental sulphur evaporated from an effusion cell in a two stage process. All the samples were found to have n-type conductivity regardless of the Bi atomic content. Measurements of conductivity as a function of temperature revealed that in the range of high temperatures the electrical transport in SnS_2 with increased Bi content films is predominantly affected by transport of free carriers in extended states of the conduction band. It was found that the energy band gap of the SnS_2 varies between 1.44 and 1.63 eV when the Bi concentration is varied between 0 and 100 %. XRD measurements allowed us to find that the SnS_2 films grow with a mixture of the SnS_2 and Bi_2S_3 phases independently of the Bi content.

Acknowledgements: This work was supported by the Universidad Nacional de Colombia and Colciencias.

References

[1] Koteeswara Reddy N, Ramakrishna Reddy K T, Physica B **368** (2005) 25–31
[2] Ramakrishna Reddya K T, Koteswara Reddya N, Miles R W, Solar Energy Materials & Solar Cells **90** (2006) 3041–3046
[3] Sanchez-Juarez A, Tiburcio-Silver A, Ortiz A, Thin Solid Films 480–481 (2005) 452–456
[4] Price L, Parkin I P, Hardy A M E and Clark R J H, Chem Mater. **11** (1999) 1792
[5] Ramakrishna K T, Purandar P, Materials Letters **56** (2002) 108
[6] Takeuchi K, Ichimura M, Arai E and Yamazaki Y, Solar Energy Mater. Solar Cells, **75** (2003) 427
[7] Bletskan D I, Journal of Ovonic Research Vol. **1**, No. 5, October 2005, 59 – 67.
[8] Swanepoel R, J. Phys. E: Sci. Instrum., Vol. **16** (1983) 1214.

Effect of preparation conditions on the properties of Cu_3BiS_3 thin films grown by a two - step process

F Mesa and G Gordillo [1]
Departamento de Física, Universidad Nacional de Colombia, Bogotá Cr.30 Nº 45-03, Colombia

E-mail: fgmesar@unal.edu.co

Abstract. Cu_3BiS_3 thin films were prepared on soda-lime glass substrates by co-evaporation of the precursors in a two-step process; for that, the metallic precursors were evaporated from a tungsten boat in presence of elemental sulfur evaporated from a tantalum effusion cell. The films were characterized by spectral transmittance, atomic force microscopy AFM and x-ray diffraction (XRD) measurements to investigate the effect of the growth conditions on the optical, morphological and structural properties. The results revealed that, independently of the deposition conditions, the films grow only in the orthorhombic Cu_3BiS_3 phase. It was also found that the Cu_3BiS_3 films present p-type conductivity, a high absorption coefficient (greater than 10^4 cm^{-1}) and an energy band gap Eg of about 1.41 eV, indicating that this compound has good properties to perform as absorbent layer in thin film solar cells.

1. Introduction

The most extensively studied polycrystalline thin film solar cells are based on $Cu(In,Ga)Se_2$ (CIGS) and CdTe as absorbent material [1]. Although high efficiency solar cells based on CIGS and CdTe have been produced, it is generally recognized that other materials may in the long term have advantages over the former. There is concern respect to the use of cadmium in both types of solar cells but the greatest concern is the supply of indium when the technology of CIGS moves to large-scale production. Current efforts are being made in various laboratories to develop new photovoltaic materials with suitable properties to be used in solar cell fabrication. The factors that should be considered in developing these new materials include: abundance of the precursor elements and low environmental impacts. Cu_3BiS_3 is one of the promising materials for low cost thin film solar cells technology thanks to its optimum energy band gap Eg and a high fundamental absorption coefficient [2, 3]. Cu, Bi and S are abundant in nature and they do not contribute to pollution during the Cu_3BiS_3 growing process.

Cu_3BiS_3 thin films have been deposited by different techniques such as: Sequential deposition of Bi_2S_3 and Cu_2S by CBD method, followed by thermal annealing in vacuum [4], reactive sputtering in one and two step processes [5], CBD (chemical bath deposition) [6], rapid polyol process from single source precursors [7] and simple ethanol-thermal route [8].

In this work, Cu_3BiS_3 thin films were grown by co-evaporation of the precursor in a two step process. The influence of the deposition conditions on the optical, morphological and structural properties was

[1] Tel./fax:571-3165000 Ext. 13017 E-mail adress: *ggordillog@unal.edu.co*

investigated through spectral transmittance, atomic force microscopy AFM and x-ray diffraction (XRD) measurements.

2. Experimental

The Cu_3BiS_3 thin films were prepared by co-evaporation of the precursor species on a glass substrate, using a system constituted by an evaporation chamber connected to a vacuum system which allows working at pressures of about 10^{-5} Torr, two tungsten boats (used to evaporate Bi and Cu respectively), a tantalum effusion cell to evaporate sulfur and a thickness monitor (Maxtec TM-400) with a quartz crystal as sensor, used to measure the evaporated elements flux. The substrate temperature was controlled with a programmable PID controller (Eurotherm 900C). The chemical composition of the Cu_3BiS_3 films is controlled by controlling the substrate temperature and the evaporated mass of the precursor species.

The deposition of the Cu_3BiS_3 films was accomplished in two stages. In the first stage a Bi_xS_y layer is grown by simultaneous evaporation of Bi and S, keeping the substrate temperature at 300° C. During this stage the flux of Bi is kept around 1 Å/s and the flux of S is kept around 4 Å/s. In the second stage the Cu_3BiS_3 compound is formed by evaporating Cu, in a sulfur environment, on the Bi_xS_y layer at a temperature around 300°C. During this stage the flux of Cu was kept at a value of about 0.8 Å/s.

The transmittance measurements were carried out with a VIS – IR Oriel spectrophotometer and the conductivity type was performed through thermoelectric power measurements. The XRD measurements were performed using the Cu Kα radiation of a Shimadzu-6000 diffraction-meter and the AFM images were obtained with a PSI AP-0100 Scanning Probe Microscope. The film thickness was measured with a Veeco Dektak 150 surface profiler.

Thermoelectric power measurements revealed that all the studied Cu_3BiS_3 films, presented p type conductivity.

3. Results and discusión

3.1 Optical characterization

Cu_3BiS_3 thin films deposited on soda-lime glass substrates were characterized through spectral transmittance measurements, in order to study the effect of the synthesis parameters on the optical constants (refractive index n, absorption coefficient α and optical gap E_g). In figure 1 are depicted typical transmission spectra of Cu_3BiS_3 films, deposited varying the Cu mass to Bi mass ratio as well as the substrate temperature. The studied Cu_3BiS_3 films were grown varying the deposition parameters as follows:

i) The mass of Cu was varied between 0.15 and 0.19 grams, keeping the substrate temperature at 300°C and the masses of Bi and S in 0.20 and 1.5 grams, respectively.
ii) The substrate temperature was varied between 280 and 320°C, keeping the amounts of Cu, Bi and S at 0.15, 0.19 and 1.5 grams, respectively.

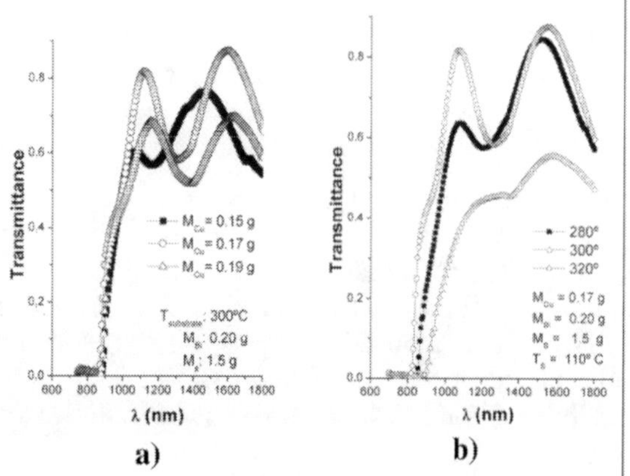

Figure 1. Spectral transmittance measurements of Cu_3BiS_3 thin films showing the effect of a) mass of Cu to mass of Bi ratio and b) substrate temperature.

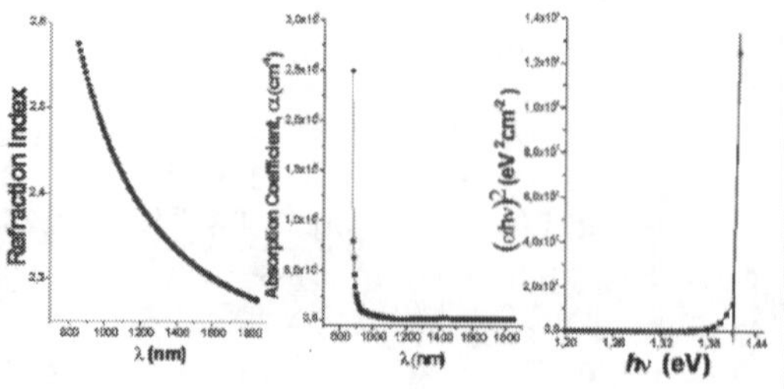

Figure 2. Curves of n vs. λ, α vs. λ and $(\alpha h\nu)^2$ vs. hν, corresponding to typical Cu_3BiS_3 thin films.

It is observed in figure 1, that the samples containing excess or deficiency in Cu or S, present low values of transmittance. This behaviour is seemingly caused by absorption of photons through states within the energy band gap associated to donor states generated by S-vacancies and by anti-site defects. Transmittances around 80% were obtained with Cu_3BiS_3 films deposited by co-evaporation of 0.19 g of Bi, 0.15 grams of Cu and 1.5 grams of S, on a substrate heated at a temperature around 300°C, in a two stage process. On the other hand, the results of figure 1 reveal that the substrate temperature also affects the transmittance of the Cu_3BiS_3 films. The transmittance decreases significantly at temperatures lower than 300°C and higher than 300 °C, probably because, under these conditions, a high density of native defects is formed in the Cu_3BiS_3 films.

In figure 2 are depicted curves of n vs. λ, α vs λ and $(\alpha h\nu)^2$ vs. hν, which were determined using the transmission spectrum and calculations based on a procedure described in detail in reference [9]. The linear behaviour of the $(\alpha h\nu)^2$ vs. hν curves indicates that the Cu_3BiS_3 thin films have a direct energy band gap Eg. The results of figure 2 also show that the Cu_3BiS_3 films have an absorption coefficient α greater than 10^4 cm^{-1} and an optical gap Eg of 1.41 eV , indicating that the Cu_3BiS_3 thin films, we have deposited in this work, have suitable properties for using them as absorber layers in solar cells.

3.2 Structural characterization

The effect of the substrate temperature and of the Cu mass to Bi mass ratio on the structural properties of the Cu_3BiS_3 films was also studied through XRD measurements. Figure 3 shows typical XRD spectra of Cu_3BiS_3 films deposited varying the substrate temperature between 280 and 320° C and the mass of Cu between 0.15 and 0.19 g (keeping the mass of Bi in 0.20 grams). The results were analyzed using the data reported in the JCPDS data base for these types of compounds and performing a theoretical simulation of the XRD pattern, with the help of the PowderCell package. This procedure allowed us to identify the phases present in the samples with a good degree of confidence. It was found that, independently of the synthesis temperature and the Cu mass to Bi mass ratio, the Cu_3BiS_3 samples grow with orthorhombic structure; the corresponding lattice constants are: a = 7.7230 Å, b = 10.395 Å and c = 6.716 Å, which are similar to those reported by Kocman [10]. No secondary, binary or ternary phases different from the Cu_3BiS_3 phase were identified.

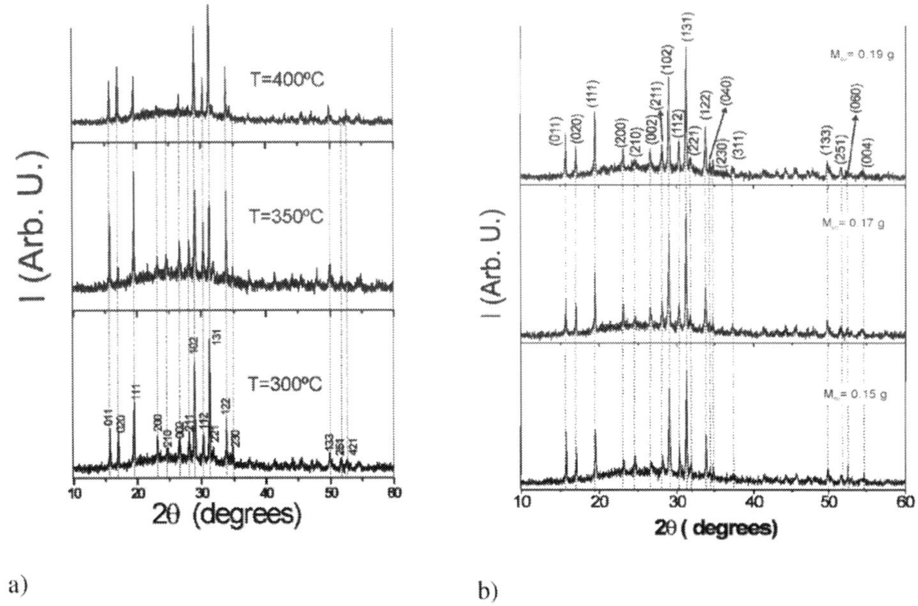

Figure 3. Typical XRD spectra of Cu_3BiS_3 films deposited: a) varying the synthesis temperature between 300 and 400 °C and b) the mass of Cu between 0.15 and 0.19 grams.

3.3 Morphological characterization

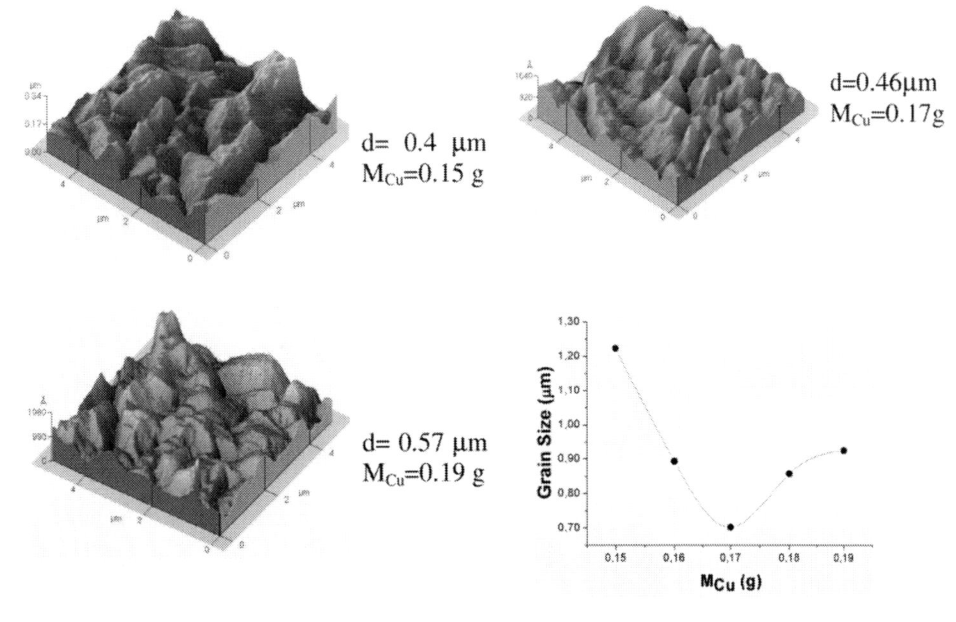

Figure 4. AFM images corresponding to Cu_3BiS_3 thin films deposited varying the mass of evaporated Cu, keeping constant the remaining deposition parameters (T_S=300°C, M_{Bi}=0.19g, M_S=2 g). The effect of the evaporated mass of Cu on the grain size is also shown.

In figure 4 are compared the AFM images taken to typical Cu_3BiS_3 thin films, deposited varying the evaporated mass of Cu between 0.15 and 0.19 grams, keeping constant the remaining deposition parameters (T_S=300°C, M_{Bi}=0.19 g, M_S=2g). Figure 4 also shows a curve of grain size vs. mass of Cu, determined from a statistical analysis carried out with the ProScan Data Acquisition software.

It is clear from the results of figure 4 that the Cu mass to Bi mass ratio significantly affects the films' morphology. In general, the grain size increases when the Cu concentration in the Cu_3BiS_3 sample decreases; the increase of the grain size observed in samples deposited using an amount of Cu greater than 0.17 grams is mainly attributed to thickness effect.

4. Conclusions

P type Cu_3BiS_3 thin films with adequate properties to be used as absorber layers in solar cells were grown using a novel procedure based on the co-evaporation of the precursor in a two step process.

It was found that, independently of the deposition conditions, the films grow only in the orthorhombic Cu_3BiS_3 phase; however the main deposition parameters significantly affect the transmittance and the morphology. It was also found that the Cu_3BiS_3 films have high absorption coefficients (greater than 10^4 cm^{-1}) and an energy band gap Eg of about 1.41 eV, indicating that this compound has good properties to perform as absorbent layer in thin film solar cells

Acknowledgements: This work was supported by the Universidad Nacional de Colombia and Colciencias

5. References

[1] Von Roedem B, Zweibel K and Ulall H S, *31st IEEE Photovoltaic Specialist Conference and Exhibition*, Lake Buena Vista, Florida , 2005, 183
[2] Schmidt J A, Suagua A E and Lescano G, *J. Chem. Termodynamics* 1998, **30**, 283-290
[3] Sonawane P S , Wani P A, Patil L A, Seth T, *Mater. Chem. Phys.* 2004, **84**, 221-227
[4] Nair P K, Huang L, Nair M T S, Hu Hailin, Meyers E A and Zingaro R A, *J. Mater. Res.*, March 1997, **12**, N° 3, 651-656
[5] Gerein N J and Haber J A, *Chem. Mater.*, 2006, **18**, 6289-6296
[6] Estrella V, Nair M.T.S and Nair P.K., Semicond. Sci. Technol., 2003, **18**, 190-194
[7] Shen G., Chen Di, Tang K. and Qian Y., *J. Cryst. Growth* 2003, **257**, 293-296
[8] Chen Di, Shen G., Tang K., Liu X., Qian Y. and Zhou G., *J. Cryst. Growth* 2003, **253**, 512-516
[9] Gracia M, Rojas F and Gordillo G, *20th European photovoltaic Solar Energy Conference*, Barcelona-España, 2005
[10] Kocman V, Nuffield E W, The crystal structure of wittichenite, Cu_3BiS_3, *Acta Crystallogr.* 1973, **B29**, 2588

Low temperature deposition: properties of SiO_2 films from TEOS and ozone by APCVD system

H Juárez[1,3], M Pacio[2], T Díaz[1], E Rosendo[1], G Garcia[1], A García[2], F Mora[1] and G Escalante[1]

[1]Centro de Investigación en Dispositivos Semiconductores, Universidad Autónoma de Puebla, 14 Sur and Av. San Claudio, San Manuel 72000, Puebla, MEXICO
[2]Ingeniería Eléctrica, Sección Electrónica del Estado Sólido, Centro de Investigación y de Estudios Avanzados del I. P. N., Av. Instituto Politécnico Nacional 2508, San Pedro Zacatenco, 07360 México, D. F. MEXICO

E-mail: hjuarez@cs.buap.mx

Abstract. An Atmospheric Pressure Chemical Vapor Deposition (APCVD) system was implemented for SiO_2 nanometric films deposition on silicon substrates. Tetraethoxysilane (TEOS) and ozone (O_3) were used and they were mixed into the APCVD system. The deposition temperatures were very low, from 125 to 250 °C and the deposition time ranged from 1 to 15 minutes. The measured thicknesses from the deposited SiO_2 films were between 5 and 300 nm. From the by Fourier-Transform Infrared (FTIR) spectra the typical absorption bands of the Si-O bond were observed and it was also observed a dependence on the vibrational modes corresponding to hydroxyl groups with the deposition temperature where the intensity of these vibrations can be related with the grade porosity grade of the films. Furthermore an analytical model has been evoked to determine the activation energy of the reactions in the surface and the gas phase in the deposit films process.

1. Introduction

Interlayer dielectric film formation technology is essential for multilevel interconnection in ULSI devices. Low temperature deposition has been required. Organosilicon materials such as tetraethoxysilane (TEOS), $Si[OCH_2CH_3]_4$ have increasingly been used as a silicon source because it provides a better film quality and excellent step coverage compared to the conventional silane method [1]. Desirable aspects of TEOS as a silicon source are safety, easy of handling, and chemical stability. However, conventional thermal CVD TEOS processes require temperatures higher than 600°C. The CVD reaction of TEOS with ozone yields silicon dioxide thin films with excellent characteristics at unusually low process temperatures (300°C to 400°C) [2-6]. However, its deposition mechanism is not well understood yet. The chemistry of this process has not been definitively characterized yet and the process therefore presents an excellent opportunity to evaluate our approaches.

In this work, SiO_2 films by APCVD using TEOS and ozone were deposited at very low temperature on silicon substrates and their optical properties were investigated. An analytical model for the deposition rate was used to obtain the activation energy of the reaction at the film.

[3] To whom any correspondence should be addressed.

2. Experimental Procedure

A schematic draw of the APCVD system is shown in figure 1 [7]. The reactor is composed by a graphite heater element into a closed quartz chamber. The deposition temperature is controlled by a k type thermocouple and a temperature controller (Eurotherm model 2416). TEOS is introduced into the reactor by flowing N_2 through a bubbler containing the precursor at 25 °C. Also an O_3/O_2 flow was introduced into the reactor. The partial pressure of the TEOS was 1.1 Torr and the ozone concentration in the oxygen was controlled to be 5%.

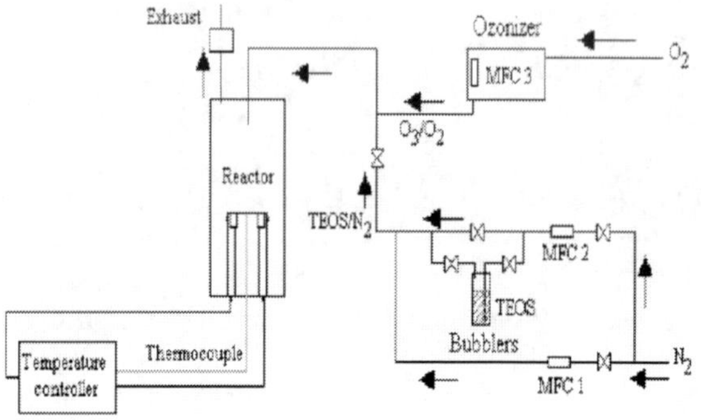

Figure 1. Schematic diagram of the APCVD system.

The SiO_2 films were deposited on n-type (100) silicon substrates of 5 Ω-cm at a pressure of 590 Torr. Prior to the deposit process, the substrates were cleaned by a standard process (RCA) to eliminate organic and inorganic moistures on their surface and the silicon wafers were introduced into the reactor. After the temperature on the substrate was stabilized the TEOS and Ozone was flowed into the chamber. The oxide films were deposited at temperatures from 125 to 250 ºC having deposition time between 1 and 15 minutes. The final film thicknesses were from 5 to 300 nm as measured from a Rudolph 439L633P ellipsometer.The films were etched in a chemical solution, composed by 15:10:300 of HF (48%): HNO_3 (70%): DI water, in volume, to determine the etch rate in the films. Film microstructure was examined by a FTIR Bruker vector 22 system.

3. Results and Discussion

The main absorbance peaks expected from SiO_2 films from an infrared spectroscopy are rocking (450 cm^{-1}), bending (800 cm^{-1}) and asymmetric stretching (1072 cm^{-1}) originated from the vibration modes of Si-O bonds. In the figure 2 a serie of absorbance spectrum are presented for SiO_2 films obtained at different temperatures and from where those main peaks are observable. Also, in some cases, additional bands corresponding to hydroxyl groups (920cm^{-1}) and other residual groups (ethoxy groups at 1395-1837 cm^{-1}) are exhibited.

When the deposition temperature was increased from 125 to 250 °C the intensity of those absorption bands decreased, this means that the quantity of residual groups incorporated in the film decreases [8]. This could signify that the film porosity is decremented principally as a consequence of the reduction in the number of the hydroxyl groups [9]. The higher etch rate values found in the films deposited at lower temperature (table 1) shown a decrease in the film density. It is known a decrease in the Si-OH bonds in the film resulted in a more dense silicon oxide film [9, 10].

The refractive index has been reported to be inversely proportional both to etch rate and the number of hydroxyl groups in the film [9-11]. Hence, the silicon oxide films obtained at low temperatures with high number of Si-OH bonds lead to low values for the refraction index. The refraction index of the as-deposited silicon oxide films increased from 1.400 to 1.461, as the deposition temperature rise from 125 to 250 °C.

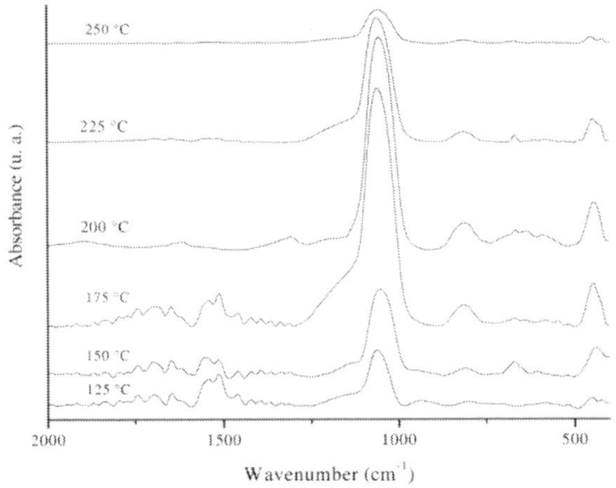

Figure 2. Absorbance spectra for SiO_2 films obtained by APCVD at different temperatures and 10 min of deposition. The flow of N_2 and O_2 were 100 and 500 sccm respectively.

Table 1. SiO_2 films TEOS-Ozone parameters.

Deposition temperature [°C]	Deposition rate [nm/min.]	Refraction index	Etch rate [nm/sec]
125	5	1.400	1.73
150	12	1.420	1.46
175	20	1.445	1.17
200	19	1.457	0.84
225	16	1.460	0.71
250	9	1.461	0.66
Thermal SiO_2	8	1.46	0.59

On the other hand, was found that the film deposition rate was dependent on the deposition temperature. From Table 1 can be observed that the minimum value for this parameter is found at 125°C. For higher temperatures it reaches a maximum and after it decreases. To explain this behavior the model of Kim and Gill [12] was taken.

The kinetic mechanism includes parasitic gas phase reactions. Thus, the deposition rate is given by [12]

$$R_{SiO_2} = k_S \frac{Q_{in}[O_3]_{in}}{V} \frac{[TEOS]_{in}}{1 + k_G [O_3]_{in}}, \qquad (1)$$

where Q_{in} is the volumetric flow rate, $[O_3]_{in}$ and $[TEOS]_{in}$ are the concentration of ozone and TEOS at the reactor input respectively, V is the reactor volume and k_S, k_G are the reaction rate constant in the surface and in gas phase respectively. The equation (1) contains the surface reaction (numerator) and the gas phase reactions (denominator). For our APCVD system the values for Q_{in}, $[O_3]_{in}$ and $[TEOS]_{in}$ were

$[TEOS]_{in} = 1.2\text{x}10^{-6}$ mol / cm^3
$[O_3]_{in} = 1.5\text{x}10^{-5}$ mol / cm^3
$Q_{in} = 26.6$ cm^3 / s
$V = 1021.4$ cm^3

Substituting those values en equation (1), it is found;

$$R_{SiO_2} = \frac{4.69x10^{-13}\, k_S}{1 + k_G \;\; 1.5x10^{-5}}, \qquad (2)$$

where k_S y k_G have the form [7]

$$k \;\alpha\; e^{-\frac{B}{RT}}, \qquad (3)$$

here, T is the temperature, R is the Boltzmann constant and B is the activation energy of the reaction.

Taking into account just the reactions in the surface, the second term in the denominator of equation (2) can be neglected and it becomes,

$$R_{SiO_2} = 4.69x10^{-13} k_S \qquad (4)$$

If the experimental values of the deposition rate from table 1 vs. the inverse of the deposition temperature are plotted; it is possible to obtain the value of the activation energy for the reactions in gas phase and on the surface. In figure 3 such characteristic is presented.

In that figure, it is observed that in the temperature range 125 to 175 °C, a straight line is obtained according to the equation (4), signifying that the deposition rate is increased with the increase in the deposition temperature because the value of the k_S in equation (3) increases with temperature.

After reach a maximum, a decrease in the deposition rate can be observed around 200 °C. This fact can be explained by the effects in gas phase reactions. At higher temperatures (>175°C), where the gas phase reactions are significant the second term in the denominator of equation (2) becomes important and therefore the theoretical deposition rate must be decrease. The deposition rate of the TEOS–Ozone process is limited by reactions in the gas phase and this cause the decrease in the deposition rate. This behavior has been also found in systems LPCVD to 30 Torr, where the maximum is located around of 400°C [7, 12, 13].

Using the experimental values of the deposition rate and numeric analysis with regard to equation (2), the values of k_S and k_G and therefore the activation energy in the surface and in gas phase were obtained. These values were 8744.0 and 27925.9 cal / mol respectively. Substituting the values in equation (1) it becomes;

$$R_{SiO_2} = \frac{4.75x10^{-4} \; e^{-\frac{8744}{RT}}}{1 + 4.5x10^{12} \; e^{-\frac{27925.9}{RT}}}. \qquad (5)$$

A plot of equation (5) is shown in figure 3. It can be seen that the experimental data fits well with the theoretical ones calculated from equation (5). A little difference observed around the maximum of the deposition rate in figure 3 can be due to a change in the predominant reactions, that is, the dominant processes are initially the reactions in the surface and later the dominant processes changes and it leads to reactions in gas phase to be dominant [7]

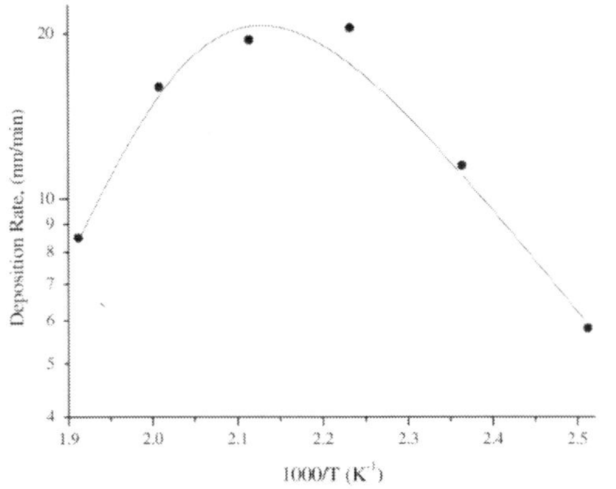

Figure 3. Deposition rate vs. inverse of the temperature for experimental values (points) and theoretical values (continuous line) from equation (5).

4. Conclusions

An APCVD system was implemented to deposit SiO_2 films on silicon substrates. The deposition temperatures ranged from 125 to 250 °C, temperatures range not reported in APCVD systems. The film thickness ranged from 5 to 300 nm. The absorption bands corresponding to the vibration modes of the SiO_2 were observed. It was also observed additional absorption bands belonging to hydroxyl groups (Si-OH bonds). It could be observed from those spectra that the composition of the silicon dioxide films changes with the film deposition temperature. Increasing the temperature, from 125 to 225 °C the quantity of hydroxyl groups incorporated in the film reduces. These produce a change in the value of the refraction index film and as consequence a decrease in the film porosity. A theoretical model was fed with parameters calculated from our APCVD system to determine the activation energy of the reactions in gas phase and on the surface. The experimental data for the film deposited rate fitted well with the ones found from that theoretical model.

5. References:

[1] Schlote J, Bugiel E, Arndt J and Wahl G 1999 SiO_2 films from tetraethoxysilane-based LPCVD: An experimental investigation of the by-product-inhibited deposition mechanism *J. Electrochem. Soc.* **146** 3415-19

[2] Fujino K, Nishimoto Y, Tokumasu N and Maeda K 1993 TEOS and ozone atmospheric pressure CVD of borophosphosilicate glass films using triethylborate and trimethylphosphate *J. Electrochem. Soc.* **140** 2922-27

[3] Kubo A, Homma T and Murao Y 1996 An SiO_2 film deposition technology using tetraethylorthosilicate and ozone for interlayer metal dielectrics *J. Electrochem. Soc.* **143** 1769-73

[4] Jung E and Gill W N 1994 A new model for low pressure chemical vapor deposition of SiO_2 films by ozone augmented tetraethoxysilane *J. Cryst. Growth* **140** 308-14

[5] Homma T, Yamaguchi R and Murao Y 1993 A room temperature chemical vapor deposition SiOF film formation technology for the interlayer in submicron multilevel interconnections *J. Electrochem. Soc.* **140** 687-92

[6] Romet S Couturier M F and Whidden T K 2001 Modeling of silicon dioxide chemical vapor deposition from tetraethoxysilane and ozone *J. Electrochem. Soc.* **148** G82-90

[7] Pacio M. 2004 *M. D. Thesis: Modeling, design and developed an APCVD system, to obtained SiO_2 nanometrics films on silicon substrates* (Puebla, México: CIDS-ICUAP, University of Puebla)

[8] Adams A. C 1983 Plasma deposition of inorganic films *Solid State Technol.* **26** 135-9

[9] Fowler B and O´Brien E 1994 Relationships between the material properties of silicon oxide films deposited by electron cyclotron resonance chemical vapor deposition and their use as an indicator of the dielectric constant *J. Vac. Sci. Technol.* B **12** 441-48

[10] Brunet-Bruneau A, Rivory J, Rafin B, Robic J Y and Chaton P 1997 Infrared ellipsometry study of evaporated SiO_2 films: Matrix densification, porosity, water sorption *J. Appl. Phys.* **83** 1330-35

[11] Herak T V, Chau T T, Thomson D J, Mejia S R, Buchanan D A and Kao K C 1989 Low-temperature deposition of silicon dioxide films from electron cyclotron resonant microwave plasmas *J. Appl. Phys.* **65** 2457-63

[12] Jung E and Gill W 1994 Analytical model for chemical vapor deposition of SiO_2 films using tetraethoxysilane and ozone *J. Cryst. Growth* **140** 315-26

[13] Ikeda Y, Numasawa Y and Sakamoto M 1990 Ozone/Organic-Source APCVD for conformal doped oxide films *J. Electron. Mat.* **19** 45-9

Journal of Physics: Conference Series **167** (2009) 012021 doi:10.1088/1742-6596/167/1/012021

Reactivity of a Pt(100) cluster modified by adsorption of a Nickel tetramer

EV Ortiz[1,2], MB Lopez[1], EA Castro[3]
[1] Centro de Investigaciones Fisicoquímicas, Teóricas y Aplicadas (CIFTA), Facultad de Ciencias Exactas y Naturales, Universidad Nacional de Catamarca, Av. Belgrano 300, (4700), Catamarca, Argentina.
[2] Facultad de Tecnología y Ciencias Aplicadas, Universidad Nacional de Catamarca, Maximio Victoria 55, (4700), Catamarca, Argentina.
[3] INIFTA, CONICET, Universidad Nacional de la Plata, Diag. 113 y 64, Suc.4, C.C. 16, (1900), La Plata, Argentina.

E-mail: mblopez@fcasuser.unca.edu.ar

Abstract. The aim of this paper is to report a study of the reactivity of Pt(100) cluster and the same system modified by a nickel tetramer towards the atomic hydrogen adsorption. This study was carried out in the framework of density functional theory which provides global and local indexes that can be used to characterize the reactivity. The analyzed reactivity descriptors were: chemical potential, chemical hardness, electrophilicity index and Fukui function. The results showed that the global reactivity descriptor predicts that the platinum cluster modified by nickel is more reactive than the pure platinum cluster and that the local Fukui function provides information about the most susceptible site to electrophilic attack in platinum cluster.

1. Introduction

There is a strong current interest in the study of supported metal nanoclusters, not only from a fundamental point of view but also because of their technological relevance in, e.g., heterogeneous catalysis or the design of nanoscale magnetic storage devices. Systems with reduced dimensions can exhibit surprising properties [1].

The increase in catalytic properties resulting from the formation of intermetallic compounds is attributed to the alteration of the electronic structure and physicochemical properties, such as work function and density of states at the Fermi level [2].

The system Pt/Ni has been widely studied in its various forms: Alloy surface [3], pseudomorphic overlayer [4] and supported nanoparticles [5]. Recent studies have revealed that the Pt-Ni alloy surface presents a considerable increase in the catalytic activity versus Pt pure compared to the reaction of oxygen reduction [6] and Ni-Pt nanoparticles which are more reactive compared to the oxidation of methanol on the surface of Pt(100) surface [7]. It is therefore of great interest to know the microscopic mechanisms responsible for the modification of the reactivity in this bimetallic system since this knowledge will allow the design of catalysts composition and specific structures.

The Density Functional Theory (DFT) has provided the basis for rigorous mathematical definitions of reactivity descriptors like chemical potential, electronegativity, chemical hardness, softness, etc., [8].

All of these are well established global indicators in chemical reactivity studies and are going to be used in this paper to characterize the reactivity of $Pt(100)_n$ cluster and $Pt(100)_nNi_m$ (n=25, m=4) toward the adsorption hydrogen atom.
Also local indicators have been introduced in the theory to explain the reactivity or selectivity at a particular site of the system. In this work, we adopted the Fukui function to analyze the reactivity of the surfaces against a hydrogen atom.

2. Theoretical background

In DFT, the chemical potential and hardness for the N-electrons system with total energy E and external potential v(r) are defined as the first and second derivatives of the energy with respect to N [8-11]:

$$\mu = \left(\frac{\partial E}{\partial N}\right)_{v(r)} = -\chi \tag{1}$$

and

$$\eta = \frac{1}{2}\left(\frac{\partial^2 E}{\partial N^2}\right)_{v(r)} = \frac{1}{2}\left(\frac{\partial \mu}{\partial N}\right)_{v(r)} \tag{2}$$

In numerical applications, μ and ηare calculated through the following approximate version of Eqs. (1) and (2), based upon the finite difference approximation and the Koopmans theorems [8],

$$\mu \approx -\frac{1}{2}(IP + EA) = \frac{1}{2}(\varepsilon_L + \varepsilon_H) \tag{3}$$

and

$$\eta \approx -\frac{1}{2}(IP - EA) = \frac{1}{2}(\varepsilon_L - \varepsilon_H) \tag{4}$$

IP is the ionization potential, EA is the electron affinity, ε_H and ε_L are the energies of the highest occupied molecular orbital (HOMO) and the lowest unoccupied molecular orbital (LUMO), respectively. The electrofilicity index is defined as [11]:

$$\omega = \frac{\mu^2}{2\eta} \tag{5}$$

This quantity is a measure of the propensity of systems to acquire electronic charge from the surroundings. Considering clusters or metallic surfaces as binding environments, ω can be useful for analyzing the interaction with small molecules.
The condensed Fukui functions are denoted as [10]:

$$f_k^+ = q_k(N+1) - q_k(N)$$
$$f_k^- = q_k(N) - q_k(N-1) \tag{6}$$

where $q_k(M)$ is the atomic electron population at atom k for either the neutral system (M=N), the cation (M=N-1), or the anion (M=N+1).

3. Computational methods

Density functional theory (DFT) as implemented in Gaussian03 [12] has been used for all the calculations. The hybrid B3LYP density functional method was used, which includes Becke´s 3-parameter nonlocal-exchange functional [13] with the correlation functional of Lee et al [14]. The effective-core-potential LANL2DZ [15] is used for the atoms of Ni and Pt and the bases 6-31G** for hydrogen atom.

The Pt surfaces with (100) crystalline plane were modeled using $Pt(100)_{25}$ clusters and Ni nanoparticles was simulated by a nickel tetramer, Ni_4, Figure 1. Since the reactivity of the surface is evaluated in terms of global and local indices based on well localized processes of charge transfer between adsorbate-substrate the use of an atomic bilayer was not considered.

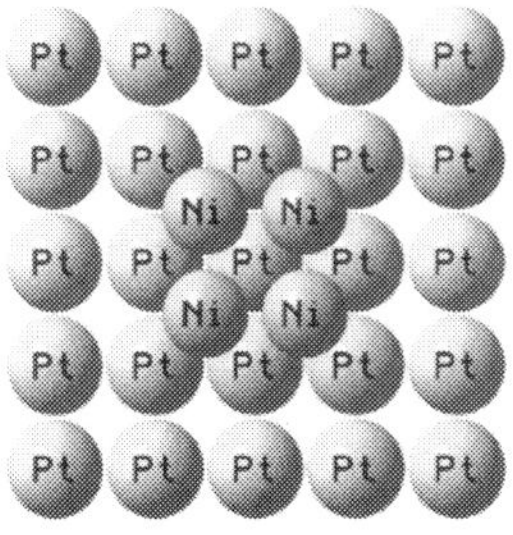

Figure 1. Pt_nNi_m (n=25, m=4) system

In all calculations the nearest neighbor Pt-Pt distance was fixed at the experimental value [16] of 0.277 nm for platinum. The optimization of the geometry of the Ni ad-atoms was carried out maintaining the geometry of the substrate fixed. The effects of relaxation of the substrate atoms were not considered.
The interaction of the hydrogen atom on Pt $(100)_n$ and Pt $(100)_nNi_m$ surfaces was optimized in different adsorption sites: on top, bridge and hollow. At each site the H-surface distance was optimized, while all other parameters of these systems were kept frozen. In all cases, harmonic frequency analysis was performed to verify that the optimized structures are local minima.
The binding energy of hydrogen atom to the M-H systems (M= $P(100)_n$, and $Pt(100)_nNi_m$) was calculated as:

$$E_{ad} = E_{M-H} - E_M - E_H \tag{7}$$

Where, E_{M-H} is the energy of hydrogen adsorbed on M system, E_M energy from the M system and E_H is the energy of hydrogen to an infinite distance from the surface, respectively.

4. Results

Table 1 shows the values of the global reactivity descriptors -chemical potential (μ), chemical hardness (η) and electrofilicity index (ω)- for the pure and modified surfaces, Pt_n and Pt_nN_m, respectively, and the binding energy (E_{ad}) of hydrogen atom on Pt_nH and modified surface, Pt_nNi_mH. The adsorption site of hydrogen atom is indicated as T (on top) and B (bridge), Figure 2. Not found adsorption on the Hollow site.

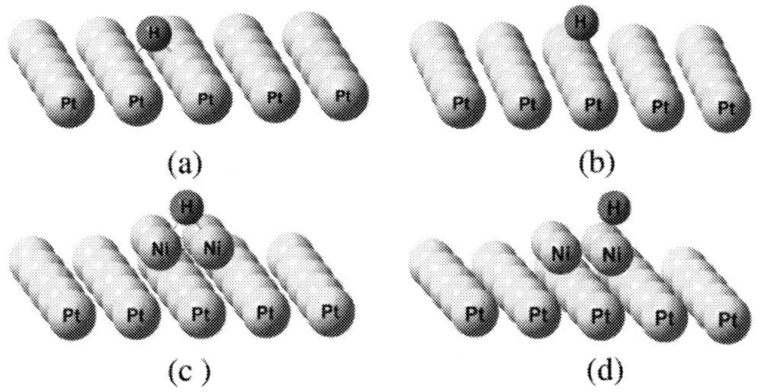

Figure 2. Hydrogen adsorption site (a) $Pt_nH(B)$, (b) $Pt_nH(T)$, (c) Pt_nNi_m H(B), (d) Pt_nNi_m H(T)

The reactivity of global indicators is an intrinsic electronic property of the system and indicates their availability for a chemical change that may occur. The chemical potential measures the escaping

Journal of Physics: Conference Series **167** (2009) 012021 doi:10.1088/1742-6596/167/1/012021

tendency of electrons from the equilibrium systems then as μ becomes more negative, it is more difficult to lose an electron but easier to gain one.

The chemical hardness can be seen as the strength of a system to charge transfer; therefore it characterizes the stability of the systems under study. The electrophilicity index is a measure of stabilization energy of the system when it is saturated with electrons which come from the external environment, i.e., it measures the tendency of a system to capture electrons.

The chemical potential, the negative of the absolute electronegativity, of the hydrogen atom is –7.20 eV and the chemical potential of Pt_n and Pt_nNi_m systems are –6.03 eV and –5.92 eV, respectively (Table 1). Therefore it can be expected that Pt_n and Pt_nNi_m systems act as a donor system and the donor Fukui function f_k^- should indicate the reactivity of different regions of Pt_n cluster.

The Pt_n system had a greater value of chemical hardness, 0.56 eV, that the Pt_nNi_m, 0.46 eV, hence the latter system is more reactive or softer. The hydrogen adsorption stabilizes Pt_n systems, being Pt_nH(T) more stable than Pt_nH(B) with a chemical hardness value equal to 0.61 eV and 0.59 eV, respectively. The same trend is observed for the modified surface but in this case the Pt_nNi_mH(B) is more stable than Pt_nNi_mH(T) with a chemical hardness value equal to 0.55 eV and 0.49 eV, respectively.

The values of the electrophilicity index confirm the tendency of the system, noting that the lower values of ω are for Pt_nH(T) and Pt_nN_mH (B).

An analysis of the binding energy of hydrogen atom in different sites covering the surface under study confirms that site energetically favorable for Pt_n cluster is on top and for Pt_nNi_m is on bridge.

Table 1. Global reactivity descriptors, chemical potential (μ), chemical hardness (η), electrophilicity index (ω) and binding energy (**E_{ad}**) for the systems: Pt_n, Pt_nNi_m, Pt_nH and Pt_nNi_mH (n = 25, m = 4).

System	μ (eV)	η (eV)	ω (eV)	E_{ad} (eV)
Pt_n	-6.03	0.56	32.41	-
Pt_nNi_m	-5.95	0.46	37.93	-
Pt_nH(T)	-6.02	0.61	29.70	-4. 64
Pt_nH (B)	-5.93	0.59	29.80	-4.45
Pt_nNi_mH(T)	-5.92	0.49	37.76	-4.53
Pt_nNi_mH (B)	-5.84	0.55	31.00	-6.56

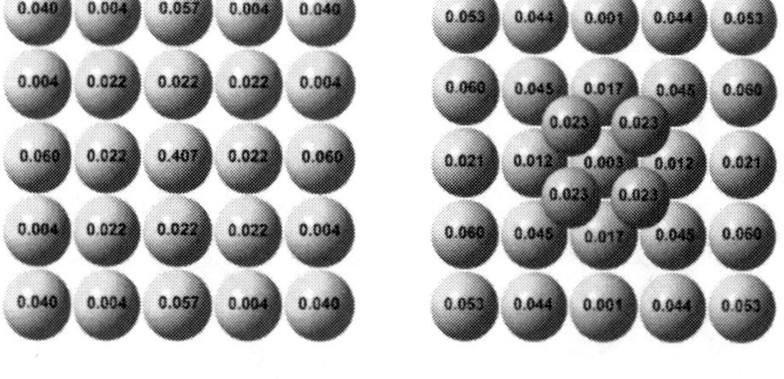

Figure 3. Fukui function (f_k^-) of Pt_n and $Pt_n Ni_m$ systems.

The Fukui function (f_k^-) allows identifying the most active sites in the surface. The greater reactivity of cluster Pt $(100)_{25}$ is located in the center with maximum value of Fukui function equal to 0.407, Figure 3. This site is the most susceptible to electrophilic attack and this result is in agreement with the previous one for hydrogen adsorption on platinum cluster, Pt_nH(T). Reactivity observed in the edge of the platinum cluster atoms can be considered artificial because the border atoms have unsaturated valences which can cause additional reactivity. The modified surface $Pt_n\ Ni_m$ presents the greater reactivity on Ni tetramer with the same value of Fukui function for each Ni atom. The four Ni atom have obviously the same probability to be attacked by electrophylic species, therefore in this case they do not provide the adsorption hydrogen site.

Future studies using larger cluster and effects of relaxation between adsorbate-substrate will allow us to make up a further analysis on the impact of the localized Fukui function to predict energetically favorable adsorption sites.

5. Conclusions

The study of the reactivity of Pt(100) cluster and of the same system modified by a nickel tetramer towards the atomic hydrogen using the global and local reactivity descriptors, allows us to arrive at the following conclusions:

-The global reactivity descriptors, chemical potential, chemical hardness and electrophilicity index confirm that the Pt cluster modified by nickel tetramer is more reactive than the pure Pt cluster.

-The local reactivity descriptor, Fukui function (f_k^-), predicts that the greater reactivity of pure Pt cluster is located in the central atom and for Pt_nNi_m surfaces in the Ni tetramer.

6. Reference

[1] Roudgar A, Groß A 2004 *Surf. Sci.* **559** L180.
[2] Greeley J, Nørskov J K, Mavrikakis M 2002 *Annu. Rev. Phys. Chem.* **53** 319.
[3] Yao K, Cheng, Y F 2007 *J Power Sourc*. **173** 96.
[4] Kitchin J R, Khan N A, Barteau M A, Chen J G, Yakshinskiy B 2003 *Surf. Sci.* **544** 295.
[5] Paulus U A, Wokaun A, Scherer G G 2002 *J. Phys. Chem. B* **106** (16) 4181.
[6] Paulus U A, Wokaun A, Scherer G G, Schmidt T J, Stamenkovic V, Markovic N M, Ross P N 2002 *Electrochimica Acta* **47** 3787.
[7] Lee S, Park K, Choi J, Kwon B, Sung Y 2002 *J. Electrochem. Soc.* **149** A1299.
[8] Parr R G, Yang W 1989 *Density Functional Theory of Atoms and Molecules* (New York, Oxford University Press).
[9] Parr R G, Yang W 1984 *J. Am. Chem. Soc.* **106** 4049.
[10] Parr R G, Von Szentpaly L, Liu S 1999 *J. Am. Chem. Soc.* **121** 1922.
[11] Parr R G, Donnelly R A, Levy M, Palke W E 1978 *J. Chem. Phys.* **68** 3801.
[12] Gaussian 03, Revision B.04.; Frisch M J; Trucks G W; Schlegel H B; Scuseria G E; Robb M A; Cheeseman J R; Montgomery Jr J A; Vreven, T; Kudin K N; Burant J C; Millam J M; Iyengar S S; Tomasi J; Barone V; Mennucci B; Cossi M; Scalmani G; Rega N; Petersson G A; Nakatsuji H; Hada M; Ehara M; Toyota K; Fukuda R; Hasegawa J; Ishida M; Nakajima T.; Honda Y; Kitao O; Nakai H; Klene M; Li X; Knox J E; Hratchian H P; Cross J B; Bakken V; Adamo C; Jaramillo J; Gomperts R; Stratmann R E; Yazyev O; Austin A J; Cammi R; Pomelli C; Ochterski J W; Ayala P Y; Morokuma K; Voth G A; Salvador P; Dannenberg J J; Zakrzewski V G; Dapprich S; Daniels A D; Strain M C; Farkas O; Malick D K; Rabuck A D; Raghavachari K; Foresman J B; Ortiz J V; Cui Q; Baboul A G; Clifford S; Cioslowski J; Stefanov B B; Liu G; Liashenko A; Piskorz P; Komaromi I; Martin R L; Fox D J; Keith T; Al-Laham M A; Peng C Y; Nanayakkara A; Challacombe M; Gill P M W; Johnson B; Chen W; Wong M W; Gonzalez C; Pople J A. 2003 *Gaussian Inc* (Wallingford CT).
[13] Becke A D 1993 *J. Chem. Phys.* **98** 5648.
[14] Yang W, Parr R G, Lee C 1988 *Phys. Rev. B* **37** 785.
[15] Hay P J, Wadt W R 1985 *J. Chem. Phys.* **82** 270, Hay P.J., Wadt W.R 1985 ibid. **82** 284.
[16] Wyckoff R W G 1965 *Crystal Structures* (second edition vol. 1) (New York: Interscience Publishers).

Identification of the additional transition around the E_0 critical point in photoreflectance spectra of GaInAsSb epitaxial films

L Tirado-Mejía[1], J J Prías-Barragán, J F Gómez, H Ariza-Calderón

Laboratorio de Optoelectrónica, Universidad del Quindío

e-mail: litirado@uniquindio.edu.co

Abstract. We present the analysis of the photoreflectance (PR) spectra line-shape of GaInAsSb films over GaSb. The studied samples were grown by liquid phase epitaxy technique. The growth parameters such as initial temperature and contact time were the same for all samples, but the precursor solution varied slightly in the As concentration. The cooling rate was different for each sample. The PR spectra were obtained at low temperature, modulated by a 632.8 nm He-Ne laser. These spectra were fitted using the third derivative lorentzian function but, for one of the samples exhibiting an additional structure around the E_0 critical point oscillation, it was also necessary to use a first derivative lorentzian function, associated to an excitonic transition. This interpretation of the new PR feature was analyzed according to X-ray diffractograms. Comparing the FWHM of the main peak in the diffractograms it was found that the sample which exhibits the excitonic transition has the best crystalline quality, and was the sample with the lower cooling rate. The excitonic energy value obtained (9.7 meV) is in agreement with the value reported in the related literature.

1. Introduction

GaInAsSb alloys are semiconductor materials with interesting applications on optoelectronic devices such as thermophotovoltaic cells, lasers, and detectors in the infrared range [1,2]. Liquid phase epitaxy (LPE) layers of $Ga_{1-x}In_xAs_ySb_{1-y}$ grown over GaSb and InAs substrates had been reported [3,4]. The selection of the substrate is according to the chosen x and y values. For layers with low x and y values the appropriate substrate is GaSb in order to obtain a good lattice match. In this type of quaternary alloys, properties such as the values of the energy band gap E_0 and lattice constant a may be tailored to the desirable performance spectral range by controlling the molar fraction of the constituent elements (x and y) [5-9]. In this work we studied by means of photoreflectance and X-ray diffraction, $Ga_{1-x}In_xAs_ySb_{1-y}$ layers grown by liquid phase epitaxy over GaSb substrates. The films are GaSb- rich. LPE growth technique is based in a slightly difference from the thermodynamic equilibrium between the solid (substrate) and the liquid solution, inducing the solid phase over the substrate by lowering the liquid solution temperature to saturate or supersaturate it. As the temperature decreases, the solid phase over the substrate appears and the molar fractions of the constituent elements in the liquid solution change due to the difference between its diffusion coefficients. And so the stoichiometry of the grown layer and its lattice parameter depend on the cooling rate and the initial and final growth

[1] litirado@uniquindio.edu.co

temperature. For material systems where it can be used, LPE has the advantage of yielding layers of accurately controlled composition and thickness, since only a few growth parameters can be varied, i. e., substrate orientation, liquid solution composition, temperature, cooling rate, and growth time [10]. The optical response of these layers is strongly related to growth parameters.

Although it is of great interest to understand the optical properties of this alloy [11,12], only a few photoreflectance (PR) measurements has been reported in the literature [13]. This is a modulated reflectivity technique in which the surface electric field of the sample is modulated by an incident chopped electromagnetic radiation. Variations on the intrinsic field produce modifications on the electronic band structure and so, in the reflectivity of the sample.

Based on the line shape, line width and energy of oscillations in the PR spectra it is possible to obtained energy values associated to electronic band structure critical points (CP's), identification of excitons, and information about electron-phonon interactions. The PR line shape is strongly related to the type of critical point. Considering the symmetry of the CP's in the band structure, it is possible to define one, two or three dimensional critical point, if the number of effective masses equal to infinite is two, one or none, respectively. In the low field regime, the PR spectra features of the different CP's could be described by Aspnes expression [14], varying some parameters according to the type of CP. We used this basic characteristic of the PR technique to identify the type of CP and the electronic transition associated to it. Muñoz et *al.* [13] present the analysis of the E_0 critical point and its behavior with temperature, but in this work we studied the additional structure present on the PR spectra and the identification of the type of critical point associated. These results are correlated with cooling rate and with crystalline quality of the samples, studied by means of X-ray diffraction patterns.

2. Experimental Procedures

The three samples under study were grown by LPE over GaSb (100) substrates by precursor quaternary liquid solutions In-rich. The solid solutions obtained have a high content of Ga and Sb. The LPE used is a conventional system with a horizontal graphite boat and a three zone furnace. The growth atmosphere was purified hydrogen from a Pt-Ag diffusion cell. The epitaxial layers, named S1, S2 and S3, were grown at 530 °C with the growth parameters presented in table 1.

Table 1. Growth parameters of GaInAsSb layers grown by liquid phase epitaxy.

SAMPLE	S1	S2	S3
X_{Ga} (%)	18.544	18.541	18.542
X_{In} (%)	57.833	57.835	57.825
X_{As} (%)	0.147	0.147	0.162
X_{Sb} (%)	23.476	23.477	23.471
Cooling rate (°C/min)	0.144	0.083	0.288
t (min)	4:51	4:50	4:30

The optical characterization was accomplished by photoreflectance technique at low temperature, with a He-Ne laser as modulating beam. The light probe used was the output of a monochromator with 32 cm of focal distance, and a 120 W QTH lamp beam. The samples holder was in a He close cycle cryostat, and the measurements were done at 12 K, with a thermoelectrically cooled PbS detector. For the structural characterization, X-ray diffraction patterns were used in the low incident angle configuration. They were recorded with a Bruker D8 Advance diffractometer, using Cu K_α radiation, and a graphite monochromator with the detector. The measurement mode was a detector scan for low incident angles. Considering that the samples are epitaxial layers, the main response in this measurement configuration is due to the film instead of the substrate. For each chosen incident angle the diffraction patterns exhibited a very few peaks mostly at high 2θ values.

3. Results and Discussion

The photoreflectance spectra at 12 K are presented in figure 1. In all of them there is a main oscillation, attributed to the E_0 critical point, corresponding to the band gap energy. A second oscillation is observed in samples S2 and S3, at lower energy. This last feature was studied using the first derivative lorentzian function, associated to excitonic transitions. The spectra were fitted using two contributions, for S2 and S3 samples, and just one contribution for S1 fitting [14]. For all samples the PR spectra were fitted with the following expression [15]:

$$\frac{\Delta R}{R} = \mathrm{Re}\left\{C_1 e^{i\theta_1}\left(E - E_{CP1} - i\Gamma_1\right)^{-2} + C_2 e^{i\theta_2}\left(E - E_{CP2} - i\Gamma_2\right)^{-2.5}\right\} \qquad (1)$$

in which C_1 and C_2 are the amplitude of each oscillation, E_{CPi} is the critical point energy, i.e. transition energy, Γ is the phenomenological broadening parameter, and θ is the phase. The first term of the right part of the equation (1) is associated to excitonic transitions, and the other term is associated to band to band transitions. The fitting curves observed in figure 1 were obtained using equation (1) with C_1=0 for S1 sample. This means that the transition responsible for the PR spectra of sample S1 is well described by a band to band critical point; instead, for samples S2 and S3 it was necessary to add a contribution (C_1≠0) corresponding to an excitonic transition. This is the expected optical response of a high crystalline quality material. The excitonic binding energy obtained for S2 was 9.7 meV; this value is in the order of the reported value for GaSb [16].

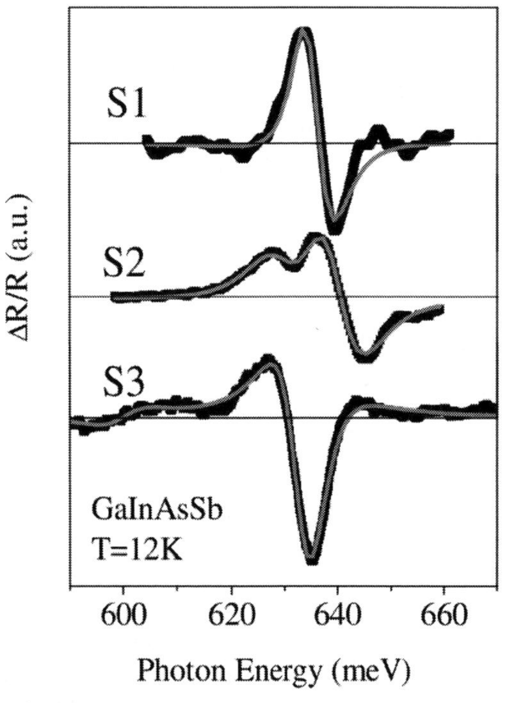

Figure 1. PR spectra at low temperature of GaInAsSb samples grown over GaSb substrates.

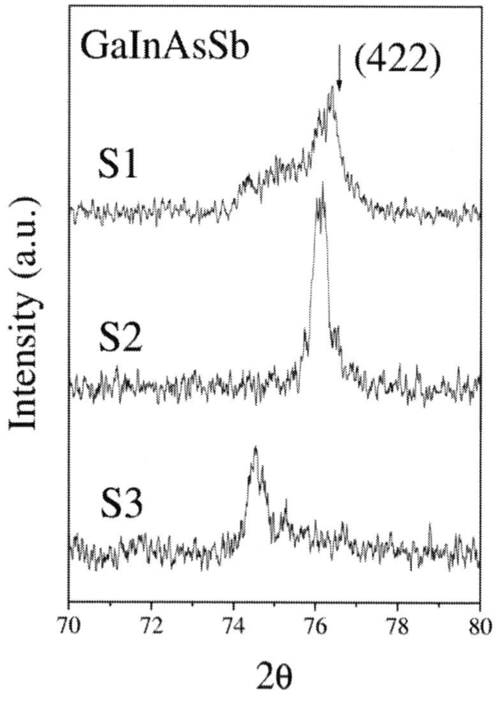

Figure 2. X ray diffraction pattern for low incident angle, of GaInAsSb samples grown over GaSb substrates.

For some semiconductor materials, PR spectrum features in the vicinity of the band to band transition have been reported. These features are associated to the line shape of a 1D critical point [17]. Equation (1) is still valid for excitonic transitions considering they form a discontinuous parabolic band, as was argue by Batz [18].

The diffraction patterns obtained in the low incident angle configuration are presented in figure 2. As the measurement mode is of low penetration depth, these patterns have exclusively film contribution, without contribution of the substrate. In the 2θ range chosen, the PDF of GaSb exhibit a diffraction maximum in 76.514° due to the (422) planes. Even though samples S2 and S3 patterns present diffraction at angles lower than GaSb, the FWHM of the peaks is small compared to the S1 pattern. This result is coherent with the optical response obtained due to the existing relation between FWHM and crystalline quality. In the other hand, the angle shift of the peak in S2 and S3 patterns is attributed to the inclusion of In and As in the GaSb matrix, changing the lattice parameter. Calculating the differences between the lattice parameters of sample S2 and GaSb substrate, the lattice mismatch is 4.5×10^{-3}. This value means that there is some distortion of the unit cell, increasing toward the generation of misfit dislocations. Instead, diffraction pattern of sample S1 has several peaks not resolved. It is possible that sample S1 has a low crystalline quality and possible present of other phases. The different optical and structural response of samples under study could be explained as a consequence of growth parameters such as cooling rate and relative molar fractions in the liquid solution. S2 is the most homogeneous and crystal.

4. Conclusions

The line shape of PR spectra contains important information about the critical points involved in the optical transitions on the semiconductor layers. The GaInAsSb samples fabricated by LPE showed differences in the optical response, influenced by the growth parameters. The analysis of X-rays and PR measurements evidenced excitonic effects on samples S2 and S3 attributed to growth conditions. Experimental results can be interpreted as a good crystalline quality of quaternary samples S2 and S3, while S1 sample exhibited a possible presence of other phases. The lattice mismatch value of S2 sample, 4×10^{-3}, which could generate a low dislocation density. The excitonic binding energy evaluated from PR measurements on S2 sample is 9.7 meV,

Acknowledgement

This work has been supported by Universidad del Quindío and the Center of Excellence for Novel Materials (CENM) under contract 043-2005 with COLCIENCIAS.

References

[1] Wang C A Shiau D A and Calawa D R 2004 *J. Crystal Growth* **261** 372
[2] Cerutti L, Boissier G, Grech P, Peron A, Angellier J, Rouillard Y, Tournié E, Genty F, Dente G C and Kaspi R 2007 *J. Cryst. Growth* **301** 967
[3] Dutt B V, Temkin H, Kolb E D and Sunder W A 1985 *Appl. Phys. Lett.* **47** 111
[4] Muhitdin A, Afrailov and Mehmet Ozer 2005 *Phys. Stat. Sol.* **2** 1393
[5] Vurgaftman I, Meyer J R, and Ram Mohan L R 2001 *J. Appl. Phys.* **89** 5815
[6] Guden M, and Piprek J. Modelling Simul 1996 *Mater. Sci. Eng.* **4** 349
[7] Díaz Reyes J, Cardona Bedoya J A, Gómez Herrera M L, Herrera Pérez J L, Riech I and Mendoza Alvarez J G 2003 *J. Phys. Condens. Matter* **15** 8941
[8] Moiseev K D, Mikhailova M P, Yu P Y, Simecek T, Hulicius E, and Oswald J 2001 *J. Phys. Condens. Matter.* **90** 2813
[9] Shim K, Rabitz H and Dutta P 2000 *J. Appl. Phys.* **88** 7157
[10] Hsieh J J 1980 *Handbook on Semiconductors* vol. 3, ed. T S Moss 415
[11] Olvera Hernández J, Olvera Cervantes J, Rojas López M, Navarro Contreras H, Vidal M A and de Anda F 2006 *J. Phys.Conf. Ser.* **28** 147

[12] Amariei A, Polychroniadis E K, Dimroth F and Bett A W 2005 *J. Cryst. Growth* **275** 1229
[13] Muñoz M, Pollak F H, Zakia M B, Patel N B, and Herrera Pérez J L 2000 *Phys. Rev. B* **62** 16600
[14] Prías-Barragán J J, Espinosa-Arbeláez D G, Álvarez G A, Tirado-Mejía L and Ariza-Calderón H 2005 *Rev. Col. Fís.* **37** 134
[15] Aspnes D E and Studna A 1973 *Phys. Rev. B* **47** 10
[16] Iyer N, Hedge S, Bajaj K K, Abul-Fald A and Michel W1993 *J. Appl. Phys.* **73** 8
[17] Ghosh S and Arora B M 1997 *J. Appl. Phys.* **81** 6968
[18] Batz B 1972 *Semiconductors and Semimetals* vol. 9, ed. Willarson R K and Beer A 37

Journal of Physics: Conference Series **167** (2009) 012023 doi:10.1088/1742-6596/167/1/012023

Multilayers of GaAs/Mn deposited on a substrate of GaAs (001)

M. Bernal-Salamanca[(1)], **A. Pulzara-Mora**[(1)], **S. Molina-Valdovinos**[(2)], **M. Meléndez-Lira**[(2)], **A. Rosales-Rivera**[(1)], **and M. López-López** [(3)]

[(1)] Laboratorio de Magnetismo y Materiales Avanzados, Universidad Nacional de Colombia, Sede Manizales, A.A. 127.

[(2)] Physics Department, Centro de Investigación y Estudios Avanzados del IPN, Av. IPN No. 2508, Apartado Postal 14-740, 07000 México D.F, México

[(3)] Centro de Fisica Aplicada y Tecnologia Avanzada, Universidad Nacional Autonóma de México, Apartado Postal 1-1010, Querétaro 76000, México

E-mail: `aopulzaram@unal.edu.co`

Abstract. In this work GaAs/Mn multilayers were deposited on GaAs (001) substrates by R.F magnetron sputtering technique, varying the deposition time (tg). Scanning electron and atomic force Microscopy studies were realized on the surface of the samples in order to determine the morphology and average roughness. X-ray diffraction spectra show that our samples tend to do amorphous. Raman spectroscopy at room temperature was employed to analyze the structural properties of the samples. We found that for a GaAs film taken as reference, the Raman spectra is dominated by the transverse (TO) and longitudinal (LO) modes located at 266 cm^{-1} and 291 cm^{-1}, respectively. However, for the GaAs/Mn multilayers the TO and LO modes decrease dramatically, and the Mn Raman modes in the range of 100 cm^{-1} and 250 cm^{-1} are evidenced. Additional new peaks located around 650 and 690 cm^{-1} are only observed for the samples with high Mn content. By using the mass reduced model we estimate that the Mn related peaks are located at 650.2 cm^{-1} and 695.2 cm^{-1}, in good agreement with the experimental data, these peaks are correlated with excitations due to $(Mn)_mAs_n$ localized structures.

1. Introduction

The relatively recent discovery of ferromagnetic III-V semiconductor-based materials with Curie temperatures (T_c) as high as 185 K has now raised interesting fundamental issues regarding the origin of ferromagnetism in materials such as $Ga_{1-x}Mn_xAs$[1-3]. The introduction of relatively small concentration of magnetic elements into nonmagnetic host semiconductor leads to dramatic change of the magneto-optic properties upon application of magnetic fields. In these III-V semiconductors, Mn^{2+} acts as an acceptor, generating free holes in the valence band. The ferromagnetism in these materials arises from the exchange interaction between these holes and the Mn^{2+} ions, and it is generally believed that there is a direct correlation between the Curie temperature (T_c) and the holes density [4]. However, a high Mn concentrations MnAs clusters can be formed. Depending on the growth conditions phase separation occurs, and either nanosize MnAs clusters are formed into the GaAs matrix or $Ga_{1-x}Mn_xAs$ regions with a low Mn concentration are generated [5]. In general, $Ga_{1-x}Mn_xAs$ thin films have been grown employing molecular beam epitaxy. However, there are few reports on deposition of $Ga_{1-x}Mn_xAs$

films by radio frequency (RF) magnetron sputtering. Sputtering deposition is one alternative that offers the possibility of deposition of $Ga_{1-x}Mn_xAs$ thin films at low cost and simple processing procedures. In this work, we report the study of structural and optical properties of GaAs/Mn multilayers grown by using RF magnetron sputtering technique. The samples were characterized by different techniques in order to determine their structural characteristics, chemical composition, surface morphology, crystal structure and optical properties.

2. Experimental Details

$Ga_{1-x}Mn_xAs$ films were grown by RF planar magnetron sputtering with two cooled targets of 1 inch diameter each one, and controlled by independent RF sources. The targets used were a (100) oriented GaAs wafer and Mn block. The growth temperature (Tg) was monitored using a thermocouple embedded in the substrate holder. No substrate bias was used, and the base pressure was of 1.2x10^{-6} Torr. The flow rate gas was controlled by using a needle valve. The films were deposited on (100) GaAs substrates located at 5 cm from targets.The substrates were cleaned by using standard methods. Before deposition the working pressure was held constant at 1.2x10^{-3} Torr, and the plasma between target and main shutter was initiated using Argon gas 99.9% purity. Then, the main shutter was opened to start the deposition alternating GaAs and Mn layers. The samples consist of three GaAs/Mn cycles, two types of samples were prepared with different GaAs and Mn deposition times:A)GaAs(15 min)/Mn(15 min) and B)GaAs(60 min)/Mn(60 min). A 30 sec. interruption time was introduced between the deposition of the GaAs and Mn layers, while the plasma is stabilized. The growth temperature was fixed at 500^oC and the working pressure was held at 1.2x10^{-3} Torr. The films surface morphology was analyzed by scanning electron microscopy (SEM) and atomic force microscopy (AFM) in non-contact mode. X-ray diffraction (XRD) using a CuK_α radiation source was used to evaluate the films crystalline structure. The structural properties were also studied by micro Raman scattering (RS) using the 632.8nm line of a He-Ne laser with a power of 20mW.

3. Results and Discussion

Figure 1 (a) and (b) show plan view SEM images (with a magnification of x 10.000) for the samples type A and B, respectively. From these images we observe that the surface of sample A is flat and smooth. However, the surface of sample B is composed of larger grains. In Fig. 1 (c) and (d), we show the AFM images on a 1x1 μm^2 scale for samples A and B, respectively. By comparing these images, we can observe that there exists a strong difference in the morphology of these samples, probably due to the thickness of the samples. For the sample A the surface is smooth with a 3 nm RMS roughness and small (9.8 nm mean height) grain size. However, for the sample B the surface is porous formed of spherically shaped grains (20.6 nm of mean height) with a value of 8.4 nm in RMS roughness [6].

To evaluate the crystalline quality of the GaAs/Mn multilayers, X-ray diffraction was performed in a θ-2θ configuration. Fig. 2(a) and (b) shows the x-ray curves of Mn bulk and GaAs film taken as references. The peaks located at 2θ = 34.8, 42.9, 47.7, 52.2, 60.4, 73.4, and 78.7 degrees in the Fig. 2(a) correspond to the crystallographic planes of β-Mn polycrystalline phase. In the Fig. 2(b) we only observed the (200)GaAs and (400)GaAs main peaks, located at 2θ = 31.7 and 66.2 degrees. For the GaAs/Mn multilayer type A (Fig.1(c)), additional to the GaAs main peaks we note peaks not well defined labeled by (*), corresponding to the Mn_4As_3 phase. However, for the GaAs/Mn sample type B (Fig.1(d)), peaks corresponding to the Ga_5Mn_8 phase appeared[7]. We observed that increasing the growth time of the individual layers of GaAs/Mn, the X-ray peaks are more defined, which may be due to an improvement of the crystalline quality of the samples caused possibly by the interdiffusion of Mn atoms into GaAs layer, or the re-evaporation of Ga and As atoms that are bonded to Mn to form GaAsMn

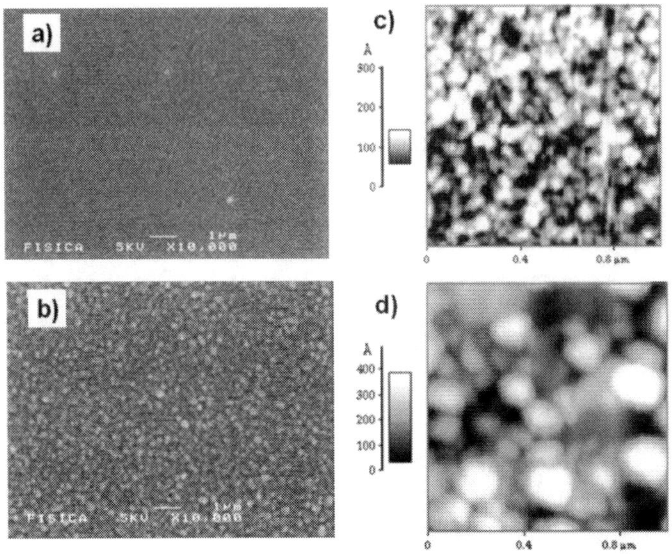

Figure 1. Plan view SEM images (left side) taken in a 1 μm scale and x10000 amplification and AFM images (right side) taken in a 1x1 μm^2 area of GaAs/Mn multilayers grown at a deposition time of: a) t_g = 15 min., b) t_g = 60 min. and 500 ^oC substrate temperature.

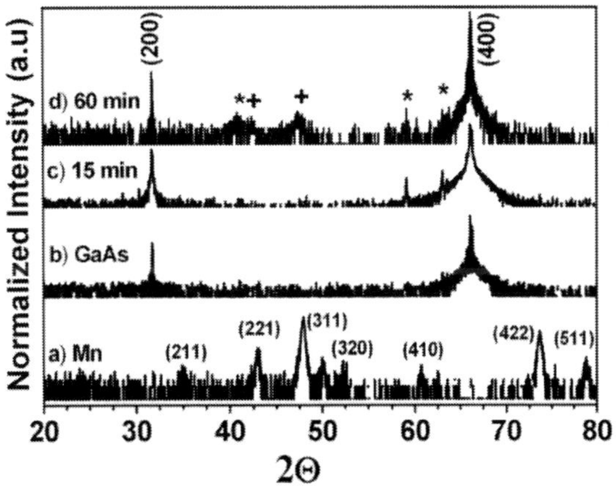

Figure 2. XRD curves of: a) Mn, b) GaAs layers taken as reference, c) and d) are the spectra of GaAs/Mn Multilayers grown by varying the deposition time per each layer: c) t_g = 15 min. and c) t_g = 60 min, and T_g = 500 ^oC substrate. The peaks labeled as (+) and (*) in this Figure denote the GaMn phase, and MnAs phase, respectively.

alloys in the interface of the layers, as suggest the Raman spectra that will be discussed below [8].

In Fig. 3(a) we show the Raman spectra of the GaAs film reference grown at $T_g = 500\ ^oC$ and $t_g = 60$ min. In this spectrum the first order Raman scattering LO and TO modes from GaAs are observed at 266 and 291 cm^{-1}, respectively. These modes present a redshift of 2 cm^{-1} probably due to stress effects [9]. Raman spectra obtained for GaAs/Mn multilayers type A and B are shown in the Fig. 3 (b) and (c), respectively. In the Raman spectra of Fig. 3

(b), is evident that the intensity of the TO and LO modes tend to disappear completely, and a broader band from 150 to 270 cm^{-1} appeared, which indicates that the zincblende structure was strongly deformed by the inclusion of the Mn atoms. This suggests that the Mn atoms affected the GaAs lattice as a whole, due to a structural disorder caused by vacancies, MnGa complex or Mn-Mn clusters formation [10,11]. In order to analyze the superimposed phonon modes, the Raman spectrum was fitted by using four Lorentzian functions. We found that the peaks located at 195, 212.7, 229.9 and 247 cm^{-1} are related to vibrational modes of Mn and As, as a consequence of the structural disorder and other built-in defects for samples with high Mn concentrations [12].

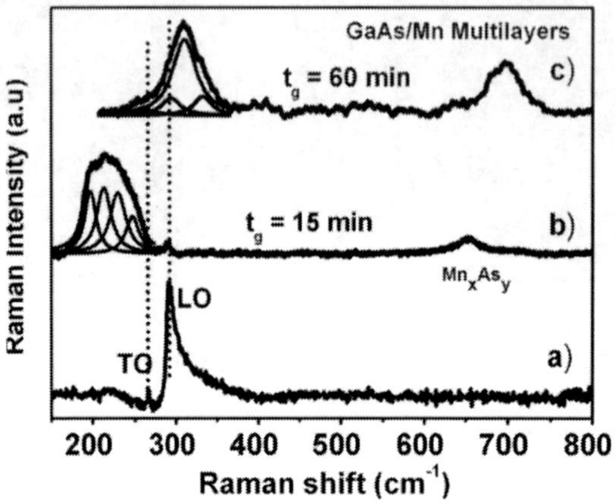

Figure 3. Room temperature Raman spectra of: a) GaAs layer taken as reference, b) and c) are the spectra of GaAs/Mn Multilayers grown by varying the deposition time per each layer: b) t_g = 15 min., c) t_g = 60 min, and T_g = 500 ^oC substrate temperature. Thicker lines denote the deconvolution of experimental data by using Gaussian functions.

On the other hand, the Raman spectrum of the Fig. 3(c) shows a broader band from 238 to 265 cm^{-1}. A more detailed analysis of this band by using four Lorentzian functions, allowed us identify the TO and LO modes related to GaAs and two additional Raman modes located at 309,8 cm^{-1} and 332 cm^{-1} associated with GaAsMn alloys. This result is in agreement with the X-ray results shown in Fig. 1(d), where a phase separation (labeled by(+)) is observed for this sample. The high Mn concentration introduced during growth at high substrate temperature allows the zincblende matrix to form MnAs clusters, and a phase separation occurs. Note that a coupled LO-phonon plasmon mode has been reported to appear in epitaxial GaMnAs layers [13], this mode could be present in sample B (Fig. 3(c)). On the other hand, new phonon modes located around 654 $cm^{-1}$and 694 cm^{-1} near to the 2LO GaAs second order mode are observed in our samples (b) and (c). These modes have been assigned to disorder activated Raman modes at the Brillouin zone boundary corresponding to the acoustic and optical phonon branches due to the incorporation of Mn-As pairs or Mn_mAs_n cluster formation on GaAs matrix [14,15].

In order to theoretical estimate the frequency of these Raman modes, we used the mass reduced model given by the following equation [16]:

$$\omega_{Mn_mAs_n} \approx \omega_{2LOGaAs}\sqrt{\frac{\mu_{GaAs}}{\mu_{Mn_mAs_n}}} \qquad (1)$$

where $\omega_{2LOGaAs}$ is the frequency of 2LO (580 cm^{-1}) phonon mode and μ_{GaAs} , μ_{MnAs} are the reduced masses of the Ga-As and Mn_mAs_n pairs, respectively. The value of the Raman frequency calculated for us using the Eq.(1) is 650.2 cm^{-1} (Mn_2As_3) and 695.2 cm^{-1} ($Mn_{12}As_3$). This value is in good agreement with the experimental value centered at 654 cm^{-1} and 696 cm^{-1} for sample A and sample B, respectively. The formation of $(GaAs)_nMn_m$ complex has been predicted theoretically by G.L. Gutsev, et al [17]. Clusters are likely formed by MnAs units because the pure Mn_m clusters possess the antiferromagnetic coupling of local excess spin densities beginning with m = 5, that drastically reduces the cluster magnetic moments. The reason for the MnAs clusters formation in the sample with high Mn content could be the existence of very small aggregates of Mn atoms in the GaAs layers, which would act as nucleation center for the clusters nucleation. The study of cluster formations is interesting, because the clustering of Mn atoms inside the GaAs host is a possible path toward increasing the Curie temperature important to technological applications in spintronics. Besides, MnAs which is a ferromagnetic metal at room temperature is on the focus of attention because it can be successfully integrated with semiconductors such as GaAs [18].

4. Conclusions

We have studied the properties of GaAs/Mn multilayers grown by R.F magnetron sputtering. X-ray diffractions show that the samples have a poor crystalline quality. In Raman scattering studies we found that in addition to some conventional modes that were also observed in GaAs, a broader band from 150 to 260 cm^{-1} appeared due to a structural disorder caused by vacancies, MnGa complex or Mn-Mn clusters formation. For the sample with high Mn content, new bands around of 650 cm^{-1} and 695 cm^{-1} are evident, probably due to a $(Mn)_mAs_n$ cluster formation. The theoretical calculations of these modes by using the mass reduced model are in good agreement with experimental data.

Acknowledgments
This work was partially supported by DIMA, COLCIENCIAS - CONACYT - México and ICTDF. The authors thank the technical assistance of R. Fragozo, A. Soto, and E. Gómez. MLL would like to thank CONACYT for the support during the sabbatical leave.

5. References

[1] T. Dietl, H. Ohno, and F Matsukura, Phys. Rev. B 63, 195205 (2001).
[2] T. Jungwirth, J. Sinova, J. Masek, J. Kucera, A.H. MacDonald, Rev. Mod. Phys. 78, 809 (2006).
[3] V. Novk, K. Olejnk, J. Wunderlich, M. Cukr, K. Vborn, A. W. Rushforth, K. W. Edmonds, R. P. Campion, B. L. Gallagher, Jairo Sinova, and T. Jungwirth, Phys. Rev. Lett 101, 077201 (2008).
[4] M. J. Seong1, S. H. Chun, Hyeonsik M. Cheong, N. Samarth, and A. Mascarenhas. Journal of the Korean Phisical Society, February 2003, pp. S124-S128.
[5] M. Bolzan, I. Bergenti, G. Rossetto, P. Zanella, V. Dediu, M. Natali. Journal of Magnetism and Magnetic Materials 316 (2007) 221-224.
[6] I.T. Yoon, T.W. Kang, D.J. Kim. Journal of Magnetism and Magnetic Materials 320 (2008) 662-665.
[7] C.H. Chen, H. Niu, C.Y. Cheng, H.H. Hsieh, S.C. Wu. Nuclear Instruments and Methods in Physics Research B (2008).
[8] D.M.G. Leite, L.F. da Silva, A.L.J. Pereira, J.H. Dias da Silva. Journal of Crystal Growth 294 (2006) 309-314.
[9] A. Pulzara Mora. Ph.D Thesis. Cinvestav, Mexico D.F, 2006.
[10] M. R. Islam, A. Hiroki and M. Yamada, Jpn. J. Appl. Phys 43, 1088,
[11] A. Koeder, S. Frank, W. Schoch, V. Avrutin, W. Limmer, K. Thonke, R. Sauer, and A. Waag. Materia condensada, 9 dec 2002. 2004.
[12] Xuelin Yanga, Jiejun Wu, Zhitao Chen, Yaobo Pan, Yan Zhang, Zhijian Yang,Tongjun Yu, Guoyi Zhang. Solid State Communications
[13] W. Limmer, M. Glunk,W. Schoch, A. K.oder, R. Kling, R. Sauer, A. Waag, Physica E 13, 589 (2002).
[14] M.R. Islam, N.F. Chen, M. Yamada. Materials Science in Semiconductor Processing 9 (2006) 184-187. 143 (2007) 236-239.
[15] G.L. Gutsev, M.D. Mochena, C.W. Bauschlicher Jr. Chemical Physics Letters 439. (2007) 95-101.
[16] X. Yang, J Wu, Z. Chen, Y. Pan, Y. Zhang, Z. Yang, T. Yu, G. Zhang. Solid State Communications 143 (2007) 236.
[17] G.L. Gutsev, M.D. Mochena, C.W. Bauschlicher Jr., J. Phys. Chem. A 110 (2006) 9758.
[18] G.L. Gutsev, M.D. Mochena, C.W. Bauschlicher Jr. Chemical Physics Letters 439. (2007) 95-101.

GaSb grown from Sn solvent at low temperatures by LPE

Compeán V. H., de Anda F., Mishurnyi V. A., Gorbatchev A. Yu.

Universidad Autónoma de San Luis Potosí, Instituto de Investigación en Comunicación Optica, Av. Karakorum 1470, Col. Lomas 4a Sec., San Luis Potosí, SLP, CP 78210, México

E-mail: fdeanda@cactus.iico.uaslp.mx

Abstract. The LPE growth of GaSb using Sn as a solvent has been studied in the temperature range 250-370 C and using liquid solutions covering a wide range of compositions. In order to find the growth conditions the phase diagram has been determined experimentally around the same temperature region. It is shown the Sn incorporates into the grown layers and that it behaves as an acceptor. The photoluminescence spectra of the grown layers with different Sn contents show characteristic peaks that can be attributed to different recombination processes.

1. Introduction

GaSb and its alloys with other III-V semiconductors are very interesting materials from the point of view of its potential application in several fields such as, thermo-photovoltaic cells, mid infrared lasers and photodetetctors for the detection and measurements of air pollutant concentrations, for medical applications, night vision, etc. Nowadays most optoelectronic devices are grown by MBE or OMCVD techniques, however, liquid Phase Epitaxy (LPE) could, still, be an adequate growth technique for some of those devices, especially those requiring thick, high purity and/or high resistivity layers. The use of Sn as the solvent for LPE growth of these materials presents some advantages that could help to meet those requirements. The solubility of GaSb in Sn is larger than in Ga [1]. Thus it is possible to grow thicker layers from Sn than from Ga. The larger solubility also allows the growth at lower temperatures and thus to get materials with less point defects. This is very interesting since it is known that the practical use of GaSb in real devices is limited by the presence of a native complex acceptor defect [2] the concentration of native defects decreases with the temperature and it has been shown that undoped GaSb grown at low temperatures changes its conductivity to N type [3]. Low growth temperatures would offer the added advantage that techniques that involve the regrowth of epitaxial layers on complex heterostructures could be applied without disturbing noticeably the heterostructures even if they comprise very thin layers. Another advantage of using a "neutral" solvent such as Sn is that if it is used to grow different GaSb based alloys, i. e. AlGaAsSb and GaInAsSb to fabricate heterostructres by LPE, the problem of mixing the current growth solution with remaining droplets of the solution used to grow the previous layer is minimized.

2. Experimental

The growth experiments and the liquidus temperature measurements were done in a semitransparent furnace under a high purity, Pd-diffused, Hydrogen fow of 0.3 lt/min. An electronic grade graphite boat of the sliding type was used in all the experiments. All the materials used to prepare the liquid solutions had a purity of 5N or higher. The liquidus temperature was determined, by observing through a microscope the liquid solution while slowly heating, as the point when the last solid disappeared. For the growth experiments the substrates were chemically polished in a solution of H2O2: HF: C4O6H6 (tartaric acid): H2O [4]. Since it is not possible to fabricate semi-insulating wafers of GaSb it is not easy to measure the free carrier concentration using standard techniques, so to determine whether Sn incorporates as a donor or acceptor, the conductivity type of the grown layers were tested by the hot probe method. The photoluminescence experiments were done using a 0.5 m monochromator, a 100 mW Ar laser for excitation and lock-in techniques to filter the signal. The setup was controlled by a PC.

3. Results and discussion

The Sn-GaSb quasi-binary phase diagram has been previously determined and some experimental points have been reported at temperatures above 300 [5 - 9] and some works about thermodynamic properties of the ternary system have been reported.

Table 1. Liquidus points of the Sn-Ga-Sb phase diagram.

T_{Liq}, K	X_{Ga}	X_{Sb}	X_{Sn}
508	0.01	0.01	0.98
520	0.013	0.013	0.974
534	0.015	0.015	0.97
546	0.018	0.018	0.964
552	0.796	0.001	0.203
557	0.02	0.02	0.96
564	0.023	0.023	0.954
566	0.99928	7.2E-4	0
566	0.997	7.0E-4	0.002
573	0.032	0.019	0.949
600	0.038	0.038	0.924
601	0.317	0.006	0.677
634	0.94	0.003	0.057
652	0.49	0.01	0.5
709	0.11	0.11	0.78

Table 1 shows the liquidus points measured in this work, when a layer was grown its conductivity type is indicated. The points corresponding to the quasi-binary part of the phase diagram are plotted in Fig. 1 together with some other points reported in the literature [5 - 9]. It can be seen that all points lie close to a continuous line. It is possible to estimate the heat of dissolution of GaSb in Sn using the approximation of ideal solutions by fitting the low temperature, low concentration data points to the equation:

$$X^{L}_{GaSb} = X_0 \exp(-\Delta H / RT)$$

The fitting gives a value of the heat of dissolution (ΔH) of the order of 8.7 kcal/mol.

All the layers grown had P type conductivity, even those grown at temperatures lower than 400 C, at these temperatures GaSb layers grown from Ga solutions in a graphite boat normally exhibit N type conductivity [3]. This is indicative that Sn is incorporated into the GaSb layers as an acceptor impurity.

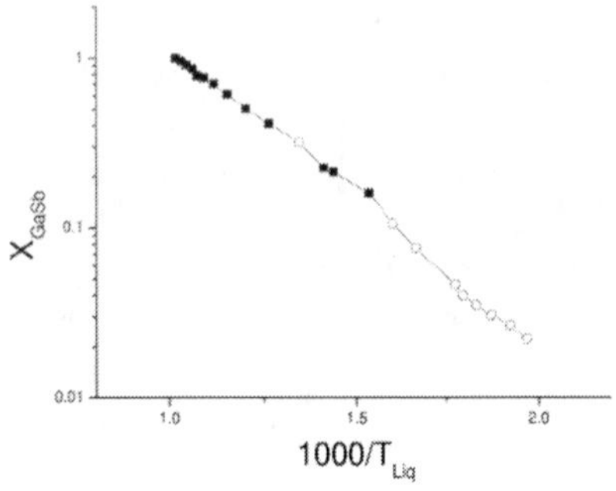

Figure 1. Liquidus of the quasi-binary phase diagram Sn-GaSb.

To confirm this observation an undoped GaSb layer was grown from a Ga solvent at 293 C and it had N type conductivity, as expected. Then other layers were grown at the same temperature but adding very small amounts of Sn to the liquid solution. It was observed that the conductivity of the layers changed to P type when the amount of Sn added was near 0.2 at. %.

The photoluminescence spectra of some of the grown samples from Ga solvent is shown in fig. 2.

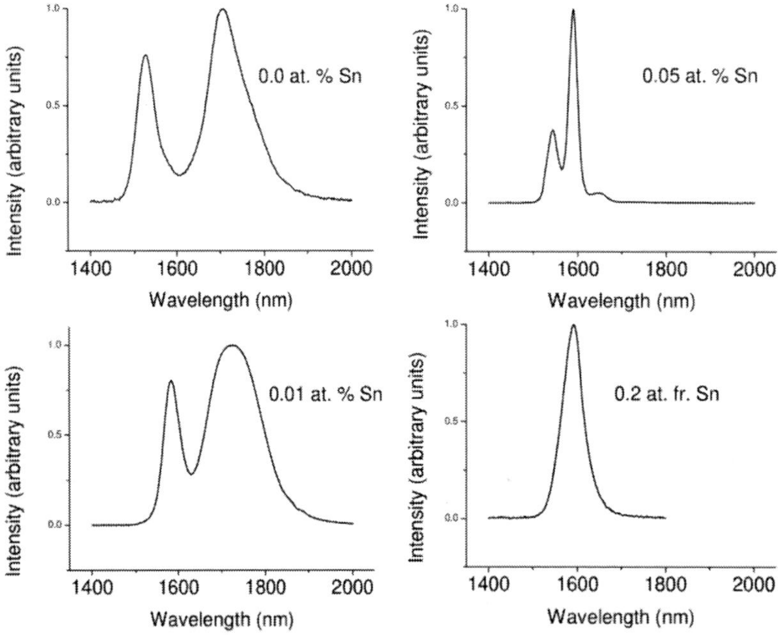

Fig. 2. Photoluminescence of layers grown from a Ga solvent doped with different amounts of Sn.

The spectra of the sample grown with no added Sn shows two peaks, one at around 1530 nm can be attributed to near band gap transitions. The emission peak at lower energies is more intense, we have not searched for the origin of this peak. When 0.01 at. % of Sn is added to the liquid solution the near band gap peak disappears and a new peak at around 1590 nm appears in the spectrum and its intensity is still less than that of the lower energy peak, this new peak can be attributed to the Sn impurity as we will see. When the quantity of Sn is raised to 0.05 at. % the lower intensity peak disappears, the peak at 1590 nm is now the main feature of the spectrum and a smaller intensity peaks appears at around 1540 nm. Finally when the Sn content of the liquid solution reaches 0.2 at. % the emission shows only the peak at around 1590 nm. This behavior is in agreement with the assumption that the 1590 peak is due to transitions involving the Sn impurities.

4. Conclusions

Some points of the Sn-Ga-Sb phase diagram have been measured. Some of these points agree with previously reported data. GaSb layers have been grown using Sn as solvent at temperatures as low as 250 C. Sn incorporates into the grown layers as an acceptor impurity and so it cannot be considered as an "ideal neutral solvent" for the growth of GaSb. The photoluminescence spectra of these samples show a peak at around 1590 nm that can be attributed to the Sn impurities.

5. Acknowledgments

This work was partially supported by CONACYT, PROMEP and FAI at UASLP.

References

[1] Chandvankar S S and Arora M M 1987 *J. of Cryst. Growth* **80** 69
[2] Dutta P S and Bhat H L 1997 *J. Appl. Phys.* **81** 5821
[3] Olvera-Hernández J, de Anda F, Navarro-Contreras H. and V A Mishurnyi 2000 *J. Cryst. Growth* **208** 27
[4] Berishev I E, de Anda F, Mishournyi V A, Olvera J, Ilyinskaya N D and Vasilyev V I 1995 *J. of the Electrochem.* Soc. **142** L189
[5] Zivkovc D, Katayama I, Kostov A and Zivkovc Z 2003 *J. Therm. Anal. Cal.* **71** 567
[6] Gerdes F and Predel B 1981 *J. Less-Common Met.* **79** 281
[7] Oelsen W and Zuhlke P 1956 *Arch. Eisenhuttenwessen,* **27** 743
[8] Oelsen W, Schurmann E, Weigt H J and Oelsen O 1956 *Arch. Eisenhuttenwessen,* **27** 487
[9] Katayama I, Fukuda Y, Hattori Y, Maruyama T 1998 *Thermochimica Acta* **314** 175

Nonequilibrium dynamics of polariton entanglement in a cluster of coupled traps

L.Quiroga[1] **and C.Tejedor**[2]

[1]Departamento de Física, Universidad de Los Andes, A.A.4976, Bogotá D.C., Colombia
[2]Departamento de Física Teórica de la Materia Condensada, Universidad Autónoma de Madrid, Cantoblanco, E-28049, Madrid, Spain

E-mail: `lquiroga@uniandes.edu.co`

Abstract. We study in detail the generation and relaxation of quantum coherences (entanglement) in a system of coupled polariton traps. By exploiting a Lie algebraic based super-operator technique we provide an analytical exact solution for the Markovian dissipative dynamics (Master equation) of such system which is valid for arbitrary cluster size, polariton-polariton interaction strength, temperature and initial state. Based on the exact solution of the Master equation at $T = 0$K, we discuss how dissipation affects the quantum entanglement dynamics of coupled polariton systems.

1. Introduction

Coupled bosonic systems appear naturally in a great variety of physical situations ranging from quantum optics, atomic/molecular physics to condensed matter physics. In the last field exciton/polariton systems are among the most notable examples. Similarly to ultracold boson atoms confined in optical lattices, trapped exciton/polariton systems are predicted to present different collective phases such as Mott insulator or superfluid phases. Experimental control of polariton trapping is rapidly improving. Among the most promising candidates are stress induced trap systems [1] and metal thin-film deposition on microcavities [2]. A proposal for the generation of polarization entangled photon-pairs from electric gate controlled polariton trapped systems has been recently put forward [3]. However, for full exploitation of the capabilities of semiconductor microcavities based entanglement, the polariton entangling power itself requires complete characterization. In the present work, we focus on polariton trapped clusters as open quantum systems evolving under nonequilibrium or dissipative conditions. The time evolution of the entanglement generated by polariton-polariton interactions jointly with dissipative effects is considered.

In order to analyze the dissipative quantum dynamics of a trapped polariton system we use a Born-Markov reduced dynamics approach. The solution of the resulting Master (Lindblad) equation (ME) describes the time evolution of the polariton system of interest. While deviations from a strictly Markovian behavior may occur in real microcavities [4], our analytical results may be used as a first step in exploring more complex relaxation dynamics in multitrap polariton systems. We restrict ourselves to the case where the system is deep in the Mott insulator or fully localized regime where polariton hopping is neglected but polariton-polariton interactions remain important. To find the full density matrix of the coupled polariton system, we use an exact Lie algebraic superoperator method for solving the ME, generalizing previous similar treatments [5, 6] to arbitrary number of coupled systems. By applying these results to a double trap, we focus our attention to noise related properties of the entanglement creation and preservation in such condensed matter highly nonlinear systems. We quantify the amount of entanglement in terms of the logarithmic negativity which has been proven to be valid as an entanglement measure for any mixed state of arbitrary bipartite systems [7]. Furthermore, this entanglement measure is easily computable having additionally an operational interpretation. We found that in the Mott regime, the polariton-polariton interaction generates naturally entanglement which is degraded by dissipation on a shorter time scale as compared with the population decay, and with an entanglement maximum being an increasing function of the initial number of polaritons. The entanglement duration time shortens as the polariton decay rates or the initial trap populations become large.

2. Theoretical method

Within the Born-Markov approach, the ME for the reduced system's density operator $\hat{\rho}$ can be written as ($\hbar = 1$)

$$\frac{d\hat{\rho}(t)}{dt} = -i[\hat{H}, \hat{\rho}(t)] - \hat{\mathcal{L}}\hat{\rho}(t) \tag{1}$$

with the first term on the right hand side describing the coherent or unitary evolution term, whereas the second term, or Lindblad term, is associated to the noisy dynamics induced by the reservoirs. Thus, we address the study of the relaxation dynamics of M coupled polariton traps described by the Hamiltonian

$$\hat{H} = \sum_{n=1}^{M} \omega_n \hat{a}_n^\dagger \hat{a}_n + \sum_{n=1}^{M}\sum_{m=1}^{M} \lambda_{n,m} \hat{a}_n^\dagger \hat{a}_n \hat{a}_m^\dagger \hat{a}_m \tag{2}$$

where $\hat{a}_n^\dagger$ ($\hat{a}_n$) are boson creation (annihilation) operators in trap n. The coupling matrix $\bar{\lambda} = \{\lambda_{n,m}\}$ is a real symmetric matrix (hereafter diagonal terms will be denoted simply as $\lambda_{n,n} = \lambda_n$). Lindblad terms, as taken to describe reservoirs acting independently on each polariton trap, are given by

$$\hat{\mathcal{L}}\hat{\rho}(t) = \sum_{n=1}^{M} \left(D_n \left[\hat{a}_n^\dagger \hat{a}_n \hat{\rho}(t) + \hat{\rho}(t) \hat{a}_n^\dagger \hat{a}_n - 2\hat{a}_n \hat{\rho}(t) \hat{a}_n^\dagger \right] + P_n \left[\hat{\rho}(t) \hat{a}_n \hat{a}_n^\dagger + \hat{a}_n \hat{a}_n^\dagger \hat{\rho}(t) - 2\hat{a}_n^\dagger \hat{\rho}(t) \hat{a}_n \right] \right) \tag{3}$$

where coefficients D_n and P_n are related to the dissipation and pumping mechanisms, respectively, for the n-th trap. In this equation, there are no cross terms of the type $\hat{a}_n^\dagger \hat{a}_m$ because we consider that decay and pumping on one trap is independent from other traps.

In order to evaluate the time evolution of the polariton traps density operator we proceed as follows. We define super-operators in the following form

$$\hat{K}_-^{(n)}\hat{O} = \hat{a}_n^\dagger \hat{O} \hat{a}_n \quad , \quad \hat{K}_+^{(n)}\hat{O} = \hat{a}_n \hat{O} \hat{a}_n^\dagger \quad , \quad \hat{K}_0^{(n)}\hat{O} = -\frac{1}{2}(\hat{a}_n^\dagger \hat{a}_n \hat{O} + \hat{O}\hat{a}_n^\dagger \hat{a}_n + \hat{O}) \tag{4}$$

and $\hat{\mathcal{N}}_n \hat{O} = [\hat{a}_n^\dagger \hat{a}_n, \hat{O}]$, where $\hat{O}$ denotes any polariton operator. Furthermore, given the fact that $[\hat{a}_n^\dagger \hat{a}_n \hat{a}_m^\dagger \hat{a}_m, \hat{\rho}(t)] = -\left[\hat{\mathcal{N}}_n \left(\hat{K}_0^{(m)} + \frac{1}{2}\right) + \hat{\mathcal{N}}_m \left(\hat{K}_0^{(n)} + \frac{1}{2}\right)\right] \hat{\rho}(t)$, the ME for the coupled polariton system, Eq.(1), can be written as

$$\frac{d\hat{\rho}(t)}{dt} = -i \sum_{n=1}^{M} \tilde{\omega}_n \hat{\mathcal{N}}_n \hat{\rho}(t) + 2 \sum_{n=1}^{M} \left[\hat{\Omega}_n \hat{K}_0^{(n)} + D_n \hat{K}_+^{(n)} + P_n \hat{K}_-^{(n)} + \frac{D_n - P_n}{2} \right] \hat{\rho}(t) \tag{5}$$

with $\tilde{\omega}_n = \omega_n - \lambda_n - \sum_{m\neq n}^{M} \lambda_{n,m}$ and $\hat{\Omega}_n = D_n + P_n + i\lambda_n\hat{\mathcal{N}}_n + i\sum_{m\neq n}^{M} \lambda_{n,m}\hat{\mathcal{N}}_m$. As the super-operators $\hat{\mathcal{N}}_n$ commute with any other super-operator, i.e. $[\hat{\mathcal{N}}_m, \hat{\mathcal{N}}_n] = [\hat{\mathcal{N}}_m, \hat{K}_{\pm}^{(n)}] = [\hat{\mathcal{N}}_m, \hat{K}_0^{(n)}] = 0$, the solution to Eq.(5) has the form

$$\hat{\rho}(t) = \prod_{n=1}^{M} e^{-i\tilde{\omega}_n\hat{\mathcal{N}}_n t} e^{2\left[\hat{\Omega}_n\hat{K}_0^{(n)} + D_n\hat{K}_+^{(n)} + P_n\hat{K}_-^{(n)} + \frac{D_n - P_n}{2}\right]t} \hat{\rho}(0) \tag{6}$$

where $\hat{\rho}(0)$ denotes the initial state. It is straightforward to demonstrate that the superoperators defined by Eq.(4) close in the su(1,1) Lie algebra

$$[\hat{K}_+^{(m)}, \hat{K}_-^{(n)}] = -2\hat{K}_0^{(n)}\delta_{m,n} \quad , \quad [\hat{K}_0^{(m)}, \hat{K}_{\pm}^{(n)}] = \pm\hat{K}_{\pm}^{(n)}\delta_{m,n} \tag{7}$$

By using the su(1,1) Lie algebra structure of the superoperators as given in Eq.(7), the ME solution expressed by Eq.(6), can finally be written in the full disentangled form

$$\hat{\rho}(t) = \prod_{n=1}^{M} e^{-i\tilde{\omega}_n\hat{\mathcal{N}}_n t} e^{(D_n - P_n)t} e^{\hat{c}_-^{(n)}(t)\hat{K}_-^{(n)}} e^{\hat{c}_0^{(n)}(t)\hat{K}_0^{(n)}} e^{\hat{c}_+^{(n)}(t)\hat{K}_+^{(n)}} \hat{\rho}(0) \tag{8}$$

The temporal coefficients are operators given by [8]

$$\begin{aligned} \hat{c}_+^{(n)}(t) &= \frac{2D_n\sinh(\hat{\Delta}_n t)}{\hat{\Delta}_n\cosh(\hat{\Delta}_n t) + \hat{\Omega}_n\sinh(\hat{\Delta}_n t)} \quad , \quad \hat{c}_-^{(n)}(t) = \frac{2P_n\sinh(\hat{\Delta}_n t)}{\hat{\Delta}_n\cosh(\hat{\Delta}_n t) + \hat{\Omega}_n\sinh(\hat{\Delta}_n t)} \\ \hat{c}_0^{(n)}(t) &= 2Ln\left[\frac{\hat{\Delta}_n\cosh(\hat{\Delta}_n t) + \hat{\Omega}_n\sinh(\hat{\Delta}_n t)}{\hat{\Delta}_n}\right] \end{aligned} \tag{9}$$

where $\hat{\Delta}_n = \left[\hat{\Omega}_n^2 - 4D_nP_n\right]^{1/2}$. In this way, polariton observables can be analytically evaluated for any number of coupled polariton traps, any combination of dissipation/pumping effects (including temperature effects) and for arbitrary initial states.

3. Entanglement nonequilibrium dynamics

Given the general solution of the ME in Eq.(8) an analytical expression can be obtained for the cluster density operator [8]. However, in the present work we focus on the zero temperature case for a two trap polariton cluster ($M = 2$). The same algebraic results could also be used to model a single trap with two polariton species, as it could be the case for a system formed by polaritons of different spin polarizations confined in the same spatial region. We consider that the system is initially prepared, by a resonant short laser pulse, in a purely separate coherent state, i.e.. $\hat{\rho}(0) = |\alpha_1, \alpha_2 >< \alpha_1, \alpha_2|$, where $|\alpha_j>$ denotes a coherent state in the j-th trap, $j = 1, 2$. The initial mean number of polaritons in trap j is given by $|\alpha_j|^2$. After this initialization step, interacting polaritons evolve in time in presence of dissipation while, in the vanishing temperature limit, incoherent pumping effects are not present ($P_n = 0$). Moreover, we take identical dissipation decay rates $D_n = \Gamma$. We take as the unit of energy the inter-trap interaction strength $\lambda_c = 1$ and consequently time is measured in units of λ_c^{-1}.

Now the question is: how to identify entanglement between two bosonic systems?. By contrast with two qubit systems for which entanglement can be unambiguously identified by the concurrence [9], for two continuous variables as it is the case here, sufficient and necessary conditions for detecting entanglement have only been established for the special case of Gaussian states [10, 11]. However, in general the polariton trap state is a mixed non-Gaussian state.

We quantify the polariton entanglement by computing the logarithmic negativity $LN(\rho) = log_2||\rho^{PT}||$, where $||\rho^{PT}||$ is the trace norm of the partial transpose density operator [7]. This latter quantity can be easily evaluated as $||\rho^{PT}|| = 2N(\rho) + 1$ where $N(\rho)$, or negativity, is the absolute value of the sum of the negative eigenvalues of ρ^{PT}. Based on the exact analytical solution for the ME given in Eq.(8), the required transpose operation on the system's density operator is trivial for any time from which the negative eigenvalues of ρ^{PT} are numerically evaluated. For any separable or unentangled state $LN(\rho) = N(\rho) = 0$. Obviously, the entanglement fully vanishes for $\lambda_c = 0$. To focus the discussion and for demonstrational purposes we consider here the case of identical traps, i.e. $\omega_1 = \omega_2$ and $\lambda_1 = \lambda_2$, which are also identically prepared $\alpha_1 = \alpha_2$.

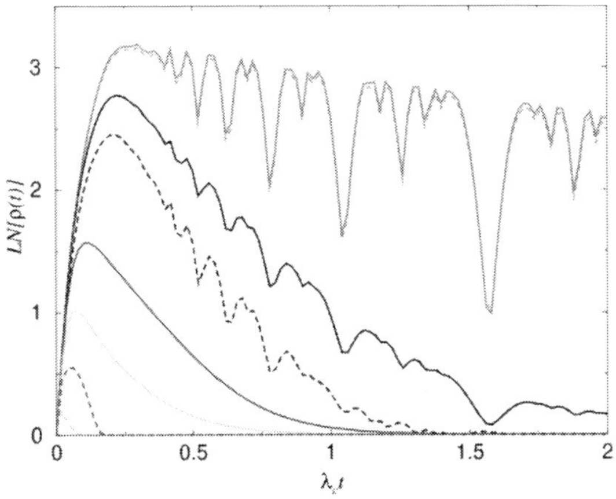

Figure 1. Logarithmic negativity as a function of time for different polariton self-interaction strengths and dissipation rates. See text for parameters.

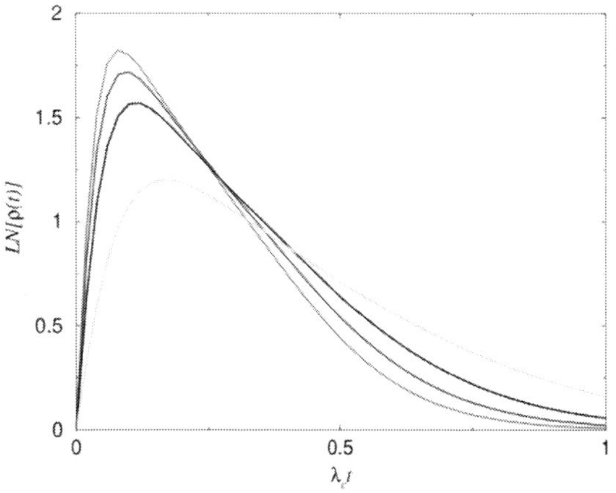

Figure 2. Logarithmic negativity as a function of time for different polariton populations. See text for parameters.

In Figure (1) we plot the time evolution of the polariton entanglement, $LN[\rho(t)]$, for a two trap cluster initially prepared with $|\alpha_j|^2 = 6$ (the amount of entanglement does not depend on the frequency ω_j of each trap). Four different values of dissipation rates $\Gamma = 0.01, 0.1, 1$ and 3 are considered and correspond in Fig.(1), from top to bottom, to colors red, black, blue and green, respectively. Additionally, for each decay rate, results for two extreme values of the ratio between the intra-trap and inter-trap interactions are displayed: $\lambda_j/\lambda_c = 1$ solid lines and $\lambda_j/\lambda_c = 10$ dashed lines (colors are still as described above). The first observation is that entanglement is immediately generated by the polariton-polariton interaction, λ_c, and it is preserved for a significant fraction of the time that polariton population in the traps is maintained. The cluster system starts its evolution in a separable state rising its entanglement degree in an almost linear way with a slope depending directly on the initial polariton population, a clear indication of how important is the effective non-linear polariton interactions for generating entanglement. At long times the system ends again in a separable state, as it should be because the traps become empty and thus the final state is nothing but the double vacuum state. At very low decay rates entanglement oscillations are visible due to the quasi-unitary evolution of inter-trap correlations governed exclusively by λ_c [8]. Note that in this limit of weak dissipation the entanglement dynamics does not depend on the intra-trap polariton interactions λ_j, as it is evident from the almost identical solid and dashed red curves in Fig.(1). As the dissipation becomes important populations and the entanglement decay faster. The polariton population follows a very simple exponential decay form given by $< \hat{a}_j^\dagger \hat{a}_j > (t) = |\alpha_j|^2 e^{-2\Gamma t}$ while for the entanglement a more pronounced dependence on the polariton self-interactions are evident. Thus, dissipation and polariton self-interactions jointly degrade significantly both the maximum attainable entanglement and the time interval where entanglement is visible.

Figure (2) shows the time evolution of $LN(\rho)$ for different polariton initial populations, peaks from top to bottom, $|\alpha_j|^2 = 10$ (red curve), $|\alpha_j|^2 = 8$ (blue curve), $|\alpha_j|^2 = 6$ (black curve) and $|\alpha_j|^2 = 3$ (green curve). Other parameters are: $\lambda_1 = \lambda_2 = -1$ and $\Gamma = 1$. It is evident that the rising time of the entanglement is shorter as the polariton population increases, emphasizing previous comments on the importance of inter-trap polariton-polariton interactions for an efficient and rapid generation of entanglement. However, on the other hand, the temporal width of the entanglement curve decreases as a function of the polariton population, indicating the high level of non-linear effects occurring in the present system.

Some numerical estimates are in order. According to Ref.[2] typical engineered trap radius range from 5 to 50 μm. Pumping pulse lasers of low intensity on the order of 1 mW give rise to a polariton density of $10^8 - 10^{10} cm^{-2}$. Thus, for experimentally achievable trap sizes, typically $10 - 100$ polaritons could be found in a single small trap. Our unit of time is fixed by the inter-trap (or inter-species) interaction strength λ_c. According to Ref.[3], $\lambda_c \sim 10 - 100\ \mu eV$, which for $\lambda_c t \sim 1$ yields to $t \sim 40\ psec$. Hence, the theoretical results we discuss above are in the lower range of experimentally accessible polariton densities and time resolution, requiring very clean microcavities samples and high quality factors Q, for entanglement detection to be observable. However, our analytical results should be indeed valuable as a starting point for more numerically involved calculations in realistic situations.

4. Conclusions

By solving exactly the Master equation for an interacting trapped polariton system we have found that polariton entanglement can be immediately produced from an initial separable multi-coherent state. The entanglement starts rising linearly with a slope depending on the initial trap populations. The entanglement maximum is an increasing function of the initial number of polaritons. However, the entanglement duration time shortens as the initial trap population becomes large. In any case polariton entanglement observation would require very good microcavities with weak dissipation rates.

LQ acknowledges MEC(Spain) for a sabbatical grant and Universidad Autónoma de Madrid-Spain for hospitality during the development of the present work. LQ also acknowledges partial financial support through the Research Project Program 2007-2008 of the Facultad de Ciencias-UniAndes. This work has been supported in part by the Spanish MEC under contract QOIT Consolider-CSD2006-0019 and by CAM under contract S-0505/ESP-0200.

References

[1] Balili R et al. 2007 *Science* **316** 1007
[2] Kim N Y et al. 2008 *Phys.Stat.Sol.* (b) **245** 1076
[3] Na N and Yamamoto Y 2008 *Preprint* quant-ph/0804.1829
[4] Rodríguez F J, Quiroga L, Tejedor C, Martín M D, Viña L and André R 2008 *Phys.Rev.* B **78** 035312
[5] Ban M 1992 *J.Math.Phys.* **33** 3213
[6] Moya-Cessa H 2006 *Phys.Reports* **432** 1
[7] Vidal G and Werner R F 2002 *Phys.Rev.* A **65** 032314
[8] Quiroga L and Tejedor C *to be published*
[9] Wootters W K 1998 *Phys.Rev.Lett.* **80** 2245
[10] Duan L-M, Giedke G, Cirac J I and Zoller P 2000 *Phys.Rev.Lett.* **84** 2722
[11] Simon R 2000 *Phys.Rev.Lett.* **84** 2726

Space-time evolution of Gaussian wave packets through superlattices containing left-handed layers

P. Pereyra[1], A Robledo-Martinez [2] and M Romero-Serrano[1]

[1] Departamento de Ciencias Básicas, Universidad Autónoma Metropolitana-Azcapotzalco, México DF, México

[2] Departamento de Energía, Universidad Autónoma Metropolitana-Azcapotzalco, México DF, México

E-mail: ppereyra@correo.azc.uam.mx

E-mail: a.robledo@mailaps.org

Abstract. We study the space-time evolution of Gaussian electromagnetic wave packets moving through $(L/R)^n$ superlattices, containing alternating layers of left and right-handed materials. We show that the time spent by the wave packet moving through arbitrary $(L/R)^n$ superlattices are well described by the phase time. We show that in the particular case where the thicknesses $d_{L,R}$ and indices $n_{L,R}$ of the layers satisfy the condition $d_L|n_L| = d_R n_R$, the usual band structure becomes a sequence of isolated and equidistant peaks with negative phase times.

1. Introduction

The possibility of having negative refraction index (left-handed) materials (with electric permittivity ϵ and magnetic permeability μ simultaneously negative), as a simple extension of the well established electromagnetic systems, has become lately an active and controversial research field, where striking and new effects are expected. Veselago[1] already predicted in 1968 some of the unique properties of EM wave propagation in left-handed media (LHM): a) that the waves appear to propagate towards the source and not away from it, b) that their group velocity is negative and, c) because waves incident on right-handed/left-handed (RH/LH) interfaces are refracted to the same side of the normal, converging and diverging lenses exchange their roles. Furthermore, Veselago proposed the constraints:

$$\frac{\partial \epsilon\omega}{\partial\omega} > 0 \qquad \frac{\partial \mu\omega}{\partial\omega} > 0 \tag{1}$$

with ϵ and μ depending on the frequency, for the energy transferred from the source to the load to be positive and to avoid causality violations. Smith and Kroll[2] on the other hand maintain that while a reversed k resembles a time-reversed propagation towards the source, the work done is nevertheless positive. Incidentally, by analyzing the implications of Eqs. (1) these authors reach the conclusion that the group velocity ought to be positive for both types of materials. A simplified approach can easily lead to the conclusion that the sign in k causes opposing group velocities.

Besides the blazing presumption of perfect lenses[3], and problems like the sign selection and directions of motion of both the energy and the electromagnetic field, it has been previously shown [4] that the transmission amplitude

$$t_L = \left[\cos\left(k_1 d_1 \cos\theta_1\right) + \frac{i}{2|n_1|n_2}\left(n_2^2\frac{\cos\theta_1}{\cos\theta_2} + n_1^2\frac{\cos\theta_2}{\cos\theta_1}\right)\sin\left(|k_1|d_1\cos\theta_1\right)\right]^{-1} = |t_L|e^{-i\theta_t} \quad (2)$$

of a single left-handed slab is just the complex conjugate of the transmission amplitude

$$t_R = \left[\cos\left(k_1 d_1 \cos\theta_1\right) - \frac{i}{2n_1 n_2}\left(n_2^2\frac{\cos\theta_1}{\cos\theta_2} + n_1^2\frac{\cos\theta_2}{\cos\theta_1}\right)\sin\left(k_1 d_1\cos\theta_1\right)\right]^{-1} = |t_L|e^{i\theta_t} \quad (3)$$

of a similar but right-handed slab[4]. An important consequence of this property is that the phase time τ of a single slab, defined as the frequency derivative of the transmission-amplitude's phase θ_t, becomes the negative of the corresponding phase time of a right-handed slab, which arises concerns of a possible causality violation. Gupta et al. studied also the transmission of electromagnetic pulses across a parallel slab of a medium where ϵ and μ are functions of the frequency ω. For the frequency range where both quantities are negative, they found a positive phase time[5]. We are interested here on this problem. As will be seen in the next section, the analysis in multilayered structures (MLS) might be more involved. It is no longer true that, except for specific cases, the transmission amplitude t_L of a MLS containing left-handed layers (LHL's) will be the complex conjugate of t_R. Therefore the phase times are not necessarily negative. But what happens when the phase time is negative?.

In this paper we will discuss the transport and transmission time of Gaussian electromagnetic wave packets through $(L/R)^n$ superlattices (LRSL's), where left-hand layers (LHL's) of refractive index $n_1 \equiv n_L = -|n_L|$ and width d_L, alternate with right-handed layers (RHL's) having refractive index $n_2 \equiv n_R$ and width d_R. We will show that in the particular case where $d_L|n_L| = d_R n_R$, the usual transmission coefficient band structure becomes a sequence of isolated and equidistant resonances (IER).

2. Transport of EM waves in an $air(L/R)^n air$ system

In a $(L/R)^n$ superlattice bounded by semi-infinite air layers, an electromagnetic wave moving across it has a total transmission amplitude t_T given by

$$t_T = \frac{1}{\alpha_T} \quad (4)$$

with (assuming $\mu_L = \mu_R \simeq \mu_o$):

$$\alpha_T = \alpha_{nr} + i\left(\frac{1+n_R^2}{2n_R}\alpha_{ni} + \frac{n_R^2-1}{2n_R}\beta_{ni}\right) \quad (5)$$

and[6]

$$\alpha_{nr} = U_n - \alpha_r U_{n-1} \qquad \alpha_{ni} = -\alpha_i U_{n-1} \quad (6)$$

$$\beta_{nr} = \beta_r U_{n-1} \qquad \beta_{ni} = \beta_i U_{n-1} \quad (7)$$

Here α_r and α_i, β_r and β_i are real and imaginary parts of

$$\alpha = e^{ik_R d_R\cos\theta_R}\left(\cos\left(k_L d_L\cos\theta_L\right) + \frac{i}{2n_L n_R}\left(n_R^2\frac{\cos\theta_L}{\cos\theta_R} + n_L^2\frac{\cos\theta_R}{\cos\theta_L}\right)\sin\left(k_L d_L\cos\theta_L\right)\right), \quad (8)$$

$$\beta = \frac{ie^{ik_R d_R \cos\theta_R}}{2n_L n_R}\left(n_L^2\frac{\cos\theta_R}{\cos\theta_L} - n_R^2\frac{\cos\theta_L}{\cos\theta_R}\right)\sin\left(k_L d_L \cos\theta_L\right), \tag{9}$$

and U_n is the Chebyshev polynomial of the second kind and order n, evaluated at α_r.

It is apparent from these results that, in a system like the one considered here, it is not as simple to conclude in general, as it was for a single slab, whether the phase time sign becomes negative or not. It depends on many factors. In the next section we will present the space-time evolution of specific Gaussian wave packets whose centroid will be defined at qualitatively different points. We will consider two examples. In the first one we will have a LRSL with $d_L = d_R$, $|n_L| \neq n_R$, and the Gaussian centroid in the photonic gap. In the second example we will have a LRSL with layer thickness and refraction parameters such that $d_L|n_L| = d_R n_R$. In this case we choose the Gaussian-wave-packet centroid coinciding with one of the isolated resonances.

3. Space-time evolution of Gaussian WP

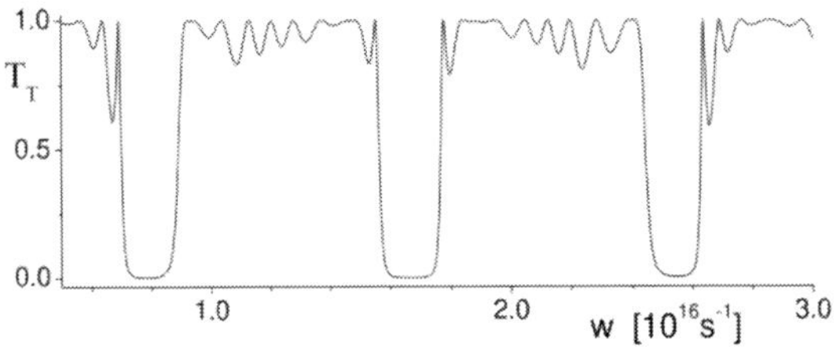

Figure 1. Transmission coefficient as function of the wave's angular frequency. The fields are assumed to move through the superlattice $(L/R)^9$ with $n_L = -2.22$, $n_R = 1.41$, $d_L = d_R = 140$nm.

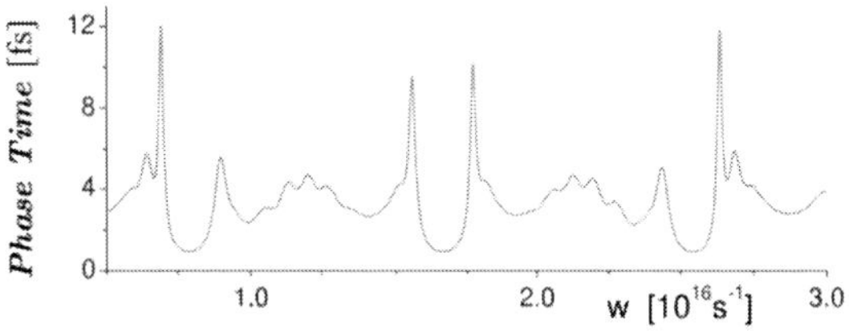

Figure 2. Phase time for the same superlattice of Figure 1, containing left-handed layers. The phase time predictions are positive.

It is well known that the superlattice phase time, generally follows the characteristic resonant band structure of the superlattice transmission coefficients[7]. However, based on the examples presented here, we can state that positive and negative phase times are likely possible. In the first example (see Figures 1 and 2), the phase time is positive for all frequencies. It will be shown that a WP defined inside the photonic gap will behave very much like in a right handed superlattice[7]. In the second example, we will have an IER LRSL, which for frequencies around

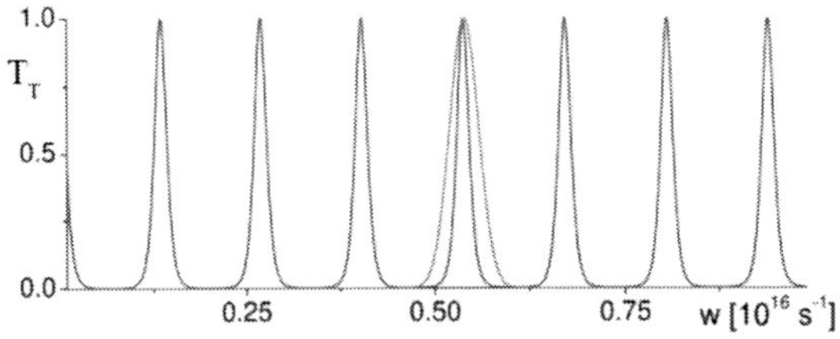

Figure 3. Transmission coefficient as function of frequency. The fields are assumed to move through the superlattice $(L/R)^9$ with $n_L = -2.22$, $n_2 = 1.41$, $d_L = d_R n_R / n_L = 316.5$nm.

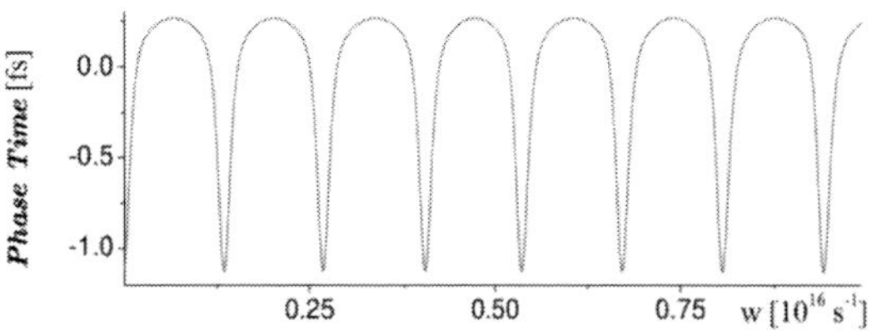

Figure 4. Phase time for the superlattice in Figure 3. The phase times obtained near the resonant frequencies are negative.

the isolated resonances, has negative phase times (see Figures 3 and 4). We will study the transmission of an electromagnetic WP through this system. At $t = 0$, we will locate the wave-packet centroid at a distance z_o from the left edge of the LRSL. Moving with the light velocity, the WP centroid reaches the left edge at $t_a = z_o/c$, therefore a snapshot of the wave packet at $t_b = 2t_a + \tau$, with τ the transmission time, should show the transmitted WP on the other side, at a distance z_o from the right edge of the IER LRSL. We will show that even that τ is negative, the electromagnetic WP at t_b will certaily appear at $z = z_o + L$, with L the IER LRSL length. We shall now discuss in detail these specific examples.

3.1. The Gaussian WP in arbitrary $(\mathrm{L/R})^n$ *superlattices*

To simplify the discussion, let us consider a WP moving to the right in the parallel polarization and with normal incidence. At $t = 0$ the centroid of the Gaussian packet is at $-z_o$. If we choose $d_L = d_R = 140$nm with $n_L = -2.22$ and $n_R = 1.41$, the transmission coefficient has the resonant bandstructure shown in figure 1, as function of the frequency w, and the phase time shown in figure 2.

As shown in Fig. 1, there is a wide photonic gap between $w = 1.6\,10^{16}\mathrm{s}^{-1}$ and $w = 1.72\,10^{16}\mathrm{s}^{-1}$. The Gaussian WP is placed at the center of this gap. In figure 5 we plot the WP at different instants of time. The phase time predicts a positive transmission time τ of 0.9223 fs for the centroid, which is much less than the time that would require to cross the LRSL were it to move at the speed of light (5.6 ps for a length of 1680 nm). Since the WP is defined almost completely inside the gap, there will be no distortion after it passes or gets reflected from the metamaterial superlattice.

In figure 5 we show snapshots of the WP at different times. In the upper frame (at $t = 0$) the WP is moving with group velocity $v_g = c$ towards the metamaterial superlattice. At $t = z_0/v_g$ the wave-packet centroid reaches the left edge of the LRSL. The WP spends some time (the transmission time τ) inside the superlattice and, at $t = z_0/v_g + \tau$ it begins to leave the LRSL. In the bottom frame the wave packet is plotted when presumably it is back to $-z_0$ or transmitted at $z = L + z_0$.

3.2. The Gaussian WP in the IER $(\mathrm{L/R})^n$ *superlattice*

A different behavior in the transport properties of metamaterial superlattices $(\mathrm{L/R})^n$ occurs when the layer-widths fulfill the relation $d_L = d_R n_R/|n_L|$. We choose $d_R n_R = \lambda_0/4$ with λ_0 the wave length at the center of the photonic gap. In figure 3 we plot the transmission coefficient, it is clear that the band structure collapses into a periodic sequence of single resonances, independent of the number of cells in the LRSL. In this case the phase time is negative for frequencies around the resonances (see figure 4). To visualize this property and to see whether we have some causality problem, we shall study the evolution of a Gaussian wave packet defined precisely at a resonance, as shown in figure 3. In figure 6 we plot the packet snapshots as the time t increases. Since the phase time prediction is negative ($\tau = -1.1129$fs), we have $z_0/v_g + \tau < z_0/v_g + \tau/2 < z_0/v_g$. As mentioned before, one would expect that at $t_a = z_0/v_g$, the WP will be arriving at the left of the LRSL, and leaving it at $t_l = z_0/v_g + \tau$. Suppose that the WP is leaving the SL at t_l, at latter times like $z_0/v_g + \tau/2 = z_0/v_g + \tau + |\tau|/2$ and $z_0/v_g = z_0/v_g + \tau + |\tau|$ the WP should be farther away. Although it is difficult to distinguish the differences between the three graphs in the middle, one can perceive that, for $t = z_0/v_g$, part of the reflected wave packet is growing at the left hand side of the superlattice as part of the reflected WP, which is much clearly seen later at $t = 2z_0/v_g + \tau$. In fact, if the phase time prediction is correct, the WP at $t = 2z_0/v_g + \tau$ should be back in the position where it was at $t = 0$ (when it is reflected), or at $z = L + z_0$ defined by the peak of the Gaussian pink curve at the right (when it is transmitted). As can be seen in the graph at the bottom, this is precisely what happens. The WP seems to travel from the far edge of the LM layers towards the source

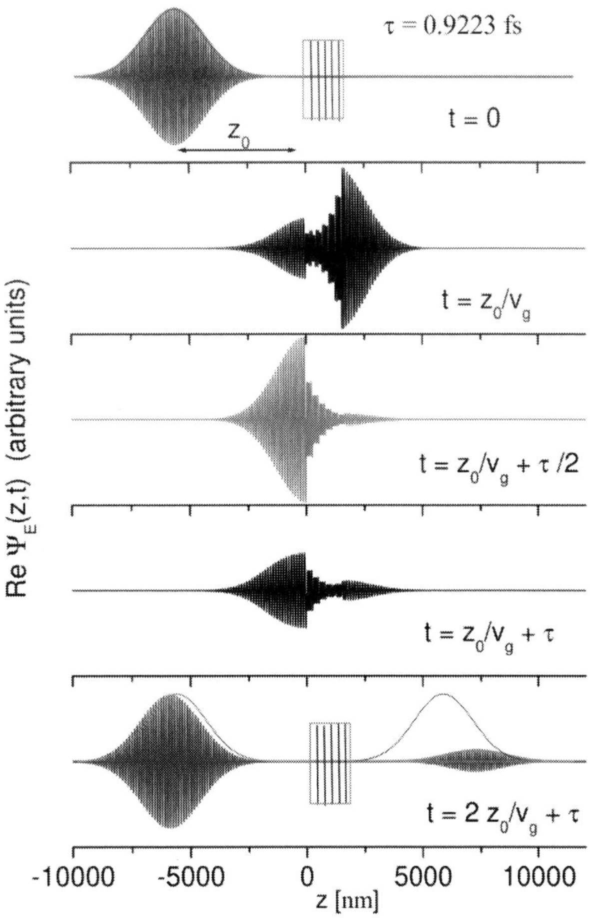

Figure 5. Time series of a WP defined in the gap with centroid at $\omega = 1.66\,10^{16}\text{s}^{-1}$. At $t = 0$s, the WP moves to the right with group velocity $v_g = c$. The phase time predicts a transmission time $\tau = 0.9223$fs through a superlattice $(\text{L/R})^6$. The WP is shown also when the centroid is presumably touching, inside and when it is presumably leaving the LRSL. At $t = 2z_0/v_g + \tau$ the reflected and the transmitted WP's should be in the position of the pink Gaussians. The black Gaussian envelope on the RHS of the bottom frame indicates the position where the WP would be were it to move with $v_g = c$. The rectangular inset shows the spatial extent of the LRSL.

before it reaches the LRSL, just as predicted by Veselago. This is the main result that we report in this paper.

4. Conclusion

We have studied the space-time evolution of Gaussian wave packets through left- and right-handed $(\text{L/R})^n$ superlattices. We have shown that the phase time describes the transmission time on these systems. We have shown also that, even though the phase time of a single LHM slab is the negative of the RHM slab phase time, the $(\text{L/R})^n$ superlattice phase times are generally positive. We have found that in the particular $(\text{L/R})^n$ superlattice with $d_L|n_L| = d_R n_R$, the usual band structure of the transmission coefficient becomes a sequence of isolated and equidistant peaks with negative phase times. Our results show that in this case no apparent

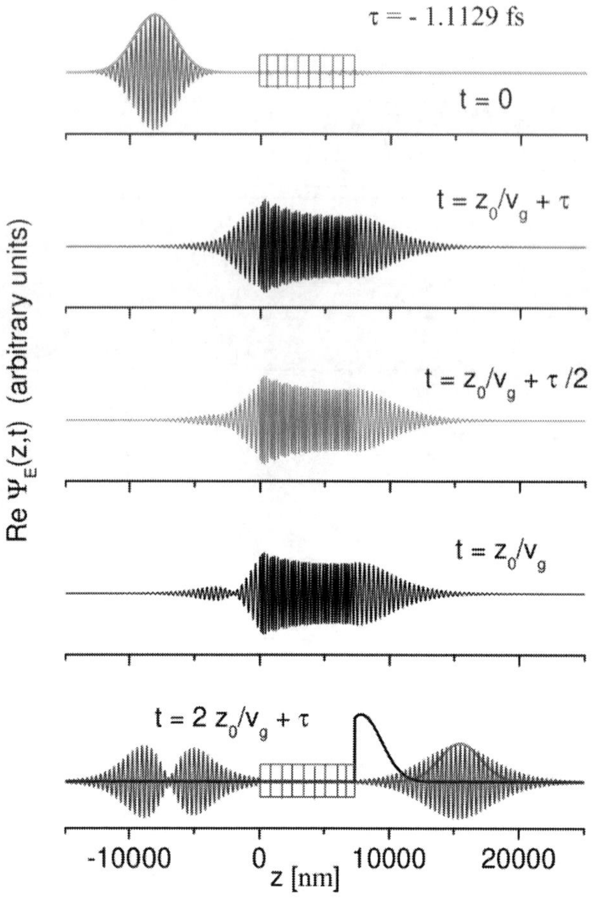

Figure 6. Time series of a WP defined in a resonance with centroid at $\omega = 5.38\,10^{16}\mathrm{s}^{-1}$. At $t = 0\mathrm{s}$, the WP moves to the right with group velocity $v_g = c$. The phase time predicts a transmission time $\tau = -1.11293\mathrm{fs}$ through a superlattice $(\mathrm{L/R})^9$. The WP is shown also at $t_a = z_0/v_g + \tau$, at $t = z_0/v_g + \tau/2$ and at $t_l = z_0/v_g$. Because of the negative sign of the phase time the sequence is inverted and it is interesting to see that at $t = 2z_0/v_g + \tau$ the transmitted WP coincides with the pink Gaussian peak at $z_o + L$.

causality violation occurs. The negative phase time of left-handed slabs deserves further research.

[1] V G Veselago 1968 *Sov. Phys. Usp.* **10** 509.
[2] D R Smith and N. Kroll 2000 *Phys. Rev. Lett.* **85** 2933.
[3] J B Pendry 2000 *Phys. Rev. Lett.* **85** 3966
[4] P Pereyra, A Robledo-Martinez and M Morales-Luna 2008 *Microelectronics Journal* **39** 394.
[5] S D Gupta, R Arun and G S Agarwal 2004 *Phys. Rev. B* **69** 113104
[6] P Pereyra 1998 *J. Phys A: Math. Gen.*, **31** 4521.
[7] P Pereyra and H P Simanjuntak 2007 *Phys. Rev. B* **75** 056604.

XIX Latin American Symposium on Solid State Physics (SLAFES XIX) IOP Publishing
Journal of Physics: Conference Series **167** (2009) 012027 doi:10.1088/1742-6596/167/1/012027

A simple model for conduction band states of nitride-based double heteroestructures

L. M. Gaggero-Sager, M. E. Mora-Ramos

Facultad de Ciencias, Universidad Autónoma del Estado de Morelos, Av. Universidad 1001, CP 62209, Cuernavaca, Morelos, Mexico

E-mail: lgaggero@uaem.mx

Abstract. In this work we propose an analytical expression for the approximate modeling of the potential energy function describing conduction band bending in III-V nitride quantum wells. It is an alternative approach to the self-consistent Poisson-Schödinger calculation. The model considers the influence of the many electron system and the built-in electric field inside the well. Hartree and exchange contributions are included along the lines of a local-density Thomas-Fermi-based theory. The effects due to the modulated doping in the barriers is also considered. We report the calculation of the energy spectrum as a function of several input parameters: alloy composition in the barriers, barrier doping concentration, and quantum well width. Our results could be of usefulness in the study of optoelectronic properties in this kind of systems.

1. Introduction

The wurtzite nitrides exhibit high values of the piezoelectric constants as well as large spontaneous polarizations due to their lack of central symmetry in the crystal unit [1, 2, 3, 4, 5]. This fact has a consequence of special interest: the formation of two-dimensional charge densities at the interfaces, when a heterostructure made of these materials is formed. This kind of charge density is induced by the electric polarization resulting from the superposition of the spontaneous and piezoelectric polarization fields of both materials across each interface.

In the particular case of heterostructures, large built-in electric fields form in wurtzite III-V nitrides thin films and multilayers. Such fields originate from heterointerface discontinuities of the macroscopic bulk polarization of the nitrides [6, 7]. It has been found that the strain and the strain-induced piezoelectric field significantly alter the subband structure [8, 9]. For instance, the optical properties of GaInN/GaN/AlGaN quantum wells are governed to a large extent by the spontaneous and piezoelectric polarization fields arising from the strongly polar nature of wurtzite nitrides. In multiple quantum wells the electric fields lead to spatially highly indirect transitions between nearest and next nearest neighbour quantum wells [10]. On the other hand, there are some works in which it is shown that of the macroscopic polarization fields can be screened via modulation doping in nitride quantum wells [11].

In this work we present a simple analytical formulation describing the one-dimensional conduction band potential profile in nitride-based quantum wells with the prototype AlGaN/GaN/AlGaN. The presence of polarization-induced interface charge densities as well as the influence of intentional doping in the barrier regions are explicitly taken into account in deriving the confining potential function. In the derivation, the one-dimensional local density Thomas-Fermi approach is used to generate the expression for the quantum well bottom profile.

2. Model and results

Let us assume that the polarization-induced surface charge densities at the interfaces of the AlGaN/GaN quantum well are given by $\sigma_1\delta(z)$ and $\sigma_2\delta(z-l_{qw})$, where l_{qw} is the width of the well region. In addition, two selectively doped regions with three-dimensional ionized donor impurity density N_d, are created in both AlGaN barriers, separated from the well interfaces via spacer layers of width d. That is, the region located at the left of $z=0$ extends from $z=-d$ to $z=-L_1-d$. The doped region located at the right of $z=l_{qw}$ extends from $z=l_{qw}+d$ to $z=L_2+l_{qw}+d$. Both doped regions will have their respective width lengths L_1 and L_2 given by the depletion approximation. Taking into account these density charge distribution in the system, the proposed form for the potential function in the system is derived using the method presented in refs. [12, 13]:

$$V(z)=\begin{cases} V_b^I, & z\leq -L_1-d;\\ \frac{4\pi}{\varepsilon}N_d(d+L_1+z)^2+V_b^I, & -L_1-d\leq z\leq -d;\\ \frac{8\pi}{\varepsilon}N_dL_1(d+\frac{L_1}{2}+z)+V_b^I, & -d\leq z\leq 0;\\ -\frac{V_{qw}^I}{\left(\frac{\eta}{z_{oI}}z+1\right)^4}-\frac{V_{qw}^{II}}{\left(\frac{\eta}{z_{oII}}(l_{qw}-z)+1\right)^4}, & 0\leq z\leq l_{qw};\\ \frac{8\pi}{\varepsilon}N_dL_2(d+\frac{L_2}{2}+z)+V_b^{II}, & l_{qw}\leq z\leq l_{qw}+d;\\ \frac{4\pi}{\varepsilon}N_d(-d+L_2-l_{qw}+z)^2+V_b^{II}, & l_{qw}+d\leq z\leq l_{qw}+d+L_2;\\ V_b^{II}, & z\geq l_{qw}+d+L_2. \end{cases} \tag{1}$$

In this expression all the quantities are expressed in effective atomic units (Rydberg energy and Bohr radius) related with the GaN. In this sense, $\varepsilon=\epsilon_{AlGaN}/\epsilon_{GaN}$ has the meaning of a relative dielectric constant, whereas;

$$\eta=\frac{2}{15\pi},\qquad z_{oj}=\left(\frac{\eta^3}{2\pi N_{eff}^j}\right)^{1/5},\qquad N_{eff}^j=\sigma_j+N_dL_j\,; \tag{2}$$

where the index j labels barrier regions I and II. The amplitudes of the Thomas-Fermi potentials that combine within the well region are given by:

$$V_{qw}^j=\frac{\eta^2}{z_{oj}^4}. \tag{3}$$

To specify the set of undetermined constants appearing in equation (1), it is necessary to impose certain restrictive conditions. In first place, we must obtain that for $z = 0$ and $z = l_{qw}$, the height of the barrier is equal to V_o, which is no other than the value for the conduction band offset in the heterostructure, provided the origin for the energy is taken at the conduction band bottom in the GaN layer. With this in mind, it is found that:

$$V_b^j = V_o - \frac{8\pi}{3} N_d L_j \left(\frac{L_j}{2} + d \right) - V_{qw}^{i \neq j} - \frac{V_{qw}^j}{\left(\frac{\eta}{z_{oj}} l_{qw} + 1 \right)^4} \tag{4}$$

The second set of restrictions are related to the charge conservation in the structure. This is attained if:

$$V'(z_{lk}^-) - V'(z_{lk}^+) = \frac{4\pi}{\varepsilon} \sigma_k, \qquad k = 1, 2\,. \tag{5}$$

The points at which the derivatives of the potential are evaluated are $z_{l1} = 0$, and $z_{l2} = l_{qw}$. This conditions determine, self-consistently, the values of the depletion lengths L_1 and L_2.

Now we discuss some properties of the potential function derived with this scheme.

- It is obvious that $\sigma_1 \neq \sigma_2$. They are going to be equal only when $l_{qw} \to \infty$. Both 2D charge densities depend upon the strain, and this changes depending on which material the substrate layer is made up, and on the value of its width.
- The value of the built-in electric field inside the well is

$$V(0^+) - V(l_{qw}^-) = -V_{qw}^I - \frac{V_{qw}^{II}}{\left(\frac{\eta}{z_{oII}} l_{qw} + 1 \right)^4} + \frac{V_{qw}^I}{\left(\frac{\eta}{z_{oI}} l_{qw} + 1 \right)^4} + V_{qw}^{II}. \tag{6}$$

This will be equals to zero when $l_{qw} \to \infty$, due to we have $\sigma_1 = \sigma_2$, $L_1 = L_2$, and $z_{oI} = z_{oII}$. In this limit we recover the potential profile of two isolated single heterostructures.

At this point it is convenient to state that this model is also valid in the case of modulation-doped double heterostructures with no polarization-induced built-in intrawell electric fields associated. In that case, we could set, for instance, $\sigma_1 = \sigma_2 = Nd_L$ $(L_1 = L_2)$ from the beginning. Under such condition, the obtained potential well profile is completely symmetric, with $V'(l_{qw}/2) = 0$.

Figure 1 shows the potential well profile, given in units of 10^{-3} eV, for a AlN/GaN/AlN double heterostructure. It has been assumed that the barriers are homogeneously doped with a donor concentration of $N_2 = 5 \times 10^{18}\,\text{cm}^{-3}$ [12]. Besides, the strain relaxation parameter is taken to be unity [13]. Three different values of the well width are considered: 100, 150, and 200 Å. It can be seen that the value of the internal electric field is significantly dependent on that parameter as predicted in equation (6).

3. Conclusions

In this work we have presented a model analytical proposal for the description of conduction band bending in nitride double heterostructures. Together with its simplicity, it provides appropriate limiting cases as well as an expression for the built-in electric field in the quantum well. A previous version of this model, applied to nitride single heterostructures have proven to be a reliable and quite accurate approach to provide very good quantitative agreements with detected single electron states in AlGaN/GaN field effect transistors [14]. Therefore, it is of interest to continue exploiting the approach here introduced. The results of that work will be published later on.

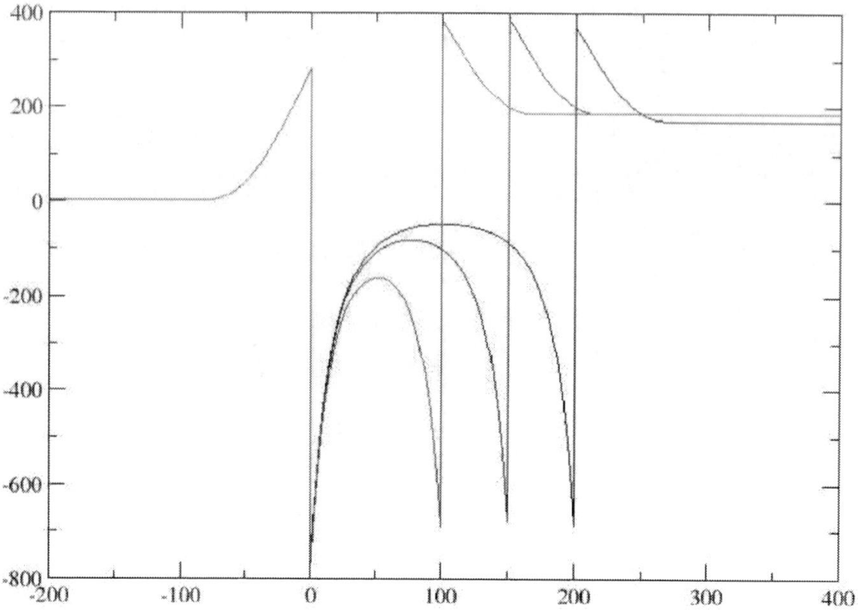

Figure 1. Potential well profile (in meV) as a function of the position (in Å) for a AlN/GaN/AlN double heterostructure, derived within the model proposed in the present article. Values of the well width included are 100, 150, and 200 Å

Acknowledgments

M.E.M.R. acknowledges support from Mexican CONACYT through Grant No. CB-80846/2007.

References

[1] Morkoç H 1999 *Nitride Semiconductors and devices* (Berlin: Springer)
[2] Bernardini F and Fiorentini V 1999 *phys. stat. solidi (b)* **216** 391
[3] Ambacher O, Dimitrov R, Stutzmann M, Foutz B E, Murphy M J, Smart J A, Shealy J R, Weimann N G, Chu K, Chumbes M, Green B, Sierakowski A J, W. J. Schaff W J and Eastman L F 1999 *phys. stat. solidi (b)* **216** 381
[4] Ambacher O, Foutz B, Smart J, Shealy J R, Weimann N G, Chu K, Murphy M, Sierakowski A J, Schaff W J, Eastman L F, Dimitrov R, Mitchell A and Stutzmann M 2000 *J. Appl. Phys.* **87** 334
[5] Foutz B E, Ambacher O, Murphy M J, Tilak V and Eastman L F 1999 *phys. stat. solidi (b)* **216** 415
[6] Fiorentini V, Bernardini F, Della Sala F, Di Carlo A and Lugli P 1999 *Phys. Rev. B* **60** 8849
[7] Coffey D and Bock N 1999 *Phys. Rev. B* **59** 5799
[8] Xiao D, Kim K W and Zavada J M 2004 *J. Appl. Phys.* **96** 723
[9] Wan S, Xia J and Chang K 2001 *J. Appl. Phys.* **90** 6210
[10] Hangleiter A 2000 *Proc. SPIE* **3944** 58
[11] Di Carlo A, Della Sala F, Lugli P, Fiorentini V and Bernardini F 2000 *Appl. Phys. Lett.* **76** 3950
[12] Gaggero-Sager L M, Mora-Ramos M E and Velasco V R 2003 *Surf. Sci.* **546** 39
[13] Mora-Ramos M E and Gaggero-Sager L M 2000 *phys. stat. solidi (b)* **220** 175
[14] Mora-Ramos M E, Tutor J and Velasco V R 2006 *J. Appl. Phys.* **100** 123708

Journal of Physics: Conference Series **167** (2009) 012028 doi:10.1088/1742-6596/167/1/012028

Tight-binding study of the hole subband structure properties of p-type delta-doped quantum wells in Si by using a Thomas-Fermi-Dirac potential

I Rodríguez-Vargas, J Madrigal-Melchor and S J Vlaev

Unidad Académica de Física, Universidad Autónoma de Zacatecas, Calzada Solidaridad Esquina Con Paseo La Bufa S/N, 98060 Zacatecas, ZAC., México.

E-mail: isaac@planck.reduaz.mx

Abstract. We present the hole subband structure of p-type delta-doped single, double, multiple and superlattice quantum wells in Si. We use the first neighbors sp^3s^* tight-binding approximation including spin for the hole level structure analysis. The parameters of the tight-binding hamiltonian were taken from Klimeck et al. [Klimeck G, Bowen R C, Boykin T B, Salazar-Lazaro C, Cwik T A and Stoica A 2000 *Superlattice. Microst.* **27** 77], first neighbors parameters that give realiable results for the valence band of Si. The calculations are based on a scheme previously proposed and applied to delta-doped quantum well systems [Vlaev S J and Gaggero-Sager L M 1998 *Phys. Rev. B* **58** 1142]. The scheme relies on the incorporation of the delta-doped quantum well potential in the diagonal terms of the tight-binding hamiltonian. We give a detail description of the delta-doped quantum well structures, this is, we study the hole subband structure behavior as a function of the impurity density, the interwell distance of the doped planes and the superlattice period. We also compare our results with the available theoretical and experimental data, obtaining a reasonable agreement.

1. Introduction

Doping plays a fundamental role in semiconductor physics due to the physical properties of semiconductor systems can be easily modified through the incorporation of impurities. Among the most important doping techniques is the so called delta-doping, which has as key features the amount of charge that can be reached and the precision of the impurity seeding (up to atomic scale). With this technique it is possible to obtain bidimensional doping densities (p_{2D}) above 10^{12} cm^{-2} and doping profile widths of tenth of angstroms, these factors causes a strong electric field that confines the charge carriers in a V-shaped potential or better known as delta-doped quantum well. Studies on p-type B δ-doped Si quantum wells have been reported [1, 2, 3, 4, 5]. Due to the technological importance of Si and the particular properties of the δ-doped systems, they appear to be of interest for application in the electronic device industry and for basic investigation as well. In this particular, the B δ-doped Si QW has been investigated from the theoretical and experimental points of view [1, 2, 3, 4, 5, 6, 7, 8, 9, 10]. It is clear that in the case of p-type systems the simultaneous consideration of the heavy and light hole bands is unavoidable because the electric charge is distributed among them. Some works [5, 6] have shown that, even at $T = 0$ K, the contribution of the split-off(so) band turns out to be important when the value of the impurity density p_{2D} is greater or equal to 2×10^{13} cm^{-2}. The reason for this effect to

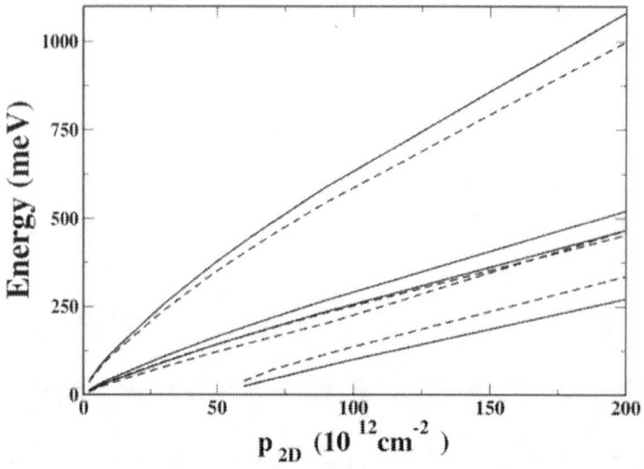

Figure 1. Hole subband structure versus doping density of SDD quantum wells in Si.

occur relies in the relative population of the different bands: for values of p_{2D} above 5×10^{13} cm^{-2}, the relative occupation number of the so_0 level is greater than the corresponding to the lh_0 level, and ranks as the second after the hh_0 population.

In the present work we study the hole subband structure of p-type delta-doped quantum wells in Si. We use a Thomas-Fermi-Dirac scheme and first neighbors sp^3s^* tight-binding approximation (including spin) for the modeling of the confining potential and the hole subband structure, respectively. We have implemented tight-binding parameters that are suitable for the description of the valence band of Si [11]. We analyze the level structure behavior of single, double and superlattice delta-doped (SDD, DDD and SLDD) quantum wells as a function of the paramount parameters in each of these systems. We find that the split-off band influences the hole subband structure, even for low impurity densities, in a considerable way. We also compare our results with the theoretical and experimental data available obtaining a reasonable agreement.

2. Mathematical method

The scheme of calculation starts by modeling the valence band profile, within the local density Thomas-Fermi-Dirac approximation. The outcome of this approach is an analytical expression for the one-dimensional potential energy function describing the band bending in a SDD quantum well considering the exchange effects [12],

$$V^*(z) = \frac{\alpha^2}{(\alpha\,|z| + z_0)^4} - \frac{2\varsigma^2(w)m_a}{\pi^2}\left[1 - \sqrt{1 + \frac{\pi^2}{\varsigma^2(w)m_a}\frac{\alpha^2}{(\alpha\,|z| + z_0)^4}}\right], \tag{1}$$

where α and z_0 are constants that come in terms of the effective masses, dielectric constant and the impurity density, and ς takes account of the coupling between the light and heavy hole bands [12].

The corresponding expression for $V^*(z)$ in DDD and SLDD is constructed with a suitable combination of the potential functions of multiple delta-wells and is incorporated to the diagonal terms of the tight-binding Hamiltonian [13],

$$TB_{ii}(n) = TB_{ii}(0) + V^*(n) \tag{2}$$

where the potential $V^*(n)$ is the potential $V^*(z)$ written in discrete coordinates, in which n numbers the atomic layers.

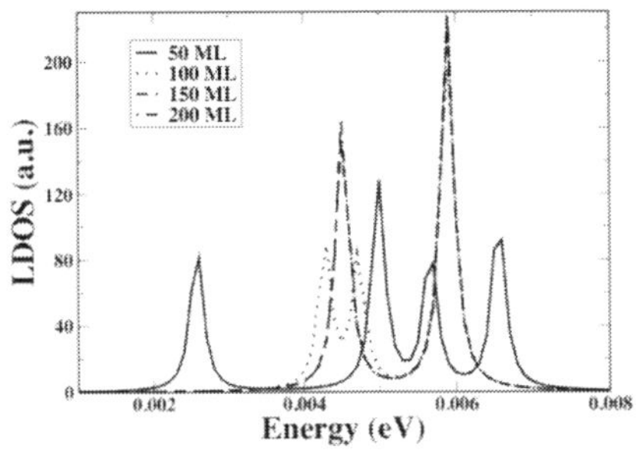

Figure 2. Local denisty of states versus energy of DDD quantum wells in Si. The doping density is fixed at 1×10^{12} cm^{-2}, while the interwell distance takes values of 50, 100, 150 and 200 ML's.

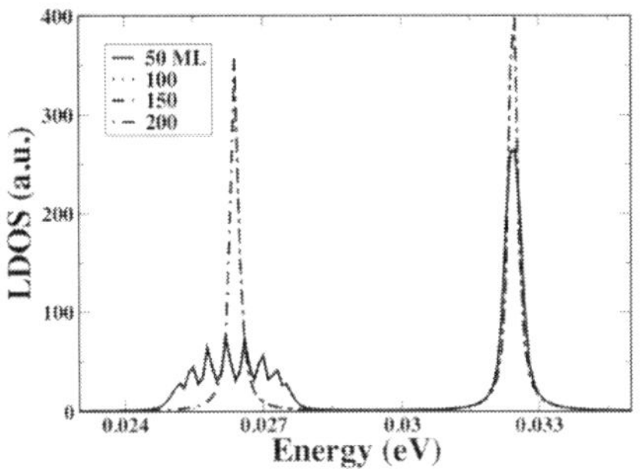

Figure 3. Local density of states versus energy of SLDD quantum wells in Si. The doping density is fixed at 8×10^{12} cm^{-2}, while the interwell distance takes values of 50, 100, 150 and 200 ML's.

3. Results and discussion

We use a set of typical input parameters for the confining potential [14]. At first, we analyze the subband hole level structure of SDD quantum wells taking into account the split-off band contribution and without it by means of the parameter m_a that appears in the kinetic term of the Thomas-Fermi energy functional. In Fig. 1 we can see the behavior of the hole subband level structure as a function of the doping density with (dashed lines) and without the split-off contribution (solid lines). From top to bottom we have the potential depth, ground heavy hole state, ground light hole state and ground split-off state, respectively. As it is possible to see from the mentioned figure the split-off contribution takes relevance in the high density limit in which energy differences from 60 meV to 80 meV are found in the hole subband level structure as compare to the calculations without the abovementioned contribution. The hole level structure behavior is the typical one expected to the delta-doped quantum well systems with a stronger confinement as the doping density is increased.

In the case of DDD we can figure out the hole level structure behavior with respect to the interwell distance through the evolution of the local density of states (LDOS). In Fig. 2 we present the LDOS of DDD quantum wells with a doping density of 1×10^{12} cm^{-2} and four interwell distances (50, 100, 150 and 200 ML). As it is clearly seen the DDD systems is practically degenerate for 150 and 200 ML meanwhile for 100 ML the ground light hole state splits in two levels with energy difference of about a 1 meV. In the case of 50 ML both the ground light hole state and the heavy hole state splits in two levels with energy differences of 2.5 meV and 1 meV, respectively. In a similar fashion the SLDD quantum well system presents a energy level splitting, however is only evident for the ground hole light state for a period width of 50 ML. The impurity density considered in Fig. 3 is 8×10^{12} cm^{-2}. In order to discern the evolution of the hole level structure, of both DDD and SLDD quantum well systems, as a function of the impurity density we have considered DDD and SLDD quantum wells with a fixed interwell distance (50 ML) and doping densities of 3, 7, 20, 60, 100 and 200 in units of 10^{12} cm^{-2} that corresponds to the top left, top middle, top right, bottom left, bottom middle and bottom right of Figs. 4 and 5, respectively. The hole level structures of the bottom part of both figures belong practically to the level structure of SDD quantum wells which means that the delta-doped quantum wells do not interact at all as a consequence of the stronger confinement experience by the carriers as the doping density is increased. On the contrary, the top part of both figures shows a clear level splitting for low doping densities due to the better sloping of the

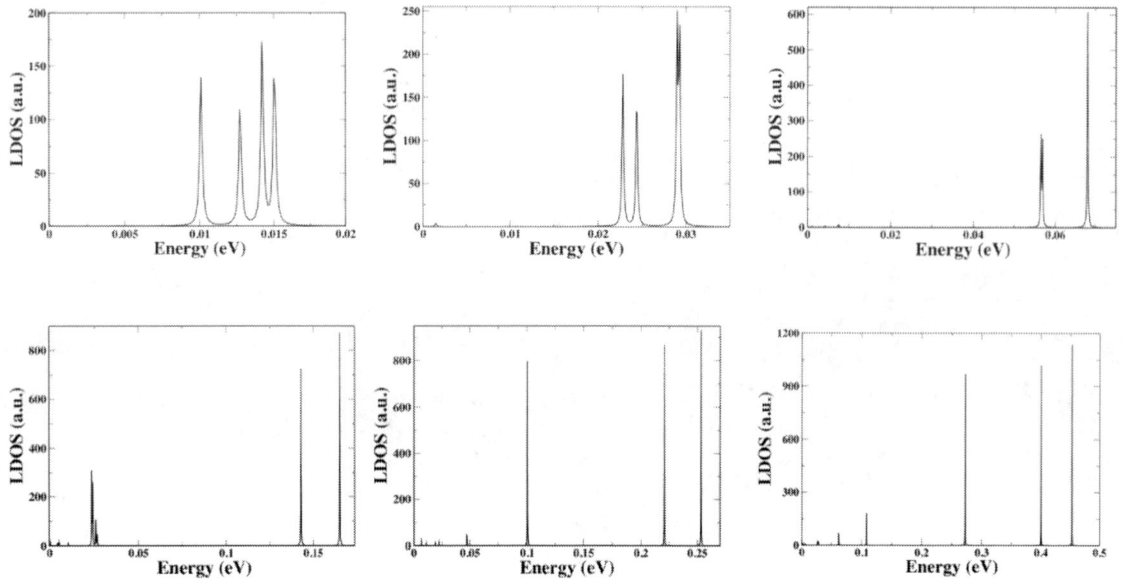

Figure 4. Local density of states versus energy of DDD quantum wells in Si with interwell distance of 50 ML and doping densities of: (Top) 3 left, 7 middle and 20 right; (Bottom) 60 left, 100 middle and 200 right. The doping density is in units of 10^{12} cm^{-2}.

hole subband levels.

Finally, in table 1 we present a comparison of our results with respect to the experimental data based on absorption and conductance measurements [3, 4, 2] as well as the theoretical data based on the $\mathbf{k}\cdot\mathbf{p}$-Envelope Function Approximation (EFA) [6, 14]. Our results agree within a 15 % for doping densities of 35 and 90 (in units of 10^{12} cm^{-2}) meanwhile for a doping density of 24 (in units of 10^{12} cm^{-2}) our approximation is frankly far from the experimental value within an error of 30 %.

Table 1. SDD hole subband energy comparison with respect to the experimental [3, 4, 2] and theoretical data [6, 14]. The doping density is in units of 10^{12} cm^{-2} and the energies in units of meV.

p_{2D}	TB	EFA	Exp.
24	79	99	110 ± 20
35	107	129	125
90	232	274	270

4. Conclusions

In summary, we have presented the hole subband level structure of quantum well systems in Si. We adopted a Thomas-Fermi-Dirac scheme for the confining potential as well as the nearest neighbors sp^{3}s* Tight-Binding approximation including spin for the hole subband level structure. We have analyzed the single, double and superlattice delta-doped quantum wells as a function of

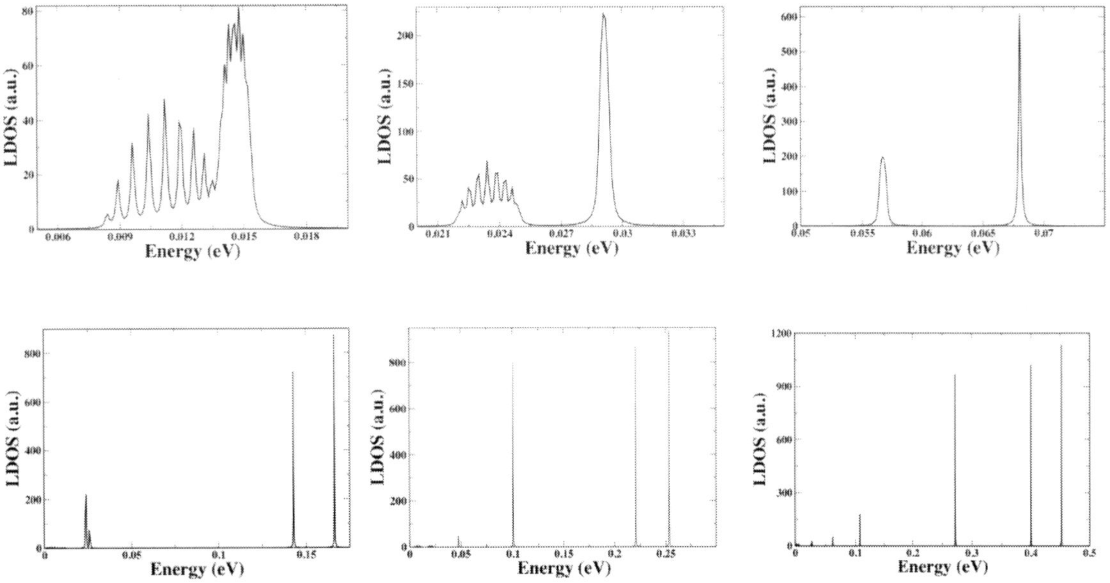

Figure 5. Local density of states versus energy of SLDD quantum wells in Si with a period of 50 ML and doping densities of: (Top) 3 left, 7 middle and 20 right; (Bottom) 60 left, 100 middle and 200 right. The doping density is in units of 10^{12} cm^{-2}.

the main parameter in each case. We also compare our results with the available experimental data obtaining a good agreement.

Acknowledgments

I.R.-V. acknowledges financial support from PROMEP-México through the program NPTC-2007.

References

[1] Gaggero-Sager L M and Pérez-Alvarez R 1997 *Appl. Phys. Lett.* **70** 212.
[2] Zhu J H, Gong D W, Zhang B, Lu F, Sheng C, Sun H H and Wang X 1995 *Phys. Rev. B* **52** 8959.
[3] Wang K L, Karunasiri R P G and Park J S 1992 *Surf. Sci.* **267** 74.
[4] Karunasiri G 1994 *Jpn. J. Appl. Phys.* **33** 2401.
[5] Siphai G M, Enderlein P, Scolfaro L M R, and Leite J R 1996 *Phys. Rev. B* **53** 9930.
[6] Gaggero-Sager L M and Mora-Ramos M E 2000 *phys. stat. sol. (b)* **220** 163.
[7] Caro J, Vink I D, Smit G D J, Rogge S and Klapwijk T M 2004 *Phys. Rev. B* **69** 125324.
[8] Chien P W, Wu S L, Lee S C and Chang S J 2002 *Mater. Chem. Phys.* **77** 426.
[9] R. Liu, A. T. S. Wee, D. H. Shen, and H. Takenaka 2004 *Surf. Interface Anal.* **36** 172.
[10] Kaesen F, Fink C, Anil K G, Hansch W, Doll T, Grabolla T, Schreiber H and Eisele I 1998 *Thin Solid Films* **336** 309.
[11] Klimeck G, Bowen R C, Boykin T B, Salazar-Lazaro C, Cwik T A and Stoica A 2000 *Superlattice. Microst.* **27** 77.
[12] Gaggero-Sager L M 2002 *phys. stat. sol. (b)* **231** 243.
[13] Vlaev S J and Gaggero-Sager L M 1998 *Phys. Rev. B* **58** 1142.
[14] Rodríguez-Vargas I and Gaggero-Sager L M 2007 *Physica B* **389** 227.

Journal of Physics: Conference Series **167** (2009) 012029 doi:10.1088/1742-6596/167/1/012029

DONOR IMPURITY IN NANOTUBE WITH TWO GaAs/GaAlAs QUANTUM WELLS: MAGNETIC FIELD EFFECTS

J D González[1], R Escorcia[1] and J Sierra-Ortega[1]

[1] Grupo de Investigación en teoría de la Materia Condensada, Universidad del Magdalena, A.A. 731, Santa Marta, Colombia

Abstract. Micro-tubes containing two GaAs/GaAlAs quantum wells (QWs) in a section of the tube layer has been fabricated and optical properties of the embedded QWs has been studied. The ground state binding energy of an off-axis donor in a cylindrical nanotube, containing two GaAs/GaAlAs quantum wells (QWs) in the presence of a uniform magnetic field is calculated as a function of the donor location as well as the density of states. A trial function for describing the asymmetric electron charge distribution is taken as a product of the combination of 1s and $2p_{x,y}$ subband wave functions and an unknown function that depends only on electron-ion separation. We found that the increasing the magnetic field the increasing the binding energy while the impurity is located in the QW1, whereas the opposite occurs when the impurity is located in the QW2. Two peaks in the curves of the binding energy, as a function of the impurity position, are also found as well as in the density of impurity states.

E-mail: jdavid0831@gmail.com

1. Introduction

Recently, Prinz et al. demonstrated that multilayer structures including strained layers are useful to form nanotubes and nanocoils [1,2]. By using lattice-mismatched epitaxial layers that rolled up when freed from the substrate due to the built-in strain and a micro-tube including two GaAs/GaAlAs quantum wells (QWs) located at positions with different types of strain was fabricated and its optical properties before and after the fabrication process were investigated by photoluminescence (PL) spectroscopy. Micro-tubes containing two GaAs/AlGaAs quantum wells (QWs) in a section of the tube layer, and also the optical properties of the embedded QWs have been studied[3]. In our previous work we show that the low lying *1s* and $2p_{x,y}$ subbands of the free electron in nanotube, become almost degenerated as the width of the repulsive core grows and the mixing of these subbands in the presence of the off-axis donor no longer should be depreciated [4].The purpose of this paper is to analyze the ground state energy of the off-axis neutral donor and the density of the impurity states in cylindrical nanotube with two GaAs/AlGaAs quantum wells with and without the presence of a magnetic field oriented along the growth axis.

2. Theory

We consider the problem of an off-axis neutral donor impurity in a cylindrical nanotube, which containing two $GaAs/Ga_{1-x}Al_xAs$ quantum wells (QWs) in the presence of an uniform magnetic field, as can see in figure 1. The material parameters m^*, the conduction-band effective mass and ε, the

[1] jdavid0831@gmail.com.

dielectric constant of the semiconductor heterostructure are assumed to be uniform throughout the heterostructure. The values of the physical parameters pertaining to GaAs (m^*=0.067 m_0, and ε =12.53, where m_0 is the free-electronmass) are used in our calculations.

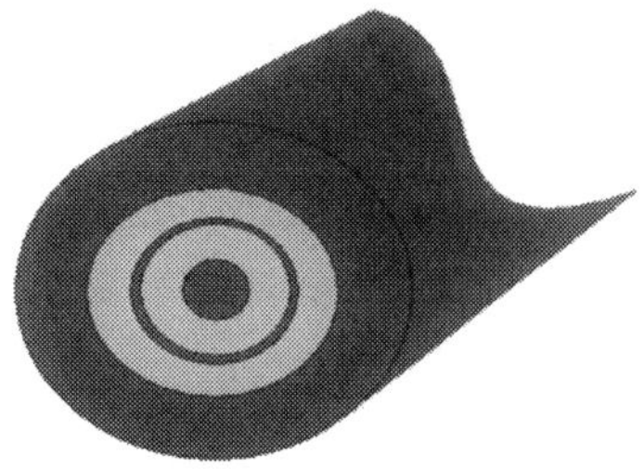

Fig.1 Cylindrical nanotube, containing two GaAs/GaAlAs quantum wells (QWs)

Within the effective-mass approach, the dimensionless Hamiltonian for the donor can be written as

$$H = H_0 - \frac{2}{|\vec{r} - \vec{\xi}|} \;;H_0 = -\nabla^2 + V(\rho) + \frac{1}{4}\gamma^2\rho^2 - i\gamma\frac{\partial}{\partial\varphi}$$

Where $\vec{r}$ and $\vec{\xi}$ are used to designate the electron and ion positions, respectively and we have introduced the effective Bohr radius $a_0^* = \varepsilon\hbar/m^*e^2$ as the unit of length, the effective Rydberg $Ry_0^* = e^2/2a_0^*\varepsilon$ as the unit of energy and $\gamma = e\hbar B/2m^*cRy^*$ as the dimensionless unit of the magnetic-field strength corresponding to the first Landau level.

To calculate the donor ground state energy, we choose a trial function as the product of a linear combination of the waves functions of the electron ground, s and the first excited, $2p_{x,y}$ states, with an unknown isotropic function $\Phi(|\vec{r} - \vec{\xi}|)$ could be interpreted as a variational wave function:

$$\psi(\vec{r}) = \left[\alpha f_{1s} + \sqrt{1-\alpha^2} f_{2p}\right]\Phi(|\vec{r} - \vec{\xi}|), \quad (2)$$

where α is a variational parameter $(-1 < \alpha < 1)$ which gives the grade of the mixing of the subbands [3]. One can see that in the particular case as $\alpha = 1$, there are not subband mixing and the trial function (2) becomes similar to the simple Bastard-type trial function. In general case, the mixing provided a decrease of the parameter α. Greater the mixing the smaller is the parameter α. In our calculations the parameter α decrease at to 0.8. Due to the inclusion of two quantum wells in the nanotube, appear the strong mixing subbands mostly in the central barrier. In this way, we must include the effect in our calculations.

For our model the wave equation for the free electron $H_o f(\vec{r}) = E_0 f(\vec{r})$ is separable and the electron wave function can be written as follows:

$$f(\vec{r}) = e^{ikz}e^{im\varphi}g(\rho), \quad (3)$$

where $m = 0, \pm1, \pm2, \ldots$ is the angular momentum in z-direction, k is the wave number corresponding to a free motion in z-direction $(-\pi < k < \pi)$ and the function $g(\rho)$ is the solution of the one-dimensional boundary value problem:

$$g''(\rho)+\frac{1}{\rho}g'(\rho)+U(\rho)g(\rho)=0;\;\; g'(0)=0;\;\; g(\infty)=0$$
$$U(\rho)=E_0-\gamma m-k^2-V(\rho)-\frac{\gamma^2\rho^2}{4}-\frac{m^2}{\rho^2} \tag{4}$$

The differential equation (4) is solved numerically using the trigonometric sweep method [5]. For modelling a nanotube containing two *GaAs/Ga*$_{1-x}$*Al*$_x$*As* quantum wells (QWs), we use the confinement potential given by the following expression:

$$V(\rho)=V_1\theta(-\rho,-R_1,W)+V_2\theta(\rho,R_2,W)-V_3\theta(\rho,R_3+W,W)+V_4\theta(\rho,R_4,W), \tag{5}$$

where:

$$\theta(z,z_0,W)=\begin{cases}0\;; & z<z_0-W\\ \left[(z-z_0)/W^2-1\right]^2; & z-W\le z<z_0\\ 1; & z\ge z_0\end{cases} \tag{6}$$

is soft-edge version of the Heaviside function, R_i, V_i represent the radii and heights of the barriers, respectively; W is a parameter related to the width of the transition region (see Fig. 2).

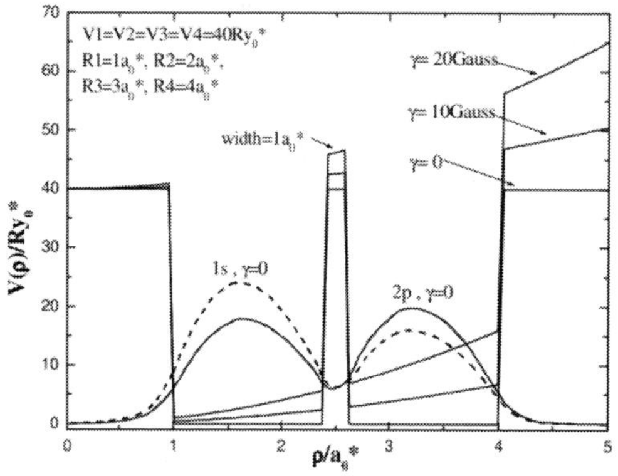

Fig. 2. Confinement Potential in cylindrical Nanotube

The solutions of the boundary value problem (4) corresponding to the bottom of the subbands $(k=0)$ with radial quantum numbers $n=0,1,2,...$ and the angular momentum m we denote as $g_{n,m}(\rho)$ and therefore in our notations the electron wave function, $f_{n,m}(\vec{r})$ and the energy $E_0(n,m)$ depend on two quantum numbers (n,m). If we assume that the off-center donor is located on the axis x, then it modifies the free electron wave functions in such way that it becomes more asymmetric in the x direction. It is reason why in the trial function (2) should be chosen the electron wave functions which gives contribution in the mixing as $f_{1s}(\vec{r})=g_{0,0}(\rho)$ for the *1s* state, and $f_{2p}(\vec{r})=g_{0,1}(\rho)\cos\varphi$ for $2p_x$ state. Starting from the variational principle and using the method described in the papers [6] one can obtain the following Euler-Lagrange equation for the correlation function:

$$-\frac{1}{J(r)}\frac{d}{dr}\left[J(r)\frac{d\Phi(r)}{dr}\right]+\left[\tilde{E}(r)-\frac{2}{r}\right]\Phi=E(D^0)\Phi \tag{7a}$$

Where $E(D^0)$ represents the energy of the neutral donor, $J(r)$ is radial part of the Jacobian and $\tilde{E}(r)$ is the averaged free electron local energy, given by the followings expressions:

$$J(r)=r^2\sum_{i,k=0,1}\alpha^{2-i-k}\left(1-\alpha^2\right)^{(i+k)/2}P_{ik}(r),\quad \tilde{E}(r)=r^2\sum_{i,k=0,1}\alpha^{2-i-k}\left(1-\alpha^2\right)^{(i+k)/2}E_0(0,i)P_{ik}(r)/J(r) \tag{7b}$$

$$P_{ik}(r) = \int_0^{2\pi}\int_0^{\pi} g_{0,0}^{2-i-k}(\tilde{\rho}) g_{0,1}^{i+k}(\tilde{\rho}) \lambda^{i+k} \sin\theta d\theta d\varphi \,;\, i,k = 0,1 \quad where \quad \lambda = 1 - r^2 \sin^2\theta \sin^2\varphi / \tilde{\rho}^2 \tag{7c}$$

$$and \quad \tilde{\rho} = \left(r^2 \sin^2\theta + \xi_\rho^{\,2} + 2r\xi_\rho \sin\theta\cos\varphi\right)^{1/2}$$

In these relations ξ_ρ is the distance from the donor position to axis. Once the functions $g_{0,0}(r)$ and $g_{0,1}(r)$ are found, the functions $P_{i,k}(r),(i,k=0,1)$ and $J(r)$ may then be calculated in straightforward way through (7b) and (7c). Finally, to define the donor energy we solve the wave Eq. 7(a) by using the trigonometric sweep method [5]. To calculate the density of the impurity states we assume that the circular cross section of the nanotube is not too small in order that we might treat the impurity positions as a continuous random variable. In this case the density of the impurity states $g(E)$ in a cylindrical nanotube is given by the following relation [4]:

$$g(E_b) = 2\pi\xi_\rho(E_b)\left|dE_b(\xi_\rho)/d\xi_\rho\right|^{-1} \tag{8}$$

In Fig. 3 and 4 we display the variation of the binding energy as a function of the distance from donor locations to the axis in a cylindrical nanotube with two square quantum well and different values of the magnetic field parallel to the axis. Additionally the different heights and small displacement to the right for minimum energy value in the second square well are shown due to increasing the magnetic field, for the lower it present a greater whereas larger are smaller, because the magnetic field provides additional confinements that tend to move the electron to the central region of the nanotube.

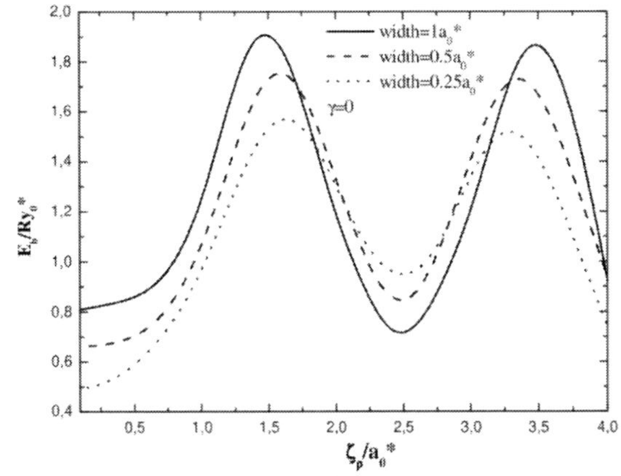

Fig.3 D^0 ground state binding energies as a function of the donor displacement ξ_ρ from the axis of the cylindrical nanotube for different central barrier's width and magnetic field $\gamma = 0$

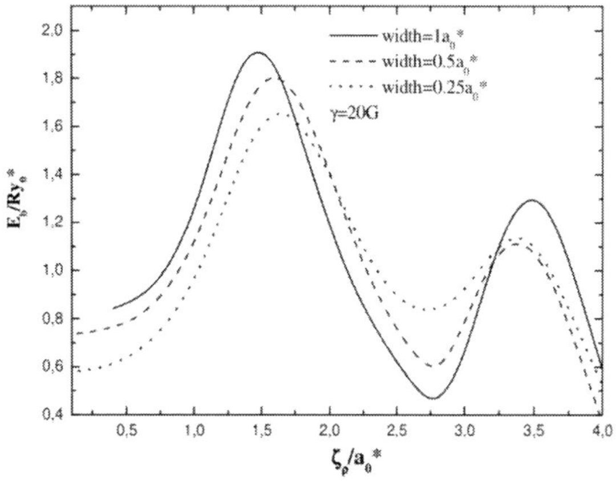

Fig.4 D^0 ground state binding energies as a function of the donor displacement ξ_ρ from the axis of the cylindrical nanotube for different central barrier's width and magnetic field $\gamma = 20G$

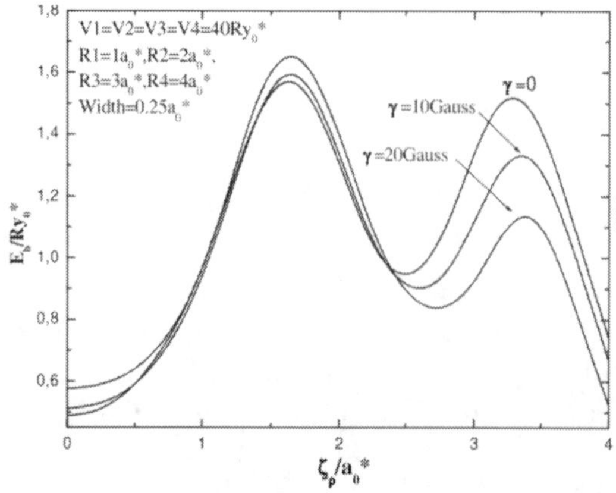

Fig.5 D^0 ground state binding energies as a function of the donor displacement ξ_ρ from the axis of the cylindrical nanotube with two square quantum well with different magnetic field values.

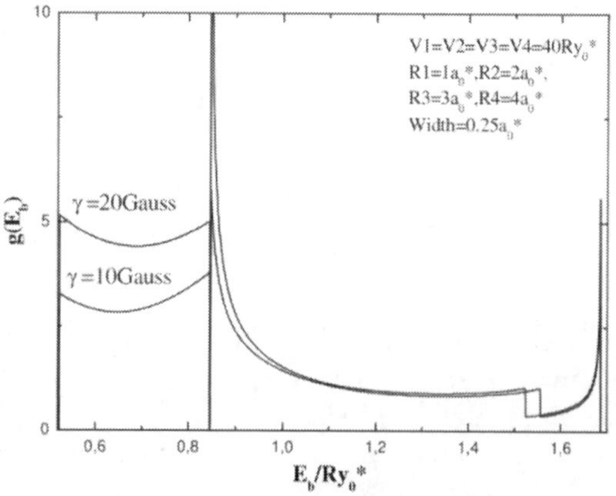

Fig.6 Density of the D^0 impurity states as a function of the binding energy in a nanotubo for different values of the magnetic field B applied along the symmetry axis.

The figure 5 shows the additional maximums of the energy due to the inclusion of the two square quantum well such behavior of the donor binding energies is also reflected in the curves of density of impurity states in nanotube. It seen from the Fig. 6 that the two well causes the appearance of the singularities near the right-side and left-side threshold of the curves of the density of the impurity states and it middle part. These singularities are due to the existence of one minimum and one maximum in the binding energies dependence where the derivate $dE(\xi_\rho)/d\xi_\rho$ is equal to zero.

3. CONCLUSION

We have presented a simple method of calculation of the binding energy for the lowest state of the off-axis D^0 donors in a nanotube with two square wells in the presence of the magnetic field applied parallel to the axis, taking into account the effect of the mixing s y p_x subbands. We found that under external magnetic field applied parallel to the axis the binding energies of donors located close to the axis increases and located far from the axis decreases. Also we present new curves that show novels energy peaks and analyze the density of the D^0 impurity states for different magnetic fields.

Acknowledgement
This work was partially financed by the Universidad Del Magdalena

4. References

[1] Prinz V Ya, Seleznev V A, Samoylov V A, Gutakovsky A K, Microelectron. Eng. **30** (1996) 439.
[2] Prinz V Ya, Seleznev V A, Gutakovsky A K, Chehovskiy A V, Preobrazenskii V V, Putyato M A, Nenasheva L A, Proceedings of the 26th International Symposium on Compound Semiconductors, Berlin, 22–26 August 1999,p. 199.
[3] Kubota K, Vaccaroa P O, Ohtania N, Hiroseb Y, Hosodab M, Aida T, Physica E **13**(2002)313–316
[4] Betancur F J, Orozco E A, Gonzalez J D and Mikhailov I D, Phys. Stat. Sol., **242** (b), 1833 (2005).
[5] F. J. Betancur, I. D. Mikhailov and L. E. Oliveira, J. Appl. Phys. D **31**, 3391 (1998)
[6] I. D. Mikhailov, F. J. Betancur, R. Escorcia and J. Sierra-Ortega, Phys. Stat. Sol., **234** (b), 590 (2002), Phys. Rev., B**67,** 156317 (2003).

Journal of Physics: Conference Series **167** (2009) 012030 doi:10.1088/1742-6596/167/1/012030

Hydrostatic-pressure-induced Γ-X mixing in delta-doped $Al_xGa_{1-x}As$

M. E. Mora-Ramos[1], **C. A. Duque**[2]

[1] Facultad de Ciencias, Universidad Autónoma del Estado de Morelos, Av. Universidad 1001, Col. Chamilpa, CP 62209 Cuernavaca, MOR., México

[2] Instituto de Física, Universidad de Antioquia, AA 1226, Medellín, Colombia

E-mail: memora@buzon.uaem.mx

Abstract. The mixing between Γ and X points in the conduction band is known to manifest in GaAs/$Al_xGa_{1-x}As$ heterostructures. Application of hydrostatic pressure leads to the variation of both the Γ- and X-energy gaps in such a way that around $P = 39$ kbar, the GaAs becomes an indirect gap one. Within a narrow interval of P around this value, both minima remain almost energetically aligned. Therefore, it can be expected that some kind of mixing between the states corresponding to the delta-quantum-wells formed at those points should manifest. In this work, we investigate the occurrence of Γ-X mixing induced by the application of hydrostatic pressure in n-type delta doped $Al_xGa_{1-x}As$, by means of a variational scheme that uses a 2×2 Hamiltonian with the off-diagonal element being a variational potential energy. Diagonal Schrödinger-like effective mass elements are written with potential functions given by the Thomas-Fermi approximation. The effect of the hydrostatic pressure is incorporated via the dependence on P of the different input parameters in the system: energy gaps, effective masses and dielectric constant. We report the variation of the ground state of the system as a function of P.

1. Introduction

Mixing of Γ and X conduction band minima has been the subject of study mainly regarding GaAs-AlAs heterostructures [1]-[9]. There are also a significant number of works dealing with the effects of the hydrostatic pressure on the optical and electronic properties in this kind of structures [10]-[15].

The main features of the spectrum of conduction band electrons in $Al_xGa_{1-x}As$ n-type delta doped quantum wells (DDQW) were recently investigated [16], within the framework of the one-dimensional local density Thomas-Fermi approximation [17]. Early experimental reports on DDQW in this material can be found in ref. [18]. On the other hand, a theoretical study of the influence of hydrostatic pressure on the electronic states of in GaAs-DDQW has shown the possibility of the formation of a high density two-dimensional electron gas (2DEG) corresponding to the DDQW states at the X minimum of the conduction band in GaAs [19]. Such approach assumed, for instance, that no pressure-induced transition to a semi-insulating state associated to the existence of deep impurity levels is present [20]. The reason for the mixing to occur is the re-accommodation of the ground state for the conduction band electrons due to the displacement of the Γ and X energy gaps as functions of P. This gives the possibility of having the single

Journal of Physics: Conference Series **167** (2009) 012030 doi:10.1088/1742-6596/167/1/012030

electron ground state formed at X located below the ground level at Γ, when P goes above of certain value.

This work is aimed at studying the mixing of Γ- and X-related electronic subbands in $Al_xGa_{1-x}As$ DDQW, induced by the application of hydrostatic pressure onto the system. Our ansatz considers that within an interval of P there must be a reciprocal influence of Γ and X DDQW states given their proximity in the energy scale. That pressure range includes the value for which the positions of the ground state levels, calculated for both minima independently, are aligned. As a consequence, the above mentioned mixing will appear. We will present some preliminary results on this subject. They are derived with the use of a simple variational scheme of calculation.

2. The Model

The electronic states in DDQW can be calculated by means of two main approaches. One of them is the self-consistent solution of coupled effective-mass-Schrödinger and Poisson equations. The second one is the use of an analytical expression for the potential well profile that arises within the formalism of the Thomas-Fermi one-dimensional local density approximation [17]. Then, with this function describing the confining potential, a single band effective mass Schrödinger equation is solved to obtain the subband structure.

In our case, as we are assuming some kind of coupling between the two minima in the conduction band, the Hamiltonian of the system should be written as a 2×2 matrix. If we use effective atomic units (Bohr radius and Rydberg energy) related with the Γ electron effective mass, it has the form:

$$H = \begin{bmatrix} h^{\Gamma} & V_{\Gamma X} \\ V_{X\Gamma} & h^{X} \end{bmatrix} \tag{1}$$

with

$$h^j = -\gamma_j \frac{d^2}{dz^2} + V_j(z), \quad j = \Gamma\,, X. \tag{2}$$

The parameter γ_j is equals to unity when $j = \Gamma$. For the X point $\gamma_X = m_\Gamma/m_X$, where m_j labels the effective masses of both minima along the direction of confinement in the structure,"z". The potential functions will be given by the expression:

$$V_j(z) = E_{Gj} - \frac{a_j^2}{(a_j|z| + z_{0j})^4}. \tag{3}$$

These one-dimensional potential functions arise from the Thomas-Fermi approximation [17]. The E_{Gj} are the corresponding energy gaps at both conduction band states. This means that the zero of the scale for the energies is taken at the bulk upper valence band edge at Γ. According to the system of units adopted we have that:

$$\begin{aligned} a_\Gamma &= \frac{2}{15\pi} \\ a_X &= \frac{2}{15\pi}\left(\frac{m_\Gamma}{m_X}\right)^{3/2} \\ z_{0\Gamma} &= \frac{2}{15\pi}\left(\frac{225\pi}{4N_{2D}}\right)^{1/5} \\ z_{0X} &= \frac{2}{15\pi}\left(\frac{225\pi}{4N_{2D}}\right)^{1/5}\left(\frac{m_\Gamma}{m_X}\right)^{1/2}, \end{aligned} \tag{4}$$

whereas N_{2D} is the two-dimensional density of ionized impurities in the delta doping spike.

The calculation of the ground state of the mixing Hamiltonian in equation (1) is performed variationally. The off-diagonal elements are set to be $V_{\Gamma X} = V_{X\Gamma} = \mathcal{V}$, which will be one of the variational parameters in the energy functional $\int \Psi^\dagger H \Psi dz$. The second variational parameter, α, appears in the choice of the two components of the trial vector wavefunction Ψ [21]:

$$\psi_\Gamma(z) = \alpha\phi_\Gamma(z) + \sqrt{1-\alpha^2}\phi_X(z)$$
$$\psi_X(z) = -\sqrt{1-\alpha^2}\phi_\Gamma(z) + \alpha\phi_X(z). \quad (5)$$

The functions $\phi_j(z)$ are proposed to have the following form:

$$\phi_j(z) = Ae^{-b_j|z|}, \text{ with } b_j = \sqrt{\mathcal{E}_{0j} - \left(E_{Gj} - \frac{a_j^2}{z_{0j}^4} + \beta_j\mathcal{V}\right)}, \quad (6)$$

where $\beta_j = \alpha^2$ if $j = \Gamma$ and $\beta_j = 1 - \alpha^2$ if $j = X$. $\mathcal{E}_{0j}$ represents the values of the ground level calculated exactly with the single-minimum Hamiltonian of equation (2).

The inclusion of the hydrostatic pressure effects is made via the dependence upon P of the main input parameters. That is, effective masses, energy gaps, and the dielectric constant. We are using the input data previously dealt with in ref. [15].

3. Results

Now, we go back to the use of traditional units. Figure 1 show the results of the variational calculation for the ground state of the mixed system with variational Hamiltonian (1). The value chose for the two-dimensional ionized impurity density is $N_{2D} = 5 \times 10^{12}\,\mathrm{cm}^{-2}$. The cases of pure GaAs (left figure) and $Al_{0.3}Ga_{0.7}As$ (right figure) are considered. Reported pressure intervals were chosen in such a way that the point of crossing between the ground levels of the unmixed DDQW at both minima is included within.

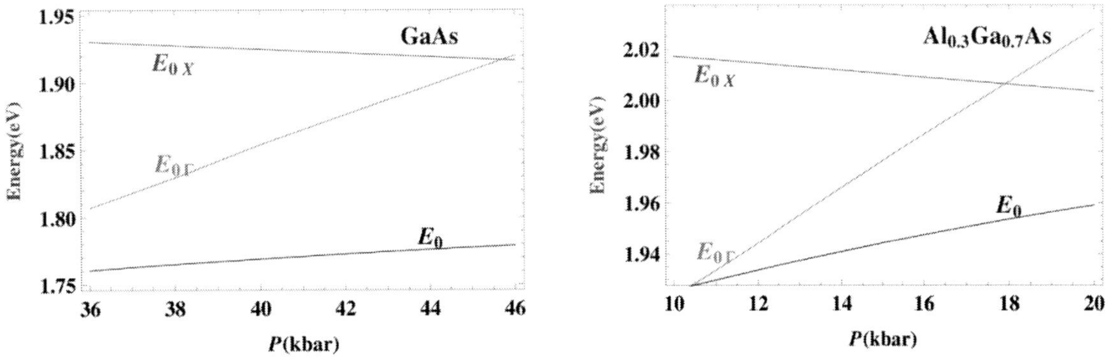

Figure 1. The ground state (in black) of the mixed $\Gamma - X$ n-type delta doped quantum well with $N_{2D} = 5 \times 10^{12}\,\mathrm{cm}^{-2}$ as a function of the hydrostatic pressure. For comparison, the corresponding independent energy ground levels at the Γ (in red) and X (in blue) points are plotted as well. Left figure corresponds to the case of GaAs. Right figure shows the results for a n-delta-doped $Al_{0.3}Ga_{0.7}As$ quantum well.

The off-diagonal variational parameter is found to depend linearly with pressure: $\mathcal{V} = \sigma P$. The values obtained for the slope, in each range of P considered are $\sigma(GaAs) = 2.83\,\mathrm{meV\,kbar}^{-1}$, and $\sigma(Al_{0.3}Ga_{0.7}As) = 2.68\,\mathrm{meV\,kbar}^{-1}$. As the calculation was performed using effective atomic units, the values resulting for the variational parameter α after minimization are constant and independent of P because the effective Rydberg is defined for each value of pressure, thus leaving the differential diagonal operators in H unaltered. That is, in effective units, both variational parameters have a fixed value, which is independent of P. When going back to "normal" units, it is necessary to multiply by $Ry^*(P)$ if the quantities involved do have units of energy. That is the case of $\mathcal{V}$. We have obtained $\alpha = 0.305$ for both cases.

It is seen that the effect of mixing is to bring down the ground state of the coupled system, when it is compared with the values $\mathcal{E}_{0j}$ calculated without the presence of hydrostatic pressure. Furthermore, the correct behaviour of E_0 as a function of pressure is detected. That is, the ground state must reproduce the a very close to $\mathcal{E}_{0\Gamma}$ when $P = 0$. In fact, there might be some kind of interaction between Γ and X minima even for null pressure; but it is always neglected. E_0 must also tend to $\mathcal{E}_{0X}$ for sufficiently high hydrostatic pressure.

However, it can be observed in the case of $Al_{0.3}Ga_{0.7}As$ that there will be a small difference between E_0 and $\mathcal{E}_{0\Gamma}$. We explain this by indicating that the choice made for the trial wavefunctions $\phi_j(z)$ is not the more accurate. They are, perhaps, the simplest option. Although they qualitatively reproduce the decaying of the electronic probability density with the distance away from the origin, these functions do not have a continuous first derivative at $z = 0$. Therefore, they can not be taken as any expansion of the eigenfunctions of the Hilbert space associated with the Hamiltonian H.

4. Conclusions

The mixing effect between Γ and X conduction band states due to the hydrostatic pressure in $Al_xGa_{1-x}As$-DDQW appears to be not negligible. Several tens of meV are the amount of the difference between the ground state of H and the eigenstates of the single-conduction-band-minimum Hamiltonians of equation (2) for the n-type DDQW. This indicates that further consideration should be taken of this phenomenon when studying the properties of delta-doped systems under pressure in such materials. A more convenient selection of the single-minimum variational trial wavefunctions is required for that purpose. It would be interesting, for instance, the inclusion of finite temperature effects. In this case, the self-consistent solution for the single minimum eigenstates would be the correct starting point for the application of the variational scheme here proposed.

Acknowledgments

M.E.M.R. acknowledges support from Mexican CONACYT through Grant No. CB-80846/2007. C.A.D. is grateful to A. Kurczyńska for her valuable comments and suggestions. C.A.D. would also like to thank the Colombian agencies CODI-Universidad de Antioquia and the Excellence Center for Novel Materials and COLCIENCIAS under contract No. 043-2005.

References

[1] Liu H C 1987 *Appl. Phys. Lett.* **51** 1019
[2] Ando T and Akera H 1989 *Phys. Rev. B* **40** 11619
[3] Pulsford N J, Nicholas R J, Dawson P, Moore K J, Duggan G and Foxon C T B 1989 *Phys. Rev. Lett.* **63** 2284
[4] Sun J P, Mains R K, Yang K and Haddad G I 1993 *J. Appl. Phys.* **74** 5053
[5] Li G H, Goñi A R, Syassen K, Brandt O and Ploog K 1994 *Phys. Rev. B* **50** 18420
[6] Foreman B A 1998 *Phy. Rev. Lett.* **81** 425
[7] di Carlo A, Pescetelli S, Kavokin A, Vladimirova M and Lugli P 1998 *Phys. Rev. B* **57** 9770
[8] Brzostowski B and Kucharczyk R 2003 *Phys. Rev. B* **67** 125305
[9] Tomić S, O'Reilly S E P, Klar P J, Grüning H, Heimbrodt W, Chen W M and Buyanova I A 2004 *Phys. Rev. B* **69** 245305
[10] Gell M A, Ninno D, Jaros M, Wolford D J, Keuch T F and Bradley J A 1987 *Phys. Rev. B* **35** 1196
[11] Shan W, Fang X M, Li D, Jiang S, Shen S C, Hou H Q, Feng W and Zhou J M 1991 *Phys. Rev. B* **43** 14615

[12] Wang D, de Andrada e Silva E A and da Cunha Lima I C 1992 *Phys. Rev. B* **46** 7304
[13] Burnett J H, Cheong H M, Paul W, Koteles E S and Elman B 1993 *Phys. Rev. B* **47** 1991
[14] Li G H, Goñi A R, Syassen K, Hou H Q, Feng W and Zhou J M 1996 *Phys. Rev. B* **54** 13820
[15] Mora-Ramos M E, López S Y and Duque C A 2008 *Eur. J. Phys. B* **62** 257
[16] Gaggero-Sager L M, Mora-Ramos M E and Olivares-Robles M A 2005 *Microel. J* **36** 416
[17] Ioriatti L 1990 *Phys. Rev.* B **41** 8340
[18] Schubert E F, Tu C W, Kopf R F, Kuo J M and Lunardi L M 1989 *Appl. Phys. Lett.* **54** 2592
[19] Mora-Ramos M E and Duque C A 2006 *Braz. J. Phys.* **36** 866
[20] Dizhur E, Voronovsky A, Kotelnikov I and Dizhur S 2006 *phys. stat. sol. (b)* **244** 453
[21] Mora-Ramos M E, López S Y and Duque C A 2008 *Physica E* **40** 1212

Effect of the hydrostatic pressure on the electron mobility in delta-doped systems

O. Oubram, M. E. Mora-Ramos, L. M. Gaggero-Sager

Facultad de Ciencias. Universidad Autónoma del Estado de Morelos. Av. Universidad 1001. CP 62209. Cuernavaca. Morelos. Mexico

E-mail: `lgaggero@uaem.mx`

Abstract. The influence of hydrostatic pressure on the electron states and low-temperature mobility in n-type GaAs δ-doped single quantum wells is studied. Values of hydrostatic pressure consider are below the so-called $\Gamma - X$ crossover, keeping all attention in the electronic properties at the Brillouin zone center. The effect of the pressure on the electron mobility is described via a relative quantity that is proportional to the ratio between $P \neq 0$ and zero pressure results. Calculation is performed using an analytical description of the potential energy function profile, based on the Thomas-Fermi approach, taking explicitly into account the dependence upon P of the main input parameters: effective masses and dielectric constant. The relative mobility increases for higher values of P. The cases of zero and finite -although small- temperature are studied, showing that the influence of T is mainly to lower the values of the relative mobility in the entire range of P considered. Numerical results are reported for a two-dimensional density of ionized impurities equals to $7.5 \times 10^{12}\,\mathrm{cm}^{-2}$.

1. Introduction

Delta-doped quantum wells (DDQW) provide the possibility of having two-dimensional conductive channels with a high density of charge carriers. In the particular case of GaAs-based systems, the n-type DDQW have been fabricated using Si atoms as donor impurities[1, 2].

Among the theoretical studies devoted to investigate the properties of the electron states in δ-doped systems one finds two main trends. In one of them, the coupled one-dimensional Schrödinger and Poisson differential equations are solved self-consistently (see for instance the references [3, 4, 5]). The other approach uses a quasi-classical scheme in which an analytical expression for the energy band potential profile is derived within the one-dimensional Thomas-Fermi local-density-functional approximation[6, 7]. Then an effective mass Schrödinger equation is solved in order to obtain the corresponding subband structure in the well.

It is known that the application of hydrostatic pressure to a crystal results in a modification of its electronic properties. This is mainly due to the deformation of the interatomic bonds[8]. There are previous reports on the influence of the hydrostatic pressure in GaAs-based n-type DDQW. In particular, a study was performed on the possibility of the formation of a high density two-dimensional electron gas associated to X-states in the conduction band, above a certain pressure threshold[9]. A different kind of investigation has shown that it may occur a pressure-induced transition to a semi-insulating state in a δ-doped system in GaAs for values of

P around 2 GPa. In this phenomenon, the participation of states within the energy gap related to DX impurity centers seem to play a fundamental role[10].

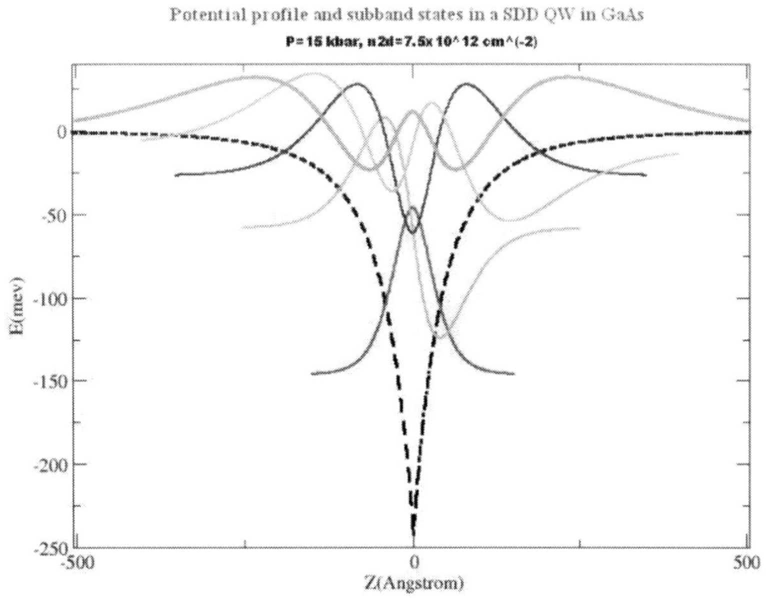

Figure 1. Potential well profile and single electron states in a single GaAs DDQW for $n_{2d} = 7.5 \times 10^{12}\,\mathrm{cm}^{-2}$.

In the present work we are aimed to study the effect of the hydrostatic pressure on the electron states in the conduction band of single DDQW of type n in GaAs by exploiting the advantages of the Thomas-Fermi approach. The obtained energy spectra will be used as a key element to calculate the relative mobility as a function of P. This is a quantity which is given by the ratio between the low temperature two-dimensional electron mobilities corresponding to the case including pressure and that without it. For the moment, we restrict ourselves to consider values of P below the point of transition from the direct to the indirect energy gap regimes, induced by pressure in GaAs. Besides, we explore the influence of finite temperature, although only limited to small values which do not threaten the validity of the approximation.

2. Model and results

As mentioned above, the calculation of the single electron states in GaAs n-type DDQW is carried out by solving a single band effective mass Schrödinger equation with a V-shaped Thomas-Fermi potential. For details of this potential function and its derivation, references [6, 7] are referred. The expression for this function is

$$V(z) - E_F = -\frac{\alpha^2}{(\alpha|z| + z_0)^4}; \tag{1}$$

where E_F represents the Fermi level in the system, whereas the parameters α and z_0 are given by:

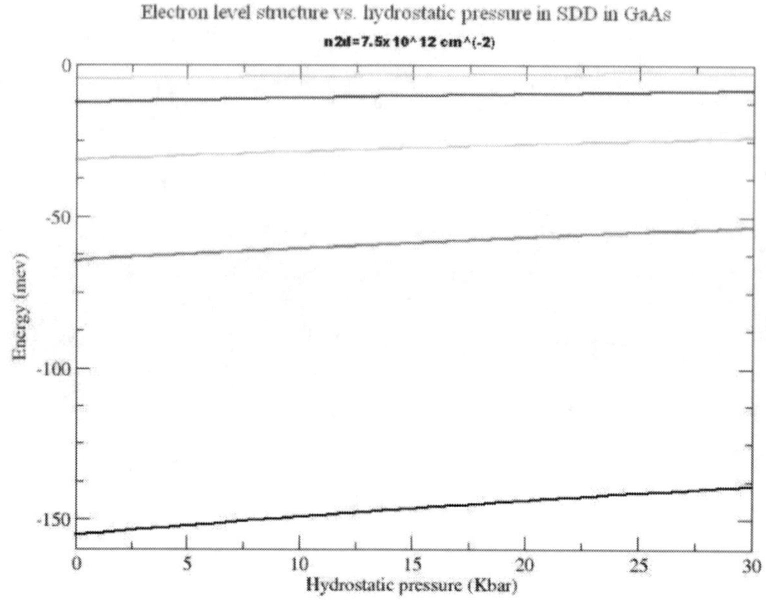

Figure 2. The variation with hydrostatic pressure of the confined energy levels in a single GaAs DDQW for $n_{2d} = 7.5 \times 10^{12}\,\mathrm{cm}^{-2}$.

$$\alpha = \frac{e^2}{\epsilon}\frac{m^{*3/2}}{15\pi\hbar^3},$$
$$z_0 = (\frac{2\epsilon^2\alpha^3}{\pi e n_{2d}})^{1/5} \quad (2)$$

Figure 1 shows a schematic representation of both the potential well profile and subband states. The results are presented for $P = 15\,\mathrm{kbar}$ and a density of ionized impurities $n_{2d} = 7.5 \times 10^{12}\,\mathrm{cm}^{-2}$. The variation of the values of each energy level confined in the DDQW as a function of P is presented in the figure 2.

The inclusion of pressure effects is made via the variation of the main input parameters upon P[11](see Table 1). For both conduction band minima, the following relation -including temperature- holds;

$$E_{\mathrm{gap}}(P) = E_1 + \beta\, P + \alpha\, T^2/(T + 204) \quad (3)$$

On the other hand, the variation of the static dielectric constant is given by

$$\epsilon = 12.74\, e^{[9.4\times10^{-5}\,(T-75.6)-1.67\times10^{-3}\,P]}, \quad (4)$$

and the corresponding effective masses are

$$m_{\Gamma-\mathrm{GaAs}} = \left[1 + \frac{2 \times 7510}{E_{gap-\Gamma-\mathrm{GaAs}}(P)} + \frac{7510}{E_{gap-\Gamma-\mathrm{GaAs}}(P) + 341}\right]^{-1} m_0, \quad (5)$$

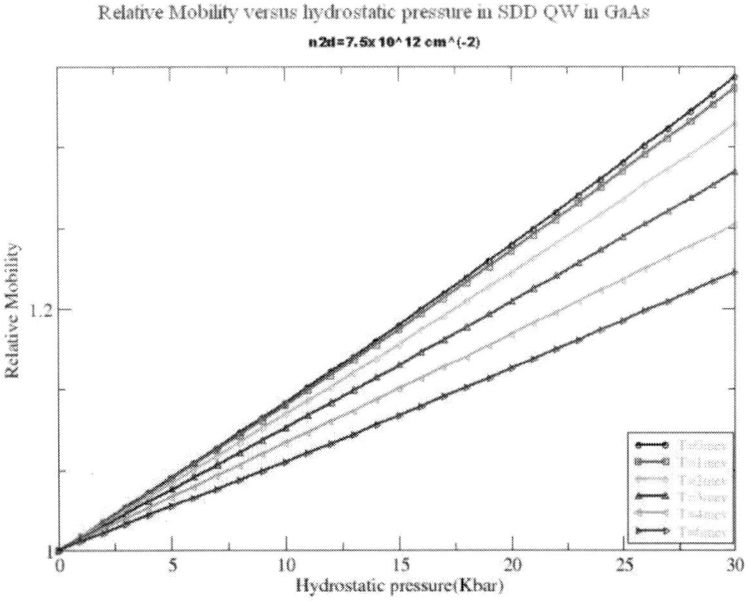

Figure 3. Relative electron mobility for a single GaAs DDQW with $n_{2d} = 7.5 \times 10^{12}\,\mathrm{cm}^{-2}$ as a function of hydrostatic pressure for different values of temperature. Reference values are taken for $P = 0$ and $T = 0$. The curves from top to bottom correspond to $T = 4, 3, 2, 1, 0\,\mathrm{meV}$, respectively.

Table 1. Parameters used in the present calculations.

	$\Gamma - \mathrm{GaAs}$	$X - \mathrm{GaAs}$
$E_1\,(\mathrm{meV})$	1519	1981
$\alpha\,(\mathrm{meV/K})$	-0.5405	-0.460
$\beta\,(\mathrm{meV/kbar})$	10.7	-1.4
$A_{(\parallel/\perp)}$		1.3/0.26

$$m_{X,(\parallel/\perp)} = (A_{(\parallel/\perp)} + 6.15 \times 10^{-3}\,P - 1.22 \times 10^{-5}\,P^2)\,m_0. \tag{6}$$

The expression for the relative mobility can be written as [12, 13]:

$$\mu_{rel} = \frac{\mu^{n\delta,P=0,T=0}}{\mu^{n\delta,P\neq 0,T\neq 0}} = \frac{\int\int_{IR^2} \rho_e^{n\delta,P=0,T=0}(z')\,\rho_{imp}^{n\delta,P=0,T=0}(z)\,|z-z'|dzdz'}{\int\int_{IR^2} \rho_e^{n\delta,P=0,T\neq 0}(z')\,\rho_{imp}^{n\delta,P=0,T\neq 0}(z)\,|z-z'|dzdz'} \tag{7}$$

Then, for calculating the corresponding relative conductivity we use

$$\sigma_{rel} = n_{rel} \cdot \mu_{rel}$$
$$n_{rel} = \frac{n_e^{P=0,T=0}}{n_e^{P\neq 0,T\neq 0}} \tag{8}$$

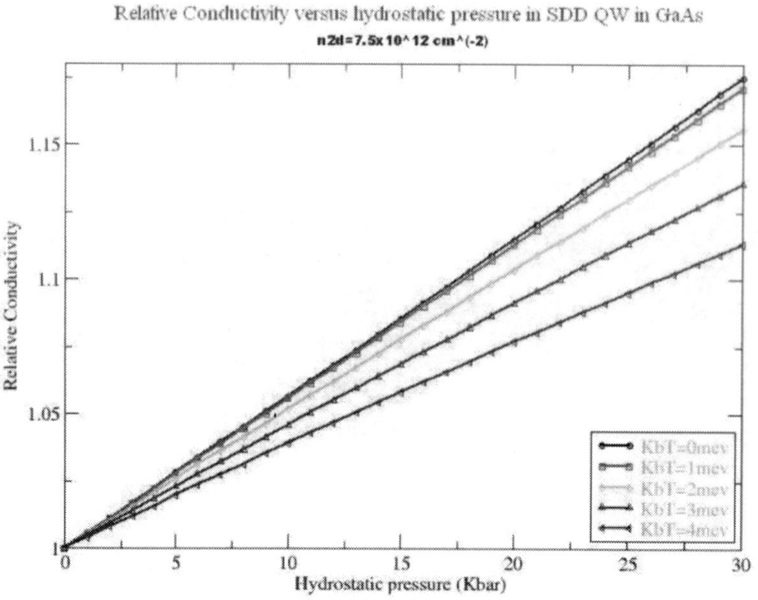

Figure 4. Relative electron conductivity for a single GaAs DDQW with $n_{2d} = 7.5 \times 10^{12}\,\text{cm}^{-2}$ as a function of hydrostatic pressure for different values of temperature. Reference values are taken for $P = 0$ and $T \neq 0$. The curves from top to bottom correspond to $T = 4, 3, 2, 1, 0\,\text{meV}$, respectively.

Figure 3 (Figure 4) shows the values of μ_{rel} (σ_{rel}),respectively, obtained for the above mentioned DDQW system. It can be seen that, when compared with zero pressure conditions, the effect of the hydrostatic pressure is to increase the mobility of the electrons in the system. The conductivity increases as well; but it does it at a slower pace. This is a result of the modulation coming from the dependence of the relative density of charge carriers upon P. On the other hand, temperature is included going from 0 to 60 K. The effect of the increment in T is to slow down the increasing of the relative mobility as a function of P. The same result can be readily seen for the relative electron conductivity from the figure 4.

References

[1] Schubert E F, Fischer A and Ploog K 1986 *IEEE Trans. Electron. Devices* **ED-33** 625
[2] Schubert E F, Tu C W, Kopf R F, Kuo J M and Lunardi L M 1989 *Appl. Phys. Lett.* **54** 2592
[3] Degani M H 1991 *Phys. Rev.* B **44** 5580
[4] Scolfaro L M R, Beliaev D, Enderlein P and Leite J R 1994 *Phys. Rev.* B **50 8699**
[5] Gaggero-Sager L M, Pérez-Alvarez R 1995 *J. Appl. Phys.* **78** 4566
[6] Ioriatti L 1990 *Phys. Rev.* B **41** 8340
[7] Gaggero-Sager L M, Mora-Ramos M E and Contreras-Solorio D A 1998 *Phys. Rev.* B **57** 6286
[8] Yu P Y and Cardona M 1999 *Fundamentals of Semiconductors, 2nd. ed* (Berlin: Springer)
[9] Mora-Ramos M E and Duque C A 2006 *Braz. J. Phys.* **36** 866
[10] Dizhur E, Voronovsky A, Kotelnikov I and Dizhur S 2007 *phys. stat. sol. (b)* **244** 453
[11] See, for example, Raigoza N, Morales A L, Montes A, Porras-Montenegro N and Duque C A 2004 *Phys. Rev. B* **69** 045323, and references therein
[12] Rodríguez-Vargas I, Gaggero-Sager L M and Velasco V R 2003 *Surf. Sci.* **537** 75
[13] Rodríguez-Vargas I and Gaggero-Sager L M 2006 *J. Appl. Phys.* **99** 033702

Journal of Physics: Conference Series **167** (2009) 012032 doi:10.1088/1742-6596/167/1/012032

D^- Energy spectrum in toroidal quantum ring

C A Gómez[1], J H Marín[†,2], W Gutiérrez[1] and L F García[1]

[1]Universidad Industrial de Santander, Bucaramanga-Colombia

[2]Universidad Nacional-Colombia, Medellín-Colombia, AA3840

Abstract. The structure of energy spectrum of the negative donor centre in a toroidal-shaped quantum ring with two different morphologies of the cross-section is analyzed. By using the adiabatic procedure we have deduced a one-dimensional wave equation with periodic conditions which describes the low-lying energy levels related to the electrons rotation around the symmetry axis. Our results are in good agreement with those previously obtained as the size of the ring cross-section tends to zero.

Introduction

Motivated by the possibility of finding new physical effects and novel technological applications to opto-electronics, many researches have begun to analyze quantum dots with diverse morphologies. Nanostructures in a toroidal-shape have become a focus of attention because their topology allow us to get special laboratories for the study the mesoscopy physics related to quantum interference effects, such as the Aharonov–Bohm effect [1,2] and persistent currents [3]. In addition, these nanostructures can be used to carry out theoretical studies aimed at finding out the effects on the few- particle energy spectrum when dimensionality is reduced up to one in a very narrow nanoring. In this sense, the two-electron exactly solvable problem was recently reported in the papers [4-5] and some devices based on InGaAs toroidal quantum rings (QRs) have been designed and realized [6–7]. One-electron into a single toroidal QR [8] or double concentric toroidal QRs [9] under the presence of the homogeneous magnetic field is the simplest system that one can study in these semiconductor nanostructures. The one-particle energy spectrum has been calculated by using the diagonalization method [8-9], while energy structure of neutral shallow donors D^0 [9-10] and D_2^+ molecular complex[11], which are also examples of one-particle systems, have been analyzed by using variational method [10] and a quasi-exact solution based on the adiabatic procedure [11-12]. As a neutral donor D^0 traps a second electron released by other neutral donor a negative-donor D^- is formed [13]. The importance of this D^- center is related to the possibility of studying the electron-electron (e-e) correlation effects under the strong confinement conditions. Taking into account that the energy spectrum of few-particle systems becomes strong depending on the quantum confinement and the morphology of the quantum dots [14], this work is concerned with the determination of the D^- center energy structure as its two electrons are constrained to be moved into a toroidal QR while the positive ion is located in any point along of the QR symmetry axis. In spite of the fact that several features of the two-electron QRs can be explained in a simple mode by using a strictly one-dimensional QR [4-5], we are interesting to explore the cross-section morphology effect on the D^- energy spectrum since the connection between of the QR shape with the e-e interaction is still subject of great research.

Physical system and theoretical framework

We consider two electrons with an effective mass m^* that are forced to move inside a nanoscopic toroidal QR whose cross-section of the area πR_t^2 may be circular with the radius R_t or a square with the side $h = \pi^{1/2} R_t$ and whose centreline, defined as a circle passing through symmetry centres of the

[†] Corresponding author: e-mail jhmarin@unal.edu.co , Phone: +57-4-4309327

torus cross-sections, has sufficiently large radius R_a $\left(R_a >> R_t\right)$. The electrons are electrically bound to the positive ion (donor centre) located on the torus symmetry axis at the distance ξ measured from its symmetry plane. Configurations and parameters of the considered models are shown in figure 1.

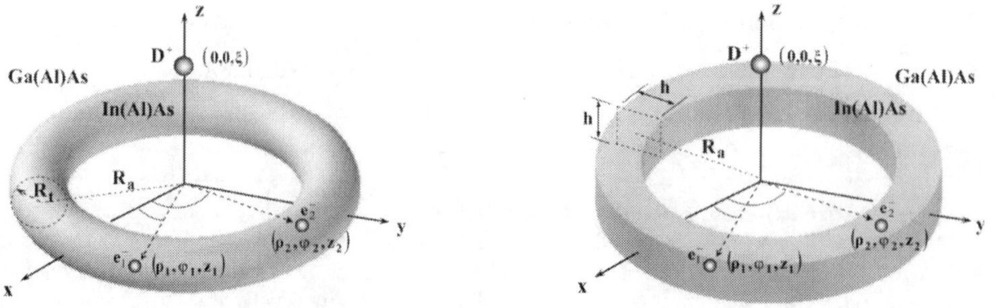

Figure 1. Configurations of negatively charged donors in a toroidal quantum rings with circular and square cross-sections

The dimensionless Hamiltonian of D^- centre in the presence of a homogeneous magnetic field $\mathbf{B}$ ($\mathbf{B} = B\hat{e}_z$) applied along the z-axis in cylindrical coordinates is given by the following expression:

$$\hat{H} = \sum_{j=1}^{2}\left(-\Delta_j - i\gamma\frac{\partial}{\partial\varphi_j} + \frac{1}{4}\gamma^2\rho_j^2 - \frac{2}{\left|\mathbf{r}_j - \boldsymbol{\xi}\right|} + V\left(\mathbf{r}_j\right)\right) + \frac{2}{\left|\mathbf{r}_1 - \mathbf{r}_2\right|} \tag{1}$$

where $\mathbf{r}_1$ and $\mathbf{r}_2$ are the vector electrons positions, $V\left(\mathbf{r}_j\right)$ is the confinement potential, equal to zero and infinite inside and outside the torous, respectively. The units of the length, the energy and the magnetic strength used in equation (1) are the effective Bohr radius, $a_0{}^* = \hbar^2\varepsilon/m^* e^2$, the effective Rydberg $Ry^* = e^2/2\varepsilon a_0{}^*$, and the first Landau level energy $\gamma = eB\hbar/2m^* cRy^*$, respectively. To describe the electrons positions let us introduce a three-dimensional orthogonal coordinate system $(\varphi, \mathbf{e}_t)$, where φ is the angle that defines the direction of the position vector $\mathbf{R}_a$ over centreline while the two-dimensional unit vector $\boldsymbol{\rho}_t$ identifies the electron position inside the given cross section. Inside the nanoring the electron position is therefore is described by

$$\mathbf{r}_j = \mathbf{R}_a(\varphi) + \boldsymbol{\rho}_t(\mathbf{e}_t) \tag{2}$$

In the case when the centreline radius R_a is much larger than the cross section radius R_t the eigenvalues of the Hamiltonian (1), can be found by using the adiabatic approximation that allows us to decouple the fast transverse electron motion from the slow rotation around z-axis [11]. The low-lying energy levels in this approximation are eigenvalues of the following effective Hamiltonian:

$$\hat{H} = -\frac{1}{R_a^2}\left(\frac{\partial^2}{\partial\varphi_1^2} + \frac{\partial^2}{\partial\varphi_2^2}\right) - i\gamma\left(\frac{\partial}{\partial\varphi_1} + \frac{\partial}{\partial\varphi_2}\right) + U_{ee}(\varphi_1 - \varphi_2) + \frac{1}{2}\gamma^2 R_a^2 - \frac{2}{\sqrt{R_a^2 + \xi^2}} + E_0 \tag{3}$$

$$U_{ee}(\varphi_1 - \varphi_2) = \left\langle f_0\left(\boldsymbol{\rho}_{t1}\right) f_0\left(\boldsymbol{\rho}_{t2}\right)\middle\| \frac{2}{\sqrt{4R_a^2\sin^2\left(\varphi_1 - \varphi_2\right) + \left(\boldsymbol{\rho}_{t1} - \boldsymbol{\rho}_{t2}\right)^2}} \middle\| f_0\left(\boldsymbol{\rho}_{t1}\right) f_0\left(\boldsymbol{\rho}_{t2}\right)\right\rangle \tag{4}$$

Here $f_0(\boldsymbol{\rho}_t)$ and E_0 are the exact wave function and the corresponding energy in an infinite barrier two-dimensional quantum well related to the electron motion along the cross-section. By using the substitution $\theta = 0.5(\varphi_1 + \varphi_2)$ and $\varphi = \varphi_1 - \varphi_2$ the two-particle Hamiltonian (4) can be separated into center-of-mass, $\hat{H}_c$ and relative $\hat{H}_r$ terms:

$$\hat{H}=\hat{H}_c+\hat{H}_r;\qquad \hat{H}_c=-\frac{1}{2R_a^2}\frac{d^2}{d\theta^2}-i\gamma\frac{d}{d\theta}+\frac{1}{2}\gamma^2R_a^2-\frac{2}{\sqrt{R_a^2+\xi^2}}+E_0;\qquad \hat{H}_r=-2\frac{d^2}{d\varphi^2}+U_{ee}(\varphi) \tag{5}$$

The eigenfunctions of the Hamiltonian (5) are $\Psi_{M,m,s}(\varphi,\theta)=\exp(iM\theta)\Phi_{m,s}(\varphi)$ being M and m the center of mass and relative angular quantum numbers ($M,m=0,\pm1,\pm2,...$) and $s=\pm$ the eigenfunction parity. The even solutions ($s=+$) correspond to singlet states and the odd solutions ($s=-$) correspond to triplet states. In our numerical work, we solve the wave equation $\hat{H}_r\Phi_{m,s}(\varphi)=E_r(m,s)\Phi_{m,s}(\varphi)$ by using the trigonometric sweep method [16].

Results and discussion

In order to check the quality of our calculation method we first calculated normalized energies $\left(\tilde{E}(M,m,s)=R_a^2\left(E_c(M)+E_r(m,s)\right)\right)$ for two-electron model $(\xi\to\infty)$ with centreline radii R_a, $4a_0$* and $20a_0$* as $R_t\to 0$ and compare them with those obtained in reference 4 for two-electron one-dimensional ring. Results in Table 1 show an excellent concordance between two sets of results.

Table N°.1 Comparison of energies $\tilde{E}(M,m,s)$ in the limit $R_t\to 0,\xi\to\infty$ with results of reference 4

	$\tilde{E}(0,0,0)$		$\tilde{E}(\pm1,1,1)$		$\tilde{E}(\pm2,0,0)$		$\tilde{E}(0,2,1)$		$\tilde{E}(\pm,1,1)$	
	$4a_0^*$	$20a_0^*$	$4a_0^*$	$20a_0^*$	$4a_0^*$	$20a_0^*$	$4a_0^*$	$20a_0^*$	$4a_0^*$	$20a_0^*$
Our results	5.18	22.40	5.68	22.90	7.18	27.54	7.91	24.40	8.41	29.54
Ref. 4	5.18	22.40	5.68	22.90	7.18	27.54	7.91	24.40	8.42	29.54

The total energies of the D^- located at the centre of torous $(\xi=0)$ with the cross-section size parameter $R_t=0.5a_0$*as functions of the torous centreline radius R_a for some low-lying levels are shown in figures 2 (for square and circular cross-sections in left- and right-side panels, respectively)

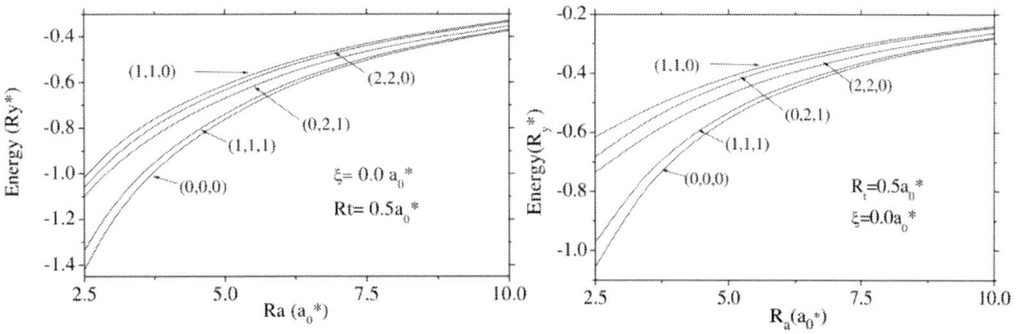

Figure 2. D^- energies of some low-lying levels as functions of the centreline radius for two different morphologies of the torous cross-sections: square (left panel) and circular (right panel).

One can observe that energies are sensitive to the morphology of the QR cross-section, although its size is very small. Comparison of energies of D^-ion shows that in a torous with square cross-section they are always below that those in torous with circular-cross section. It is due to the fact that the averaged separation between electrons is larger and the effective potential given by (4) is deeper in torous with square cross-section than circular one. As a result, the experimental observation of the D^-centre is more probable in torous with square cross-section than in circular one, since the the

D^- ground state energy is lower and the system is more stable against the dissociation. Also one can see that the greater the ring centreline radius the greater is the D^- energy and the smaller are gaps between different levels, independently of the cross section morphology. It is due to the fact that the contributions in the total energy of terms corresponding to the centre of mass and relative motion are inversely proportional to the centreline radius, M/R_a^2 and m/R_a^2, respectively, both tends to zero as the radius increases. On the other hand, one can see, as the centreline radius decreases, the energies for all levels monotonically decrease whereas the ground state energy tends the well known value of $1.055Ry^*$ [15] which corresponds to an energy value reported for the D^- in bulk. It is due to the Coulomb energy term is strongly dominated by attraction between the electrons and the positive ion.

In figure 3, we display the evolution of the D^- energy levels with the positive ion position along the z-axis for a QR with square cross-section. Remarkable evolutions of these curves are the result of the strong competition between electron-electron repulsion and the electron-ion attraction energy terms. As the positive ion is beyond $12a_0*$ from the centre of the toroidal ring, the attractive potential energy term becomes depreciable while the electron-electron repulsion term increases because the electrons are constrained to move into the ring. This is the reason by which all energy levels in figure 3 have positive values as $\xi > 12a_0*$. In the region $\xi < 12a_0*$, one can observe a substantial increase of the total energy accompanied of a notable variation in the slope of the curves, while the energy values for all levels become positive only from $\xi > 12a_0*$. From this value, the energy undergoes a significant reduction due to the dominance of electron-electron interaction. .

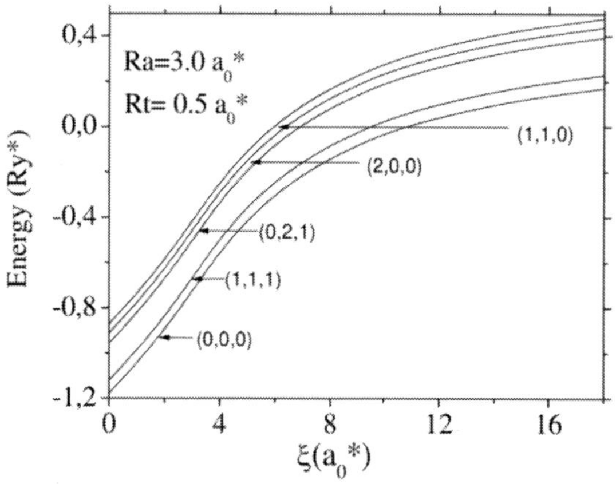

Figure 3. Energies of some low-lying levels of on-axis D^- in a quantum ring with square cross-section as function of the distance between the positive ion position to the center of ring.

The dependence of some energies levels $E(M,m,s)$ with $|M| \le 3$ on the magnetic field strength for a toroidal QR with mean radius $R = 4a_0*$ and square cross-section of side $1a_0*$ is shown in figure 4 for three different values of positive ion position, 0.0 (left-panel), 5.0 (center panel) and $50a_0*$ (right panel), respectively. From these panels, we can see once again that the greater the positive ion distance the greater is the energy values and the presence of the magnetic field yields the splitting of the energy levels whose angular momentum M has different signs. The behaviour of these curves is the result of the strong competition between the paramagnetic and diamagnetic terms in the Hamiltonian (5). The contribution of diamagnetic term is quadratic with the magnetic field strength while the paramagnetic field is linear. For this reason, as the magnetic field strength is small, all the curves are quasi-linear with a constant slope, being positive for M>0 and negative for M<0, and they takes a parabolic shape for large values of the magnetic field strength which makes that the all curves with M<0 take a minima for values of γ equal to $|M|/R^2$ a very narrow cross-section. This fact is responsible of the well known Aharonov-Bohm effect [1-2] which corresponds to the oscillatory behaviour of the lowest level energy as measure as the magnetic strength growing with period given by $1/R^2$. In the present case, where the mean radius is equal to $4a_0*$ this period is equal $0.0625a_0*$

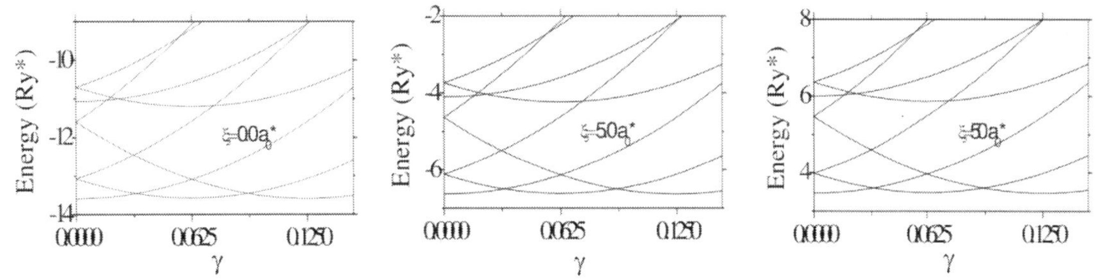

Figure 4 D^- energy levels as a function of the magnetic field strength in a toroidal quantum ring with mean radius $R = 4a_0*$ and square cross-section of side $R_t = 1a_0*$

Conclusions

In short, the negative charged donor energy levels in very narrow toroidal QRs with two different shapes are investigated. We have shown that D^- energy structure is dependent on the QR's cross-section morphology. Finally, at the limit cases when the positive ion position tends to infinity or the size of the cross-section tends to zero, our results are an excellent concordance with the exact solutions.

Acknowledgements

This work was financed by the Industrial University of Santander (UIS) through the Dirección General de Investigaciones (DIF Ciencias, Cod. Cod. 5124) and the Excellence Center of Novel Materials ECNM, under Contract No. 043-2005 and the Cod. No. 1102-05-16923 subscribed with COLCIENCIAS. We thank I. D. Mikhailov for helpful discussions.

References

[1] Aharonov Y and Bohm D 1959 Phys. Rev. **115** 485
[2] Viefers S, Koskinen P, Singha Deo P and Manninen M 2004 *Physica* E **21** 1
[3] Ihna T; Fuhrera A, Heinzela T, Ensslina K, Wegscheiderb W; and Bichlerc M 2003 *Physica* E **16** 83
[4] Zhu J-L, Dai Z and Hu X 2003 Phys. Rev. B **68** 045324
[5] Marín J H, García F and Mikhailov I D 2006 Braz. Jour. of Phys. **36**(3B) 940
[6] Lorke A , Bohm S and Wegscheider W 2003 Superlattices Microstruct. **33** 347
[7] Mendach S, Schumacher O, Heyn C, Schnull S, Welsch H and Hansen W 2004 Physica E **23** 274.
[8] Encinosa M 2005 Physica E **28** 209
[9] Chen G-Y, Chen Y-N and Chuu D-S 2007 Solid. State Commun. **143** 515
[10] García F, Marín J H, Paredes H and Mikhailov I D 2005 Phys. Stat. Sol.(c) **2** 3630
[11] Gómez C A, García L F, W. Gutiérrez W and Marín J H 2008 Microelectronics Journal **39** 1279
[12] Betancur J, Gutiérrez W and Pina J, 2007 Physica B **396** 12
[13] Mikhailov I D, Marín J H, and García F 2002 Phys. Stat. Sol. (b) **242** 1636
[14] Jacak L, Hawrylak P and Wójs A 1997 *Quantum Dots* (Berlin: Springer)
[15] Bethe H A and Salpeter E E 1957 *Quantum Mechanics of one-and two-Electron Atoms* (Berlin: Springer-Verlag)
[16] Betancur F J, Mikhailov I D and Oliveira L E 1998 J. Phys. D: Appl. Phys. **31** 3391

Optical response of confined excitons in GaInAsSb/GaSb Quantum Dots heterostructures

R Sánchez –Cano[1], L Tirado-Mejía[2], N Porras-Montenegro[3], G Fonthal[2], H Ariza-Calderón[2]

[1]Departamento de Física, Universidad Autónoma de Occidente, A.A. 2790, Cali, Colombia
[2]Laboratorio de Optoelectrónica, Universidad del Quindío, A.A. 4603 Armenia, Colombia
[3]Departamento de Física, Universidad del Valle, A.A. 25360, Cali, Colombia

E-mail: rsanchez40@gmail.com

Abstract. The narrow–gap $Ga_{1-x}In_xAs_ySb_{1-y}$ compounds are suitable materials for heterostructure devices operating in the infrared wavelength range. In these compounds grown by liquid phase epitaxy over GaSb single crystals, for x and y values in the range of 0.10 to 0.14 for both variables, the photoluminescence optical response at 12K is blue-shifted by 20 meV related to the photoreflectance response. We believe this behavior is due to possible higher electronic confinement in some places of the heterostructure, possibly formed in the interface during the growth process. In order to explain this behavior, in this work we study the exciton recombination energy in spherical Quantum Dots (QDs) on $Ga_{1-x}In_xAs_ySb_{1-y}$/GaSb, using the variational procedure within the effective-mass approximation and considering an electron in a Type I band alignment formed by two semiconductors with similar parabolic conduction bands. Our results are in good agreement with recent experimental results.

1. Introduction

In ternary III-V compounds semiconductor alloys $A_xB_{1-x}D$ or AC_yD_{1-y} type the band gap energy and the lattice parameter are generally both function of a single composition parameter, so they cannot be selected independently. In the quaternary $A_{1-x}B_xC_yD_{1-y}$ semiconductor alloys type, on the other hand, the two composition parameters, x and y, allowed the band gap and the lattice parameter to be selected independently within the constraints of a given substrate system. Quaternary alloys are appropriate materials for heterostructure devices because they provide a natural means of tuning the lattice parameter. In particular, in the quaternary alloy GaInAsSb it is possible to tune the magnitude of band gap so as to work in the infrared wavelength range [1-5]. The samples used in this work were grown by the liquid phase epitaxy (LPE) technique; details of growing conditions have been published elsewhere [6]. In the growth process the substrate and the liquid solution are put together at a temperature below the *liquidus* temperature. As temperature decreases, the solid phase from the liquid solution will appear. If substrate and liquid solution are in contact at a temperature higher than the equilibrium temperature the liquid solution will dilute the substrate due to its subsaturate condition. And as the temperature decreases again, the formation of the layer is induced. Is in this process where quantum dots could be formed in the interface.

The optical characterization of a GaSb sample and two epitaxial films of $Ga_{1-x}In_xAs_ySb_{1-y}$, lattice matched to GaSb, were carried out by photoreflectance (PR) and photoluminescence techniques (PL) [7]. PL measurements were carried out in different zones of the samples exhibiting the same energy position. This behavior was also observed for PR measures.

A lot of attention has been dedicated to the study of low-dimensional hetereostructures such as GaAs/GaAlAs, InGaAsP/InP and CdS [8-11] but there is a small number of works on the novel quaternary hetereostructures.

We present a calculation of the binding energy (E_{1b}) and the ground state energy levels of heavy-exciton (E_X) in spherical quantum dots (QDs) with a finite height potential for dot radii in the range of 5 – 20 nm and several values of Indium concentration x on $Ga_{1-x}In_xAs_ySb_{1-y}$ in the range of 0.10 to 0.14. The number 1 in the sub index corresponds to a ground state. We use the variational procedure within the effective-mass approximation, and we considered that electrons and holes are both confined within the same layers of $Ga_{1-x}In_xAs_ySb_{1-y}$. Excitons energies are calculated and compared with experimental data. These comparisons show that higher electronic confinements, such as QDs, are formed during the growth process of the film by LPE technique, in some zones of the heterostructure.

2. Theory

In this analysis it is considered an exciton, formed by an electron - hole pair, in a quantum dot system consisting of GaSb/GaInAsSb/GaSb such that the electrons and holes are both confined within the same layers of GaInAsSb.

In the effective-mass model treatment the Hamiltonian we use to describe the exciton localized in a quantum dot is given by:

$$H = -\frac{\hbar^2}{2m_e}\nabla_e^2 - \frac{\hbar^2}{2m_h}\nabla_h^2 - \frac{e^2}{\varepsilon_b(x)\left|r_e - r_h\right|} + V_e(r_e) + V_h(r_h) \qquad (1)$$

where $\varepsilon_b(x)$ is the static dielectric constant, m_e (m_h) is the effective mass of electron (hole) in $Ga_{1-x}In_xAs_ySb_{1-y}$ spherical quantum dot, r_e (r_h) is the electron (hole) position relative to the center of the spherical QD and, $V_e(r_e)$ and $V_h(r_h)$ are the barriers height of the confining potential for electrons and holes respectively. The confinement potentials for the electron and hole, in the Hamiltonian of the system are given by:

$$V_e(r_e) = \begin{cases} 0 & r_e, r_h < R \\ \Delta E_C & r_e, r_h > R \end{cases} \quad ; \quad V_h(r_h) = \begin{cases} 0 & r_e, r_h < R \\ \Delta E_V & r_e, r_h > R \end{cases} \qquad (2)$$

where R is the radius of the spherical QD, and ΔE_C and ΔE_V are the conduction and valence band offsets respectively.

The conduction and valence band offsets, ΔE_C and ΔE_V, in the equation (2) are given by:

$$\Delta E_C = \chi_{(GaInAsSb)} - \chi_{(GaSb)} \quad ; \quad \Delta E_V = \Delta E_g - \Delta E_C \qquad (3)$$

where $\chi_{(GaInAsSb)}$ and $\chi_{(GaInAsSb)}$ are the electron affinity of the semiconductors, and ΔE_g is the band gap change.

We followed a variational approach to solve the Hamiltonian expressed in equation (1). Taking into account the spherical confining geometry, $V_e(r_e)$, $V_h(r_h)$ confinement potentials, and Coulomb interaction between the electron and hole we take the product of 1s-hydrogenic wave function as the trial wave function for the exciton [8-11]. The ground – state wave function is given by:

$$\psi_{(r_e,r_h)} = \begin{cases} A(R-\alpha r_e)(R-\alpha r_h)e^{-\alpha(r_e+r_h)} & r_e, r_h \le R \\ \dfrac{A(1-\alpha)^2 R^4}{r_e r_h} e^{-2(\alpha-\beta)R-\beta(r_e+r_h)} & r_e, r_h \ge R \end{cases} \qquad (4)$$

where A is the normalization constant of the wave function, and α, β are the variational parameters to be determined after minimization of the total energy (E). With the boundary conditions:

$$\frac{1}{\mu_{in}^{*}}\frac{1}{\psi_{in}}\frac{\partial\psi_{in}}{\partial r_t}\Big|_{t=R} = \frac{1}{\mu_{out}^{*}}\frac{1}{\psi_{out}}\frac{\partial\psi_{out}}{\partial r_t}\Big|_{t=R} \tag{5}$$

where the subscript t is depending the electron or hole case and μ_{in}^{*}, μ_{out}^{*} are the reduced effective mass of the exciton inside and outside the spherical QD; we can express β by means of α as:

$$\beta = \frac{\gamma[\alpha + \alpha(1-\alpha)R] + \alpha - 1}{(1-\alpha)R} \quad ; \quad \gamma = \frac{\mu_{out}^{*}}{\mu_{in}^{*}} \tag{6}$$

The exciton energy E_X is found by minimizing the total energy E

$$\left\langle \psi_{(r_e,r_h)} \middle| H(\alpha,x,R) \middle| \psi_{(r_e,r_h)} \right\rangle = E \left\langle \psi_{(r_e,r_h)} \middle| \psi_{(r_e,r_h)} \right\rangle \tag{7}$$

The variational binding energies of the 1s-excitons states (E_{1b}) are obtained by subtracting from the lowest electron (E_{1e}) and hole subbands energies (E_{1h}), the ground-state exciton energy (E_X) expressed in equation (7). These subbands energies, E_{1e} and E_{1h}, are determined solving the transcendental equations numerically for finite spherical quantum wells. From these results it can be found the critical radius value required for the existence of a bound state [12].

3. Results and Discussion

The exciton binding energy, E_{1b}, presents a maximum when plotted versus the radius of the exciton. E_{1b} increases as R is reduced until it reaches a maximum, and then decreases rapidly, as presented in figure 1. The value of R at which E_{1b} reaches a maximum is smaller for larger x. Furthermore, we found that $\langle R \rangle \approx$ 15nm corresponds to the average radius for which the exciton binding energy is a maximum in the complete range of Indium concentration ($0.11 \leq x \leq 0.15$).

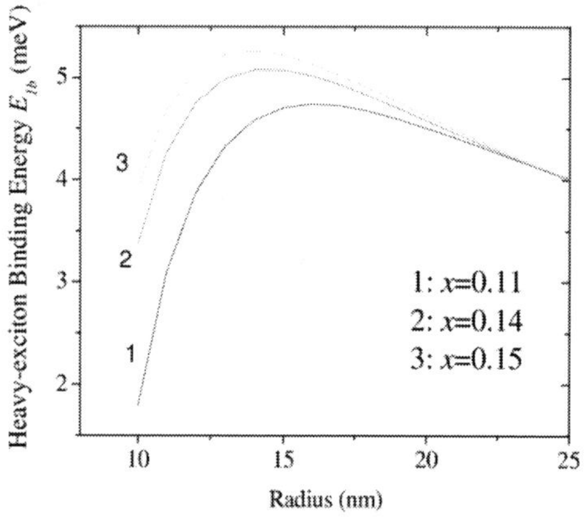

Figure 1. Binding energy of heavy-exciton of the ground state E_{1b}, in a spherical GaSb/ $Ga_{1-x}In_xAs_ySb_{1-y}$-GaSb quantum dot as a function of the radius for three different Indium concentration: x=0.11, x=0.14, x=0.15.

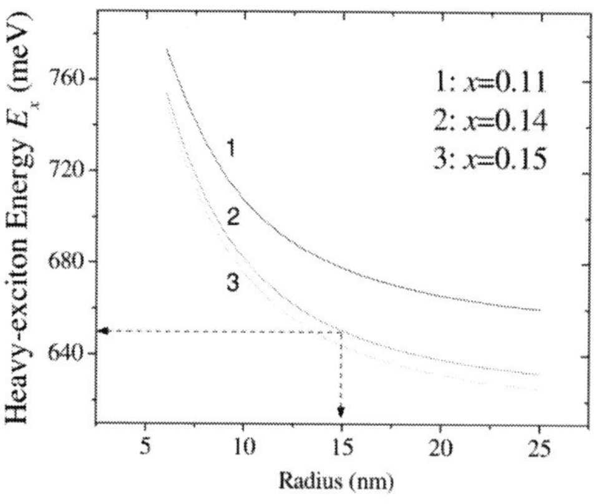

Figure 2. Heavy-exciton energy of ground state in a spherical GaSb-$Ga_{1-x}In_xAs_ySb_{1-y}$/GaSb quantum dot as a function of the radius for three different Indium concentration: x=0.11, x=0.14, x=0.15.

In figure 2 it is observed that the exciton energy increases as the radius decreases and approximates to a limiting value when the radius approaches to the critical radius. It corresponds to the particle radius

for which the electron is not confined (R ≈ 6nm), while the hole is still confined. We found that for <R> ≈ 15nm and $x = 0.14$ the exciton energy is $E_X = 0.65$eV. This value for the exciton energy is in agreement with the experimental photoluminescence response showed in figure 3. In materials where these quantum structures are present, PL spectra exhibit transitions in which they are involved. The difference between the band to band and QDs emission energy will be the responsible for the anomalous behavior presented by PL emission energy compared to PR oscillation energy.
In figure 3 we show results obtained by PR and PL techniques [3] for a GaInAsSb sample. The photoluminescence optical response at 12K is blue-shifted by 20 meV related to the photoreflectance response. This behavior was not observed in GaSb single crystals as shown in figure 4.

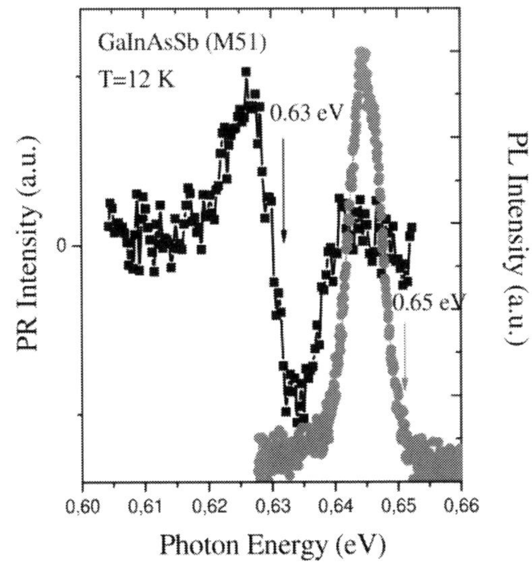

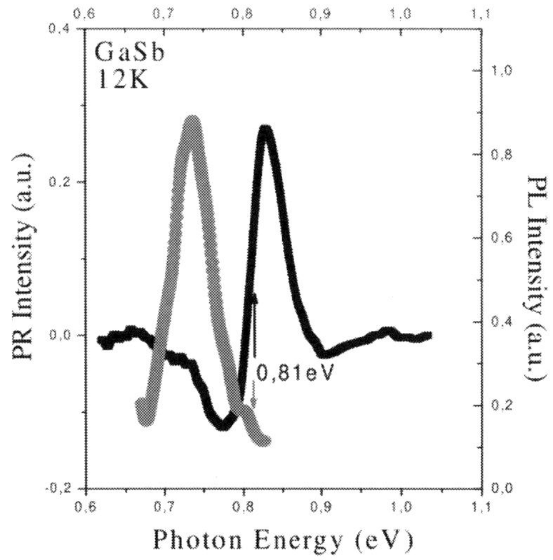

Figure 3. Photoreflectance (solid squares) and photoluminescence (solid circles) spectra for sample GaInAsSb M51 measured at 12K; the photoluminescence peak is shift 20 meV related to the photoreflectance response [3].

Figure 4. Photoreflectance (solid squares) and photoluminescence (solid circles) spectra for GaSb single crystal measured at 12K; the photoluminescence peak and photoreflectance response are both in 0.81eV [3].

4. Conclusions

We have determined the ground-state energy and binding energy of heavy-excitons confined in GaSb-$Ga_{1-x}In_xAs_ySb_{1-y}$-GaSb spherical quantum dots using variational procedure within the effective-mass approximation as a function of the quantum dot radius for several values of the Indium concentration (x).

We compared our theoretical results with the available experimental data and found good agreement. These comparisons show that higher electronic confinements, such as QDs, are formed during the process of growth of the film by LPE technique in some places of the heterostructure. This indicates that the assumption of finite barrier height and the introduction of the reduced effective mass of the exciton inside and outside the spherical quantum dot would be essential to achieve a good agreement between theory and experiment.

Acknowledgments
This work was supported by Universidad del Quindío, Universidad Autónoma de Occidente in Cali-Colombia, and the Center of Excellence for Novel Materials CENM under contract 0043-2005 with COLCIENCIAS.

References

[1] Shim K, Rabittz H and Dutta P 2000 *J. Appl. Phys.* **88** 7157
[2] Andreev I A, Il'inskaya N D, Kunitsyna E V, MikhaÏlova M P and Yakovlev Yu P 2003 Semiconductors. **37** 949
[3] Stoyanov N D, MikhaÏlova M P, AndreÏchuk O V, K D Moiseev, Andreev I A, Afrailov M A and Yakovlev Yu P 2001 Semiconductors. **35** 453
[4] Popov A A, Sherstnev V V, and Yakovlev Yu P 1998 Tech. Phys. Lett. **24** 73
[5] Ru G, Zheng Y and Li A. J 1995 Appl. Phys. **77** 6721
[6] Tirado-Mejía L, Ramírez J G, Gómez M E and Ariza-Calderón H 2006 *Brazilian J. Phys.* **36** 1070
[7] Tirado-Mejía L and Fonthal G, 2007 Private communication Laboratorio de Optoelectrónica, Universidad del Quindío, A.A. 460 Armenia Colombia
[8] Marín J L, Riera R and Cruz S A 1998 *J. Phys.: Condens. Matter* **10** 1349
[9] Porras-Montenegro N and Pérez-Merchancano S T 1992 *Phys. Rev. B* **46** 9780
[10] Laheld U E H, Pedersen F B and Hemmer P C 1995 *Phys. Rev. B* **52** 2697
[11] Greene R L, Bajaj K K and Phelps D E 1984 *Phys. Rev. B* **29** 1807
[12] Davydov A S 1965 *Quantum Mechanics*, Pergamo Press, 130

Exciton in type-II quantum dot

J Sierra-Ortega[1], R A Escorcia[1], and I D Mikhailov[2]
[1] Universidad del Magdalena, *A. A. 731, Santa Marta, Colombia*
[2] Universidad Industrial de Santander, *A. A. 678, Bucaramanga, Colombia*

E-mail: jsierraortega@gmail.com

Abstract We study the quantum-size effect and the influence of the external magnetic field on the exciton ground state energy in the type–II InP quantum disk, lens and pyramid deposited on a wetting layer and embedded in a GaInP matrix. We show that the charge distribution over and below quantum dot and wetting layer induced by trapped exciton strongly depends on the quantum dot morphology and the strength of the magnetic field.

1. Introduction

The study of the behaviour of Wannier excitons in both type-I and type-II self-assembled quantum dots (QDs) has attracted considerable attention in recent years. In type-I structures, such as *InAs/GaAs* or *InAlAs/AlGaAs*, both carriers, which forms exciton, are located inside the quantum dots, while in type-II structures such as *InP/GaInP* or *GaSb/GaAs* inside the dots can be located only electron or only hole. Separate localization of the carriers in type-II axially symmetrical heterostructures can lead to interesting effects such as persistent currents, oscillations of the exciton ground state energy, and crossovers of the energy levels [1]. At the same time, analysis realized recently shows that the modification of the density distribution in two-dimensional structures with broken symmetry produces a partial or total quenching of these effects for low-lying energy levels [2]. On the other hand, according to the results obtained previously for type-I QDs in form of the disk [3], lens, and pyramid [4] the three-dimensional density distribution of carriers in QDs is modified due to an appreciable tunnelling of carriers toward the matrix through the upper and lower QD's junctions. As the electron is lighter than the hole it's tunnelling toward the matrix is stronger than one of the electron. Therefore in the regions above and below the type-I QD with trapped exciton the negative charge exceeds the positive one, while in the central region of the QD the charge distribution is inverted. In this work we present results of similar calculations of the charge distributions, provided by the trapped exciton around and inside type-II QDs with different morphologies.

2. Theoretical model

We consider a model of axially symmetrical quantum-well structure in the form of thin InP layer whose thickness d depends on the distance ρ from the axis as:

$$d(\rho)=d_b+d_0\left[1-(\rho/R_0)^n\right]^{1/n}\vartheta(R_0-\rho) \quad (1)$$

Here $\vartheta(x)=0$ for $x<0$ and $\vartheta(x)=1$ for $x>0$, d_b is the wetting layer (WL) thickness, R_0 and d_0 are

[1] Corresponding author: Tel.: +575 4308684; fax: +575 4213011. (J. Sierra-Ortega)

the QD base radius and maximum height over the WL, respectively. The structure is embedded in GaInP matrix. The QD morphology according to relation (1) is controlled by means of the integer shape-generating parameter n which is equal to 1, 2 or tends to infinity for conical pyramid-like, lens-like and disk-like geometrical shapes, respectively. For sake of the mathematical convenience and, besides, in accordance with the experiment, we assume, that the height-to-base radius aspect ratio in all cases is small, i.e. $(d_0+d_b)/R_0 << 1$. In our model type-II quantum dot the position vectors $\mathbf{r}_i$ of the electron sited mainly inside the layer and of the hole sited mainly outside the layer (figure 1) are given in cylindrical coordinates by the components $(\rho_i, z_i, \vartheta_i)$. Here and in what follows indices $i=e,h$ correspond to the electron and the hole with the masses m_e, m_h, respectively.

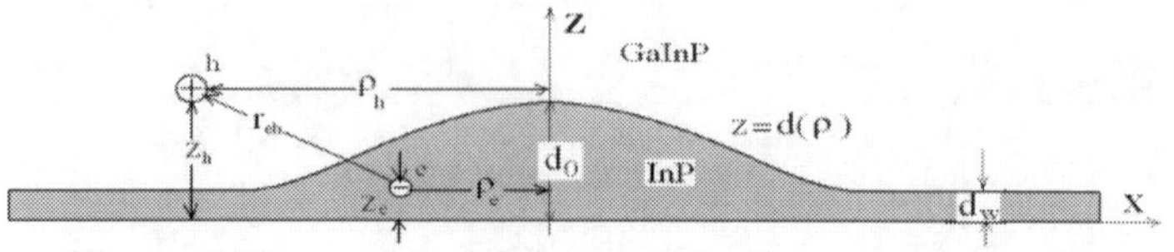

Figure 1 The model of the exciton in II-type quantum dot

Below we use as the unit of length the exciton effective Bohr radius $a_0^*=\hbar^2\varepsilon/\mu e^2$, as the unit of energy the effective Rydberg $Ry^*=e^2/2\varepsilon a_0^*$, and $\gamma=e\hbar B/2\mu c Ry^*$ as the unit of the strength of the magnetic field B applied along the symmetry axis, being $\mu=m_e m_h/(m_e+m_h)$ the reduced mass, e the electron charge, and ε the dielectric constant. In the effective-mass approximation the dimensionless Hamiltonian of the exciton corresponding to its ground state can be written as:

$$\hat{H}=-\sum_{i=e,h}\left[\eta_i\left(\frac{1}{\rho_i}\frac{\partial}{\partial\rho_i}\rho_i\frac{\partial}{\partial\rho_i}+\frac{\partial^2}{\partial z_i^2}+\frac{\gamma^2\rho_i^2}{4}\right)+V_i(\rho_i,z_i)\right]-2/r_{eh} \tag{2}$$

Here $V_e(\rho_e,z_e)=0$ and $V_h(\rho_h,z_h)=V_{h0}>0$ if $0<z_i<d(\rho_i)$, $i=e,h$, otherwise $V_e(\rho_e,z_e)=V_{eo}>0$ and $V_h(\rho_h,z_h)=0$, $\eta_e=\mu/m_e$, $\eta_h=\mu/m_h$, and r_{eh} is the electron-hole separation. In our model the hole is retained close to the QD only due to the attraction to the electron trapped inside the layer and therefore the typical scale for the distance between them is the effective Bohr radius. This value can be assured by the presence of the factor $\exp(-r_{eh})$ in the ground state wave function. It is the reason why we choose the trial function for the trapped exciton as follows:

$$\Psi(\mathbf{r_e},\mathbf{r_h})=\chi(r_{eh})f_e(\rho_e,z_e)\Phi(\rho_h),\quad \chi(r_{eh})=\exp(-r_{eh}) \tag{3}$$

Here an unknown envelope function $\Phi(\rho_h)$ describes a modification of the Slater orbital due to the presence of the repulsive $V_h(\rho_h,z_h)$ and attractive, $\gamma^2\rho_h^2/4$ potentials in the Hamiltonian (2) and $f_e(\rho_e,z_e)$ is the electron ground state wave function in QD, the exact solution of the equation:

$$\left(-\eta_e\Delta_e+V_e(\rho_e,z_e)+\frac{\eta_e\gamma^2\rho_e^2}{4}\right)f_e(\rho_e,z_e)=E_e f_e(\rho_e,z_e) \tag{4}$$

Taking into consideration that the structures under consideration have a small the height-to-base radius aspect ratio we found the solution of the equation (4) corresponding to the lowest energy E_e by using the procedure corresponding to the adiabatic approximation, described in reference [4]. The exciton energy of the ground state E_X is found by minimizing the ratio:

$$E_X=\left\langle\chi(r_{eh})f_e(\rho_e,z_e)\Phi(\rho_h)\middle|\hat{H}\middle|\chi(r_{eh})f_e(\rho_e,z_e)\Phi(\rho_h)\right\rangle/\left\langle\chi^2(r_{eh})f_e^2(\rho_e,z_e)\Phi^2(\rho_h)\right\rangle \tag{5}$$

Taking the functional derivative with respect to Φ gives a wave equation of the form:

$$-\frac{\eta_h}{J(\rho)}\frac{d}{d\rho}J(\rho)\frac{d\Phi(\rho)}{d\rho}+\tilde{V}(\rho)\Phi(\rho)=-E_b\Phi(\rho);\quad E_b=E_e-E_X, \tag{6}$$

where E_b is the exciton binding energy, $J(\rho)$ and $\tilde{V}(\rho)$ are the Jacobian and the effective potential energy as functions of the distance from the hole position to the axis, respectively, defined as:

$$J(\rho)=\left\langle \chi^2 f_e^2 \delta(\rho_h-\rho)\right\rangle;\quad \tilde{V}(\rho)=-1+\eta_h\gamma^2\rho^2/4+\left\langle V_h(\rho_h,z_h)\chi^2 f_e^2\delta(\rho_h-\rho)\right\rangle \tag{7}$$

The presence in these expressions of delta-function, $\delta(\rho_h-\rho)$ allows us to simplify essentially the calculation of the correspondent dependencies. By using the substitution $\Phi(\rho)=u(\rho)\sqrt{\rho/J(\rho)}$ the equation (6) is reduced to a form typical for two-dimensional central force problem

$$-\frac{\eta_h}{\rho}\frac{d}{d\rho}\rho\frac{du(\rho)}{d\rho}+V_{eff}(\rho)u(\rho)=E_X u(\rho);\, V_{eff}(\rho)=E_e+\tilde{V}+0.5\left(w'-0.5w^2\right);\, w(\rho)=J'(\rho)/J((\rho)) \tag{8}$$

Here V_{eff} is the effective potential which controls the in-plane motion of the hole around the QD. The potential curves V_{eff} in the disk, the lens and the conical pyramid with the base radius 8nm, the thickness of the WL 2nm and the greatest QD height over the WL 3nm for the zero-magnetic field case are presented in Fig. 2. It is seen that the effective potentials strongly depends on the QDs morphology.

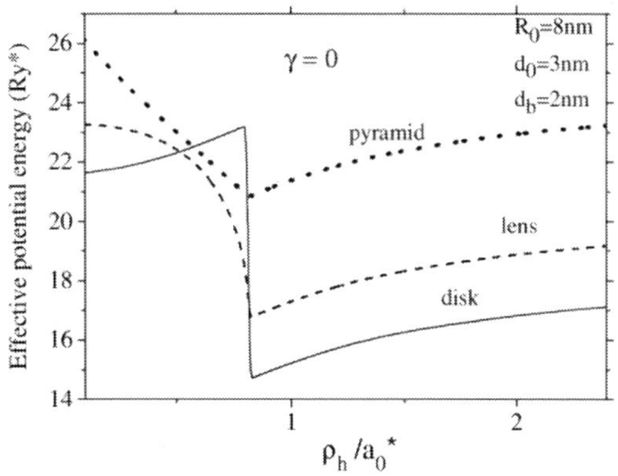

Figure 2 Effective potential curves for the in-plane motion of the hole in QDs with different profiles

3. Results and discussion

We solve one-dimensional wave equations (8) by using the trigonometric sweep method described in Ref. [3]. In our calculations we took the parameters: $m_e=0.077m_0, m_h=0.6m_0, V_e=250meV$, $V_h=-50meV$, and $\varepsilon=12.61$, which are typical for InP/GaInP system [5]. In order to understand the effect of the QD morphology on the exciton ground state energy we first calculated the density of the charge distribution inside and around the QD with trapped exciton, defined as the difference between the hole and the electron density distributions... It is clear that the change of the InP layer morphology should provide a variation of the averaged separation between the electron and the hole. For example, if in the disk-shaped QD the electron can be displaced inside the QD along the layer relatively easy, following to the motion of the hole, in the case of the conically-shaped QD the electron is mainly localized close to the axis. Therefore the mean distance between the electron and the hole in a conically-shaped QD is larger and the binding energy is inferior than those in the disk. In Fig. 3 we present contour plots, which correspond to the level lines of the charge density distributions along a cross section in the middle of the InP/GaInP quantum disk, lens and conical pyramid with trapped exciton. The left- and right-side parts of the figure correspond to the models of QDs without and with WL, respectively. It is seen that in contrast to the distribution obtained previously for the type-I QD in reference [4], the density of the radial charge distribution inside QDs is generally negative while outside QDs it is positive. Also, one can see that the region of the positive charge over and below QDs is extended toward axis while the morphology of the QD is changed from disk-shaped to conical-shaped. The existence of the WL leads to a lowering of the effective barrier height for the in-plane electron motion reinforcing significantly it's tunnelling in the lateral direction, resulting in a stronger charging of the peripheral regions in the lateral direction that can be seen in right-side plots of Fig. 3.

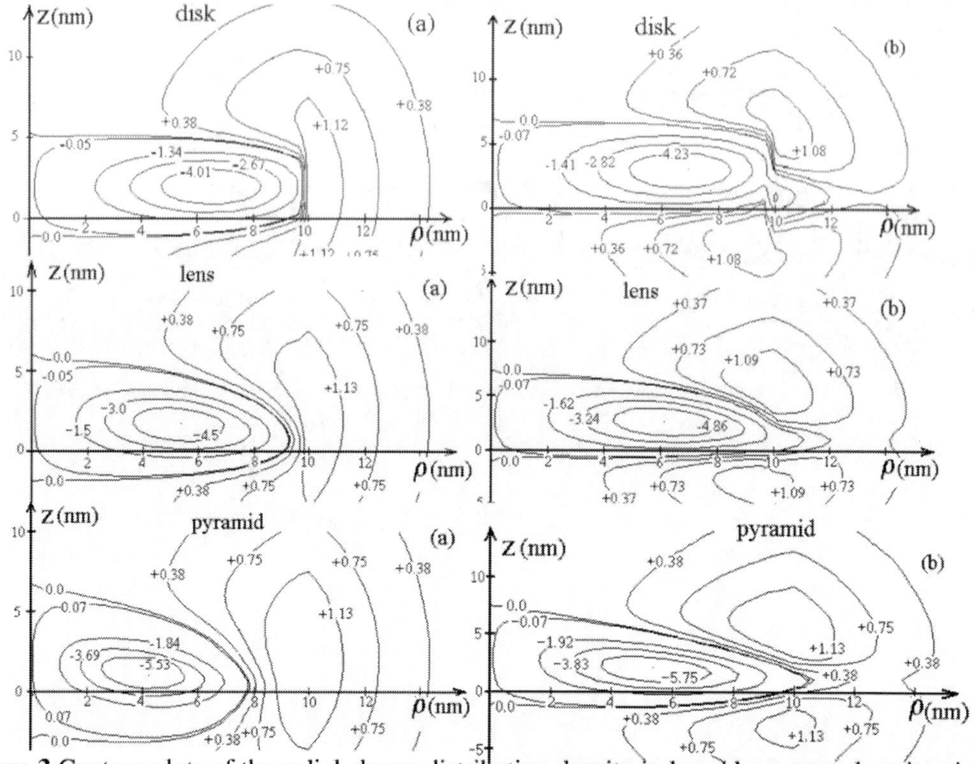

Figure 3 Contour plots of the radial charge distribution density induced by captured exciton in a plane trough the axis of symmetry of the InP/GaInP quantum disk, lens and conical pyramid with the base radii 10nm and the heights 4nm, and with 0 (left-side) and 2nm (right-side) thickness of the WL.

Comparing contour plots in Fig. 3 with those from reference [4] for type-I QDs one can observe that the charging of the central y peripheral regions in the type-II QDs is essentially stronger than one in the type-I QDs. It is due to the fact that the regions of the carriers' localization in the type-I QDs are essentially overlapped while in the type-II QDs they are strongly separated

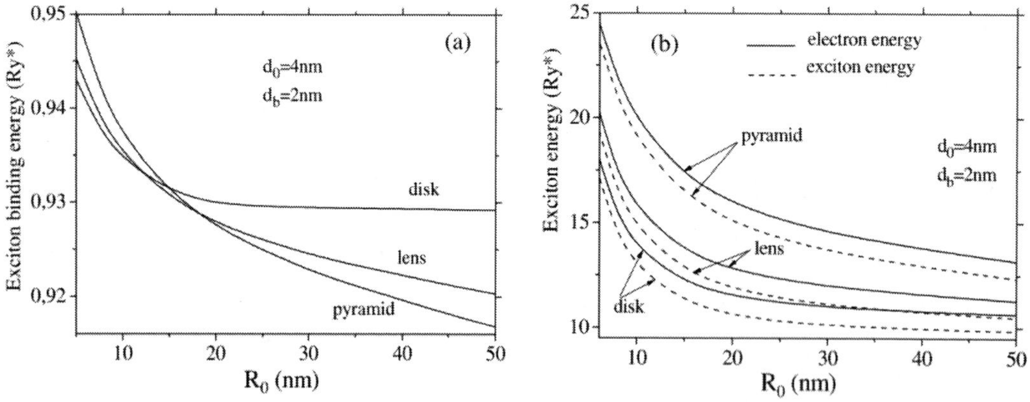

Figure 4 Exciton binding (a) and total (b) energies versus radii R_0 in QDs with different profiles.

In Fig. 4 we display the exciton binding and total energies dependencies on the base radii of the disk, the lens and the conical pyramid with height 4nm over WL, whose thickness is 2nm. It is interesting to note that for small QD radii $(R_0 < 15nm)$ the largest binding energy has the exciton in the conical pyramid followed by the lens and disk, whereas for large QD radii this order is inverted. It is due to the fact that the motion of electron is mainly restricted in QD close to axis and that the largest confinement is provided in the conical pyramid followed by the lens, and the disk. Therefore, as the QD radius is greater than the effective Bohr radius (about 15nm), the largest separation between the hole and the electron is in the conical pyramid. The order is reversed for $R_0 < 15nm$. One can see in Fig. 4(b) that the total energy dependencies in QDs with different shapes are rather similar being the energy in the conical pyramid for all base radii higher than those in the lens and in the lens higher than in the disk. In Fig. 5 we display the exciton binding energies dependences on the magnetic field strength for QDs with different shapes. It is seen that in all cases the exciton binding energy increases and that this increase is larger in the pyramid and the lens. The magnetic field affects mainly the density distribution of the hole, displacing its peak toward heterostructure axis, decreasing in this way the electron-hole separation and increasing the binding energy. Such modification of the distribution under magnetic field is less pronounced in disk-shaped QD.

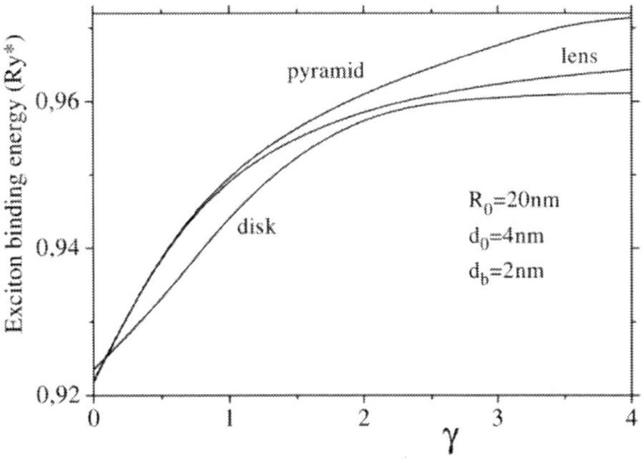

Figure 5 Exciton binding energy as a function of the magnetic field in QDs with different profiles

4. Conclusions

In order to study the effect of the morphology on the charge distribution in type-II quantum dots with trapped exciton we propose a simple trial function for calculating the charge distribution and the ground state energy of the exciton in flat disk-shaped, lens-shaped, and cone-shaped InP/GaInP quantum dots with different base radii, heights and wetting layer thicknesses. We present novel contour plots of the radial charge distribution density induced by captured exciton in InP/GaInP quantum dots with different morphology. It is found a strong influence of the morphology on the charge distribution inside, over and below QD with trapped exciton and on the exciton energy dependencies on the QD radius and the magnetic field strength.

Acknowledgements

This work was financed by the Industrial University of Santander (UIS) through the Dirección General de Investigaciones (DIF Ciencias, Cod. 5124) and the Excellence Centre of Novel Materials –ECNM, under Contract No. 043-2005 and the Cod. No. 1102-05-16923 subscribed with COLCIENCIAS.

5. References

[1] A.O. Govorov; A.V. Kalameitsev, R. Warburton, K. Karrai, S.E. Ulloa 2002 *Physica* E 13 297
[2] M. H. Degani, M. Z. Maialle, G. Medeiros-Ribeiro, and E. Ribeiro *Phys. Rev.* 2008 B **78** 75322
[3] Janssens K L, Peeters F M and Schweigert V A 2001 *Phys. Rev.* B **63** 205311
[4] Marín J H, Mikhailov I. D., and García L. F. 2007 Physica B **398** 135
[5] Janssens K L, Partoens B., and Peeters F M 2001 *Phys. Rev.* B **64** 155324

Journal of Physics: Conference Series **167** (2009) 012035 doi:10.1088/1742-6596/167/1/012035

Analysis of multiferroic properties in $BiMnO_3$ thin films

M. Grizalez[1], **G.A. Mendoza**[2], **and P. Prieto**[3]

[1] Universidad de la Amazonia, Florencia, Colombia
[2] Department of Physics, Universidad Nacional de Colombia, Bogotá, Colombia
[3] Center of Excellence on Novel Materials - CENM, Colombia

E-mail: pprieto@calima.univalle.edu.co

Abstract. Textured $BiMnO_3$ [111] thin films on $SrTiO_3$ (100) and $Pt/TiO_2/SiO_2$ substrates were grown via r.f. magnetron sputtering (13.56 MHz). The XRD spectra confirmed a monoclinic structure and high-quality textured films for the $BiMnO_3$ films. The films grown on $SrTiO_3$ (100) showed higher crystalline quality than those developed on $Pt/TiO_2/SiO_2$. Through optimized oxygen pressure of 5×10^{-2} mbar during the r.f. sputtering deposition, the crystalline orientation of the $BiMnO_3$ film was improved with respect to the previously reported value of 2×10^{-1} mbar. The values of spontaneous polarization (P_s), remnant polarization (P_r), and coercive field (F_c) from ferroelectric hysteresis loops (P–E) at different temperatures were also obtained. Samples with higher crystalline order revealed better dielectric properties (high P_s and P_r values and a low F_c). For films on both types of substrates, the ferroelectric behavior was found to persist up to 400K. Measurements at higher temperatures were difficult to obtain given the increased conductivity of the films. Magnetic hysteresis loops from 5K to 120K were obtained for $BiMnO_3$ films grown on $SrTiO_3$ and $Pt/TiO_2/SiO_2$ substrates. The results suggested that the coexistence of the magnetic and electric phases persists up to 120K.

1. Introduction

The term multiferroic has been coined to describe materials in which two or all three ferroic properties, namely, ferroelectricity, ferromagnetism, and ferroelasticity coexist in the same phase [1]. This indicates that these materials have spontaneous polarization, magnetization, and strain and that these order parameters can be regulated by the application of electric fields and/or magnetic fields, and/or by using mechanical stress. These compounds have gained renewed and ever-increasing research interest in the last three years [2]. Particularly, for magnetoelectric materials, it is possible to find a coupling between the magnetic and ferroelectric properties through the magnetoelectric effect. Those substances have stirred scientific, as well as research interest, given the numerous potential applications they can offer based on the mutual control of magnetic and electric orderings. The number of ferroelectromagnets is sparse and is reduced dramatically to a few cases, owing to the incompatibility between magnetism and ferroelectricity. Most ferromagnetic oxides have a symmetrical center and do not allow electric polarization, while most ferroelectric oxides consist of transition metal ions without the seed of magnetism, i.e., active electrons, *d*. Hence, only a few multiferroics have been reported to date [3]; although the history of studies on magnetoelectric materials dates to the 1950s with Smolenski *et al.* [4,5]. Recently, there has been renewed interest in the simple perovskite $BiMnO_3$ as a multiferroic material. Moreira dos Santos *et al.* reported on the synthesis of $BiMnO_3$ thin films, demonstrating the coexistence of ferromagnetic and ferroelectric properties [6]. This was also reported by Eerenstein *et al.* [7]. Recent theoretical calculations also suggest the likelihood of both ferromagnetic and ferroelectric characteristics, because of the covalent bonding between the bismuth and oxygen atoms [8]. These properties make this material potentially interesting for technological applications and in studying magnetoelectric interactions. An interesting characteristic of $BiMnO_3$ is that it can only be synthesized in bulk at high pressures of at least 6 GPa and high temperatures around 1100K [9], making it a difficult material for research. One way to facilitate research of such a compound would be its stabilization as a high-quality thin film. Through optimized oxygen pressure of 5×10^{-2} mbar during the r.f. sputtering deposition, the crystalline orientation of the $BiMnO_3/SrTiO_3$ film can be improved with respect to the earlier value of 2×10^{-1} mbar reported in another paper published by the authors [10]. In this study, we conducted structural, magnetic, and ferroelectric characterization of $BiMnO_3/SrTiO_3$ and $BiMnO_3/Pt/TiO_2/SiO_2/Si$ films.

2. Experimental Details

$BiMnO_3$ thin films were deposited by employing the r.f. magnetron sputtering technique onto $SrTiO_3$ (100): Nb 0.1% and $Pt/TiO_2/SiO_2/Si$ substrates in O_2 atmosphere. Ceramic targets were prepared by a solid-state reaction from a stoichiometric mixture of Bi_2O_3 and MnO_2. This mixture was pre-reacted in atmospheric air at 973K for 24 h. To obtain a denser target, polyvinyl butyral (PVB) was added to the reacted powder of 2.0% concentration in weight. The PVB was later removed after 5 h of annealing at 773K in air. Subsequently, the pellet was sintered at 1060K in a hermetic furnace in atmospheric air for 12 h to obtain a compact dark-gray pellet with a density in the order of 80% of the theoretical value with a 1-inch diameter. The deposition process commences with a long period of pre-sputtering (about one day) to avoid chamber pollution and target poisoning: beginning by heating the substrate to the desired temperature of 1120K at the system base pressure of 6.4×10^{-4} mbar, depositing the film for 1 h by using an optimized oxygen pressure of 5×10^{-2} and 5×10^{-3} mbar for $SrTiO_3$ and $Pt/TiO_2/SiO_2/Si$ substrates, respectively, carrying out a thermal treatment in an oxygen atmosphere for about 15 min and cooling the substrate to room temperature in the deposition gas atmosphere. The thickness of the films was estimated from the deposition rate at 100 nm.

A PANalytical X'Pert PRO X-ray diffraction system was used with Cu K_α radiation at room temperature. Polarization measurements of $BiMnO_3$ thin films on $SrTiO_3$ and $Pt/TiO_2/SiO_2/Si$ substrates were conducted by using an RT66 test system (Radiant Technologies). The cooling system and temperature controller used for ferroelectric measurements was a Cryodyne 22C model with LTS series. The system to collect data used for ferroelectric measurements was the precision LC analyzer with the VISION software from Radiant Technologies. Ferroelectric hysteresis loop measurements were conducted on the $BiMnO_3$ thin films in a temperature range between 100K and 400K. We used a mask to deposit circular top electrodes with diameters between 0.1 and 0.5 mm on the films. For the 1-mm thick target pellets, the dielectric losses or dissipation factor, $\tan\,\delta$, as conductivity, were measured as a function of temperature from room temperature to 870K. An HP 428A LCR was used at 10 selected values of frequencies between 20 Hz and 1 MHz.

3. Results and Discussion

X-ray diffraction was used to analyze the crystal structure of the thin films, indicating a monoclinic structure, where the (-113) peak was reported as the 100% peak (PDF 01-0071-5450). This indicates high-quality textured $BiMnO_3/SrTiO_3$ [111] films (Figure 1a). For $BiMnO_3/Pt/TiO_2/SiO_2/Si$ (Figure 1b), the XRD showed that the intensity of the (222) peak is 3 times higher than the (-113) peak, suggesting light textured growth along the [111] orientation.

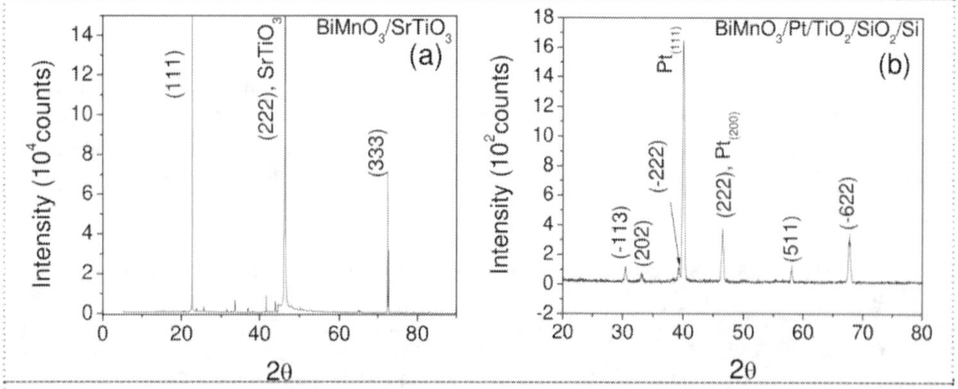

Figure 1. XRD diffraction pattern for (a) $BiMnO_3/SrTiO_3$ (b) $BiMnO_3/Pt/TiO_2/SiO_2/Si$

Typical ferroelectric hysteresis loops (P–E) of the thin films are given in Figures 2 and 3. Values for spontaneous polarization (P_s), remnant polarization (P_r), and coercive field (F_c) are presented in Table 1. As expected, the bulk material presents larger values of F_c and smaller values of P_s and P_r than the film samples. The sample with higher crystalline order ($BiMnO_3/TiSrO_3$) has better dielectric properties (high P_s and P_r values and a low F_c). For both films on $TiSrO_3$ and $Pt/TiO_2/SiO_2/Si$ substrates, the P_s and P_r values are improved by reducing temperature. For bulk material, no remarkable effect of temperature was observed. The P–E hysteresis loop persists up to 400K, but it is difficult to have good measurements at higher temperatures, given the increasing conductivity of the sample.

Table 1. Characteristic values for $BiMnO_3$

Sample $BiMnO_3$	T(K)	P_s ($\mu C/cm^2 \pm 0.01$)	P_r ($\mu C/cm^2 \pm 0.01$)	F_c ($kV/cm \pm 0.02$)
Bulk	220	0.25	0.12	4.42
Bulk	300	0.23	0.11	4.65
On $Pt/TiO_2/SiO_2/Si$	105	1.51	1.07	2.24
On $SrTiO_3$	122	2.21	1.50	2.12
On $SrTiO_3$	300	0.70	0.50	4.21

Magnetic hysteresis loops for $BiMnO_3$ on $SrTiO_3$ and $Pt/TiO_2/SiO_2/Si$ at 5K are shown in Figure 4. From magnetization, *M*, vs. applied field, *H*, curves, the values of M_s and H_c were obtained. These values, along with the values corresponding to $BiMnO_3$ developed at $P = 2\times10^{-1}$ mbar [10], are given in Table 2. The M_s and H_c are influenced by the crystalline quality of the films, i.e., the value of M_s increases when the crystalline qualities increase; while the value of H_c decreases with low crystalline qualities of the film. Magnetic hysteresis loops from 5 to 120K were obtained for $BiMnO_3$ films developed on $SrTiO_3$ and $Pt/TiO_2/SiO_2$ substrates (Fig. 4).

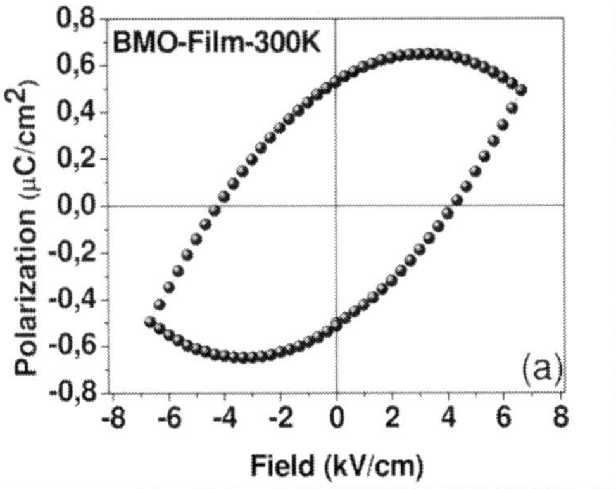

Figure 2. P-E hysteresis loop at 300K of $BiMnO_3/SrTiO_3$ film.

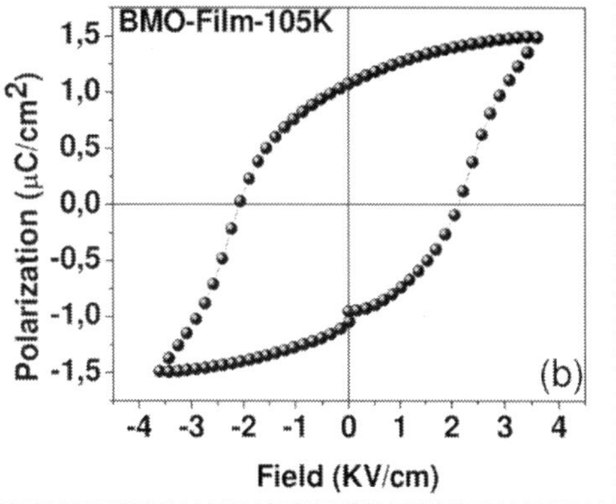

Figure 3. P-E hysteresis loop at 105K of $BiMnO_3/Pt/TiO_2/SiO_2/Si$ film.

Table 2. Magnetic properties of $BiMnO_3$ at 5K

Sample $BiMnO_3$	Oxygen Pressure (mbar)	M_s (μ_B/Mn site)	H_c (Oe)
[10]$BiMnO_3/SrTiO_3$	2×10^{-1}	2.78	50
$BiMnO_3/SrTiO_3$	5×10^{-2}	2.85	28
$BiMnO_3/$ $Pt/TiO_2/SiO_2/Si$	5×10^{-3}	2.75	38

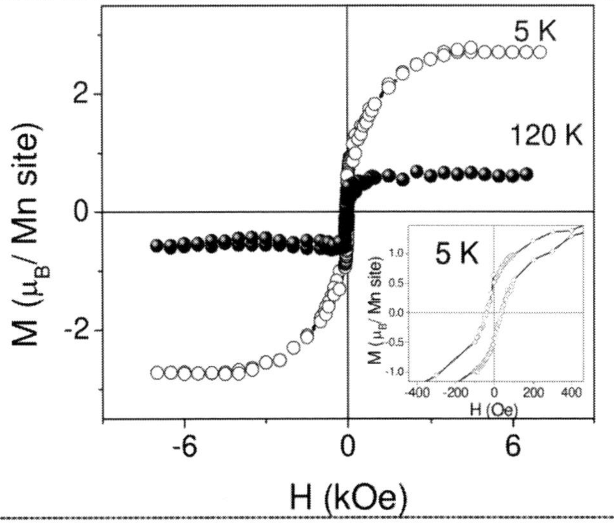

Figure 4. Magnetization (M vs H) hysteresis loops at 5K and 120K on $BiMnO_3/SrTiO_3$ grown in oxygen pressure of 5×10^{-2} mbar. Inset displays H_c ~ 28 Oe.

Figure 5 shows conductivity as being independent of frequency below 673K. For middle frequencies (5000-10000 Hz), conductivity enhances strongly from 720 K: and for a frequency of 1000 Hz, the conductivity increases in the range of 720–820K and then decreases. However, for small frequencies, we did not observe any remarkable changes in the conductivity. This complex behavior of BMO suggests that this material starts to break down if it is heated at the temperature range of 770K–920K, and that the path to decomposition is very complex with several metastable phases. In Figure 6, the dissipation factor, tan(δ) vs. temperature, shows possible ferroelectric–paraelectric transition around 630K. The magnitude of tan(δ) diminishes when the frequency is increased. We also observed a shift to higher temperatures when the frequency was increased. This may be attributed to the fact that conductivity increases when the temperature is increased.

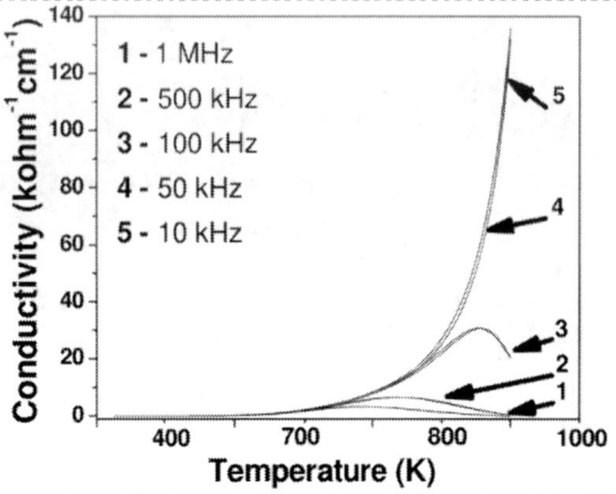

Figure 5. Conductivity vs. Temperature for a $BiMnO_3$ bulk sample.

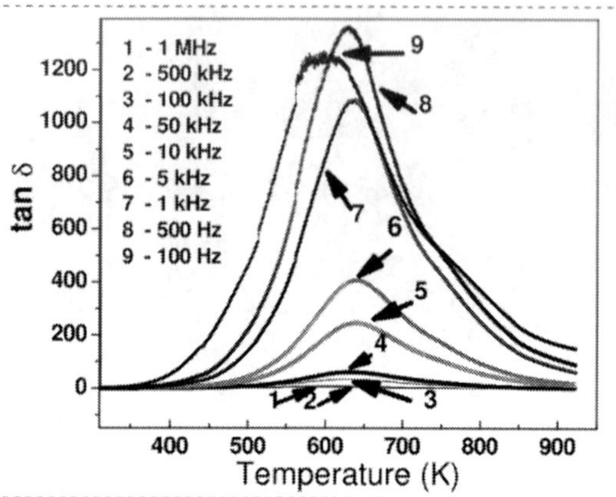

Figure 6. Dissipation factor (tan δ) vs. temperature for a $BiMnO_3$ bulk sample.

4. Conclusions

By optimizing the oxygen partial pressure in an r.f. sputtering deposition, we have obtained high-quality textured $BiMnO_3$ thin films on $SrTiO_3$ and $Pt/TiSO_2/SiO_2$. A ferroelectric-paraelectric transition around 630K for the $BiMnO_3$ bulk has also been indicated. For both samples, films and bulk, the P_s and P_r values have been improved by reducing the temperature, while no remarkable effect of temperature was observed on the bulk material. The M_s and H_c were observed to be influenced by the crystalline quality of the films, i.e., the M_s increases when the crystalline qualities increase, while the H_c decreases with low crystalline quality of the film. Furthermore, magnetic hysteresis loops from 5K to 120K were obtained for $BiMnO_3$ films grown on $SrTiO_3$ and $Pt/TiO_2/SiO_2$ substrates. Thus, the results obtained suggest that the coexistence of the magnetic and electric phases persists up to 120K.

Acknowledgments: This work was supported by the *Centro de Excelencia en Nuevos Materiales* - CENM (www.cenm.org) under Colciencias contract 043-2005.

References

[1] Hill N A, Filippetti, A 2002 *J. of Mag. and Mag.Mat.* **242-245** 976-979.
[2] Chi Z H, Xiao C J, Feng S M, Li F Y, Jin C Q, Wang X H, Chen Z, Li T 2005 *J. Appl. Phys.* **98** 103519.
[3] Hill N A 2000 *J. Phys. Chem.* B, **104** 6694.
[4] Khomskii D I 2006 *J. of Mag. and Mag.Mat* **37** 000-0001.
[5] Smolenskii G A, Chupis I E 1982 *Usp. Fiz. Nauk* **137** 415–448.
[6] Moreira dos Santos A, Parashar S, Raju A R, Zhao Y S, Cheetham A K, Rao C N R 2002 *Sol. State Comm.* **122** 49-52.
[7] Eerenstein W, Morrison F D, Scott J F, Mathur N D 2005 *App. Phys. Let.* **87** 101906.
[8] Hill N A, Rabe K M 1999, *Phys. Rev.* B **59** 8759.
[9] Moreira dos Santos A, Cheetham A K, Tian W, Pan X, Jia Y, Murphy N J, Lettieri J, Schlom D G 2004 *App. Phys. Let.* **84** 91-93.
[10] Grizalez M, Delgado E, Gómez M E, Prieto P 2007 *Phys. Stat. Solidi* C **4** 4203-4208.

Journal of Physics: Conference Series **167** (2009) 012036 doi:10.1088/1742-6596/167/1/012036

Resistance switching in silver - manganite contacts

F. Gomez-Marlasca and P. Levy [1]
Materia Condensada, GIA, GAIANN, CAC –CNEA,

and

Instituto de Nanociencia y Nanotecnología, CNEA,
Gral Paz 1499 (1650) San Martín, Pcia. Buenos Aires, Argentina

E-mail: levy@cnea.gov.ar

Abstract. We investigate the electric pulse induced resistance switching in a transition metal oxide-metal contact at room temperature - a non volatile, reversible and multilevel memory device. Using a simple multiterminal configuration, we find that the complementary effect -in which the contact resistance of each pulsed electrode displays variations of opposite sign- is strongly influenced by the history of the pulsing procedure. Loops performed by varying the magnitude and sign of the stimulus at each pulsed electrode allow to disentangle their sole contribution at different stages of the process. Electromigration of oxygen ions and vacancies is discussed as participating at the core of the underlying mechanisms for resistance switching.

1.Introduction

Electric pulse induced resistance switching (RS) is the focus of intense basic research nowadays. [1,2] This non-volatile electrically-induced and electrically-sensed memory effect has definite technological interest due to fast switching and scalability. Although RS has been observed in many different metal-oxide couples, a significant part of the community focusing this scenario is presently devoted to the study of transition metal oxides contacted through metal electrodes. [3,4,5] Among several families, perovskite based structures are being systematically studied [6] due to additional attractive features based on multifunctionality.[7] Manganites, cuprates and titanates exhibit bipolar type RS in which electrochemical migration of oxygen ions and vacancies is regarded as the driving mechanism. [1,8] In this case, RS takes place at the interface between the metal electrode and the oxide, in an interface – type path, as described thoroughly by A.Sawa [2] and references therein.

Evidence towards an oxygen diffusion mechanism by means of electric transport measurements [9] were recently obtained following a Hysteresis Switching Loop (HSL) procedure. The feasibility to induce related reversible switching at the nanoscale by means of conducting atomic force microscopy probe measurements has been recently shown. [10] Here, we take advantage from a simple multiterminal configuration design to systematically explore -under different pulsing protocols including HSL- differences between electrical behavior of complementary pulsed electrodes, i.e. each single electrical contact that suffered the pulsing discharge. We show that unveiled differences are directly related to the electric field established along each metal-oxide junction. Electromigration of oxygen ions and accumulation of vacancies are shown to play a role at the core of the underlying mechanism of the RS effect.

[1] To whom any correspondence should be addressed.

2.Experimental

We performed a systematic study on a manganite sample, with hand painted milimeter sized silver paste electrical contacts. As schematically shown in the Inset of Figure 1, electric pulses and bias current are always injected through terminals A and D, while voltage can be (simultaneously) measured at AB, BC and CD contacts by means of floating voltmeters. All experiments and data to be shown here were performed at room temperature under ambient pressure and magnetic field, on a polycrystalline sample of $La_{0.325}Pr_{0.300}Ca_{0.375}MnO_3$ (LPCMO) with average grain size 2 microns. Growth details, as well as temperature and intermediate range magnetic field dependences of the RS phenomena on LPCMO were previously reported. [7,11,12]

3.Results and Discussion

Initially, all contacts displayed comparable resistance values, as expected for a symmetric device, i.e. one in which all electrodes are made with the same metal and having similar dimensions. After repeatedly pulsing, a history dependant forming asymmetry was induced, hereafter we call it the "formed sample". Figure 1 exhibits panels displaying a) the pulsing sequence; b) AB, c) BC, d) CD measurements on each pair of contacts. Pulsing consisted of groups of 10 single pulses of 2 ms time width and 3.2 V amplitude. "Positive" means that positive carriers are injected through terminal A, producing a positive voltage drop across AB electrodes. After pulsing, nonvolatile voltage from each pair of electrodes were acquired sequentially using a small bias current (+1mA).

As observed in Figure 1, robust switching between Low Resistance (LR) and High Resistance (HR) nonvolatile states is obtained at electrodes AB and CD upon pulsing with alternate polarities. Note that a positive stimulus switches AB electrode to LR, and switches -in a complementary way [11]- the CD contacts to HR. Complementary electrodes are defined to be sole electrical contacts that suffered the pulsing discharge, A and D in this case. The complementary character of the voltage drop at AB and CD terminals refers to the sign of the change [11], i.e. it is not a numerical cancel of values, in agreement with (but seldom discussed at) previous reports in the area using a multiterminal approach. Resistance of the LPCMO sample (~0.08 ohm), as measured through BC terminals (standard 4 probe method) was unaffected by pulsing at our detection level.

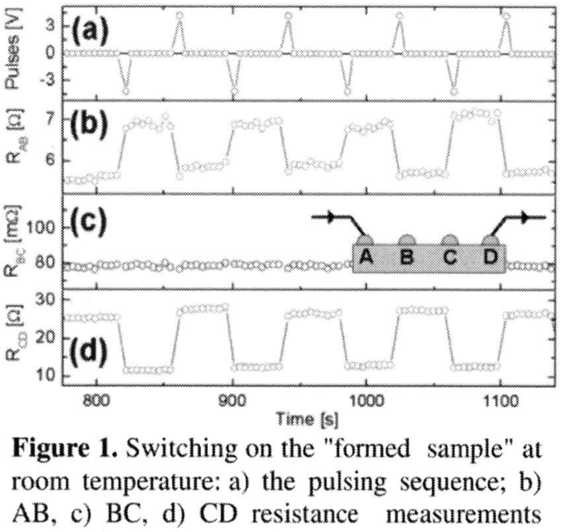

Figure 1. Switching on the "formed sample" at room temperature: a) the pulsing sequence; b) AB, c) BC, d) CD resistance measurements at each pair of electrodes. Inset: schematics of the multiterminal configuration.

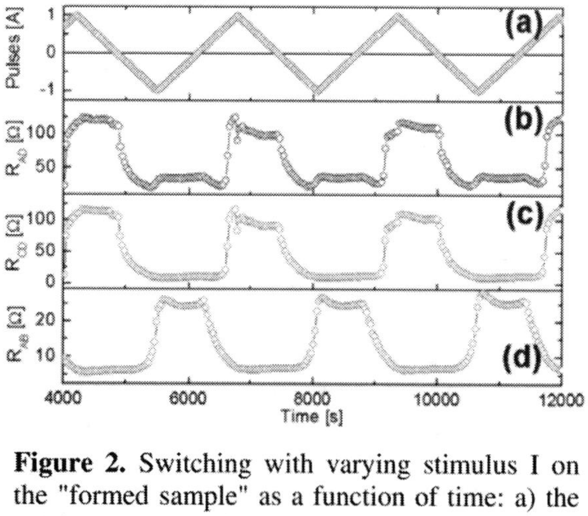

Figure 2. Switching with varying stimulus I on the "formed sample" as a function of time: a) the pulsing sequence; b) AB, c) CD, d) AD resistance measurements performed after pulsing.

The next experiment corresponds to an electrical pulsing protocol which scans in a loop mode the current during pulsing, the procedure hereafter referred to as HSL. The initial state for this HSL is unambiguously defined after repeatedly cycling (3 or more times) the stimulus –a single pulse of 2 ms time width- within + /- 1A amplitude with 10 V as limiting voltage. Then, the starting point of the loop is defined by injecting the stimulus at -1 A. The stimulus is then increased/decreased in steps of +/- 0.05 A, so as to describe a loop from -1 A to +1A, and then back to -1A. Remanent (i.e. nonvolatile) voltages at each pair of electrodes as well as the stimulus are depicted in Figure 2, as a function of time. Heating effects are expected to be present during pulsing. However, note that even for the highest I pulsing value (1A) the current density injected remains below 100 A/cm^2. While pulsing time was maintained small (~ few ms), no tracking of heating was evidenced on nonvolatile resistance values (see Figures 1 and 2).

The HSL response is depicted in Figure 3 as a function of the stimulus. Following the electrode AB, at the beginning of the sequence the HR state at -1A is sustained when I pulsing is decreased, reaching I = 0 with practically the same initial remanent value (see path "I" in Figure 3). Surprisingly, after I changes sign, its value decreases, reaching smoothly the LR state at about +1 A (path "II"). On the other hand, the complementary contact CD switches from LR to HR state in a sudden way and only after the stimulus I surpasses +0.5 A. Upon decreasing I from +1 A towards –1 A, the voltage at CD changes smoothly starting as soon as I changes sign, while AB switches steeply only after –0.5 A are surpassed (path "IV").

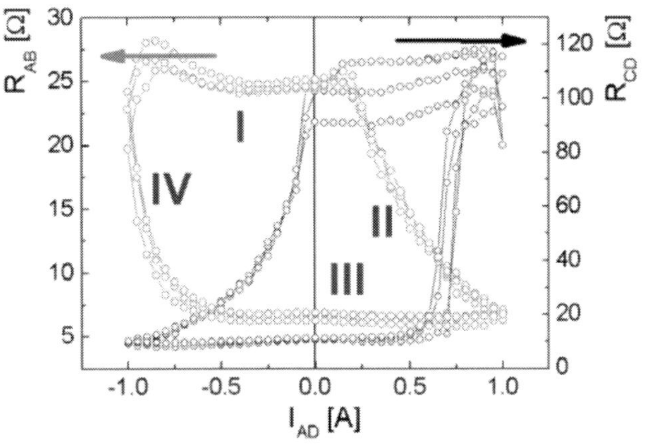

Figure 3. Non volatile voltage measured after pulsing following the Hysteresis Switching Loop procedure on the "formed sample", as a function of pulsing stimulus I.

The HSL, a distinctive fingerprint of each pulsed contact, is quite unexpected, evidencing a new aspect of complementary behavior. As seen on Figure 3, although pulsing is simultaneously applied to electrodes A and D, the response is not the same at both pulsed contacts, i.e. RS is obtained with different features at AB and CD electrodes, depending on pulsing history. Remarkably, it is possible to induce RS in one electrode and not in the other. This fact signals that complementarity [11] can be circumvented if certain pulsing protocols are applied. Interestingly, control of multilevel nonvolatile states emerge naturally upon performing the HSL experimental sequence.

When a positive pulse is applied at terminals AD, negative oxygen ions (O--) are ejected from the vicinity of the D electrode leaving the interface with vacancies and therefore interrupting and disturbing electrical transport. Pulsing with decreasing positive values leaves unaffected the HR state as these vacancies remain unfilled. Only when the polarity of the stimulus is changed, a smooth and gradual filling of vacancies with mobile O-- starts. Conversely: starting with the D electrode plenty of O--, when negative pulsing decreases towards zero, no net change is observed until the positive threshold electric field for detrapping O-- is attained. The observed difference in trapping-detrapping process is the disclosed origin of the RS mechanism.

Next, we applied a direct current of 1 A for 17 seconds (positive polarity). This protocol was a priori expected to produce a severe electric injure to the metal-oxide junction due to self heating at contacts. After this procedure, the "17sec-1A" sample still exhibited switching behavior similar to the previously described "formed" sample. Contact resistance figures appear to be drastically lower. The RS performance of the contacts AB and CD is now much more *dissimilar* as compared to the behaviour before the "17sec-1A" procedure. As expected, the intrinsic resistivity of LPCMO is unaffected by the electromigration process.

With the aim of measuring the dynamic resistance (i.e. resistance during pulsing) the HSL was performed using a single pulse having a time width of 60 ms. Under this protocol, the temperature at the contacts raises periodically due to self heating. The decay of their resistance during pulsing reveals the semiconducting like dependence of the tandem metal-oxide interface plus intrinsic LPCMO resistance. Though the maximum density current produces negligible volume heating, dissipative contact power at the interface (~ 100 W/cm^2) is undoubtedly high and local heating cannot not be avoided. We used the last voltage value obtained before the end of the pulse as a measure of the resistance during pulsing. After pulsing, the thermal decay occurring for some few seconds could be tracked with a small bias current (data not shown). Thus, in order to allow heat drift, event pulsing was separated by 30 sec intervals. Obtained data for electrodes CD on the "17sec-1A" sample is shown in Figure 4 as a function of time.

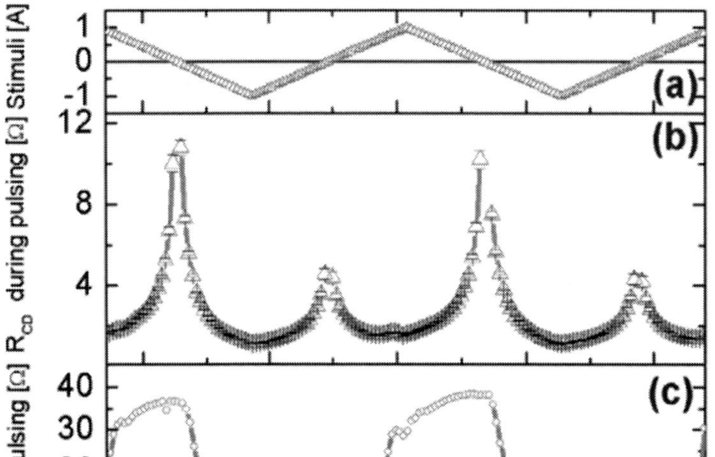

Figure 4. Resistance at electrodes CD on the "17sec-1A" sample as a function of elapsed time: a) Pulsing secuence; b) R_{CD} *during* pulsing (i.e., V_{CD}/I_{AD}); c) R_{CD} *after* pulsing (i.e., V_{CD}/I_{bias}).

Upon performing the HSL procedure on the "17sec-1A" sample (see Figure 5.a, path "I" starting at I=-1A) the remanent value of R_{AB} does not switch even after reversing and increasing the polarity of the stimulus up to + 0.40 A, when the RS is initiated. Both vertical edges of this HSL (paths "II" and "IV") display similar features being almost parallel with smooth corners, thus determining a HSL quite different to the one obtained on the "formed" sample (see Figure 3). On the other hand, as shown in Figure 5.b, the electrodes CD (remarkably displaying a higher contact resistance than the AB contact) exhibited a HSL which appears (at least in its shape appearance) unaffected by the "17sec-1A" process, and resembles the initial HSL of the CD electrodes.

This difference is undoubtedly related to the resistance level at each contact, which determines the effective electric field at the oxide metal junction when pulsing is performed. Note that the R_{AB} value and the amplitude of the HSL are both low, as a consequence of the electromigration produced on the "17sec-1A" sample. The difference above discussed for the HSL of AB and CD electrodes is also apparent studying the dynamic (i.e.volatile) resistance values measured at AB and CD *during* pulsing (see Figures 5.c and 5.d). These V –I curves exhibit non linear V-I behavior and a two level resistance state. While the curve for the AB contact is almost non hysteretical, data for the CD contact shows a clear jump from the LR to the HR state at + 0.6 A with hysteresis upon cycling the stimulus in a figure-of-eight mode, evidencing the nonsymmetrical transport behavior related to polarity dependant effects.

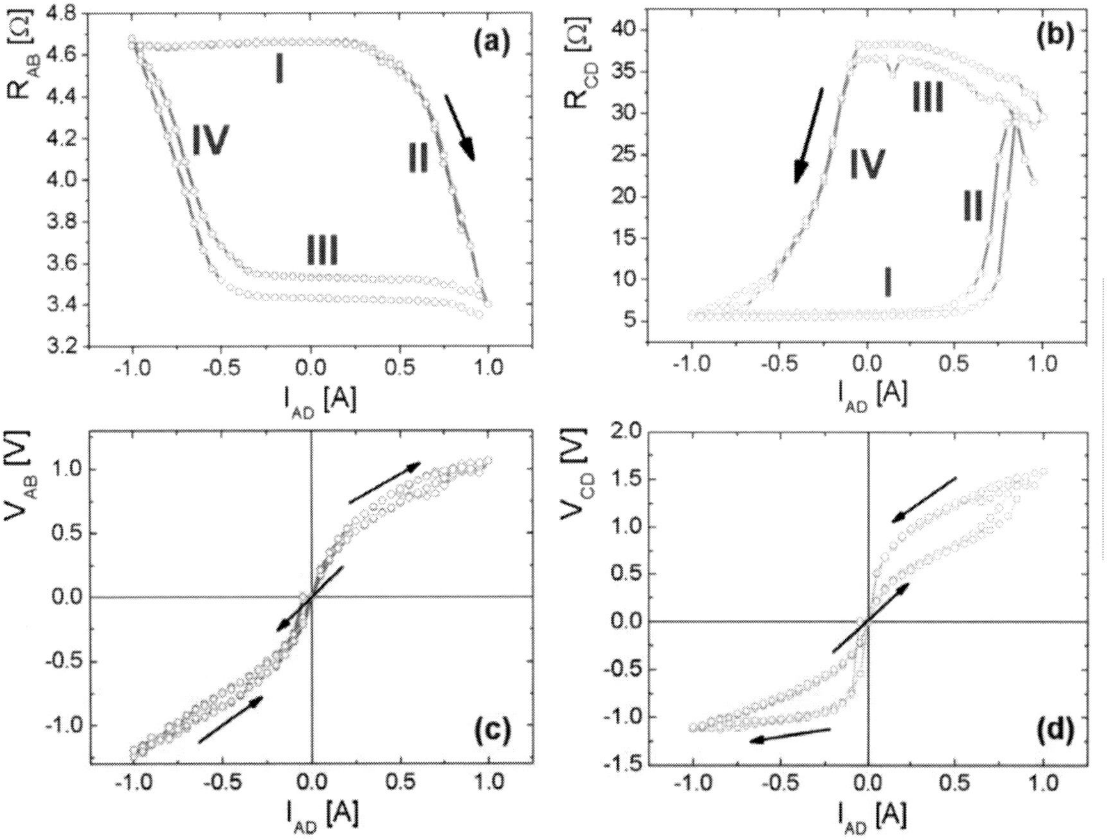

Figure 5. Comparison of volatile and nonvolatile states on the "17sec-1A" sample as a function of pulsing stimulus I_{AD}: a) R_{AB} *after* pulsing; b) R_{CD} *after* pulsing; c) V_{AB} *during* pulsing; d) V_{CD} *during* pulsing.

Switching is produced at the D electrode when a positive pulsing stimulus (+ 0.6 A in this case) is applied through AD terminals, as negative oxygen ions are detraped leaving a HR level at D ascribed to the presence of vacancies at the metal – oxide interface. This oxygen depleted state at D is easily converted to LR level soon upon polarity of the stimulus changes. Note that the negative pulsing in this part of the loop has very low incremental resistance (i.e. $\Delta V / \Delta I \sim 0.1$ ohms) related to the easy movement of O-- ions towards empty oxygen vacancies. These results and analysis are in full agreement with previous reports on $La_{0.66}Ca_{0.33}MnO_3$ / Ag contacts. [13] Moreover, with the above described approach and analysis, qualitative understanding is provided for HSL performed with decreased stimulus threshold due to intercalation of ferroelectric layers, as very recently reported. [14]

4.Conclusions

In conclusion, using very simple electrical transport techniques and protocols, we disclosed the core of the mechanism behind resistive switching in silver manganite contacts. We anticipate that similar procedure applied to different metal – transition metal oxide couples could disentangle plenty of data published since pulsing emerged as a possible nonvolatile memory for applications. We speculate that using metals with higher work functions [1] will produce steeper transitions at higher voltages, and that the use of oxides with higher oxygen mobility or an increase of the working temperature will degrade switching and retention capabilities. Besides, an open route for downscaling memory cells is envisaged, as the overall effect still relies on interface type conducting paths.

Acknowledgments
We acknowledge A.G. Leyva for sample preparation, M. Quintero for contacting electrodes. Support from the MeMO and EPICO initiatives is acknowledged. PL is member of CONICET.

References

[1] G.Meijer, Science 319 (2008) 1625
[2] A.Sawa et al., Materials Today, 11, (2008) 28
[3] J.Jameson et al., Appl. Phys. Lett. 91 (2007) 112101
[4] S. Seo et. al. Appl. Phys. Lett. 85, (2004) 5655; S. Seo et al. ibid vol86 (2005) 093509.
[5] J. Yang et al., Nature Nanotechnology 3, 429 (2008)
[6] M.Hamaguchi et al., Appl. Phys. Lett. 88, 142508 (2006)
[7] M.Quintero et al., Appl. Phys. Lett. 86 (2005) 242102
[8] Z.Szot et al., Nature Materials. 5 (2006) 312
[9] Y.Nian et al., Phys. Rev. Lett.98 (2007) 146403
[10] C. Cen et al., Nature Materials 7, 298 (2008)
[11] M.Quintero et al., Phys. Rev. Lett. 98 (2007) 116601
[12] M.Quintero, PhD Thesis (in spanish avalilable at http://www.tandar.cnea.gov.ar/doctorado/Tesis/Quintero.pdf)
[13] M. Belogolovskii et al., Low Temp. Phys. 28 (2002) 391
[14] Z. Xing et al., Appl. Phys. Lett. 91 (2007) 052106

EPR, μ -Raman and Crystallographic Properties of Spinel Type $ZnCr_2O_4$

C. Vargas-Hernández*, **O. Almanza and J. F.Jurado**
Universidad Nacional de Colombia, Manizales-Colombia
Laboratorio de Propiedades Opticas de Materiales-POM
E-mail: *`Corresponding author: cvargash@unal.edu.co`

Abstract. Structural, vibrational and electron paramagnetic resonance (EPR) analysis for compound $ZnCr_2O_4$ are shown in this work. These types of materials are used in technological applications as humidity sensors and piezoelectric devices. The compound was obtained by mean of solid state reaction technique from binary precursors ZnO and Cr_2O_3. After three thermal treatments the sample structure was monitoring using X ray diffraction (XRD), the spinel cubic phase has been indexed within $O_h^7(Fd3m)$ spatial group. It is observed normal spinel phase. Micro-Raman analysis revealed bands for normal vibration modes of Zn and Cr atoms in tetrahedral and octahedral environments formed by oxygen atoms at approximately 400 and 900 cm^{-1}, respectively. Bands around 941 cm^{-1} are associated possibly to vacancies in the tetrahedral and octahedral sites due to interaction between Zn and Cr ions. EPR signal from 150 to 300 K isothermals indicates a transition between inverse spinel to normal spinel type in a central field around 3350 G. A signal at approximately 3400 G corresponding to the Cr^{+3} in tetrahedral sites is observed near the central field.

1. Introduction

The spinels are face-centered cubic structures, general chemical formula AB_2X_4, (A = Zn, Cd, Fe, Cu or Ge, B = Cr, Al, Co, Ga or In, and X can be O, S, Se or Te), presented as a compact arrangement of X ions in which interstices are occupied by A or B cations forming a cubic cell. Zinc chromium oxide $ZnCr_2O_4$ face-centered cubic structure, $O_h^7(Fd3m)$ spatial group compounds are found within these spinel families, with tetrahedral sites occupied by Zn atoms and octahedral sites occupied by Cr atoms. When this occurs, it is said to be a normal-type spinel. When the cationic distribution is inverted , an inverse spinel configuration is obtained. Due to vacancies, the crystalline field is modified and this allowing the occupation of tetrahedral sites by Cr atoms,this produces a mixture of configurations between normal and inverse spinel, which can exist at low temperatures. Since the 1930's, ACr_2X_4 chromite spinel structures have been studied with great interest [1-3]. Different methods have been used to explain transport phenomena in these types of materials and enormous efforts have been made during the last decades to reconcile explanations of their magnetic and electrical properties. Because this type of spinel varies in its magnetic and transport properties when it comes into contact with gaseous atmospheres, it has been used as the basis for sensory devices, especially with regard to humidity [4]. A large variety of spinel-type families have been extensively studied, including the MCr_2O_4 (M=Mg, Mn, Zn) type, as to which the correlation between the substitute ion and the physical properties of the final compound has been reported. This paper presents results of structural and vibrational analysis through Raman microscopy, XRD and EPR for the $ZnCr_2O_4$ compound obtained by solid reaction, as well as the mixture of configurations between normal and inverse spinel at low temperature. Vibrational spectroscopy has been widely used because it provides information about local structural features of inorganic solids. Group Theory analysis predicts the following normal vibration modes for spatial group $O_h^7(Fd3m)$: $T = A_{1g}(R)+E_g(R)+F_{1g}(in)+3F_{2g}(R)+2A_{2u}(in+2E_u(in)+4F_{1u}(IR)+2F_{2u}(in)$ [5]. There are five Raman active modes, according to reports and previous calculations. The EPR signal was adjusted with curves Lorentzian. The zero field interaction is the D term and it is denominated the parameter of crystalline field, S is the spin and $\vec{H}$ is the applied magnetic field. In the case of a spinel with octahedral and tetrahedral sites the Hamiltonian $\hat{H}$ is [6-8] :

$$\hat{H} = g\beta\vec{H} \bullet \vec{S} + \vec{S} \bullet \vec{D}_{Oct} \bullet \vec{S} + \vec{S} \bullet \vec{D}_{Tetra} \bullet \vec{S} \tag{1}$$

Where D_{oct} and D_{Tetra}, have tetrahedral and octahedral symmetry, respectively. g is the gyromagnetic factor, β is the Bohr magneton , and D is symmetric and it is a function of the orbital matrix elements with excited crystal field states, which takes the form:

$$D = \begin{pmatrix} -D/3+E & 0 & 0 \\ 0 & D/3+E & 0 \\ 0 & 0 & 2/3D \end{pmatrix} \tag{2}$$

For chromium, $S = 3/2$, I = 0, J = 2S +1, consequently, matrix dimensionality is four, because $m_s = \pm 1$. The transitions are E_1-E_2, E_2- E_3 and E_3-E_4, and the transition energy values take the form of:

$$\begin{aligned} E_1 - E_2 &= 2G + d_{22} + 2d_{33} - d_{11} + \frac{3}{8G}\left(d_{13}^2 + d_{23}^2\right) \\ E_2 - E_3 &= 2G - \frac{3}{G}\left(d_{13}^2 + d_{13}^2\right) + \frac{3}{8G}\left((d_{22} + 2d_{33})^2 + 4d_{12}^2\right) \\ E_3 - E_4 &= 2G - (d_{22} + 2d_{33} - d_{11}) + \frac{3}{8G}\left(d_{13}^2 + d_{23}^2\right) \end{aligned} \tag{3}$$

with $G = \frac{1}{2}g\beta H$ Experimental curves of the spinel type $ZnCr_2O_4$ EPR signal are Lorentzian and their functional first derivative expression is:

$$L_y(H_j) = \frac{16y_m\left(\frac{H_j - H_0}{1/2\Delta H_{pp}}\right)}{\left[3 + \left(\frac{H_j - H_0}{1/2\Delta H_{pp}}\right)^2\right]^2} \tag{4}$$

The experimental EPR curves were adjusted using the equation (4) in order to obtain the transition energy. With equations (1-3) the value of D is determined in both octahedral and tetrahedral sites. The g factor in the calculations was obtained using $g = 2.0023 - \frac{\lambda}{\Delta}$. Where 2.0023 is the value of g for the free electron and λ is the spin-orbit coupling with $\Delta = 10D$, and Δ is the usual notation for the separation between these energy levels, and are used for octahedral and tetrahedral sites.

2. Experimental Procedure

The $ZnCr_2O_4$ compound was sintered from the binary compounds ZnO (99.9%) and Cr_2O_3 (99.9 %) in stoichiometric ratios through a solid-state reaction technique. The precursors were mixed and macerated for three hours. The final compound's stoichiometric ratios were calculated from the expression:

$$ZnO + Cr_2O_3 \underset{\Delta Q}{\Rightarrow} F(Zn,O) + G(Cr,O) + ZnCr_2O_4 \tag{5}$$

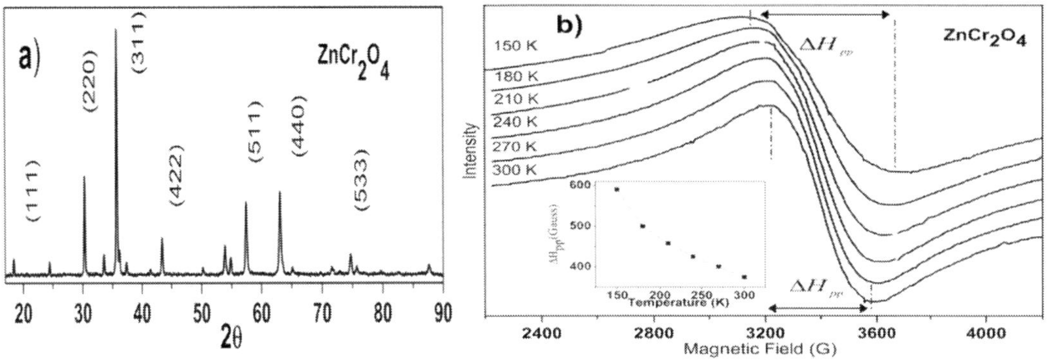

Figure 1. Spinel $ZnCr_2O_4$ **a)** XRD after three thermal treatments **b)**EPR Signal

where the F and G parameters correspond to non-reactant binary phases or remnants of the initial heat treatment, implying that a second and third treatment with the respective maceration to eliminate them is required. The F(Zn, O) compound phase crystallizes into two structures, the Zincblende in the cubic system and Wurtzite in the hexagonal system. The G (Cr, O) phase crystallizes in the cubic CrO_3 system and hexagonal Cr_2O_3 system. The post maceration mixture was heat treated to a temperature of $750°C$, gradually reducing the temperature until it stabilized at room temperature, avoiding residual strain in the final product. The optimum conditions for obtaining zinc chromium oxide $ZnCr_2O_4$ required sinterization at $750°C$ for 24h and three thermal treatments, in order to obtain normal spinel structure. For more details of sample preparation, see references [2-3]. XRD, μ-Raman and EPR techniques at room temperature were employed for the analysis. X-ray diffraction patterns were obtained with CuK radiation using Rigaku Miniflex II difractometer (λ =1.540562). For Raman measurements a confocal LabRamHR Horiba Jobin Yvon high resolution with a source of monochromatic radiation of 473 nm were used. Electronic Paramagnetic Resonance Spectra (EPR) was obtained by means of a BRUKER EPS 300 spectrometer working in the X band. Samples were measured between 150 and 300 K, with the help of an ERAIIIVT variable temperature accessory with $\pm$ 0.5 K precision range. EPR measurements were taken on a 3350 G central field with a wide sweep of 2000 G, a modulation frequency of 100 Hz and a microwave frequency of 9.44 GHz.

3. Results and Discussion

Figure 1a shows the XRD spectrum of spinel $ZnCr_2O_4$ where peaks characteristic of the normal type structure are indexed. Residual binary phases comprise 2% of the end product obtained. Figure 2 illustrates the Raman spectra of precursor ZnO and Cr_2O_3. The Cr_2O_3 reagent used was of low crystallinity but the $750°C$ for 24 hours process used was adequate for crystallization. Maceration was insufficient for spinel formation but the homogenization process is favored by the diffusion at high temperature. To obtain the spinel phase, it is required three thermal treatments using the same parameters. The Figure 2b Raman spectrum show characteristic bands for normal vibration modes of Zn and Cr atoms in tetrahedral and octahedral environments formed by oxygen atoms around 400 and 900 cm^{-1}, respectively. The bands around 941 cm^{-1} are associated with vacancies in tetrahedral and octahedral sites due to the interaction between the Zn and Cr atoms and may play an important role in surface gas absorption. The oxygen vacancies are generated because the sintering process is carried out at high temperature. The EPR signal line shapes reveal an asymmetry due to the existence of two signals from the environment surrounding the paramagnetic ion because the crystalline field depends on the environment and the size of the ion located in the tetrahedral and octahedral sites. Measurements of EPR as a function of temperature indicate that the occupation-distribution of chromium and zinc ions

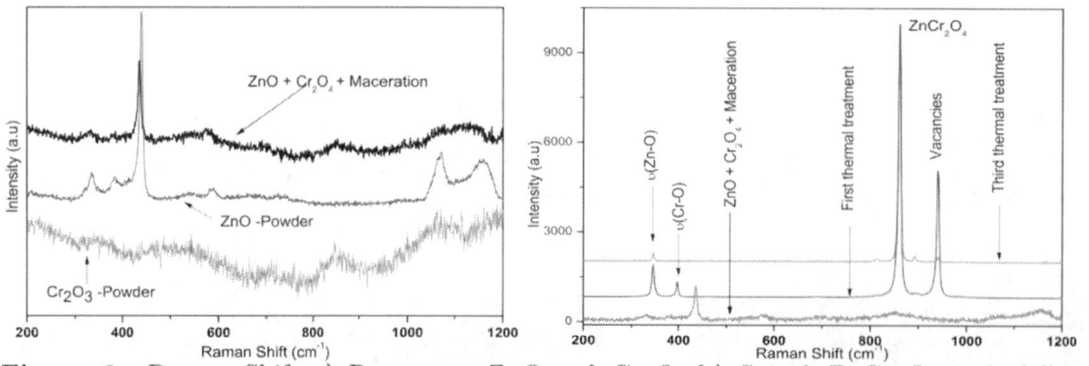

Figure 2. Raman Shift **a)** Precursors ZnO and Cr_2O_3 **b)** Spinel $ZnCr_2O_4$ with different thermal treatments

at low temperatures is configured as a partially inverse spinel. A gyromagnetic factor change from 1.941 to 1.985 is observed by EPR at temperatures around 210 K suggesting that there is competition between the Zn ($Zn^{+2} = 0.74$ Å) and Cr ($Cr^{+3} = 0.64$ Å) within the unit cell in octahedral sites formed by ions O ($O^{-2} = 1.45$ Å). A signal at approximately 3350 G corresponding to the Cr^{+3} in tetrahedral sites is observed near the central field. Lorentzian curve adjustments display normalized $D_{tetra} > D_{oct}$ values, which may be due to larger sized octahedral sites and because of minor overlap in wave functions indicating lower inter-spin coupling. Consequently there is a smaller crystalline field for octahedral coordination. In fitting the curves, the value $D_{tetra}/D_{tetra} = 1.89$ was obtained because tetrahedral sites size increased and octahedral sites size decreased and unit cells became smaller, resulting in greater direct overlap of Cr^{+3} ion wave functions. This behavior is influenced by vacancies that are generated in the sintering process which modify the O^{-2} ion environment and the respective distance between Cr ions [3].

4. Conclusions

The crystalline field D parameter obtained for tetrahedral and octahedral sites has the following relationship: $D_{tetra}/D_{tetra} = 1.89$. The $ZnCr_2O_4$ spinel at low temperatures has a distribution "realignment" of Zn and Cr ions (favored due to the existence of vacancies) occurs in octahedral and tetrahedral sites, resulting in a partially inverse spinel configuration around 210K. The vacancies generated by the thermal treatments (941 cm^{-1}) at high temperatures it have favored at low temperatures the competition between ions of Zn and Cr by octahedral sites .

Acknowledgments

This work was supported financially by the *DIMA* (Universidad Nacional de Colombia). We thank to the laboratory of chemistry of the Universidad Nacional de Colombia-Manizales.

References

[1] L. Mancicl, Z. Marinkovic, P. Vulic, C. Moral and O.Milosevic, Sensors 2003, 3, 415-423.
[2] C. Vargas-Hernandez. Tesis de Magister. Universidad Nacional, Bogot. I/1994.
[3] T.F.W.Barth and E. Posnjak, Zeits, f, Krist. 82,325 8, (1932).
[4] E. J. W. Verwey and J. H. de Boer, Rec. Trav. Chim. Pays. Bas, 55,531(1936)
[5] W. B. White and B. A. Deangelis, Interpretation of the vibrational spectra of spinels. spectrochim. Acta A. 1967, 23, 985-995.
[6] C. P. Poole and H. Farach, The theory of magnetic resonance, Wiley-Interscience, 1972
[7] J. Beauvais et al. Solid state Commun. Vol.79, N 5, Pag. 435. (1991).
[8] S. A. Oliver, H. H. Hamdeh, and J. C. Ho, Phys. Rev. B, Vol. 60, N5, 3400(1999).
[9] Z. Wang, P. Lazor, S. K. Saxena, and Gilberto Artioli, Journal of Solid State Chemistry 165, 165170 (2002).

Spectroscopy analysis of spiropyran-merocyanine molecular transformation

R Delgado Macuil[1], M Rojas Lopez[1], A Orduña Díaz[1], V Camacho Pernas[2]

[1] Centro de Investigación en Biotecnología Aplicada Del IPN, Tepetitla Tlaxcala México C.P. 90700

[2] Universidad Politécnica de Puebla, Puebla México.

E-mail: rdmacuil@yahoo.com.mx

Abstract. Spiropyrans materials are very studied for their application in transduction systems for biosensors. In this work the protocol of generation of multi and mono layers films and the efficiency of the transduction system are analyzed. The analyses were based on UV/Vis and Infrared vibracional spectroscopy in Attenuated Total Reflexion mode (ATR). The films were deposited in glass substrates, using (3-Aminopropyl) trimethoxysilano like linker. The UV/Vis spectra show that the deposition protocol allows the generation of a homogenous film from the first layer. Similar results were observed in ATR. The efficiency of the films, when analyzing them after light UV irradiation, is greater in the samples with two and three layers of deposit, since its absorption peaks are most intense in UV as in ATR. The efficiency falls from the fourth deposit layer. In ATR the efficiency is assumed by the vibration of the NO bond at 1338 cm^{-1}.

1. Introduction

Photochromism is the process of inducing color changes in a medium by incident electromagnetic radiation. Many different compounds have been found to exhibit photochromism; some of them are ones called spiropyrans, which have been the most extensively photochromic material studied over the past decades, since the discovery of the spiropyrans photochromic reactions in 1952 by Fisher and Hirshberg [1].

In particular, spiropyran (SP) has a great potential for application in optical devices, for example as data storage films and waveguides [2, 3] mainly, because of their physical and chemical properties could be controlled by illumination. More recently these molecules have been proposed for use in photoswitching of protein activity (biosensors transducers) [4,5].

When the SP is dissolved in solvents with low polarity, a colorless solution was obtained; when the solvent polarity is increased, colored solution was obtained. In fact solvents polarity not only affect to the color solution, the molecular structure is affected too [6]. Even in polar solvents certain spiropyrans in equilibrium with their trans-merocyanine forms were found.

In general, SP absorbs in the ultra violet (UV) region and not in the visible region. Upon UV illumination, the SP colorless isomer undergoes heterolytic cleavage of the NO bond to form the colored isomer (merocyanine). The coloration change (colorless to blue) in the sample is the physical

observation of the molecular transformation. This process can be reversed by visible illumination, see figure 1 [7].
The radical group NO would be changed to obtain different absorption peaks for the merocyanine state. The main feature of these materials is the several cycles of conversion and reconversion process that could be done. Each cycle limits or decreases the optical properties of the sample, by the photodegradation processes.

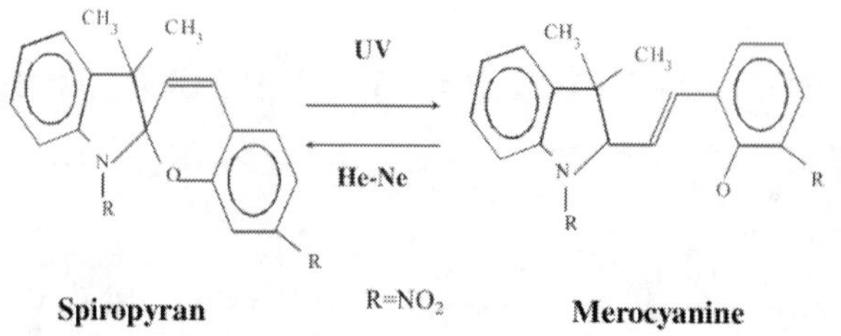

Figure 1. Schematic molecular structure of the sample before UV illumination (i.e. λ= 355 nm) and after isomerization. The process is reversible under visible illumination (i.e. λ= 633 nm).

Crucial for all the applications where the spiropyrans have found, it is a detailed understanding of the transitions kinetics between the spiropyran (SP) to merocyanine (MC) forms. The conversion of SP to MC occurs via a photochemical route involving ultraviolet photons. The transition from SP to MC has been studied and determined to occur on the picosecond to nanosecond time scale, but it depends of the source of transformation applied. For this, it is necessary to analyze the molecular behavior of these kinds of structures. In this context, vibrational spectroscopies would seem to be appropriated to study them, since they provide subtle details about molecular structure in liquids, gas or solid state. For example, Infrared [2, 7-10] and Raman [8,10] spectroscopic studies were done to observe the conformational and intermolecular interactions between the polymeric base and the spiropyran. Other works analyzed the solvent polarity effects into spiropyran molecule, using only the absorption spectrum of the structure [11-13].
Attempts to detect the main structural changes occurring in the ring-opening/closing process in spiropyrans, using IR vibrational spectroscopy, were reported by Schiele and Arnold for a series of indolinonitrospiropyrans [14] and later by Guiliano et al. for benzothiazolinonitrospiropyrans [15].
In this work, two spectroscopic techniques were used to study the ring-opening process of spiropyran to merocyanine in an acetone solution. The infrared vibrational analysis was done via the Total Attenuated Reflexion mode (ATR) whereas the UV/Vis spectroscopy was used as preliminary analysis of the opening process. The UV-Vis shows the changes in the visible region due to the molecular transformation. Infrared spectra show the conversion process from spiropyran to merocyanine in two ways: by the new peaks generation and by the changes in the line form of the peak form due to the opening process.

2. Materials and methods

2.1. Sample preparation

The spiropyran used in this work was 1',3'-Dihydro-8-methoxy-1',3',3'-trimethyl-6-nitrospiro[2H-1-benzopyran-2,2'-(2H) indole], 97%], which was dissolved in 40 ml of ethanol. The functionalized substrate was prepared in an aqueous solution with 1% mM of 3-Aminopropyltrimethoxysilane, silane is the linker between corning crystal and spiropyran molecule. Figure 1 show the spiropyran molecular structure used in this work and their corresponding merocyanine form after UV illumination. For multilayer generation, a solution of 10 ml of pyridine and 90 ml of dimethilformamide was used as linker between the spiropyran multilayers. The spiropyran was added to the functionalized substrate by substrate immersion with a frequency of 10 times/minute, into the spiropyran solution for two different times; a half and one hour. So for two layers generation, the substrate was immersed once in the pyridine/dimethilformamide solution and twice for a film of three spiropyran layers and so on.

2.2. UV/Visible spectroscopy

The absorption of spiropyran – merocyanine films, was performed in a Genesis UV/Vis spectrometer. First the spiropyran functionalized films were measured and after that the films were exposure for 10 minutes to UV irradiation inside a GelDoc Transilluminator with a 365 nm lamp; the power intensity of the 365 lamp was 500 μW/cm^2. A functionalized substrate was used as baseline in all the measurements.

2.3 Infrared spectroscopy

An Infrared spectrometer FTIR Bruker Vertex 70 in Attenuated Total Reflexion (ATR) sampling mode was employed. A diamond ATR crystal of one reflection was used during the measurements. The silane spectrum was used as baseline and the acquisition time was 30 seconds to obtain the final vibration spectra. For the molecular transformation analysis, functionalized films were measured first and after the films were exposure for 10 minutes to UV irradiation inside a GelDoc Transilluminator. For the conversion kinetics analysis, the spiropyrans films were irradiated during 20 minutes with a 4 watts Blacklight fluorescent tube used in billets identification, and 20 spectra were obtained.

3. Results and Discussion

3.1 UV/Vis Spectroscopy

Figure 2 shows the UV/Vis absorption spectra of the spiropyran functionalized films, preparated for a half and one hour. The spectra show a peak associated to the spiropyran state in the ultraviolet region at 370 nm and another less intense peak at 560 nm associated to the merocyanine state. The beginning of the ring-opening process in the colorless samples, could be due to two factors: the solvent polarity used in solutions and to the deposition protocol (mechanical movement at low frequency).

In both figures it is possible to observe that films with two, three and four layers have the more intense absorption peaks. The sample with five layers has the low intensity peaks; in fact the best sample, in both deposition times, is the sample with two layers. This behaviour could be associated to the molecular opening inhibition due to large films deposited.

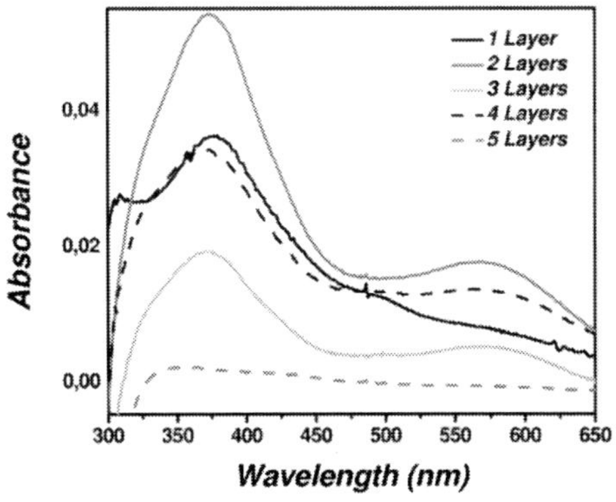

a

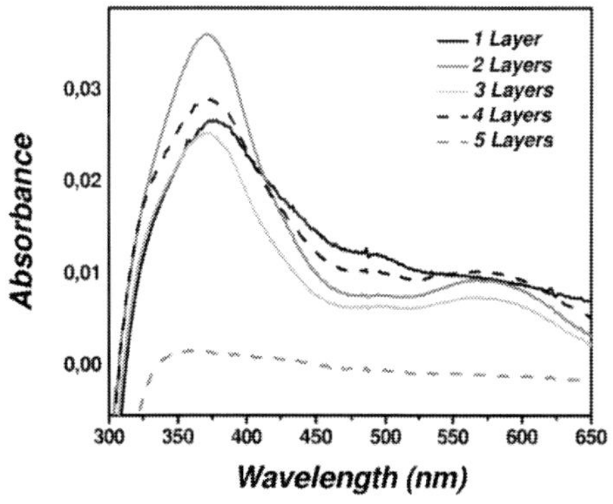

b

Figure 2. UV/Vis absorption spectra of samples at a).- half hour and b).- one hour; of immersing in the spiropyran solution.

After that, the spiropyran functionalized films were illuminated with UV radiation, by 10 minutes in the GelDoc Transilluminator. Even no colour is present in the samples due to the thickness of the deposited films, UV/Vis spectra show the increasing of the absorption peaks in the samples of two, three and four layers. Figure 3 shows this behavior for two and four layers films; the merocyanine spectra show a more intense absorption peak in the merocyanine state. In fact the presence of the peak at 560 nm is more evident. These results are in agreement with other works [3,8]. The intensity increasing for the peak at 370 nm, is due to a better absorption efficiency when the conversion is induced.

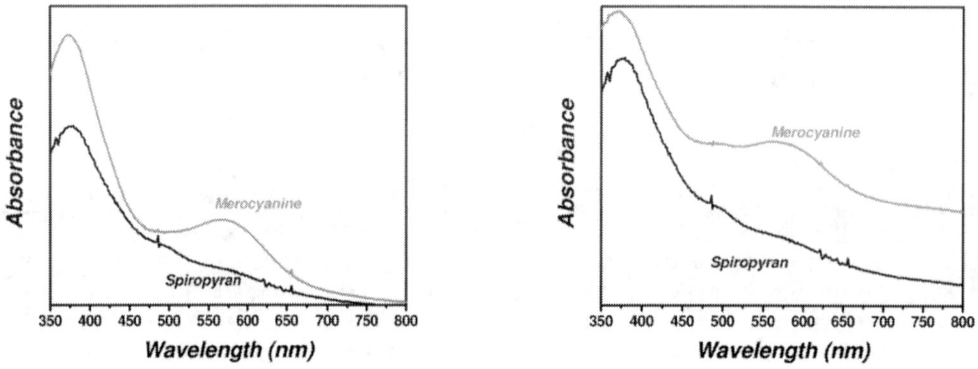

a c

Figure 3. UV/Vis spectra of samples before and after UV irradiation.
a) Two layers, b) Four layers.

3.2 Infrared Spectroscopy

Figure 4 shows FTIR results in the finger print region (1250-1650 cm^{-1}). The peaks are associated to the molecular bonds for the merocyanine state. Infrared absorption presents an increase in the intensity of the peaks associated to the NO bond (at 1338 and 1520 cm^{-1}) and the generation of new peaks associated to the bonds: CN, CH bending (symmetric and asymmetric) and to the aromatic ring; when the ring-opening process occurs. In the CH region (data not showed), the mainly difference is the disappearing of the peak at 3025 cm^{-1}, which is due to the breaking of the CO or CN bond when the ring-opening process occurs.

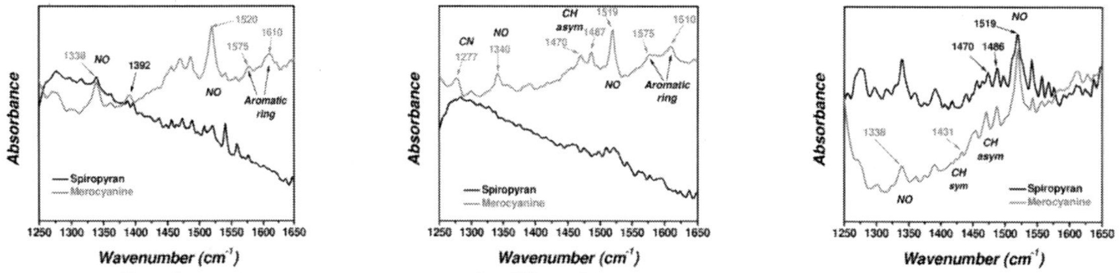

a.- Two layers b.- Three layers c.- Four layers

Figure 4. Infrared spectra of samples before and after UV irradiation.

Finally in figure 5, the FTIR absorption associated to the kinetics of the ring opening process is presented. The spiropyran films were irradiated in a continuous way during 20 minutes with a 4 watts Blacklight fluorescent tube. The samples were measured by ATR mode and the spectra were acquired on real time ever minute. This figure shows only seven spectra of the all kinetics for a clear lecture. The generation or transformation of the peaks observed at 1 minute is very clear, especially in the region 1300 to 1700 cm^{-1}. Due to the low intensity of the source of irradiation, changes are more notorious after 10 minutes of sample irradiation.

Table 1 summarizes the frequency values of the FTIR peaks and their assignment, observed in figure 5when the ring-opening process occurs.

The FTIR kinetics analysis could be performed when another UV source, with large power.

Table 1. Frequency values of the FTIR peaks observed in the kinetics of the ring opening processes.

Figure number	***Wavenumber***	***assignment***	***Figure number***	***Wavenumber***	***assignment***
1	1338	NO	4	1527	NO
2	1376 and 1395	CH bending sym.	5	1558 and 1618	Aromatic ring
3	1456 and 1473	CH bending asym.	6	1650	C=C

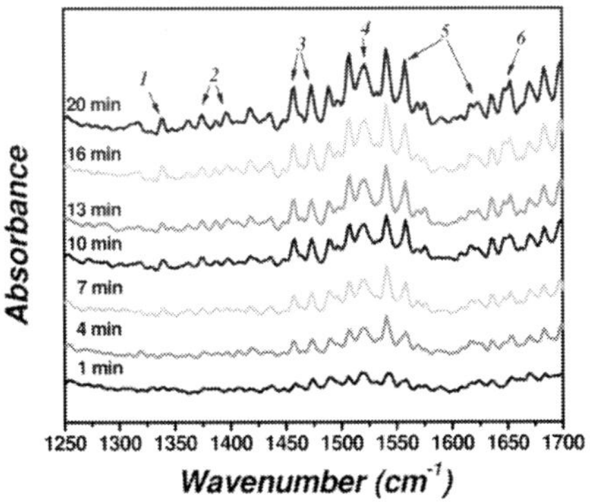

Figure 5. Kinetics of the ring opening process by FTIR of the sample with three layers.

4. Conclusions

In this work an analysis of the ring-opening process of spiropyran to merocyanine state was done by UV/VIS and Infrared spectroscopies. UV/VIS shows that the spiropyran functionalized films with two and three layers present the best response. A peak at 560 nm associated only to merocyanine appears. The vibrational response confirms these results obtained by UV/VIS. In the finger print region, infrared spectra show changes in intensity of the peaks associated to the bonds: NO, CN, CH and to the aromatic ring; when the ring-opening process occurs. The FTIR kinetics analysis show how the ring opening process is done, the increasing and generation of peaks could be observed better after 10 minutes of UV irradiation.

5. References

[1] Fisher E and Hirshberg Y 1952 J. Chem. Soc. 4522.
[2] Bobrovsky A, Boiko N and Shibaev V 1999 *Adv. Mat.* **11** 1025.
[3] Ramos García R, Delgado Macuil R, Iturbe Castillo D, González de los Santos E and Soriano Corral F 2003 *Opt. Quant. Elect.* **35** 641.
[4] Berkovic G, Krongauz V and Weiss V 2000 Chem. Rev. **100** (5) 1741.
[5] Wilner I 1997 Acc. Chem. Res. **30** 347.
[6] Takayanagi M, Nakata M, Ozaki Y, Iriyama K and Tasumi M 1997 *J. Mol. Strut.* **407** 85.
[7] Delgado-Macuil R, Rojas-Lopez M, Gayou V L, Orduña Díaz A and Diaz-Reyes J 2007 *Mat. Characterization* **58** 771.
[8] Crano R Guglielmetti 1999 *Organic Photochromic and Thermochromic Compunds Vol 2* (New York: Kluwer Academic) Chapters 1-2.
[9] Hoon L, Jaung Y and Jeong S 2002 *Bull. Korean Chem. Soc.* **23** 1045.
[10] Takayanagi M, Nakata M, Ozaki Y, Iriyama K and Tasumi M 1997 *J. Mol. Struct.* **407** 239.
[11] Chibisov A and Gorner H, 1997 *J. Phys. Chem.* A **101** 4305.
[12] Khimenko V, Chibisov A and Gorner H, 1997 *J. Phys. Chem.* A **101** 7304.
[13] Wojtyk J, Wasey A, Kazmaier P, Hoz S and Buncel E 2000 *J. Phys. Chem.* A **104** 9046.
[14] Schiele C and Arnold G 1967 *Tetrahedron Lett.* **13** 1191.
[15] Guiliano M, Davin-Pretelli E, Mille G, Chouteau J and Guglielmetti R 1978 Helv. Chim. Act. **61** 1072.

Journal of Physics: Conference Series **167** (2009) 012039 doi:10.1088/1742-6596/167/1/012039

Electric anisotropy in high density polyethylene + carbon black composites induced by mechanical deformation

E. Vigueras-Santiago, S. Hernández-López, M.A. Camacho-López and O. Lara-Sanjuan

Laboratorio de Investigación y Desarrollo de Materiales Avanzados (LIDMA). Facultad de Química, UAEM. Paseo Colón esq. con Paseo Tollocan, s/n. C.P. 50000, Toluca, Estado de México, México.

E-mail: eviguerass@uaemex.mx

Abstract. High density polyethylene + carbon black composites with electrical anisotropy was studied. Electrical anisotropy was induced by uniaxial mechanical deformation and injection moulding. We show that anisotropy depends on the carbon black concentration and percentage deformation. Resistivity had the highest anisotropy resistivity around the percolation threshold. Perpendicular resistivity showed two magnitude orders higher than parallel resistivity for injected samples, whereas resistivity showed an inverse behaviour for 100% tensile samples. Both directions were set respect to the deformation axe. Anisotropy could be explained in terms of the molecular deformation (alignment) of the polymer chains as a response of the deformation process originating a redistribution of the carbon black particles in both directions. Alignment of the polymer chains was evidenced by polarized Raman spectroscopy.

1. Introduction

Actually, the study and synthesis of novel conductive polymer composite (CPC) with carbon black (CB) are of the great interest due to its broad practical applications [1]. In particular is possible to modify the electrical resistivity in at least 11 magnitude orders without important changes in the CPC density [2]. Furthermore, they have the ability of significantly change their electrical resistivity by the stimulus of external parameters as temperature, solvents, pressure, AC electric field, etc [1,3]. Percolation theory is the most cited description for explaining the electric conduction in this type of systems [4]. However, such a description doesn't consider fundamental questions related with physical or chemical interaction among conductive particles and polymer matrix which take place during the CPC preparation. The optimum develop of these materials is an experimental work for each pair polymer-conductive particles. Is well known that both: appropriated disaggregating and dispersion of the CB particles produce the building of conduction networks [5]. Then the electrical properties of the PCP depend of the processing parameters [5]. In this work we are interested in study the changes in electrical resistivity of high density polyethylene-based CPC with high structure CB. For such propose, CPC in a width composition range of CB were prepared. The main goal is to show the possibility of controllably induces electrical anisotropy in this type of CPC [6,7]. The anisotropy could be explained in terms of a change in spatial configuration of the CB network chains during the uniaxial polymer matrix stretching, induced mainly by the molecular orientation of HDPE polymer chains along of the tensile axe [8,9].

2. Experimental section

HDPE YUZEX® - 8800 and Vulcan XC72-CB donated by Cabot, Co. were used as received. CPCs were obtained by melting mixing using a home made Banbury micromixer at 175°C, 55 rpm for 30 min at compositions of 1 to 20% wt/wt. CPCs were processed to cylinders of 0.5 cm radio and 0.5 cm high by thermomechanic molding following the methodology described in [3]. Silver paste SPI was deposited on both faces of the cylinders as electrical contacts. The electrical resistivity was measured using a Keithley electrometer 6517A . The electrical resistivity error was 6%.

For the anisotropy study, CPCs were processing to bars of 5 x 5 x 50 mm by injection molding. After, these bars were stretched by tension at a deformation rate of 1 cm/mm at 80°C using home made equipment. Molecular orientation as a deformation function was monitored by polarized light microRaman spectroscopy, using a Micro Raman Labram HR-800 de Jobin Yvon-Horiba, a He-Ne (λ=632.8 nm) polarized laser and an Olympus BX4 microscope. The relative intensities (ri) of the symmetric C-C vibrations from the main chain, with a typical frequency on 1129 cm^{-1} were compared with the respective asymmetric ones which appear at 1060 cm^{-1} for compressed, injected and stretched HDPE samples.

3. Results and discussion

The resistivity changes as a function of CB composition are shown in Figure 1. The squared curve corresponds to cylinder samples. The circle and triangle symbol curves correspond to the injected bars samples which were measured in both directions: parallel (circles) and perpendicular (triangles), as set up in Figure 2. According to our results, we could observe that conductive networks building depend on the processing method. In particular, resistivity values for injected (bar) samples were higher than those of compressed (cylinder) samples as shown in Figure 1. Differences between both directions of injected samples and thermomechanical molded samples could due to a slight orientation of the HDPE chains during injection processing.

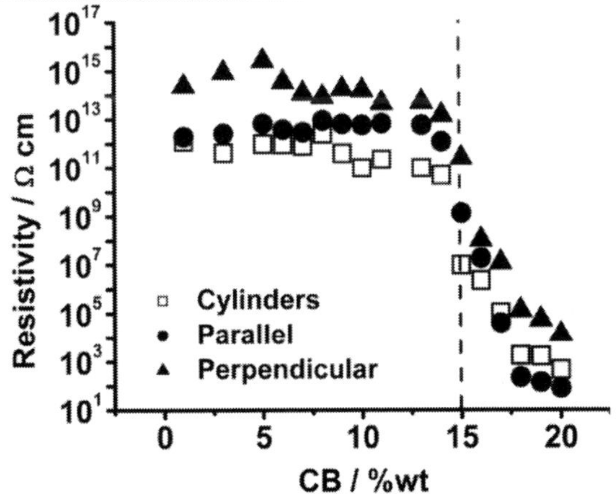

Figure 1. Influence of the processing method (compression molding and injection) on the HDPE+CB CPC's resistivity.

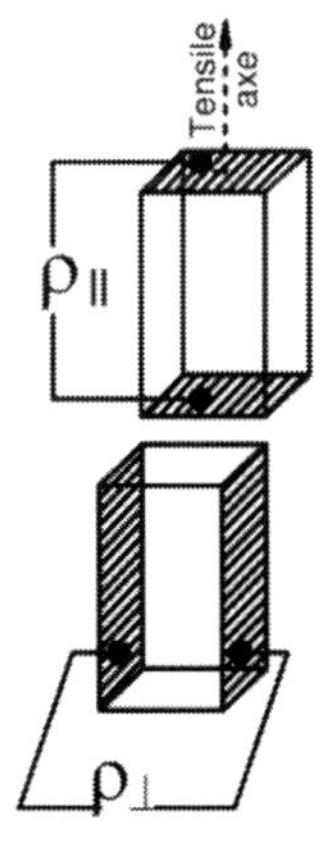

Figure 2. Configuration set up for resistivity measurements: a) parallel resistivities were measured trough the transversal sections perpendicular to tensile axe b) perpendicular resistivities were evaluated between 2 lateral opposite bar faces.

Figure 3 shows the electrical resistivity as a function of the deformation percentage for a 15% CB-CPC bar. Perpendicular direction of a no stretched sample had 2 magnitude order higher resistivity than parallel one (10^9 and 10^7 Ω.cm, respectively, Figure 1). As it was stretched the resistivity diminished at least 5 magnitude orders in both directions, but an inversion of the resistivity values was detected at 100% of deformation At his point, a parallel direction is more resistive than perpendicular one (10^4 and 10^2 Ω.cm, respectively). For deformations larger than 100%, the resistivity lightly increases for both directions until they reach almost the same value at 250% deformation.

The explanation for those facts are that as the deformation percentage increases, the polymer chains increases their orientation modifying at the same time the electrical network configuration of CB conductive particles. Then electrical resistivity of a stretched bar is modified in several magnitude orders in both directions. The resistivity behavior of HDPE+CB CPC under stretched deformation presented a different behavior in comparison with rubber-based CPC as polybutadiene, polyisoprene and silicon, for example, where an increasing in the length is accomplished of a size diminishing in the lateral direction. Consequently an increase of resistivity along of the parallel direction is observed while along the perpendicular direction the resistivity decreases. For a stretched HDPE+CB CPC, a number of the contacts in the conductive pathway increase in both directions. This could be consequence of the better packing of the chains polymer along the stretching axe diminishing the volume surrounded the particles and bringing near the CB particles (more dense material). This effect could increases the contacts among them and the possibility to build more conductive networks in both directions.

Raman spectroscopy allowed evidencing the polymer chain orientation after the injection and tension processes. Comparing the intensities of the symmetric C-C vibrations which appear at 1129 cm^{-1} with the asymmetric ones at 1060 cm^{-1} was possible to establish an orientation difference as shown in Figure 4 [10]. For the samples processed by compression molding, the Raman intensity correlation was $ri = 1.33$ in any direction, curve(c) in Figure 4. Whereas for the injected samples that correlation shown lightly variations: 1.36 and 1.17 for parallel and perpendicular directions, respectively. For the tensed sample variation on relationship between symmetric and asymmetric C-C vibrations were significant as shown in Figure 4.

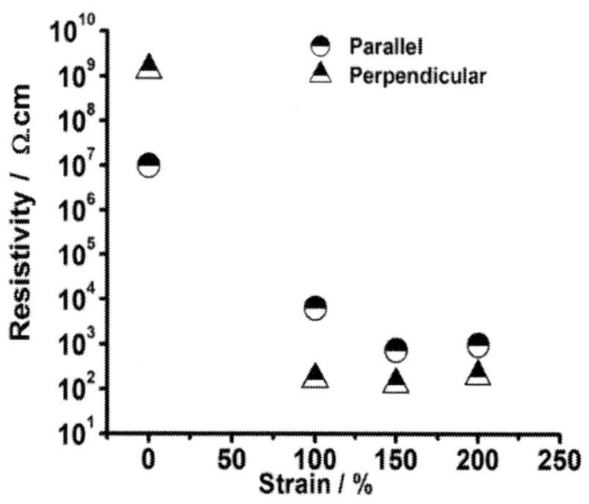

Figure 3. Resistivity dependence of deformation percentage for a 15%CB wt sample.

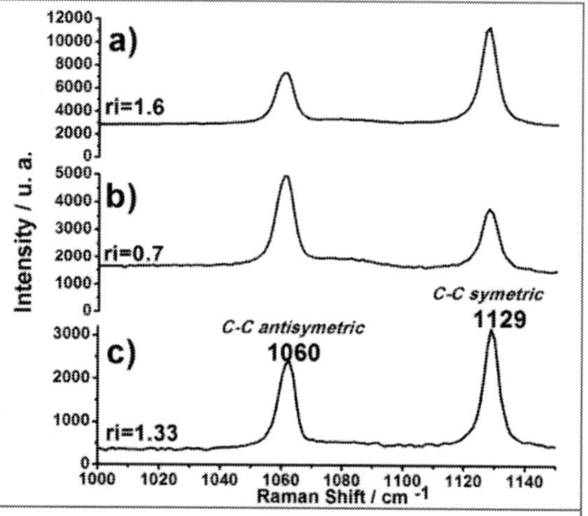

Figure 4. a) and b) Raman spectra of a 100% stretched HDPE, a) parallel, b) perpendicular and c) HDPE unoriented sample

4. Conclusions

In this work we showed that polymer composites of HDPE + CB are materials with a possible electrical anisotropy. Such anisotropy can be induced by uniaxial deformation. The electrical anisotropic effects were higher for CB concentrations near to threshold percolation, detecting differences between parallel and perpendicular directions of at least 2 magnitude orders. The explanation of the electrical anisotropy is described in terms of the percolation theory criteria, considering that the CB particles connectivity is modified and improved by the packing and the alignment of the polymer chains into the HDPE semicrystalline matrix.

5. Acknowledgements

Financial support by CONACyT-NSF Ref: J110.403/2007 is gratefully acknowledged.

References

[1] Sartore L, Penco M, Della Sciucca S, Borsarini G, and Ferrari V 2005 New carbon black composite vapor detectors based on multifunctional polymers *Sensors and Actuators B: Chemical* 111
[2] Donnet J B, Bansal R. L, Wang M.-J (Eds.) 1993 Carbon Black New York Dekker
[3] San Juan-Farfán R, Hernández-López S, Martínez-Barrera G, Camacho-López M.A, Vigueras-Santiago E 2005 Electrical characterization of polystyrene-carbon black composites. *Phys. Stat. Sol. (c)* 2 3762
[4] Stauffer D, Aharony A 1994 Introduction to percolation Taylor and Francis, London
[5] Zhang W, Dehgheni —Sanij A A 2007 Carbon Based Conductive Polymer Composites. *Matter Sci.* 42 3408
[6] Carmona F, El Amarte A 1987 Anisotropic electrical conductivity in heterogeneus solids wiyh cylindrical conducting inclusions *Physical Review B* 35 3284
[7] Skal A, Grebnev I 1992 Percolation and transport coefficients of biaxial anisotropic composites *J. Phys. Condens. Matter.* 4 1521
[8] Peterlin A 1971 Molecular model of drawing polyethylene and polypropylene *J. Mater. Sci.* 6 490
[9] Yui H, Wu G, Sano H, Sumita M, Kino K 2006 Morphology and electrical conductivity of injection molded polypropylene /carbon black composites with addition of high density polyethylene *Polymer* 49 3599
[10] Citra M J, Chase D B, Ikeda R M, Gardner K H 1995 Molecular orientation of high density polyethylene fibers characterized by polarized raman spectroscopy *Macromolecules* 28 4007

Thermoelectric effect in nano-scaled lanthanides doped ZnO

Otal E. H.[1], Schaeuble N.[2], Aguirre M. H[2], Canepa H.R.[1], Walsöe de Reca N. E.[1]

1 _ Centro de Investigación en Sólidos, CITEFA, San Juan Bautista de La Salle 4397 (B1603ALO) Villa Martelli, Buenos Aires, Argentina

2 _ Solid State Chemistry and Catalysis, Empa, Swiss Federal Laboratories for Materials Testing and Research, Ueberlandstrasse 129, CH-8600 Duebendorf, Switzerland

canepa@citefa.gov.ar

myriam.aguirre@empa.ch

Abstract. Start Nano-scaled ZnO with 1% Er doping was prepared by soft chemistry methods. The synthesis was carried out in anhydrous polar solvent to achieve a crystal size of a few nanometers. Resulting particles were processed as precipitates or multi layer films. Structural characterization was evaluated by X-Ray diffraction and transmission and scanning electron microscopy . In the case of films, UV-Vis characterization was made. The thermoelectrical properties of ZnO:Er were evaluated and compared with a typical good thermoelectric material ZnO:Al. Both materials have also shown high Seebeck coefficients and they can be considered as potential compounds for thermoelectric conversion.

1. Introduction

The ZnO is a common conductive oxide with a wide band gap of 3.5 eV. The electrical properties of ZnO are exploited in gas sensors [1] and varistors [2, 3]. Furthermore, materials derived from ZnO have been investigated as potential candidates for high-performance n-type oxide thermoelectric materials. The thermoelectric properties of undoped and doped ZnO (e.g. ZnO:Al, ZnO:Sb, ZnO:Pr) are well studied [4-13]. The best thermoelectric properties were obtained with Al-doped ZnO by Ohtaki and Tsubota et al. [4,7]. Fujishiro et al. reported the effect of microstructural control on thermoelectric properties of hot-pressed Al-doped ZnO [14]. After hot pressing, dense products of polycrystalline $Zn_{0.99}Al_{0.01}O$, containing nanosized grains with sizes below 1 μm in diameter were obtained. The power factor $S^2\sigma$ of the nanostructured $Zn_{0.99}Al_{0.01}O$ prepared by hot pressing was higher in the measured temperature range (300°C to 800 °C) than that of polycrystalline $Zn_{0.99}Al_{0.01}O$ prepared by normal sintering at 1400 °C. Enhanced thermoelectric performance of nanostructured ZnO were also reported by Ohtaki et al. [15]. Highly dispersed nanosized closed pores (nanovoids) are revealed to be effective to substantially enhance the thermoelectric performance of bulk sintered body of n-type Al-doped ZnO, resulting in a dimensionless figure-of-merit of ZT = 0.65 at 1250 K. Rare earths cations are interesting because of their large ionic radii and trivalent oxidation state. Metals of the d-group are usually introduced into ZnO matrix, occupying substitution or interstitial site. Due to its mayor ionic radii, lanthanides are not easily introduced and phase segregation is produced [16]. The route reported by Schwartz [17] uses polar aprotic solvents to solvate cations in solution and obtain nanostructured ZnO. The results of its modification to obtain erbium doped ZnO are shown below.

1. Experimental

The Erbium-doped samples were prepared by soft chemistry process: Zinc acetate dihydrate and Erbium acetate hydrate were dissolved in dimethylformamide and basified by drop wise of a KOH ethanolic solution under vigorous agitation. Once the base was aggregated, the colloids were precipitated by ethyl acetate and centrifuged to separate the solid from the supernatant. To eliminate residues of solvent, particles were suspended in ethanol, precipitated twice in ethyl acetate and dried in vacuum.

Films were prepared by spin coating from the final solution and dried at 800K for 5 minutes. Multilayer films were obtained by this proceeding. For optical characterization, fused quartz (GM associates Inc) was used as substrate and for microscopy characterization silicon wafers were used.

XRD was measured from dried samples. A PW 3710 Philips diffractometer was used to do these measurements. Instrumental broadening was corrected by LaB_6 standard. Samples and standard were measured in Bragg Brentano configuration in the range 10°-100° with 0.02°/step.

For TEM analysis, powder samples suspended in alcohol were deposited in holey carbon copper grid. SEM images were obtained from films deposited over Silicon. The TEM and SEM analysis were performed in a Philips CM30 at 300 kV and a Karl Zeiss FESEM DSM 982 GEMINI, respectively.

Spectrometric measurements were performed in a multi-wavelength diode array HP 8453 spectrophotometer (Hewlett Packard, Palo Alto, CA, USA) controlled through the Chem-Station software package. Samples were measured perpendicular to incident light in a modified sample holder. Quartz substrate without ZnO film was used as blank.

Electrical conductivity and Seebeck coefficient were measured simultaneously in the temperature range of $330\ K \leq T \leq 1273\ K$, in pressed powder pellets prepared by sintering at 1272 K for 1h. The electrical resistivity ρ was measured by the DC four-probe technique by Ozawa Science equipment: two platinum contacts placed at the ends of the bar-shaped sample supply the current and two platinum wires looped around the pellet measure the potential difference. The Seebeck coefficient was measured in the following way: a well defined temperature gradient was applied to the sample by cooling one end with cold air and the voltage difference generated by the sample was recorded. The temperature gradient was measured by two thermocouples (Pt / Pt + 13 \% Rh) attached to the platinum layer contacts. The Seebeck coefficient of the sample was finally obtained after extracting the contribution of the thermocouple ($S_{sample} = S_{measured}-S_{Pt}$).

2. Results and discussion

The XRD of solids showed a broadening due to the particle size (figure 1). The peak broadening showed a small anisotropy in the a-axis direction. The average particle size over the first peaks, corrected by instrumental broadening, was 10nm. This size is bigger than Bohr radius for ZnO [18] and in consequence nanoscale constituents would not introduce quantum-confinement effects to enhance the power factor $S^2\sigma$.

Samples showed a uniform growth and thickness was increased by increasing the number of deposited layers. This was confirmed by UV-Vis spectroscopy. The absorbance of the films deposited on fused quartz exhibited a linear correlation with number of layers, indicating a linear increase of deposited mass. An approximate thickness of 60nm/layer was obtained.

Films conserved the same nanostructure after deposited and dried, as can be observed in SEM image in figure 2. Most of particles remain in the nanometric size after depositing while a few needlesgrothw over the surface but keeping the nano-scale range

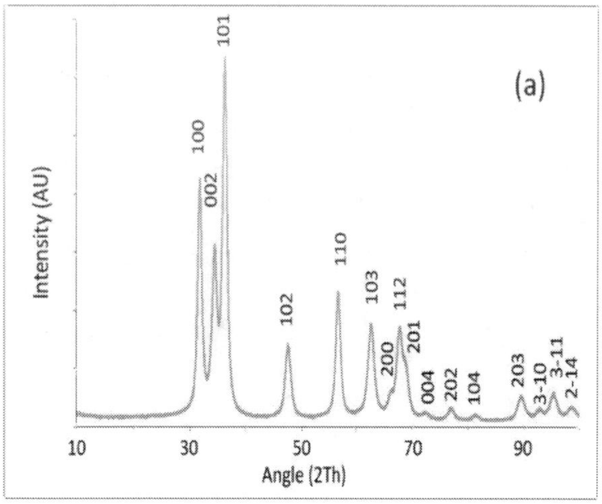

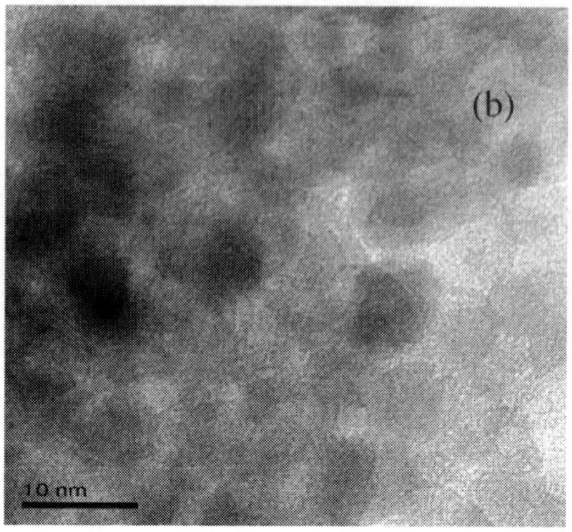

Fig 1 – a) XRD pattern diffraction for erbium doped ZnO sample. b) HRTEM micrographs for erbium doped ZnO before precipitated. The peak broadening due to the nanosized crystals was confirmed by TEM.

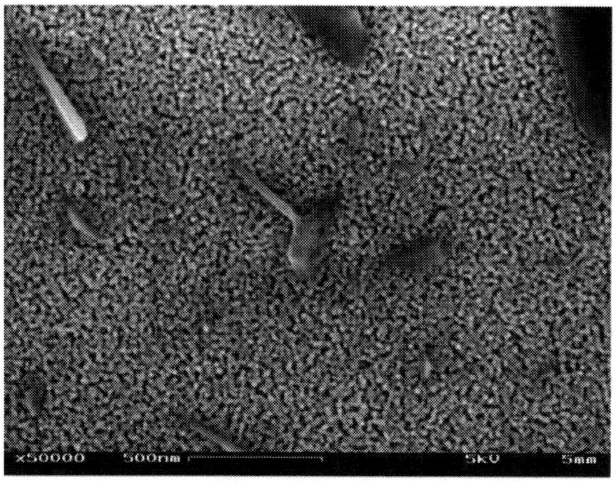

Figure 2 - Scanning microscopy of the planar view of a film.

The Seebeck coefficient and electrical conductivity were summarized as a function of temperature in figure 3 (a) and (b) , respectively. The Seebeck coefficient measurements for ZnO:Er showed a value of -510mW/K , while ZnO:Al is reported to be -550 mW/K [19] . This values shows a p-type behavior of ZnO:Er and comparable thermoelectric properties with ZnO:Al. The conductivity of the sample shows a resistance of 4kΩ.cm at 475K and 1Ω.cm at 1200K. A transition from insulator to semiconductor is observed. This abrupt change can be attributed to modifications in electronic level population produced by changing in erbium location sites, although many others phenomena could contribute to this like oxygen loss, zinc valence change and crystal cell parameters reduction. Studies of photoluminescence, XAS (XANES, EXAFS and XEOL) and XRD, of samples calcinated at different temperatures, are being performed which will help to obtain more information about this process. Preliminary results exhibited changes in luminescence spectra, showing a transition from a disordered site to a more crystalline and symmetric one, which seems to support our supposition.

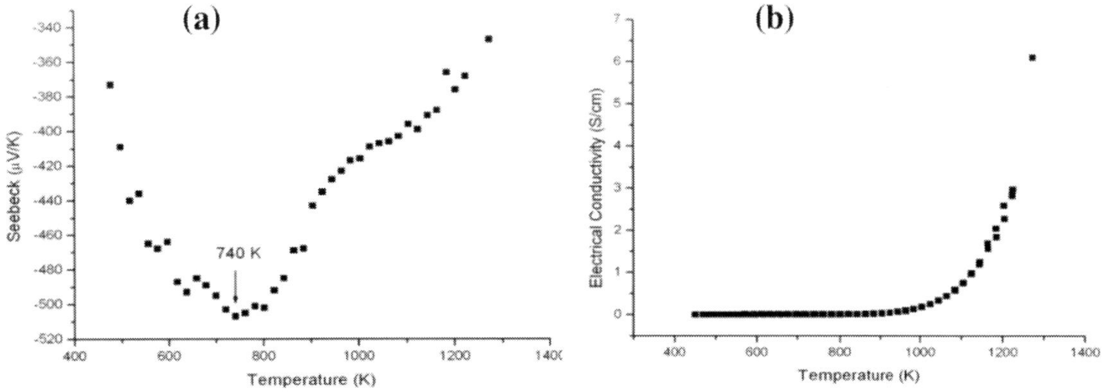

Figure 3 – a) Temperature dependence of Seebeck coefficient. At 740K a minimum is observed, b) Temperature dependence of electrical conductivity.

3. Conclusions

Efficient erbium introduction into ZnO was achieved by wet chemistry methods.

High Seebeck coefficient was obtained for ZnO:Er. This value was compared with Seebeck coefficient of typical good thermoelectric material.

An abrupt change in electrical conductivity was found at 790K, probably related to changes in erbium environment with temperature. Studies of photoluminescence, XAS (XANES, EXAFS and XEOL) and XRD of samples calcinated at different temperatures are being performed in order to obtain more information about this process

Acknowledgments

The authors are indebted toYPF Foundation, CITEFA, CONICET and EMPA for financial support that enabled them to accomplish this work.

References

[1] S.-N. Bai, T.-Y. Tseng, Journal of Applied Physics 74 (1), pp. 695-703 (1993).
[2] Bok-Hee Lee, Sung-Man Kang , Current Applied Physics (6), 5, pp 844-851 (2006).
[3] Y.-M. Chiang, W.D. Kingery, , L.M. Levinson, Journal of Applied Physics 53 (3), pp. 1765-1768 (1982).
[4] Tsubota, T., Ohtaki, M., Eguchi, K., Arai, H. , J. Mater. Chem. 7 (1), 85 - 90 (1997).

Journal of Physics: Conference Series **167** (2009) 012040 doi:10.1088/1742-6596/167/1/012040

[5] Tsubota, T., Ohtaki, M., Eguchi, K., Arai, H., International Conference on Thermoelectrics, ICT, Proceedings, pp. 240-243 (1997).
[6] Ohtaki, M., Tsubota, T., Eguchi, K., Arai, H., Journal of Applied Physics 79 (3), pp. 1816-1818, (1996).
[7] Ohtaki, Michitaka, Tsubota, Toshiki, Eguchi, Koichi, International Conference on Thermoelectrics, ICT, Proceedings, pp. 610-613 (1998).
[8] K. Park, J.K. Seong, S. Nahm, J. of Alloys and Compounds, 455 (1-2), 331-335 (2008).
[9] Tanaka Y., Ifuku T., Tsuchida K., Kato A., Journal of Material Science Letters 16, 155 - 157 (1997).
[10] Kaga, H., Kinemuchi, Y., Tanaka, S., Makiya, A., Kato, Z., Uematsu, K., Watari, K., Japanese Journal of Applied Physics, Part 2: Letters, Volume 45, Issue 42-45, 10 November 2006, Pages L1212-L1214.
[11] Inoue, Y., Okamoto, Y., Morimoto, J., Journal of Materials Science, Volume 43, Issue 1, Pages 368-377 (2008).
[12] Inoue, Y., Okamoto, M., Kawahara, T., Okamoto, Y., Morimoto, J. Materials Transactions, Volume 46, Issue 7, Pages 1470-1475 (2005).
[13] Fuda, K., Sugiyama, S., Materials Research Society Symposium Proceedings Volume 886, Pages 473-478 (2006).
[14] Fujishiro Y., Miyata, M., Awano, M., Maeda, K., J. Am. Ceram. Soc. 86 (12) 2063 - 2066 (2003).
[15] Ohtaki, M., Hayashi, R., International Conference on Thermoelectrics, ICT, Proceedings, art. no. 4133287, pp. 276-279 (2006).
[16] Mustafa D.,Biggemann D., Wu J., Coffer J. L. and Tessler L.R., Superlattices and MicrostructuresVolume 42, Issues 1-6, July-December 2007, Pages 403-408
[17] Schwartz, D. A.; Norberg, N. S.; Nguyen, Q. P.; Parker, J. M.; Gamelin, D. R., Journal of the American Chemical Society ; 2003; 125(43); 13205-13218
[18] Brus L.E. Journal ofPhysical Chemistry. 1986, 90, 2555-2560
[19] Schäuble N. – Diploma Thesis. EMPA – Zurich, Switzerland – July 2008

Short range investigation of sub-micron zirconia particles

M C Caracoche, J A Martínez

Departamento de Física, IFLP, Facultad de Ciencias Exactas, CICPBA, Universidad Nacional de La Plata, Argentina

P C Rivas

IFLP-CONICET, Facultad de Ciencias Agrarias y Forestales, Universidad Nacional de La Plata, Argentina

F Bondioli, V Cannillo

Dipartimento di Ingegniria dei Materiali e dell'Ambiente, Facoltà di Ingegneria, Università di Modena e Reggio Emilia, Italy

A M Ferrari

Dipartimento di Scienza a Metodi dell' Ingegneria, Università di Modena e Reggio Emilia, Italy.

E-mail: cristina@fisica.unlp.edu.ar.

Abstract. The Perturbed Angular Correlations technique was used to determine the configurations around Zirconium ions and their thermal behavior in non-aggregated sub-micron zirconia spherical particles. Three residues containing- Zr surroundings were determined for the non-crystalline starting particles, which were identified under the assumption of a certain chemical reactions sequence during synthesis. While the one made up mainly by hydroxyl groups was common to both samples, the two involving mainly organic residues were particle size dependent. Upon crystallization, both samples stabilized in the t′- and t- tetragonal forms and the Xc-cubic form but their amounts and temperatures of appearance were different. On heating, the structure of the smaller particles became gradually monoclinic achieving total degradation upon the subsequent cooling to RT.

1. Introduction

A growing demand for improved performance of ceramics in loadbearing and functional applications has raised the importance of chemical purification and processing of the starting powders. In particular, ceramic powders with high purity consisting of sub-micron spherical, unagglomerated, monosized particles are desirable for making high quality advanced ceramics. In the present paper, configurations around Zr(Hf probes) in spherical sub-micron zirconia particles and their thermal evolution have been determined using the Perturbed Angular Correlation (PAC) nuclear technique. From PAC spectra, the quadrupole parameters of the electric field gradient (EFG), i.e. the quadrupole frequency, the asymmetry parameter and the frequency distribution width, are obtained. Due to the extremely localized character of the EFG ($\propto r^{-3}$), several non-equivalent lattice sites as well as their relative fractions can be determined. Hyperfine interactions of equilibrium and metastable phases of zirconia are already known [1-2].

Journal of Physics: Conference Series **167** (2009) 012041 doi:10.1088/1742-6596/167/1/012041

2. Samples preparation and characterization

Zirconia particles were obtained by the sol-gel method using tetra-*n*-propylzirconate (TPZ), capronic acid (CA) and ethanol (EtOH). Synthesis was performed by slow addition of an alcoholic solution containing CA (0.016 M) and TPZ (0.1M), previously reacted for 30 min at room temperature, to a mixture of EtOH and water (50:1 molar ratio). Particle precipitation was carried out at 40 °C for 35 min under vigorous magnetic stirring. After evaporation of volatiles (12h at 100 °C), particles showing different sizes and size distribution were obtained depending on the water concentration (0.35 M or 0.6 M) [3]. Particle morphology was examined by transmission electron microscopy, TEM (JEM 2010). To estimate the particle size distribution, image analysis was carried out on TEM micrographs. The specific surface area of the powders was obtained by the B.E.T. method (Gemini 2360). Residual carbon content was determined using a Carlo Erba CHNS-O analyzer. X-ray diffraction (XRD) data were collected at room temperature by a Philips 3710 diffractometer over the 20-80° 2θ range. The diffractometric thermal behavior of the powders was investigated using a Philips X'Pert Pro in the 20-1000°C temperature range. The bulk thermal behavior of the synthesized samples was studied by thermogravimetric and differential thermal analyses in air at a heating rate of 10 °C/min, using a simultaneous TG/DTA apparatus (Netzsch STA409). For the PAC experiment, Hf probes were produced by thermal neutron irradiation of the samples. Measurements were carried out at RT and at increasing temperatures up to 1000 °C. A final RT measurement was also performed.

3. Results

Average particle diameters determined out of TEM images, specific surface areas and residual

Table I. Results obtained by TEM, BET and Elemental analysis.

Sample	TEM mean particle diameter (nm)	B.E.T. specific surface area (m^2/g)	Carbon content (wt%)
SMALL	220 ± 40	102.9	7.2
BIG	550 ± 60	9.6	10.8

carbon content of the powders are reported in Table I. TEM micrographs revealed non aggregated spherical sub-micron particles and narrow size distributions for both samples. Particles prepared with the lower water concentration exhibited greater sizes. According to their relative mean particle sizes, the two powder samples investigated were named SMALL and BIG.

DTA curves (figure 1) evidence the organic residues elimination (a double peak for BIG) over

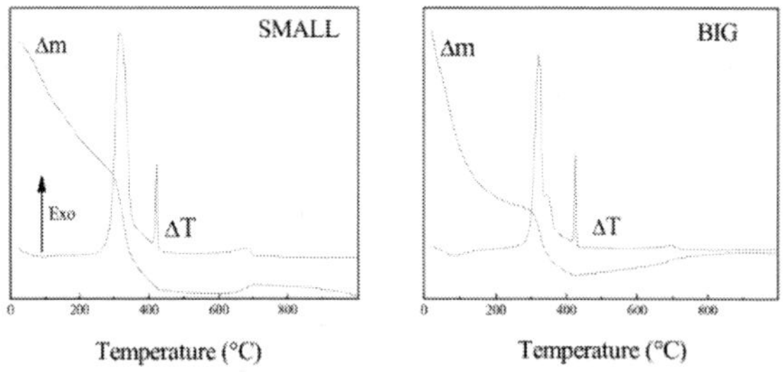

Figure 1. TG-DTA results obtained for both samples.

300 °C and the exothermic crystallization peak at about 430 °C. Close to 700 °C a change in the baseline can be observed too. Except for the crystallization peak, DTA signals are simultaneous with TG mass changes. Mass losses up to crystallization sum up 23 and 33% for SMALL and BIG respectively, in similar proportions to the carbon content determinations (see Table I). Over 600 °C, both samples show an apparent mass gain. The starting diffractograms revealed the rather amorphous nature of the obtained particles. The crystalline oxygen defective tetragonal structure $ZrO_{1.95}$ in both samples was present at 500°C. At 600°C, the existence of a small amount of monoclinic phase in SMALL could be observed. Results at 700°C showed that the

defective tetragonal structure had transformed to the regular tetragonal ZrO_2. Figure 2 displays selected spin rotation curves and in figure 3 the thermal evolution of the relative fractions of the hyperfine components for BIG and SMALL is shown. PAC results for SMALL indicate, from RT to 400 °C, the presence of two interactions, hereinafter named α (full up triangle) and β (full down triangle), being α clearly dominant.

Between 500 and 650 °C the interactions describing the defective tetragonal t'-form [1] (open up triangle) and the cubic Xc-form [3] (open down triangle), replace the former ones. At higher temperatures it is seen that the t'-form has disappeared to give rise to the tetragonal t-form (open circle) and to increasing amounts of both the Xc-form and of an interaction which depicts a

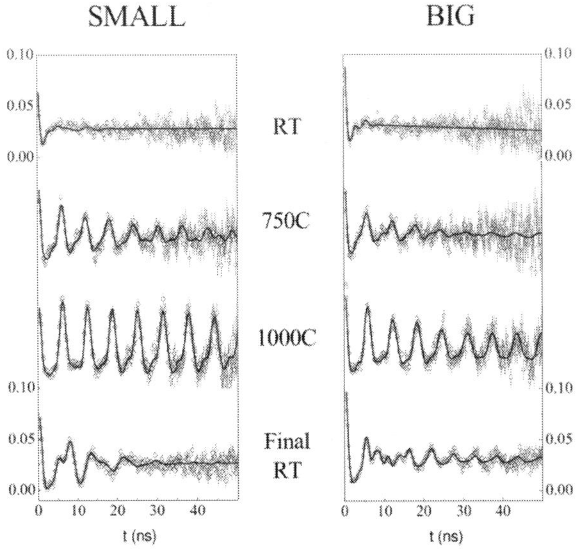

Figure 2. Selected PAC spectra obtained at indicated temperatures.

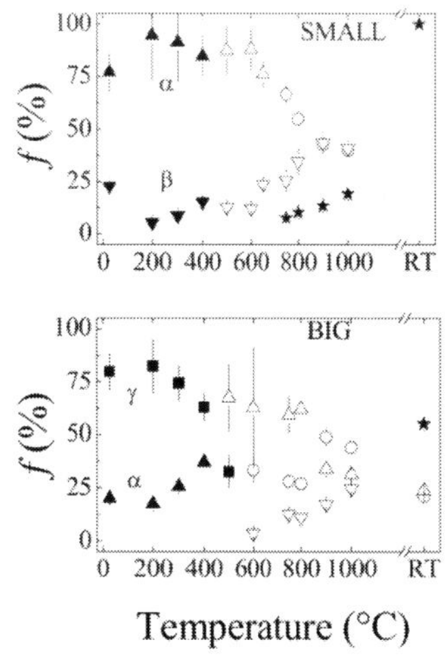

Figure 3. Relative fractions of the different hyperfine components.

somewhat distorted monoclinic phase (full star). At 1000 °C the sample is described by equal amounts of the tetragonal t- and cubic Xc-forms and 15% of the monoclinic polymorph. Upon cooling to RT the sample has become entirely monoclinic. PAC results in BIG show the presence of the α interaction accompanied by a great amount of a new one hereinafter called γ (full square). At 500 °C α has disappeared, γ is clearly decreasing and the appearance of a great amount of the t'-form is evidenced. The t- and Xc-forms are present from 600 °C on and over 800 °C they grow at the expense of the defective t'-form. Cooling down from 1000°C, the sample is half monoclinic and half tetragonal, exhibiting nearly equal amounts of both t- and t'-tetragonal configurations.

4. Discussion

In order to discuss PAC results in the as–prepared samples and their thermal evolution, a simple model for the reactions occurred during synthesis will be assumed. According to Lerot et al.,[4] in the alcoholic solution of CA and TZP some substitutions of $(OC_3H_7)^-$ by $(OOCC_5H_{11})^-$ groups could have taken place. Since the complete replacement was avoided due to the very low CA/TZP ratio employed, most probably the following reaction occurred:

$$2\,Zr(OC_3H_7)_4 + 3C_5H_{11}COOH \rightarrow Zr(OC_3H_7)_3(OOCC_5H_{11}) + Zr(OC_3H_7)_2(OOCC_5H_{11})_2 + 3\,C_3H_7OH$$

Lerot has also reported that the subsequent water addition causes the hydrolysation of exclusively the alkoxide groups. Thus, assuming that water has hydrolysed mainly zirconium-alkoxy

Journal of Physics: Conference Series **167** (2009) 012041 doi:10.1088/1742-6596/167/1/012041

bonds belonging to the less acidified species, this process would have led for SMALL (abundant water) to the species:

$$Zr(OH)_3(OOCC_5H_{11}) + Zr(OH)_2(OOCC_5H_{11})_2$$

and for BIG (scarce water) to the species:

$$Zr(OH)_3(OOCC_5H_{11}) + Zr(OH)(OC_3H_7)(OOCC_5H_{11})_2.$$

Accordingly, PAC results evidence that the as-prepared SMALL and BIG samples exhibit a common interaction (α), the major amount of which was found in SMALL. Thus, it has been assumed that this interaction describes the configurations around Zr of the hydroxyl rich and common to both samples, $Zr(OH)_3(OOCC_5H_{11})$ species. In turn, β in SMALL would describe the configurations in the $Zr(OH)_2(OOCC_5H_{11})_2$ species and γ in BIG, would describe configurations typical of the $Zr(OH)(OC_3H_7)(OOCC_5H_{11})_2$ species, the richest one in organic groups. The higher carbon content determined in BIG supports the previous interpretation. On heating, α, β and γ interactions disappear over a thermal range fairly coincident with that over which the firing and crystallization of organic residues is revealed. Additionally, the two different carbon chains modeled for BIG match the firing of two different organic residues as suggested by the DTA. Regarding the crystalline forms, even when the t′-, t- and Xc- forms appear in both SMALL and BIG, their temperatures of appearance and thermal evolution are somewhat different. Relative to stabilization, some evidences must be pointed out. Upon crystallization the t'-form is predominant in both samples, as revealed by XRD and PAC results at 500ºC. But distinctly, it disappears above 650°C in SMALL and is always present in BIG though reduced to a minor contribution above 800°C. The decrease in the amount of the t′- form is known to take place by the filling of oxygen vacancies [5] through atmospheric oxygen absorption, in agreement with the mass gain determined by TG and the change of the XRD pattern between 500 and 700°C. Knowing the role played by the surface in chemical reactivity, the particles of larger specific area, SMALL, are more proclive to absorb the oxygen that heals the stabilizing defects, thus leading to the t'- to t- destabilizing transformation. Once the regular t- form has appeared, the ordinary thermally induced- crystallite growth and stress elimination are most probably the cause for the occurrence of the distorted monoclinic form. This latter form can be thought as a monoclinic structure involving some types of defects as vacancies or residual hydroxyl groups that, as is occurs in the tetragonal phase[6,7], do not produce important crystallographic modifications. Finally, the complete degradation of SMALL when cooling down to RT is the expected consequence of the important amount of monoclinic phase preexisting at 1000°C. On the other hand, the lower reactivity in BIG seems to impede the oxygen absorption that causes the removal of t'-form. Thus, cooling down to RT, the formation of the monoclinic phase is hardened and only half of the sample becomes degraded.

5. Conclusions

The highly localized PAC technique allowed the determination of three different Zr surroundings in the non-crystalline starting structures. Assuming a plausible chemical sequence during synthesis, the one common to both samples was thought as involving mainly hydroxyl groups. The other two resulted from the presence of different organic residues and were present each one in each sample. In the bigger spheres, these organics rich configurations were predominant and probably consisting of two varieties of Zr-ligand bonds. Crystallization led to the formation of both tetragonal t′- and t- forms and the cubic Xc- form. The temperature of appearance, phase content and thermal behavior of the crystalline products depended on the different residues containing- precursors. The smaller particles degraded to the monoclinic phase gradually when heating and totally on cooling. On the contrary, the lower specific surface area of the bigger spheres favored the permanence of the oxygen defective t′- form on heating, thus leading to a partial degradation on cooling. According to these results, it seems that the smaller particles are less suitable for load bearing applications for which the tetragonal phase permanence is decisively required.

References

[1] Rodríguez A M, Caracoche M C, Rivas P C, Pasquevich A F and Mintzer S R 2001 *J. Am. Ceram. Soc.* **84** 188-192.

[2] Figueroa S, Desimoni J, Rivas P C, Caracoche M C and de Sanctis O 2006 *J. Am. Ceram. Soc.* **89** 3759-3764.

[3] Bondioli F, Cannillo V, Fabbri E and Messori M 2006 *Polimery* **51** 794-798.

[4] Lerot L, Legrand F and De Bruycker P 1991 *J. Mater. Sci.* **26** 2353-2358.

[5] Caruso R, Benavídez E, de Sanctis O, Caracoche M C, Rivas P C, Cervera M, Caneiro A and Serquis A 1997 *J. Mater. Res.* **12** 2594-2601.

[6] Gómez R, López T, Bokhimi X, Muñoz E, Boldú J L and Novaro O 1998 *J. Sol-Gel Sci. and Tech.* **11** 309-319.

[7] Torres-García E, Paláiz-Barranco A, Vázquez-Ramos C and Fuentes G A 2001 *J. Mater. Res.* **16** 2209-2212.

Synthesis and characterization of cobaltite nanotubes for solid-oxide fuel cell cathodes

F Napolitano, L Baqué, H Troiani, M Granada and A Serquis

Instituto Balseiro-Centro Atómico Bariloche and CONICET, San Carlos de Bariloche, Argentina

E-mail: aserquis@cab.cnea.gov.ar

Abstract. $La_{1-x}Sr_xCo_{1-y}Fe_yO_{3-\delta}$ oxides are good candidates for solid oxide fuel cell (SOFC) cathodes because these materials present high ionic and electronic conductivity, and compatibility with Cerium Gadolinium Oxide (CGO) electrolytes allowing a lower operation temperature. In this work, we report the synthesis of $La_{0.4}Sr_{0.6}Co_{0.8}Fe_{0.2}O_{3-\delta}$ (LSCF) nanotubes prepared by a porous polycarbonate membrane approach, obtaining different microstructures depending on sintering conditions. The structure and morphology of the nanotubes and deposited films were characterized by X-ray diffraction, transmission and scanning microscopy. Finally, we obtained nanostructured films of vertically aligned LSCF tubes deposited over the whole surface of CGO pellets with diameter up to 2.5cm in a direct and single step process.

1. Introduction

Solid oxide fuel cells (SOFC) are electrochemical devices whose efficiency is theorically very high for power generation, and have become a great candidate for future energy generation due to its possible fuel flexibility [1]. Their main disadvantage is the high work temperature (800-1000°C), which brings new challenges and difficulties with materials and structure stabilities. Therefore a high effort is currently made to lower the work temperature without loosing the SOFC benefits in the so called intermediate temperature SOFC (IT-SOFC) [1, 2]. For that purpose a lot of research is made to find new materials or to improve the performance of the different cell components: cathode, anode, electrolyte and interconnection materials. In the cathode case, it is necessary to know which of the oxygen reduction steps is the rate determining step that may be present due to mayor losses coming from lowering work temperature [3].

On one hand, the use of mixed ionic and electronics conductors like $La_{1-x}Sr_xCo_{1-y}Fe_yO_{3-\delta}$ (LSCF) as cathode material, to replace the simple electronics conductors, allows that the oxygen reduction reaction may take place over all cathode surface and not be limited to triple phase boundary region [4]. On the other hand, the microstructure also has a strong influence in cathode performance, which may be substantially improved using nanostructures [5, 6]. It is interesting to identify the rate determining step by modelling the electrochemical response of cathodes with different micro or nanostructures. The system complexity involves several associated parameters such as porosity, grain size, thickness, tortuosity, etc.

Hollow and solid cylindrical nanometric structures of mixed valent manganese oxide-based compounds have been obtained by Leyva *et al* by wetting sacrificial porous substrates [7][8]. Sacanel *et al* synthesized LSC and LSCF nanotubes by denitration process [9], Bellino *et al* painted electrolyte surfaces with an ink made with LSC nanotubes (previously obtained by pore wetting technique) and attached them with a high temperature thermal treatment [6].

In this work, the synthesis and characterization of vertically aligned tubular nanostructures of LSCF (from now on nanotubes) were deposited on cerium gadollinium oxide pellets. The synthesis was made through an acetate approach and polymeric porous membrane acting as mould. Then, a nanostructured electrode of simple and well defined geometry could be obtained which may allow a system analysis by simple models.

2. Experimental

$La_{0.4}Sr_{0.6}Co_{0.8}Fe_{0.2}O_{3-\delta}$ tube structures were prepared by the acetate route [10] and porous polycarbonate membrane (characteristic pore diameters of 200, 400 and 800nm) acting as a mould or template [8]. Stoichiometric amount of $SrCO_3$, $Co(CH_3COO)_2{\cdot}4H_2O$, $Fe(CH_3COO)_2$ and La_2O_3 were mixed in acetic acid solvent (0.1M) under reflux and magnetic stirring. The polycarbonate membrane pores were filled with the acetate solution by two different methods: capillarity and vacuum. In the first one the membrane was submerged in the solution and left in ultrasound for 5 minutes to help filling the pores by capillarity effect. In the later one, the membrane was placed over a silicate filter and the solution was forced to pass through the membrane making vacuum at the other side of the filter.

Each membrane, with its pores filled with the acetate solution, was pretreated in a microwave oven with exposure times ranging from 30 to 300 seconds. Then it was positioned over a $Ce_{0.9}Gd_{0.1}O_{2-\delta}$ (CGO) pellet (up to 2.5 cm of diameter) acting as substrate and fired at 400°C for 2 hours and then to 800°C for 6 hours to complete the synthesis of the perovskite phase.

The obtained tube structures were characterized by X-ray diffraction (XRD) with a Philips PW 1730/10 difractometer with Cu K_α radiation using a low background Si monocrystal substrate. The morphology was characterized by scanning (SEM) and transmission (TEM) electron microscope (Philips 515 and Philips CM-200UT, respectively). Dark field, bright field and high resolution images acquired with TEM were used to determine grain size distributions and interplanar distances.

3. Results and Discussion

Figure 1 shows SEM images corresponding to samples made with (a) ultrasound and (b and c) vacuum with different pore filling methods. Nanotubes made with the vacuum method have thicker walls, better defined surface and endings than those prepared with ultrasound method. The ultra-thin wall tubes prepared with ultrasound-capillarity method can be noted by the transparency of some tubes at Fig. 1(a). The complete filling of the membrane pores is not

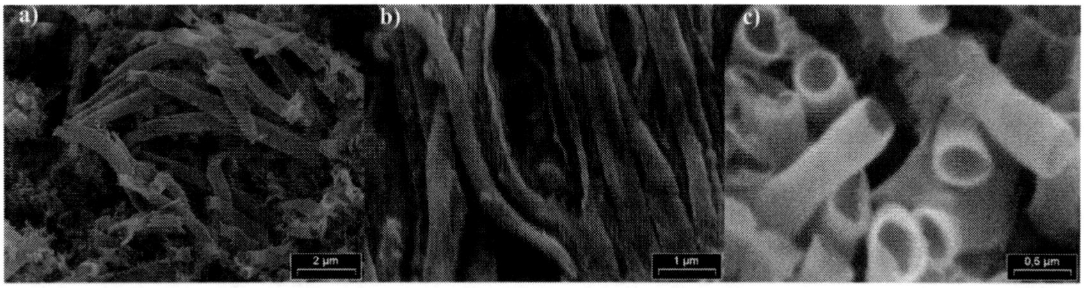

Figure 1. SEM images of nanostructures formed with a) capillarity filling method, b) vacuum method with home-made equipment and c) vacuum method with commercial filtering system adapted to the used membranes.

Journal of Physics: Conference Series **167** (2009) 012042 doi:10.1088/1742-6596/167/1/012042

Table 1. Crystallite size and residual structural tension calculated from XRD.

Sample	Crystallite Size (nm)	Residual strain (%)
bulk	30 ± 1	0.38 ± 0.02
400nm-60s	19 ± 1	0.54 ± 0.02
400nm-120s	21 ± 1	0.49 ± 0.02
800nm-30s	22 ± 1	0.47 ± 0.02
800nm-60s	21 ± 1	0.50 ± 0.02

assured using this method, while the vacuum system forces the solution to pass through the pores. It can also been observed that commercial vacuum system allows to obtain more homogeneous walls thickness with less external roughness than home-made equipment.

SEM images were used to characterize the morphology and the presence of hollow nanostructures for all characteristic pore size used. The microwave exposure time influence is shown in Fig. 2(a) for samples made with the commercial vacuum filling method. Measurements showed a decrease of tube diameters with higher exposure time but these results may be hindered by the measurement dispersion. The shrinkage rate (deduced from the final diameters as a function of the initial pore size in Fig. 2(b)) was higher for bigger pore membrane size due to more available space for diameter reduction in larger pores. The wall thickness was not significantly changed by microwave exposure times while it is reduced when using smaller pore diameters (e.g. walls thickness measured from SEM images were ~100nm for 800nm samples and ~60nm for 400nm samples).

A comparison of XRD analysis corresponding to a nanotubes sample and to a bulk sample (prepared as a pressed pellet of solution crystallized powder), with the same final thermal treatment, is displayed in Figure 3. Additional reflections, corresponding to precursors oxides, were observed only in the powder sample diffractogram meaning that the used temperature was not enough for complete LSCF phase formation. On the contrary, the microwave pretreatment

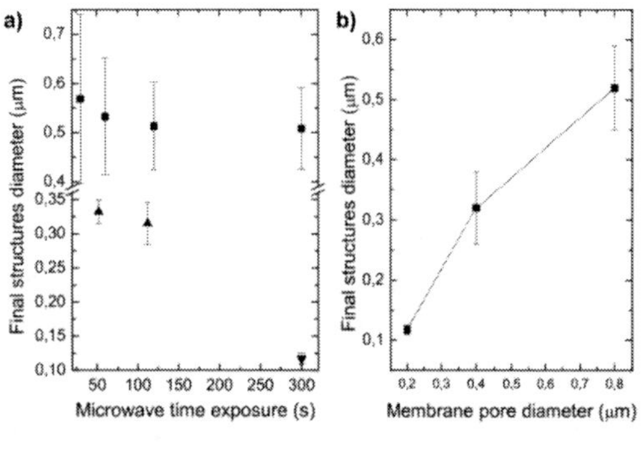

Figure 2. a) Final tube diameters for several synthesis conditions (■ 800nm, ▲ 400nm, ▼ 200nm) and b) structure shrinkage for different characteristic diameters of template.

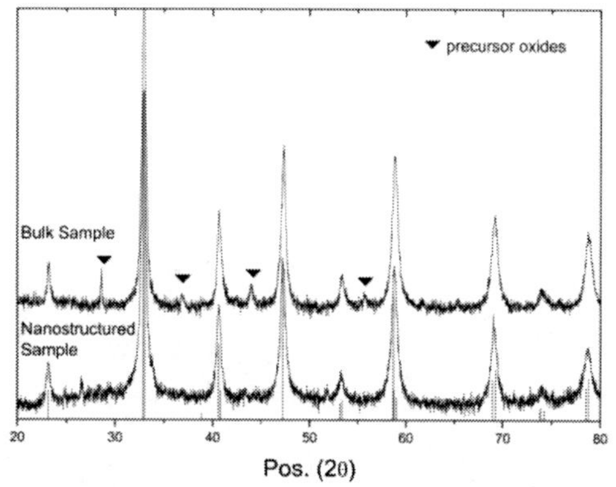

Figure 3. Bulk sample and nanostructured film X-ray diffractograms with the same final thermal treatment. Perovskite phase reflections are indicated with vertical lines.

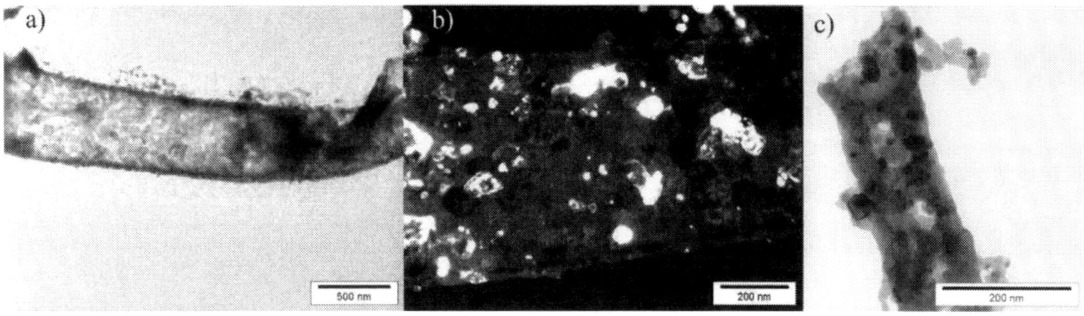

Figure 4. a) 800nm sample TEM bright field image used for individual structure characterization, b) dark field image of same sample used for measurement of crystalline single domain and c) 200nm sample bright field TEM image.

helps to reduce the synthesis temperature of the perovskite phase in the nanostructured film where only LSCF peaks were observed.

Crystallite sizes and residual structural tensions were calculated for nanotubes made with all pore sizes and also for the bulk sample using the 47.3° reflection of the x-ray diffractogram (see Table 1). All the nanostructured samples showed similar results but their residual strain were significantly higher than one corresponding to the bulk sample and the crystallite size were lower which may be explained by the confinement produced by tube walls. TEM images were acquired from colloidal suspension of 200 and 800nm pore diameter samples. From Figure 4 it can be seen that tube walls are more irregular than observed at SEM images (difference due the better resolution of TEM), diameter and wall thickness have been measured for single tubes obtaining: (550 ± 30)nm and (58 ± 7)nm, respectively for 800nm diameter template, (145 ± 10)nm and (41 ± 7)nm for 200nm diameter template. Diameters showed similar values in comparison with SEM images measurements. TEM diameter values measured along the axis direction on individual tube had a 5% dispersion. Wall thickness measurements from TEM images resulted on smaller values than those obtained from SEM images due to better resolution, and similar values were found between 200 and 800nm samples.

Grain size distribution has been obtained from dark field TEM images of both samples, resulting a mean value of 30nm (ranging from 12 to 85nm) for 800nm sample and 14nm (from 4

Figure 5. a) Final sample of nanotube monolayer full covering a 2.5cm diameter CGO pellet. b) SEM image of long range distribution of nanotubes deposited over the CGO pellet. c) detail of layer surface showing the alignment between nanotubes.

to 70nm) for 200nm sample, these values are similar to those found for crystallite size at XRD for 800 and 200nm template samples. Planar length of (2.74 ± 0.01)Å and (3.78 ± 0.06)Å were obtained from HRTEM images which can be identified with 110 and 012 planes of LSCF phase rhombohedral structure [11].

Once the synthesis process was optimized, vertically aligned nanotubes monolayer films were obtained, full covering CGO pellets of diameters up to 2.5cm (Figure 5). This optimization for LSCF tubes deposition film requires the use of commercial vacuum method for pore mould filling, drying the membrane surface before the thermal pre-treatment (microwave exposure time of 60 second), and final synthesis onto CGO pellet at 800°C.

4. Conclusions

Hollow tubular structures were synthesized through an acetate solution approach and polycarbonate porous mould technique for several nominal pore diameters (200, 400 and 800nm). Morphological characterization of these structures was made through images acquired by SEM and TEM. We observed that the best tubular structure formations were reached with commercial vacuum filling method and the measurements of final tube diameters result on decreasing values with higher microwave exposure times. These values were lower than characteristic mould pore size and the rate of shrinkage was higher for bigger pores. Diameter dispersion along tubes axis was between 6 and 9% and similar wall thickness were measured for samples made by different mould sizes.

X-ray diffractograms showed perovskite phase formation at 800°C (final thermal treatment temperature) which is lower than the temperature necessary for bulk samples. Lower crystallite size and higher tension were calculated from x-ray diffractograms for nanotubes than those for a bulk sample due to tube walls confinement. Distribution mean values of grain sizes obtained at dark field TEM images are comparable with XRD average crystallite size.

Nanostructured films of vertically aligned LSCF tubes were deposited over the whole surface of CGO pellets with diameter up to 2.5cm at the same time of final thermal treatment in a direct and single step process. LSCF films deposited over the two sides of CGO pellets, forming symmetrical cells, will allow to carry out the study of electrode reaction through the electrochemical impedance spectroscopy technique.

5. Acknowledgments

Gabriela Leyva is acknowledged for useful suggestions. F.N. thanks to Ernesto Scerbo and Carlos Cotaro for SEM observations and also to Carlos Ayala for the vacuum equipment. This work was funded by University of Cuyo, CONICET and ANPCyT.

References

[1] Singhal S C 1999 *Proc. 6th Int. Symp. on SOFC* (Honolulu:The Electrochemical Society) pp 97280
[2] Steele B C H and Heinzel A 2001 *Nature*, **414**, 345
Brandon N P, Skinner S and Steele B C H 2003, *Annu. Rev. Mater. Res.*, **33**, 183
[3] Ivers-Tiffee E, Weber A and Herbstritt D 2001 *Journal of the European Ceramic Society* **21** 1805
[4] Teraoka Y, Zhang H M, Furukawa S and Yamazoe N 1985 *Chem. Lett.* 1743
Teraoka Y, Zhang H M, Okamoto K and Yamazoe N 1988 *Mater. Res. Bull.* **23** 51
Taia L-W, Nasrallahb M M, Andersonc H U, Sparlind D M and Sehlin S R 1995 *Solid State Ionics* **76** 259
Taia L-W, Nasrallahb M M, Andersonc H U, Sparlind D M and Sehlin S R 1995 *Solid State Ionics* **76** 273
[5] Baqué L, Caneiro A, Moreno M S and Serquis A 2008 *Electrochemistry Communications In press*
[6] Bellino M, Sacanell J, Lamas D, Leyva A G and Walsöe de Reca N E 2007 *J. Am. Chem. Soc.* **129** 3066
[7] Leyva A G et al 2004 *Journal of Solid State Chemistry* **177** 3949
[8] Leyva A G, Stoliar P, Rosenbusch M, Levy P, Curiale J, Troiani H and Sanchez R 2004 *Physica* B **354** 158
[9] Sacanell J, Bellino M, Lamas D and Leyva A G 2007 *Physica* B **398** 341
[10] Xia Y, Armstrong T, Prado F and Manthiram A 2000 *Solid State Ionics* **130** 81
[11] ten Elshof J and Boeijsma J 1996 *Powder Diffraction* **11** 28

Study of $LnBaCo_2O_{6-\delta}$ (Ln = Pr, Nd, Sm and Gd) double perovskites as new cathode material for IT-SOFC

E. Chavez, M. Mueller, L. Mogni[1], A. Caneiro

Centro Atómico Bariloche-CNEA, Instituto Balseiro. Av. Bustillo 9500, S. C. de Bariloche 8400, Argentina

E-mail: [1]mogni@cab.cnea.gov.ar

Abstract. Oxides with double perovskites structures of general composition $LnBaCo_2O_{6-\delta}$ (Ln = Pr, Nd, Sm and Gd) were synthesized by solid state reaction with the purpose to evaluate new materials to be used as cathodes in intermediate temperature solid oxide fuel cell (IT-SOFC). A preliminary study about electrochemical properties was performed by impedance spectroscopy between 500 and 800 °C under atmosphere of pure O_2. Symmetrical cells were obtained by spray deposition of $LnBaCo_2O_{6-\delta}$ (Ln = Pr, Nd, Sm and Gd) at both sides of a dense ceramic electrolyte. The impedance spectroscopy measurements as a function of temperatures show a hysteresis loop which could be associated to a tetragonal/orthorhombic phase transition. The existence of this transition was corroborated by high temperature X-Ray diffraction and Differential Scanning Calorimetry measurements.

1. Introduction

Layered cobaltites with general formula $LnBaCo_2O_{6-\delta}$ (Ln = rare earth, $0 \leq \delta \leq 1$) have been extensively studied due to their attractive magnetic and transport properties at low temperatures (below 150 °C) [1, 2, 3]. These cobaltites present two main structural features [3, 4]: The first one is the perfect order in alternating (0 0 1) layers of the Ln and Ba ions due to great differences between the ionic radii of them. The second structural feature is due to the fact that the oxygen vacancies are located at the Ln layers with a great tendency to form ordered patterns. This results in an order coexistence of Co ions in octahedral (CoO6) and pyramidal (CoO5) environments. Then, the magnetic and transport properties are highly dependent on the oxygen content.

At high temperatures cobaltites are interesting due to their high mixed conductivity. This property make these materials attractive for applications as gas sensors, oxidation catalysts, or materials for fuel cells. In the last years, the trend to low the operating temperature of Solid Oxide Fuel Cells (SOFCs) to the intermediate temperature range (500-700 °C) has become one of the main SOFC research goals. In this range of temperatures, the contribution of the cathode to the total resistance of the cell usually becomes the dominant contribution. This is due to the high activation energy associated to the oxygen reduction reaction taking place in the cathode side.

Recent studies demonstrated that layered cobaltites, $LnBaCo_2O_{6-\delta}$ with Ln = Gd and Pr [5, 6], could be identified as a potential cathode materials for intermediate-temperature

SOFC (IT-SOFC) due to their excellent oxygen transport properties, i.e. high oxygen surface exchange coefficient, reasonable oxide ionic diffusivity and high electronic conductivity. However, an order-disorder phase transition recently reported for these compound around 500 °C [6, 7] could affect the electrochemical performances of this family of compounds. This order-disorder phase transition involves a rearrangement of oxygen vacancies going from the low-temperature one-dimensional distribution alternating filled and empty chains of oxygen along the a axis (orthorhombic symmetry O, Pmmm space group) to the high-temperature two-dimensional distribution of vacancies in (0 0 1) layers, i.e. in the LnOx plane (tetragonal symmetry T, P4/mmm space group).

In this work we present preliminary results of the effect of Ln cation (Ln = Pr, Nd, Sm, Gd) on the structural and electrochemical properties of $LnBaCo_2O_{6-\delta}$ compounds. The electrochemical performance was evaluated by means of impedance spectroscopy while the phase transition was investigated by high temperature X-ray diffraction (HT-XRD) and differential scanning calorimetry (DSC).

2. Experimental

$LnBaCo_2O_{6-\delta}$ (Ln = Pr, Nd, Sm and Gd) compounds were obtained by solid state reaction at 1100 °C using stoichiometric amounts of lantanides oxides, Co_3O_4 and $BaCO_3$. The presence of single-phase materials was verified in the four samples by room temperature XRD using a Philips PW1700 diffractometer with CuK_α radiation and a graphite monochromator. The samples display Orthorhombic (Pmmm) or retain the high temperature Tetragonal (P4/mmm) symmetry depending on the cooling rates. Figure 1 shows the typical diffraction patterns of these compound for both symmetries.

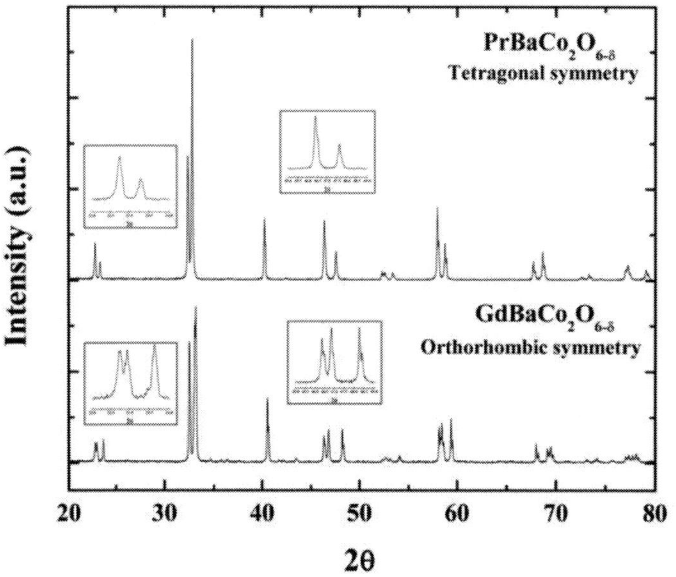

Figure 1. Characteristic XRD diagram of tetragonal and orthorhombic phases.

Symmetrical $LnBaCo_2O_{6-\delta}/Ce_{0.9}Gd_{0.1}O_{2-x}/LnBaCo_2O_{6-\delta}$ cells were prepared by spray deposition of double perovskites paint on both sides of the $Ce_{0.9}Gd_{0.1}O_{2-x}$ (CGO) dense

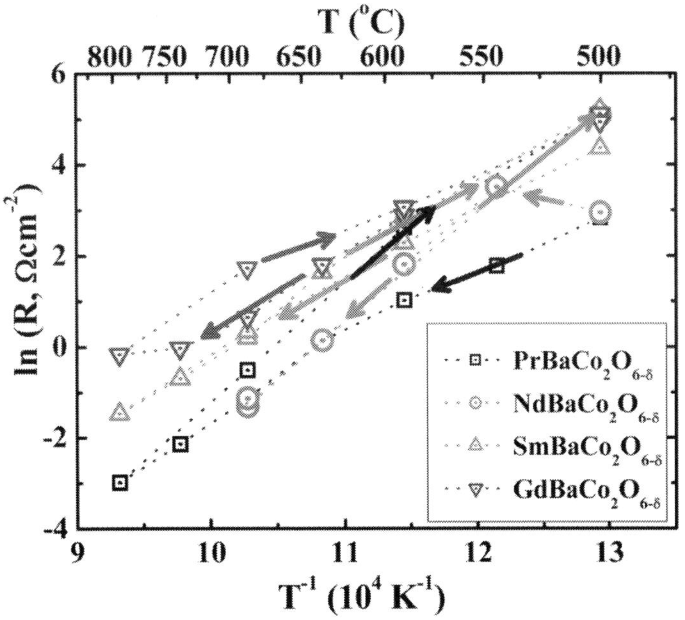

Figure 2. Arrhenius plot for the total electrode resistance of polarization R_P by $LnBaCo_2O_{6-\delta}$ (Ln = Pr, Nd, Sm and Gd) between 500-800 °C.

electrolyte. Then, these films were sintered at 1000 °C for 1 h. The structural characterization of the as-prepared film was performed by XRD before and after the electrochemical measurements. No significant differences were found between them. The uniform morphology and composition of as-deposited electrode films were checked after the electrochemical measurements using scanning electron microscope (SEM, Philips 515) and energy dispersive spectroscopy (EDS, Genesis 2000).

The electrochemical characterization was performed by Electrochemical Impedance Spectroscopy (EIS) measurements under pure O_2 atmosphere at temperatures (T) from 500 to 800 °C using a frequency response analyzer (FRA) coupled to an AUTOLAB potenciostat. Measurements were performed increasing and decreasing T in the 1 MHz-0.1 Hz frequency range with null voltage bias. The amplitud of the applied voltage signal was 10 mV.

The HT-XRD studies were carried out using an Anton Paar camera coupled to the diffractometer. The measurement were performed from room temperature to 900 °C between $20 \leq 2\theta \leq 80$ ° under static air.

The thermal behavior of the samples was studied using a differential scanning calorimeter (Modulated DSC 2910 TA Instruments). Measurements were done under air atmospheres at a heating rate of 5 °C/min between room temperature and 550 °C.

3. Results and Discussions

3.1. Electrochemical characterization

The EIS data shown in fig 2 indicate an hysteretic behavior of the electrode polarization resistances R_P vs. T above 500 °C. These results disagree with that observed by Tarancón el al. [6], who found hysteresis in their electrical conductivity measurements but not in the impedance measurements performed in air between 400 and 800 °C. We assume that this behavior should be associated to the order-disorder phase transition since our XRD data after EIS measurements show no differences with those obtained before, excluding chemical reactivity between electrode/electrolyte. Despite the increase in the polarization resistance, these R_P values are lower than those of mostly conventional cathode materials and decrease as the ionic radii of Ln increases: $R_P^{Pr} \leq R_P^{Nd} < R_P^{Sm} \leq R_P^{Gd}$.

3.2. HT-XRD study

The change of symmetry from orthorhombic Pmmm to tetragonal P4/mmm was followed trough the splitting of the peak around $2\theta = 22.5°$ (see figure 3). We selected these peaks in order to avoid the diffraction peak of Pt holder at $2\theta \sim 47°$. This transition takes place at 350-400, $\sim$ 400, 375-400 and around 450- 500 °C for Ln = Pr, Nd, Sm, and Gd, respectively. For Pr compound we found a transition temperature different to that reported by Streule et al. [7] (503 °C). However, in that case the transition was study in a sample prepared with an oxygen content of $6 - \delta \simeq 5.48$ and the NPD measurements were carried out under Ar atmosphere in order to maintain the $6 - \delta$ value constant. By other hand, we found for the Gd compound a similar transition temperature than that reported by Tarancón et al. [6] under comparable experimental conditions.

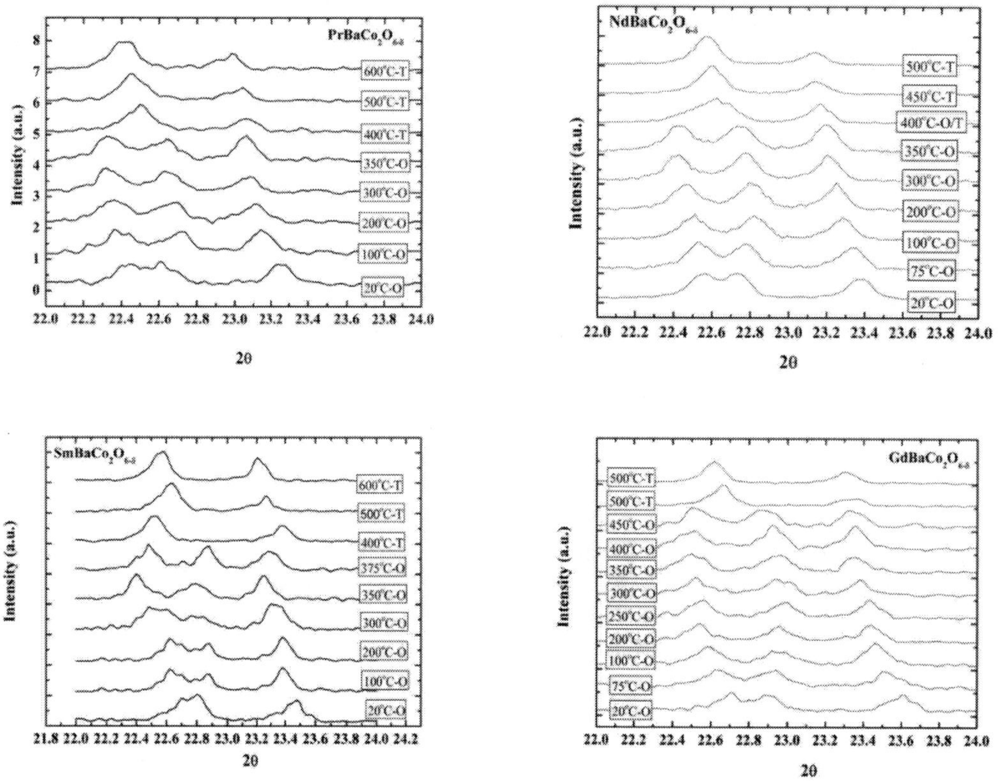

Figure 3. High temperature XRD evolution with T showing the orthorhombic-tetragonal (O/T) phase transition for $LnBaCo_2O_{6-\delta}$ (Ln = Pr, Nd, Sm and Gd).

However, previous studies performed at room temperature in samples with different oxygen content indicate that the order-disorder phase transition depends on the oxygen stoichiometry. Thus, at $6 - \delta < 5.45$, $GdBaCo_2O_{6-\delta}$ exhibits tetragonal symmetry while above it ($5.45 \leq 6 - \delta \leq 5.60$) becomes orthorhombic [2]. Therefore, it seems that this phase transition can be achieved either, by controlling temperature or oxygen content. With this in mind, it seems reasonable to observe a different behavior between our EIS measurements and those of Tarancón et al [6]. Our measurements were performed under pure O_2, then it is reasonable to expect for our samples a higher oxygen content than that of Tarancón's sample. Therefore, the stability of the orthorhombic symmetry should be shifted to higher temperatures and hampering the electrochemical performance.

3.3. Thermal analysis

In figure 4 are shown the heat flows as a function of T obtained for 5 °C/min heating rate for $LnBaCo_2O_{6-\delta}$ (Ln = Pr, Nd, Sm and Gd)compounds.

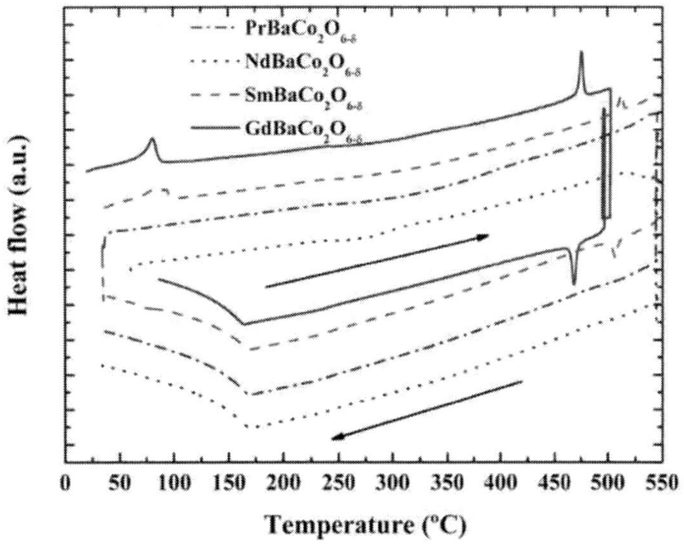

Figure 4. Thermal analysis by DSC with heating rate of °5 C/min for $LnBaCo_2O_{6-\delta}$ (Ln = Pr, Nd, Sm and Gd).

The DSC measurements didn't show the O/T transition observed by HT-XRD for Pr and Nd samples (see fig. 4). Besides, the transition temperatures for Sm sample (512-508 °C) is higher than that of Gd sample (476-468 °C). This results disagree with those observed by HT-XRD. This disagreement may be related to different experimental conditions between DSC and HT-XRD. DSC is a dynamic technique while our HT-XRD data are collected at constant temperatures.

The enthalpy change associated to the O/T transition are ΔH = 0.85 and 3.78 J/g for Sm and Gd compounds, respectively. The value obtained for Gd compound is in agreement with that of Tarancón et al. (3.9 J/g) [6]. The ΔH values decreases as we move right to left in periodic table (Sm and Gd samples). Therefore, it should be expected lower ΔH values for Nd and Pr samples, with the consequent decrease of the peak until it disappears. Thus, this fact will explain the lack of the DSC peak for Nd and Pr samples.

Besides, we want to emphasize that this order-disorder transitions seems to depend on thermal history of the samples. Thus, if samples are fast cooled, it's possible to retain the tetragonal phase at room temperature and, a subsequent fast heating avoid the orthorhombic phase formation. We think that these behaviors indicate a kinetics effect over the O/T transition which is related not only with the oxygen content and temperature, but also to the oxygen diffusion inside of each materials.

4. Conclusion

The HT-XRD and DSC studies show an order-disorder phase transition (O/T) at temperatures around 500 °C that strongly depends on the thermal history and the oxygen content of the samples.

The R_P values vary according to $PrBaCo_2O_{6-\delta} \leq NdBaCo_2O_{6-\delta} < SmBaCo_2O_{6-\delta} < GdBaCo_2O_{6-\delta}$. Then, the ionic radii of Ln seems to play a role for improving the electrochemical properties. However, the R_P values show a hysteresis loop that could be associated to the O/T transitions and could compromise the electrochemical performance in SOFC applications.

Acknowledgments

The authors want to acknowledge Institute Balseiro, Institute Sabato, CONICET, ANPCyT, and UnCuyo by financial support. We also want to acknowledge the technical support of E. Scerbo, C. Cotaro and M. Esquivel.

References

[1] C. Martin, A. Maignan, D. Pelloquin, N. Nguyen, B. Raveau, 1997 *Appl. Phys.Lett.* **71** 1421.
A. Maignan, V. Caignaert, B. Raveau, D. Khomskii, G. Sawatzky, 2004 *Phys. Rev. Lett* **93** 26401.
S. Streule, A. Podlesnyak, J. Mesot, M. Medarde, K. Conder, E Pomjakushina, E. Mitberg and V. Kozhevnikov, 2005 *J. Phys.: Condens. Matter.* **17** 3317.
I. O. Troyanchuk, N. V. Kasper, D. D. Khalyavin, H. Szymczak, R. Szymczak, and M. Baran, 1998 *Phys. Rev. Lett.* **80**, 3380.
C. Frontera, J. L. García-Muñoz, A. Llobet, and M. A. G. Aranda, 2002 *Phys. Rev.* B **65**, 180405 R.
T. Vogt, P. M. Woodward, P. Karen, B. A. Hunter, P. Henning, and A. R. Moodenbaugh, 2000 *Phys. Rev. Lett.* **84**, 2969.

[2] A.A. Taskin, A.N. Lavrov, Y. Ando, 2005 *Phys. Rev.* B **71** 134414.

[3] A. Maignan, C. Martin, D. Pelloquin, N. Nguyen and B. Raveau, 1999 *J. Solid State Chem.* , **142**, 247.

[4] C. Frontera, A. Caneiro, A. E. Carrillo, J. Oró-Solé, and J. L. García-Muñoz, 2005 *Chem. Mater.* **17**, 5439.

[5] A. Tarancón, A. Morata, G. Dezanneau, S.J. Skinner, J.A. Kilner, S. Estrade, F. Hernández-Ramírez, F. Peiró and J.R. Morante, 2007 *J. Power Sources* **174** 255.
A.A. Taskin, A.N. Lavrov, Y. Ando, 2005 *Appl. Phys. Lett.* **86** 91910.
A. Tarancón, S.J. Skinner, R.J. Chater, F. Hernández-Ramírez, J.A. Kilner, 2007 *J. Mater. Chem.* **17** 3175.
A. Chang, S. J. Skinner, J.A. Kilner, 2006 *Solid State Ionics* **177** 2009.
G. Kim, S. Wang, A.J. Jacobson, L. Reimus, P. Brodersen and C.A. Mims, 2007, *J. Mater. Chem.* **17** 2500.

[6] A. Tarancón, D. Marrero-López, J. Peña-Martínez, J.C. Ruiz-Morales, P. Núñez, 2008, *Solid State Ionics* **179** 611.

[7] S. Streule, A. Podlensyak, D. Sheptyakov, E. Pomjakushina, M. Stingaciu, K. Conder, M. Medarde, M.V. Patrakeev, I.A. Leonidov, V.L. Kozhevnikov, J. Mesot, 2006 *Phys. Rev.* B **73** 94203.
S. Streule, A. Podlensyak, E. Pomjakushina, K. Conder, D. Sheptyakov, M. Medarde, J. Mesot, 2006 *Physica* B **378-380** 539.

Journal of Physics: Conference Series 167 (2009) 012044 doi:10.1088/1742-6596/167/1/012044

Dependence on the growth direction of the strain in AlGaSb alloys

M Rojas-López[1], E Momox-Beristain[2], R Delgado-Macuil[1], V L Gayou[1] A Orduña-Díaz[1], B Salazar- Hernández[3], A G Rodríguez [4],

[1] *CIBA-Tlaxcala, IPN, Tlaxcala, Tlax. México*
[2] *FC-BUAP, Puebla, Pue. México*
[3] *CIICAp-UAEM, Cuernavaca, Mor. México*
[4] *IICO-UASLP, San Luis Potosí, S.L.P. México*

E-mail: marlonrl@yahoo.com.mx

Abstract. High resolution x-ray diffraction profiles were obtained from $Al_xGa_{1-x}Sb$ layers grown on (001) and (111) GaSb substrates. The out of plane lattice parameter, was estimated directly from the symmetrical diffractions for (001) and (111) alloys. These results show that all the layers are strained, and those grown on (001) GaSb are slightly more strained than the corresponding layers grown on (111) GaSb. This difference is explained by the dependence of the strain ratio on growth direction. The out of plane lattice parameter as a function of Al content is higher than the corresponding bulk lattice parameter of $Al_xGa_{1-x}Sb$ layers obtained with Vegard's law. Also, the perpendicular and the in-plane lattice parameter expected for pseudomorphic alloys, was estimated from the strain ratios, assuming an elastic deformation and using the EDX alloy composition to interpolate the elastic constants C_{ij}. This estimation also shows that almost all the layers are fully strained.

1. Introduction

$Al_xGa_{1-x}Sb$ compound is an important material for infrared, optical and high speed devices in the spectral region from 1.3 to 6.5 μm [1-2]. This material also has been used for low-noise and high-speed avalanche photodiodes [3], as well as injection lasers made of AlGaAsSb and GaInAsSb lattice matched to either GaSb or InAs substrates emitting in the wavelength range of 2-4 μm [4]. These devices have been fabricated primarily from layers grown by molecular beam epitaxy (MBE), liquid phase epitaxy (LPE) and by organometallic vapor phase epitaxy (OMVPE). However, the growth of AlSb-containing compounds by OMVPE presents some problems as the high level of carbon, a p-type impurity that is incorporated with methyl-based precursors and has hindered n-type doping [5]. For this reason, this material turns out to be interesting to the scientific community and it is by that it is required to investigate more about the properties of this kind of III-V semiconductor. In this work, it is performed an analysis of vibrational properties of the ternary $Al_xGa_{1-x}Sb$ alloy grown by LPE on GaSb substrates with (001) and (111) crystallographic orientation, using Raman spectroscopy. The out of plane and in-plane lattice parameters are determined by High Resolution X-ray Diffraction (HRXRD) for both (001) and (111) orientations.

2. Experimental

Ternary $Al_xGa_{1-x}Sb$ layers were grown by Liquid Phase Epitaxy (LPE) at 400 °C on (001) and (111) GaSb substrates. HRXRD rocking curves were obtained using a Philips Diffraction System that employs x-ray radiation (1.541 Å). Energy dispersive x-ray spectroscopy (EDX) was employed to estimate alloy composition of the ternary $Al_xGa_{1-x}Sb$ layers by using a JEOL microprobe system. Raman scattering measurements were performed at room temperature using the 5145 Å line of an Ar laser at normal incidence for excitation. Scattered light was filtered with a notch filter Kaiser and analyzed using a single gratting monocromator SPEX, coupled to a CCD detector cooled to 140 K using liquid nitrogen.

3. Results and discussion

Figure 1 shows rocking curves of $Al_xGa_{1-x}Sb$ layers grown on (001) GaSb substrates at several alloy compositions (x=0.06, 0.08 and 0.15). Two prominent peaks are observed, corresponding to the x-ray reflection from both, the GaSb substrate and from the ternary layer. Figure 2 shows similar curves from ternary layers grown on (111) GaSb substrates (x=0.16, 0.20, 0.36). For both crystallographic orientations (001) and (111) of the substrate, the diffraction profiles were obtained from the (004) and (444) reflections respectively.

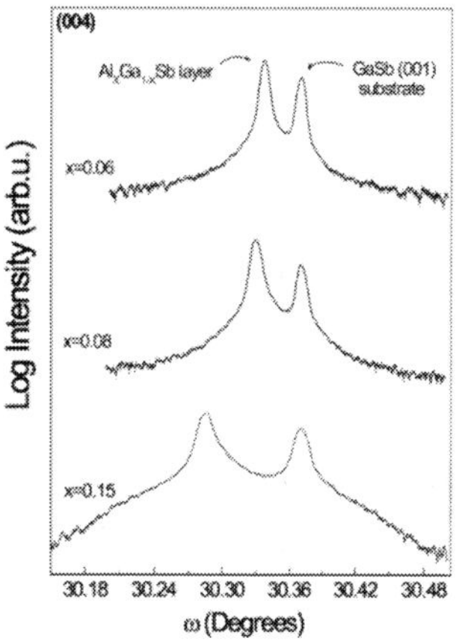

Figure 1. X-ray rocking curves of $Al_xGa_{1-x}Sb$ layers grown on (001) GaSb substrates. The diffraction profiles were obtained from the (004) reflection.

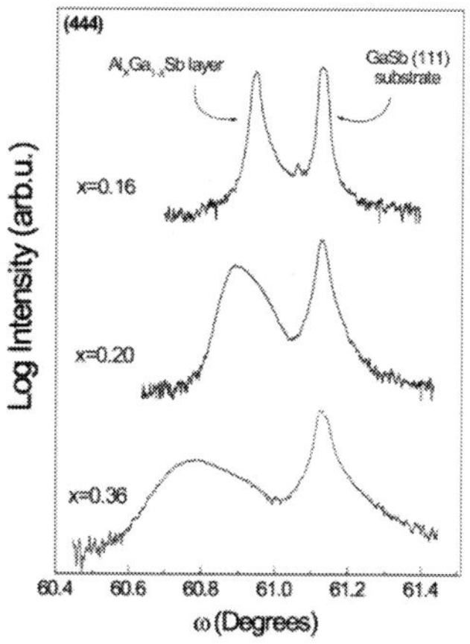

Figure 2. X-ray rocking curves of $Al_xGa_{1-x}Sb$ layers grown on (111) GaSb substrates. The diffraction profiles were obtained from the (444) reflection.

In both cases the relative lattice mismatch is of the order of ($\sim 10^{-3}$). The EDX composition and lattice mismatch data for (001) and (111) ternary layers are summarized in Table I. The lattice parameters obtained from the rocking curves are shown in figure 3. The out of plane lattice parameter ($a_\perp$), obtained directly from the symmetrical diffractions is given by full squares (triangles) for (001) ((111)) alloys. The solid line gives the bulk lattice parameter of $Al_xGa_{1-x}Sb$ layers accordingly to Vegard's law. It is clearly observed that all the layers are strained, and those grown on (001) GaSb are slightly more strained than the corresponding layers grown on (111) GaSb. This difference is explained by the dependence of the strain ratio ($\lambda = \varepsilon_\perp / \varepsilon_\parallel$) on growth direction [6,7].

Table 1. Composition and lattice mismatch data of the (001) and (111) $Al_xGa_{1-x}Sb$ epitaxial layers.

Sample number	GaSb substrate orientation	Al content, x EDX	Lattice mismatch (10^{-3})
GaSb	(111)	0.00	---
M19	(111)	0.05	0.48
M8	(111)	0.13	1.35
M7	(111)	0.16	1.74
M11	(111)	0.20	2.22
M12	(111)	0.25	2.61
M15	(111)	0.36	3.40
M3	(001)	0.06	1.19
M4	(001)	0.08	1.49
M6	(001)	0.15	2.68

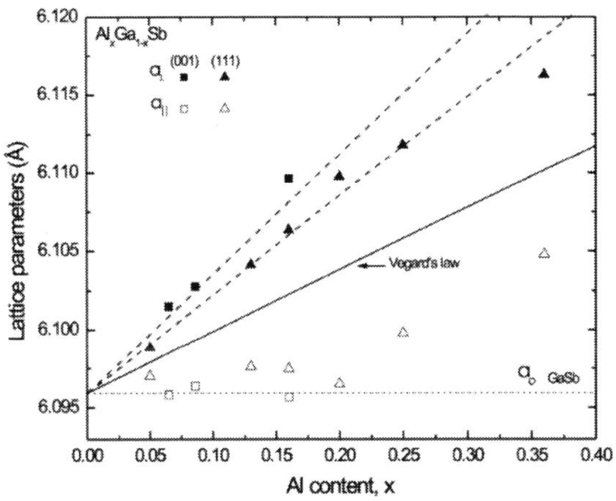

Figure 3. In-plane (open symbols) and out of plane (full symbols) lattice parameters of $Al_xGa_{1-x}Sb$ layers grown on (001) and (111) GaSb substrates. The dashed lines give the expected out of plane parameters for pseudomorphic layers.

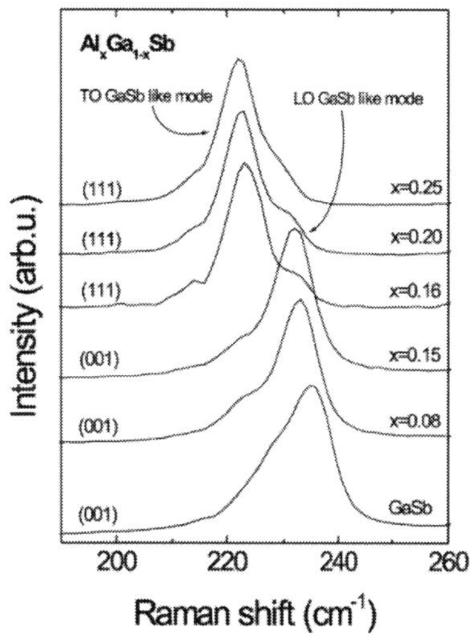

Figure 4. Raman spectra of $Al_xGa_{1-x}Sb$ layers with (001) and (111) crystalline orientations. It is noticeable the dominant TO or LO mode in each case.

The dashed lines give the perpendicular lattice parameter $a_{\perp}$ expected for pseudomorphic alloys as a function of Al content, determined from the ratios λ_{001} and λ_{111} [6,7]. It can be seen that almost all the layers for both orientations are fully strained. The in-plane lattice parameters ($a_{\|}$), were determined from the strain ratios, assuming an elastic deformation and using the EDX composition to interpolate the elastic constants C_{ij}. The behavior of $a_{\|}$ with Al content, also shows that almost all the layers are fully strained, except for the (111) grown layers with x = 0.25 and 0.36 which are partially relaxed.

Figure 4 shows Raman scattering results of $Al_xGa_{1-x}Sb$ for the two crystallographic orientations (001) and (111) of the GaSb substrate. In this figure is depicted the GaSb like mode of this ternary system. This mode can also be deconvoluted in TO and LO GaSb like modes. As it is well known the selection rules for the (001) orientation in the crystal allow only the observation of the LO mode, whereas for the (111) orientation it is allowed the observation of both TO and LO like modes. However, although both modes are allowed for the (111) orientation, the TO GaSb like mode domains in intensity over the LO GaSb like mode. On the other hand, as expected for the (001) orientation, the LO GaSb like mode domains in intensity over the TO GaSb like mode. Figure 5 shows the Raman scattering spectra of the AlSb like mode from the $Al_xGa_{1-x}Sb$ alloy. In this case TO and LO AlSb like modes have similar intensities for the (001) crystallographic orientation, whereas for the (111) orientation the TO AlSb like mode is slightly more intense with respect to the LO AlSb like mode. It is noticeable the presence of a very low intensity band at 338 cm^{-1} associated to the LO AlSb mode of the bulk binary compound [8]. This band appears only for the (111) orientation and may suggest the formation of regions rich in AlSb.

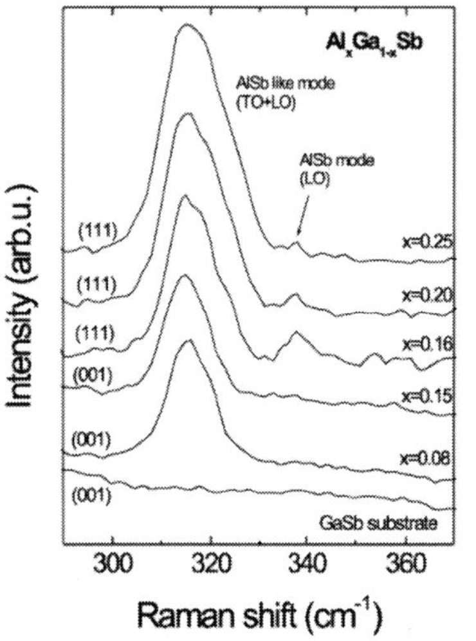

Figure 5. Raman spectra of the AlSb like mode observed in $Al_xGa_{1-x}Sb$ alloys having a (111) crystalline orientation. It is observed a blueshift of the vibration frequency as the Al content increases.

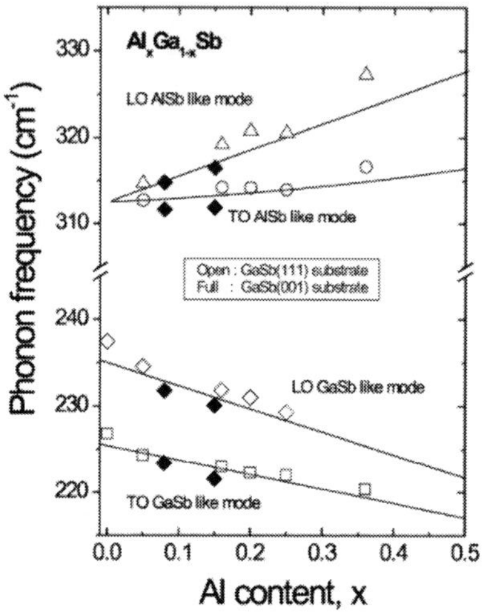

Figure 6. Experimental Raman frequency values (symbols). Continuous lines represent the experimental far infrared reflectance measurements reported in ref [9].

The compositional dispersion of TO and LO phonon frequencies of both GaSb and AlSb like modes measured in the $Al_xGa_{1-x}Sb$ alloy can be observed in Fig. 6. The symbols are the experimental results obtained from the deconvolution of the Raman spectra. GaSb like and AlSb like modes show a clear splitting in frequency shift as the Al content varies. The TO and LO phonon frequencies of each binarylike mode redshift and their splitting decreases as the volume fraction of the corresponding binary constituent in the ternary layer decreases. The dependence of the phonon frequencies on the alloy composition and the LO-TO splittings have been analyzed in several works [8, 9, 10]. The modified random element isodisplacement model (MREI) has shown to be one of the most successful models for treating the optic phonons in mixed crystals [10, 11]. The principal feature of the model is the inclusion of a large second neighbor force constant that reflects the interaction among the alloy atoms on one of the sublattices. Variations of the REI model include strong second neighbor interactions through local field coupling that is based on specific microscopic models for the infrared effective change [12]. Continuous line in Fig. 6 is a fit of the experimental TO and LO phonon frequencies, as a function of alloy composition measured by far infrared reflectance [9]. It is observed a good agreement of the experimental results with the ones previously reported, principally for the GaSb like mode, in which the (001) and (111) crystallographic orientation and residual strain of the layer do not seem to modify the phonon frequency of this mode. In contrast to the GaSb like mode, the AlSb like mode for the (001) orientation shows Raman frequencies slightly higher than the corresponding to (111) orientation. This fact suggests that the strain in the ternary layer is not fully relaxed, consistent with the HRXRD results, in which the perpendicular lattice parameters have a higher value for the (001) orientation of the layer. It is well known that Raman shifts to higher frequencies are associated to biaxial compression of the lattice constant of the layer caused by the substrate. In this case biaxial compression of the GaSb substrate causes a stretching of the perpendicular component of the lattice constant, principally in the (001) orientation, as is indicated in Fig. 3. However in the GaSb like mode this effect is almost no detected whereas for AlSb like mode is more evident, and it could be attributed to a much higher degree of disorder in the AlSb sublattice as a consequence of the low Al content of the layer [13]. In this form HRXRD and Raman techniques have shown to be able to measure differences in lattice mismatch and strain in $Al_xGa_{1-x}Sb$ layers with (001) and (111) crystallographic orientation.

4. Conclusions

$Al_xGa_{1-x}Sb$ epitaxial layers were grown at several Al contents by using the liquid phase epitaxy method at 400 °C on (001) and (111) GaSb substrates. The layer composition was measured directly with EDX obtaining values between x=0.05 and x=0.36. HRXRD results show that most of the studied layers are fully strained for both orientations. Perpendicular strain in (001) layers is higher than in (111) grown layers. Raman scattering measurements were used to measure the frequency shift and relative intensity of the optical GaSb like and AlSb like modes. The spectra show a clear splitting in frequency shift as the Al content varies. The TO and LO phonon frequencies of GaSb like mode shows a shift to low energies and their splitting decreases as the volume fraction of the corresponding binary constituent in the ternary layer decreases. It was observed a slight shift to low energy of the TO and LO GaSb like modes in (001) layers, suggesting that the ternary layer is not fully relaxed, consistent with HRXRD measurements.

References

[1] Hong K. Choi (Ed.), Long-Wavelength Infrared Semiconductor Lasers, John Wiley and Sons, New Jersey, 2004.
[2] M. Mehta, A. Jallipalli, J. Tatebayashi, M.N. Kutty, A. Albrecht, G. Balakrishnan, L.R. Dawson D.L. Huffaker, Photonics Technology Letters, IEEE **19** (2007) 1628.
[3] K. Abu El-Rub, C.H. Grein, M.E. Flatte, H. Ehrenreich, J. Appl. Phys. **92** (2002) 3771.
[4] A. Salhi, Y. Rouillard, A. Pérona A, P. Grech, Semicond. Sci. Tech. **19** (2004) 260.
[5] C.A. Wang, J. Cryst. Growth **170** (1997) 725.
[6] E. Anastassakis, J. Appl. Phys. **68** (1990) 4561.
[7] A. Navarro-Quezada, A.G. Rodríguez, M.A. Vidal, H. Navarro-Contreras, J. Crystal Growth **291** (2006) 340.
[8] S. G. Choi, S. K. Srivastava, C. J. Palmstrom, Physics of Semiconductors **772** (2005) 125.
[9] R. Ferrini, M. Galli, G. Guizzetti, M. Patrini, A. Bosacchi, S. Franchi, R. Magnanini, Phys. Rev. B **56** (1997) 7549.
[10] J. Olvera-Hernandez, J. Olvera-Cervantes, M. Rojas-López, H. Navarro-Contreras, M.A. Vidal, F. de Anda. J. Phys. Conf. Ser. **28** (2006) 147.
[11] V. Vorlicek, K.D. Moiseev, M.P. Mikhailova, Yu. P. Yakolev, E. Hulicius, T. Simecek, Cryst. Res. Technol. **37** (2002) 259.
[12] G. Lucovsky, R.M. Martin, E. Burstein, Phys. Rev. B **4** (1971) 1367.
[13] R. Cuscó, L. Artús, K.W. Benz, J. Phys. C **7** (1995) 7069.

Preparation and electrical characterization of the compound $CuAgGeSe_3$

M A Villarreal[1], L M de Chalbaud, B J Fernández, A Velásquez-Velásquez and M Pirela

Laboratorio de Temperaturas Bajas, Centro de Estudios en Semiconductores, Departamento de Física, Universidad de Los Andes, Apartado de Correos Nº 1, La Hechicera, Mérida 5251, Venezuela

E-mail: mavu@ula.ve, clio@ula.ve

Abstract. This work reports the synthesis and electrical characterization of the compound $CuAgGeSe_3$. This material was synthesized by direct melting of the constituent elements, in their stoiochiometric ratio inside an evacuated quartz ampoule. The chemical analysis (EDX) confirmed the 1:1:1:3 stoichiometric ratios for the compound. The differential thermal analysis showed the existence of a principal phase that melts at 558°C and a second phase at 636°C. The X-ray powder diffraction analysis indicated that the compound crystallizes in the monoclinic system, space group Cc, with unit cell parameters: a = 6.776(0) Å, b = 11.901(5) Å, c = 6.772(0) and β = 108.2(0)°. The study of the electrical properties was realized in the temperature range from 80 to 300 K and under a magnetic field of 14 kG. Employing the Mott transition model, we were able to obtain the temperature dependence of the resistivity and we estimated that the activation energy is 25.3 meV in the low temperatures region. The mobility temperature dependence is analyzed by taking into account the scattering of charge carriers by acoustic phonons, polar optic phonons and thermally activated hopping. From the analysis, the activation energy is estimated to be around 38 meV and the characteristic temperature of the phonons is estimated to be around 400 K.

1. Introduction

Preliminary studies [1-8] of the family I_2-IV-VI_3 compounds show a high optical-acoustical efficiency for the infrared region application. These materials belong to two possible diamond-like structure families which derive from order 3 of the binary II-VI with two kinds of cations [9]. The derivation is made respecting the established rules for diamond-like compound of normal valence [2] and the substitution of the cations establishes four electrons for each atom and eight for each anion. Several previous structural studies of Cu_2GeSe_3 showed lots of controversy in regards to the unit cell. By means of a photographic study of a polycrystalline sample, Rivet [10] reports a sphalerite structure with a = 5.445 Å in this compound. In this work we found that the structure of the $CuAgGeSe_3$ is consistent with a monoclinic subcell of space group *C1c1* (9) determined by the strongest group reflections. Velásquez et al [11] reported the structure and cell parameters of Ag_2GeSe_3 which also belongs to the monoclinic system.

[1] Corresponding author: e-mail: mavu@ula.ve Phone: + 58 274 2401322, Fax: + 58 274 2401286

Previous studies of the electrical properties [12] point out that Cu_2GeSe_3 exhibits semiconductor behaviour at liquid nitrogen and a metallic one at room temperature. In order to determine the quality as a semiconductor, we enterprise in this work the synthesis and electrical characterization of the compound $CuAgGeSe_3$.

2. Experimental

The compound $CuAgGeSe_3$ was synthesized by direct fusion of the component elements weighted in the stequiometric proportions 1: 1: 1: 3. The differential thermal analysis was made in a Perkin Elmer DTA-7 equipment. The chemical analysis of the $CuAgGeSe_3$ was made in a KEVEX EDX Delta-3 model equipment connected to a Hitachi S-2500 scanning electron microscope. In order to study the structural characteristics of $CuAgGeSe_3$ diffraction methods were employed in polycrystalline samples. The diffraction data of X-rays were taken at room temperature in a Siemens D5005 difractometer equipped with CuKα (λ = 1.54059Å at 40 KV and 30 mA) radiation in configuration Θ/ Θ, with a diffracted graphite beam monochromator. The angle (2Θ) measuring rank was of 10°-100°, the scanning was of 0.020 and the counting time was 45 s/step.

Employing the Van der Pauw technique the resistivity and the Hall Effect were measured from 80 to 300 K under a magnetic field of 14 kG. Indium solder contacts were attached onto a clean and polished surface that was previously flux etched. The contacts were completely ohmic as deduced from the I-V characteristic taken from 1 to 20 mA, at 80 K and at room temperature.

3. Results and Discussion

Three different regions of the ingot were scanned and the average atomic percentages Cu (15.4%), Ag (11.6%), Ge (17.5%) and Se (55.5%), gave an atomic ratio close to the ideal value 1:1:1:3.

The thermogram of $CuAgGeSe_3$ is presented in the figure 1, it shows a well defined affluent tip at 558°C corresponding to his fusion point and another one of less intensity which can be described to a small secondary phase.

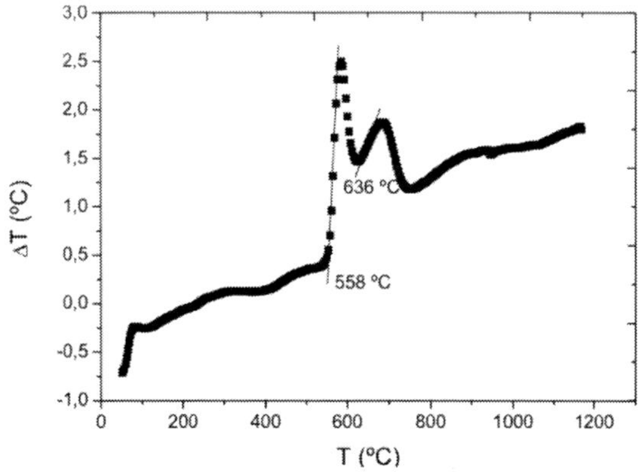

Figure 1. Thermogram of the compound $CuAgGeSe_3$

In figure 2 is given the X-ray powder diffraction pattern. The refinement of the unit cell was made with the program NBS* AIDS83, obtaining the following cell parameters: a = 6.7760 Å, b = 11.9015 Å, c = 6.7720 Å and β = 108.20°. Likewise as in the thermogram, in the X-ray diffraction it is also noticeable a secondary phase. A low intensity $GeSe_3$ phase is reported for Cu_2GeSe_3 in the reference file (PDF 71-117).

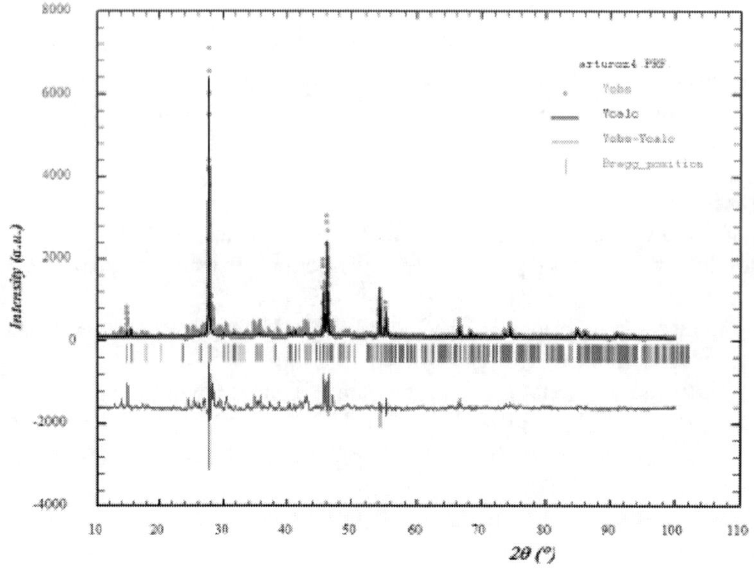

Figure 2. X-ray powder diffraction pattern. Generate with FULLprof Suite program.

The material $CuAgGeSe_3$ together with Cu_2GeSe_3 and Ag_2GeSe_3 all of them posses a common novel sphalerite superstructure. In the structure of $CuAgGeSe_3$ each tetrahedral is bonded to four anions, and conversely, each selenium atom is also tetrahedral coordinated to four cations. This bonding scheme is consistent with the sphalerite-type subcell, a comparison between the refined atomic coordinate and the atom coordinate derived from sphalerite structure confirms the close structural relationship.

Table 1. Atomic coordinates and occupation factor, derived from the Rietveld refinement.

Atoms	Oxid	Wyck	x	y	z	foc	Site
Cu1	+1	4a	0.6222(6)	0.4254(1)	0.2443(1)	0.5	1
Ag1	+1	4a	0.1446(0)	0.2447(5)	0.2369(8)	0.5	1
Ge1	+4	4a	0.6381(4)	0.0822(7)	0.2179(1)	0.5	1
Se1	-2	4a	0.2571(9)	0.4189(9)	0.1033(9)	0.35	1
Se2	-2	4a	0.7909(4)	0.4412(0)	0.6419(9)	0.18	1
Se3	-2	4a	0.7711(5)	0.2543(5)	0.1185(0)	0.18	1

Figure 3 shows the resistivity as a function of the inverse of temperature. At low temperatures below 170 K, the compound presents a behavior characteristic of an extrinsic semiconductor, meanwhile, at high temperatures between 170 and 300K the material behaves as a metal. This semiconductor-metal shift is known as a Mott transition [13] and it is ascribed to a shallow impurity band near the valence band. The intrinsic region is never reached; for this reason, a model similar to Friszsche's model was used to fit the data. A semiconductor-like and a metallike expression were combined to obtain:

$$\rho = \rho_0 \frac{T}{T_0} + \rho_1 e^{E_1 / K_B T} \tag{1}$$

In this formula we have assumed only one activation energy E_1 and $T_0 = 273K$ as a fixed parameter.

The value for the activation energy obtained from a fit to the resistivity in the low temperature region was E_1= 25.3 meV which corresponds to a 0°C resistivity of $\rho_0 = 5.314\,\Omega.cm$ and a resistivity $\rho_1 = 0.126\,\Omega.cm$, used has fitting parameters.

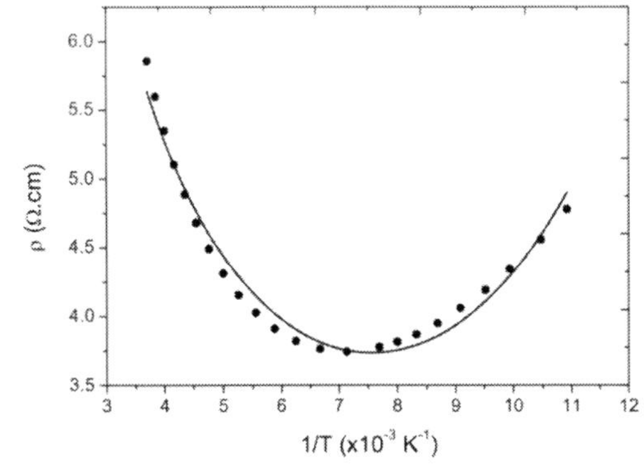

Figure 3. Resistivity of the compound $CuAgGeSe_3$.

The mobility as a function of the inverse of temperature is shown in figure 5. This behavior is usually well explained using standard scattering processes. Thus, at low temperatures, according to the Cutler-Mott model [14] neutral and ionized impurities dominate. The Erginsoy and Brooks-Herring models are not suitable. At high temperatures, acoustic phonons ruled by the usual acoustic deformation potential model and optical polar phonons ruled by a phenomenological model of the form $\mu_{op} \propto (e^{\theta/T} - 1)$ are appropriate. To fit the data we used the Matthiessen rule.

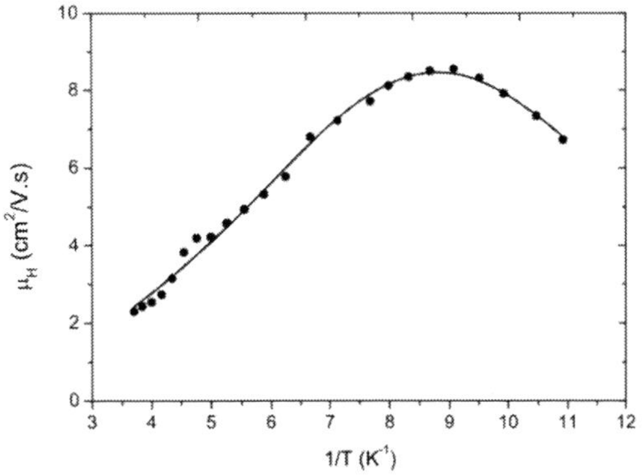

Figure 5. Mobility of the compound $CuAgGeSe_3$

To avoid parameter dependency, we define: $C_{ll} = \rho v_{sl}^2$, $C_{ac} = m^{*5/2} E_{ac}^2$ and $A = x_d / m^{*3/2}$. Thus, using, C_{ac}, μ_0, W, A and θ as adjustable parameters, the following values are obtained: activation energy $W = (38 \pm 2)\, meV$, hopping factor $\mu_0 = (8{,}65 \pm 0{,}43)$, phonon characteristic temperature $\theta = (400 \pm 8)\, K$ and elastic constant $C_{ll} = 1.7x10^{12} \pm 5.1x10^{10}\, dynes/cm^2$. The acoustic deformation potential and dielectric constant they cannot be known due to the fact that the effective mass of the compound is not known.

As regards to the electrical resistivity and mobility we found a close behaviour between Cu_2GeSe_3 and $CuAgGeSe_3$. However, in comparison to Ag_2GeSe_3 a different electrical behaviour has been reported [15].

Acknowledgements
This work was supported by the CDCHT of the Universidad de Los Andes through Project No. NURR-C-301-01-05-C.

References

[1] Berger L I and Prochukhan V 1969 *Ternary Diamond-Like Semiconductors*. Consultans Bureau, New York
[2] Parthé E and Garín J 1972 *Monastsh. Chem.* **102** 1197
[3] Rogarocha E P, Melikhova A N and Panasenco M N 1975 *Neorg. Mater.* **11** 839
[4] Fernández B J, Henao J A and Delgado J M 1996 *Cryst. Res. Technol.* **31** 65
[5] Villarreal M A, Fenández B J, Pirela M and Velásquez-Velásquez A 2003 *Rev. Mex. Fis.* **49 S3** 198-200
[6] De Chalbaud L M, Díaz G, Delgado J M, Mora A E and Sagredo V 1997 *Mater. Res. Bull.* **32** 1371
[7] Olekseyuk L D, Piskach L V, Parasyuk O V, Melñiyk O M and Lyskovetz 2000 *J. Alloys Compd.***298** 203
[8] Delgado G E, Mora A J, Pirela M, Velásquez-Velásquez A, Villarreal M A and Fernández B J 2004 *Phys. Stat. Sol. (a)* **201** 2900-2904
[9] Parthé E, in: Westbrook J H, Fleischer R L (Eds.) 1995 *Intermetallic Compounds, Principles and Applications*. Jhon Wiley & Sons (Chichester, UK)
[10] Rivet J 1965 *Ann. Chim.* **10** 243
[11] Velásquez-Velásquez A, Belandria A, Fernández B J, Avila Godoy R, Delgado G and Acosta-Najarro G D 2000 *Phys. Stat. Sol. (b)* **220** 683-686
[12] Villarreal M A, Fernández B J, Pirela M and Velásquez-Velásquez A 2007 *Rev. Mex. Fis.***S53** (7) 303-306
[13] Mott N F 1978 *Physics Today* **31** 42
[14] Cutler M and Mott N F 1969 *Phys. Rev.* **181** 1336
[15] Velásquez-Velásquez A 2005 *Tesis de Maestría* Postgrado de Materia Condensada, Facultad de Ciencias, Universidad de Los Andes, Mérida-Venezuela

Nickel-induced crystallization of amorphous silicon

J A Schmidt[1,2], N Budini[2], P Rinaldi[2], R D Arce[1,2] and R H Buitrago[1,2]

[1] INTEC (CONICET-UNL), Güemes 3450, S3000GLN Santa Fe, Argentina
[2] FIQ (UNL), Santiago del Estero 2829, S3000AOM Santa Fe, Argentina

jschmidt@intec.unl.edu.ar

Abstract. The nickel-induced crystallization of hydrogenated amorphous silicon (a-Si:H) is used to obtain large grained polycrystalline silicon thin films on glass substrates. a-Si:H is deposited by plasma enhanced chemical vapour deposition at 200 °C, preparing intrinsic and slightly p-doped samples. Each sample was divided in several pieces, over which increasing Ni concentrations were sputtered. Two crystallization methods are compared, conventional furnace annealing (CFA) and rapid thermal annealing (RTA). The crystallization was followed by optical microscopy and scanning electron microscopy observations, X-ray diffraction, and reflectance measurements in the UV region. The large grain sizes obtained – larger than 100 μm for the samples crystallized by CFA – are very encouraging for the preparation of low-cost thin film polycrystalline silicon solar cells.

1. Introduction

After many years of research and development, the hydrogenated amorphous silicon (a-Si:H) technology acquired a degree of maturity that allows to deposit thin uniform films over large area glass substrates with a high velocity. However, problems with the stability and electrical quality of the material could not be overcome [1]. Recently, there has been a renewed interest in a-Si as a precursor to obtain polycrystalline silicon (poly-Si) for different applications like active-matrix liquid crystal displays, active-matrix organic light emitting diodes, and solar cells [2-5].

Among the different methods that allow a solid phase transformation from the amorphous to the polycrystalline state we have the conventional furnace annealing (CFA) and the rapid thermal annealing (RTA). CFA uses a conventional oven to heat the samples, and allows the crystallization of a-Si (non-hydrogenated, prepared by evaporation) at a temperature of 600 °C in 48 h, providing a material with a grain size in the order of 1.5 μm [6]. RTA uses infrared radiation as a heating source, and has the advantage of the high heating speed (up to 60 °C/s) that reduces the crystallization time. Besides, the radiation can be applied in pulses that heat the sample without heating the glass substrate (which is transparent to the infrared radiation). However, the grain size obtained in the crystallization of a-Si is also in the range of a few micrometers.

In an effort to reduce the crystallization temperature and crystallization time, and to increase the grain size, the effect of Nickel as an inductor of crystallization has been studied [2,3,7,8]. The addition of very small amounts of Ni (less than a monolayer) to the a-Si surface has a beneficial effect on the crystallization. Although there is an agreement about the formation of the nickel silicide $NiSi_2$, which acts as a seed for the growth of crystalline silicon [9], the details of the crystallization mechanism are still under debate.

In this work, the Ni-induced crystallization of a-Si:H samples is studied by using the CFA and RTA methods. The results of both thermal treatments for a series of samples covered with different Ni concentrations are compared. A thermal process that induces a solid phase transformation from an amorphous to a polycrystalline material with the largest possible grain size is explored.

2. Experimental details

Samples were prepared on Schott AF-37 glass in a capacitively coupled Plasma Enhanced Chemical Vapor Deposition (PECVD) reactor. A relatively high frequency of 50 MHz and a power density of 120 mW/cm^2 were used, being the distance between electrodes of 1.1 cm. Intrinsic a-Si:H samples from pure silane (SiH_4), and slightly p-type samples adding some diborane (B_2H_6), were deposited without any hydrogen dilution. The substrate temperature was 200 °C and the gas pressure was 600 μbar, leading to a deposition speed of 15-20 Å/sec. Nickel was deposited on top of the films by dc sputtering using a low power density of 95 mW/cm^2 and an Ar pressure of 70 mTorr. The Ni concentration was determined on test samples by atomic absorption spectroscopy [10], and varied between 8.7×10^{13} and 1.5×10^{13} atoms/cm^2, which means between 5 and 1 % of a monolayer.

Both CFA and RTA were performed at atmospheric pressure under nitrogen flow. CFA started with a dehydrogenation step at 400 °C for 24 h, followed by an annealing at 550 °C for different periods. The deposition of silicon by PECVD from SiH_4 gas leads to a-Si:H, a material with a hydrogen content in the range of 10%. The direct heating of this material to 550 °C causes a violent hydrogen effusion that damages the samples. Therefore, a previous dehydrogenation step is needed. Some of the samples were dehydrogenated in the conventional furnace, and then crystallized by RTA. In this case, the samples were a) heated from room temperature to 350 °C in 300 sec, b) heated from 350 to 630 °C in 90 sec, c) kept at 630 °C for 60 sec, and d) cooled from 630 to 350 °C in 90 sec; repeating the sequence b) - c) - d) until full crystallization. Other samples were also dehydrogenated in the RTA furnace, applying a gradual heating from 350 to 550 °C in 1800 sec between steps a) and b). The number of pulses required to fully crystallize the samples varied between 6 and 48 depending on the thickness of the samples and the Ni concentration.

Crystallization was controlled through optical microscopy, scanning electron microscopy, reflectance in the UV and X-ray diffraction.

3. Results and discussion

Figure 1 a) is a photograph, taken with an optical microscope in reflection mode, of a partially crystallized p-type 100 nm thick sample. The sample was covered with a Ni density of 4.4×10^{13} at/cm^2, which is less than 3% of a monolayer (1.61×10^{15} at/cm^2). The crystallization was performed by CFA at 550 °C for 24 h, after a dehydrogenation step as described in section 2.

A distribution of circular grains can be seen in figure 1, with an average diameter around 100 μm. In the lower-right corner it can be seen that the disk-shaped grains have collided, giving straight grain boundaries. Figure 1 b) is a detailed view of the previous one, taken with a higher magnification. A dense circular grain of ~ 150 μm – and still growing – can be appreciated. In the centre of the circle a dark spot can be seen, which is attributed to the $NiSi_2$ nucleus. It is known that the first stage of the crystallization process is the diffusion of Ni atoms to form a $NiSi_2$ precipitate, which should reach a critical size for the crystallization to proceed [9]. This $NiSi_2$ precipitates can be formed at as low as 350 °C, in the initial stages of thermal annealing, and act as seeds for the growth of c-Si. In our case, the dehydrogenation step at 400 °C for 24 h is enough to permit the Ni diffusion to form the silicide. $NiSi_2$ forms octahedral precipitates having eight {111} faces. The small lattice mismatch (0.4%) between $NiSi_2$(111) and Si(111) facilitates the formation of epitaxial c-Si on the {111} faces of the precipitates. In situ studies revealed that single-crystal Si needles grow in <111> directions from the $NiSi_2$ precipitates [9,11]. Only <110> oriented $NiSi_2$ precipitates, which have four {111} faces perpendicular to the a-Si surface, can lead to a long-range growth of crystallites. Depending on the thickness of the samples a mixture of vertical and lateral growth may occur, so it is important to study the crystallization of samples having different thicknesses.

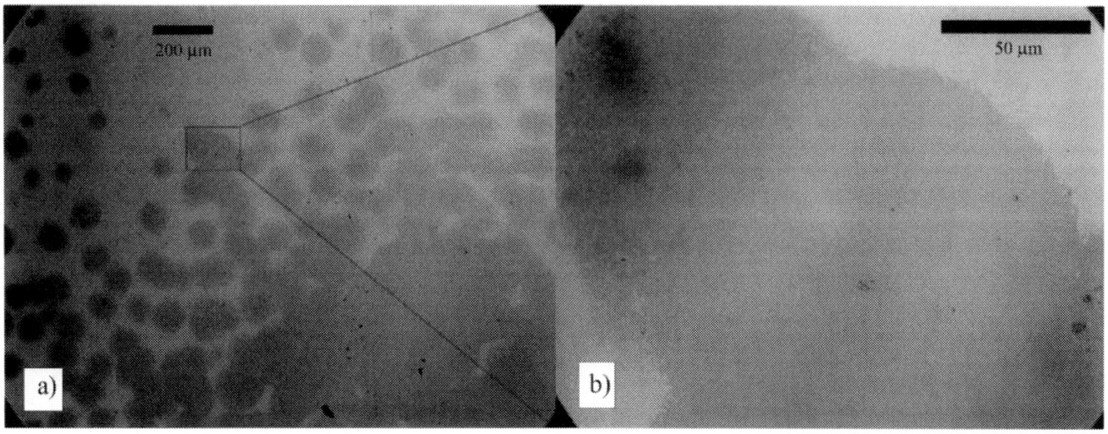

Figure 1. a) Reflection optical microscope image of a p-type sample annealed at 550 °C for 24 h by CFA, and b) detail of a region of the sample, showing a growing grain.

Figure 2 is a SEM image of a partially crystallized intrinsic layer (0.3 µm thick), focused on a grain boundary region. Inside the grains, some small pockets of amorphous material (etched away by a Secco solution) can still be appreciated. At the grain growing front, a network structure of branched crystallites can be seen. The width of the needlelike crystallites is about 100 nm. The primary and secondary arms' directions of the needles have the regular angles of 70 or 110° (the so-called bi-directional needle network structure) [12]. This indicates that the crystal exhibits a preferential <110> orientation with respect to the film surface normal. The preferential orientation is confirmed by the X-rays diffraction spectrum shown in figure 3. From these data the relative integrated intensities 100:5 for the (111) and (220) peaks are obtained, which is a higher ratio than in previous reports [13]. The strong orientation of the grains is important for practical applications, since it promotes the uniformity of the material properties over the dimensions of the device.

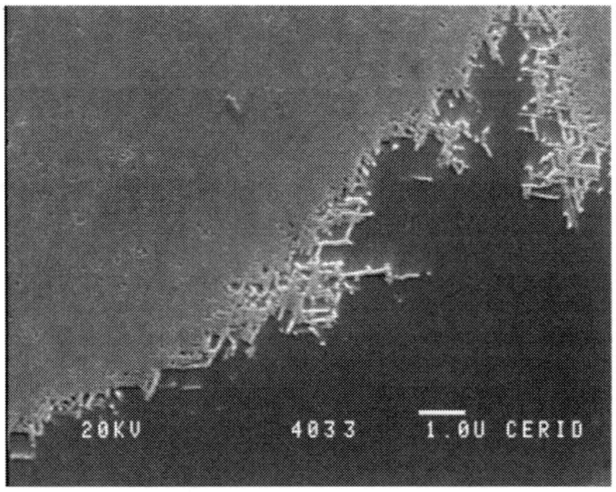

Figure 2. SEM image of a partially crystallized intrinsic sample, where a grain growing front can be seen. The length of the bar is 1 µm.

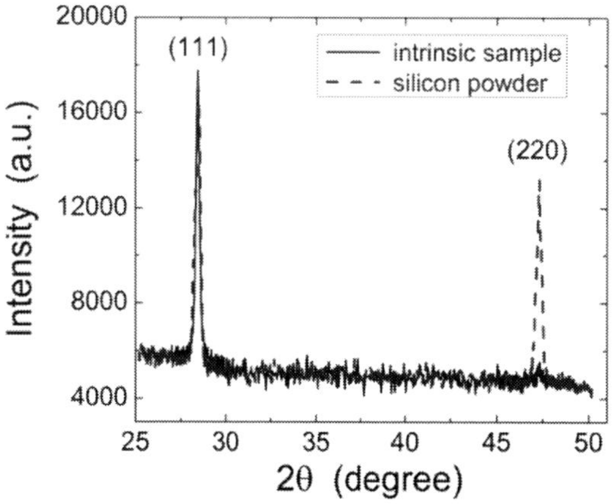

Figure 3. X-ray diffraction patterns of the intrinsic sample (solid line) and of crystalline silicon powder (dashed line).

Figure 4 is a photograph, taken with an optical microscope in reflection mode, of the same film shown in figure 1, but now covered with a Ni density of 2.9×10^{13} at/cm^2. Both the dehydrogenation and the crystallization were performed by RTA, applying 18 heating pulses as described in section 2. The sample can be seen partially crystallized, with a mixture of some disk-shaped grains (~ 30 µm diameter) and some small grains of irregular borders. The same result was obtained on samples that were dehydrogenated by CFA and then crystallized by RTA. When these samples were fully crystallized, after the application of ~ 36 heating pulses, a mixture of large and small grains was also observed. Large grains are ascribed to the Ni-induced crystallization, while small grains come from the spontaneous solid phase crystallization of a-Si at this elevated temperature of 630 °C. Even when the sample reaches this temperature only for periods of 60 sec, it seems to be enough to promote the spontaneous crystallization of a-Si. Therefore, a lower peak temperature should be tried, which will be the subject of further research.

Figure 5 shows the UV-light reflectance spectra of a polished single crystalline silicon wafer (dashed line), of the sample of figure 1 after full crystallization by CFA (solid line), and of the sample of figure 4 after full crystallization by RTA (dotted line). The CFA sample shows a high degree of crystallinity, evidenced by the height of the peak at ~ 270 nm. The RTA sample, on the other hand, exhibits a lower crystallinity due to the defective phase present at the boundaries of the small grains. These results are in agreement with the optical microscope observations of figures 1 and 4.

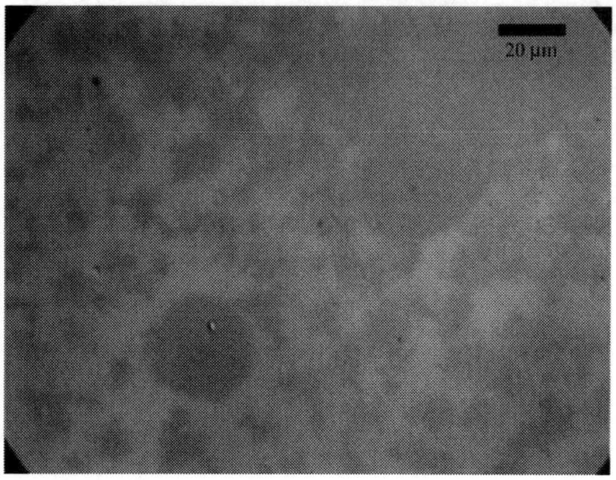

Figure 4. Reflection optical microscope image of a p-type sample annealed by RTA.

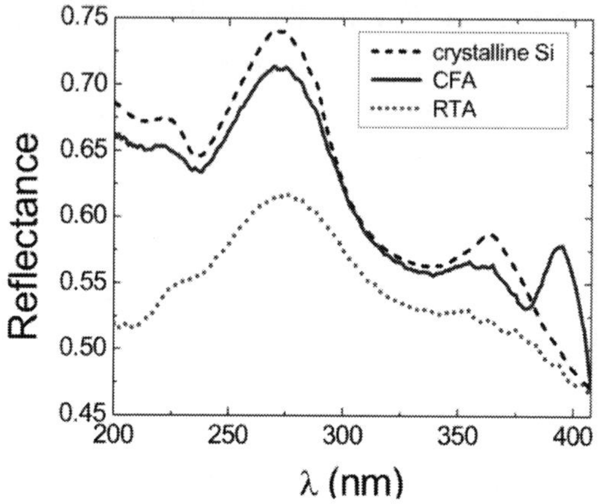

Figure 5. UV-light reflectance spectrum of a crystalline silicon wafer, a sample crystallized by CFA and a sample crystallized by RTA.

4. Conclusion

The Ni-induced crystallization of a-Si by using the CFA and RTA methods has been studied. Although RTA has the great advantage of its high speed, the parameters still have to be adjusted to obtain a uniform distribution of grain sizes. CFA has proven to be a convenient method to crystallize a-Si, leading to disk-shaped grains with a diameter in excess of 100 µm. This is a very encouraging result towards the possible application of these films to semiconductor devices like solar cells.

Acknowledgments

This work was supported by ANPCyT (Projects 22-20267 and 22-25749), CONICET (Project PIP 5246) and UNL (Project CAI+D 28-158).

References

[1] Schropp R E I and Zeman M 1998 *Amorphous and Microcrystalline Silicon Solar Cells: Modeling, Materials and Device Technology* (Boston: Kluwer Academic Publishers).
[2] Choi J H, Cheon J H, Kim S K and Jang J 2005 *Displays* **26** 137.
[3] Maramatsu S-I, Minagawa Y, Oka F, Sasaki T and Yazawa Y 2002 *Sol. En. Mater. and Solar Cells* **74** 275.
[4] Mahan A H, Ahrenkiel S P, Schropp R E I, Li H and Ginley D S 2008 *Thin Solid Films* **516** 529.

Journal of Physics: Conference Series **167** (2009) 012046 doi:10.1088/1742-6596/167/1/012046

[5] Nast O and Hartmann A J 2000 *J. Appl. Phys.* **88** 716.

[6] Aberle A G 2006 *Thin Solid Films* **511 – 512** 26.

[7] Kim K H, Nathan A and Jang J 2008 *J. Non-Cryst. Solids* **354** 2341.

[8] Pereira L, Martins R M S, Schell N, Fortunato E and Martins R 2006 *Thin Solid Films* **511-512** 275; Pereira L, Martins R M S, Schell N, Fortunato E and Martins R 2007 *Thin Solid Films* **516** 104.

[9] C Hayzelden and J L Batstone 1993 *J. Appl. Phys.* **73** 8279.

[10] M B Sperling and B Welz 1999 *Atomic Absorption Spectrometry* (Weinheim: Wiley-VCH).

[11] M Miyasaka, K Makihira, T Asano, E Polychroniadis and J Stoemenos 2002 *Appl. Phys. Lett.* **80** 944.

[12] J-S Ahn, Y-G Yoon and S-K Joo 2006 *J. Cryst. Growth* **290** 379.

[13] Choi J H, Kim D Y , Kim S S, Park S J, Jang J 2003 *Thin Solid Films* **440** 1, Yoon S Y, Park S J, Kim K H, Jang J, Kim C O 2000 *J. Appl. Phys.* **87** 609.

Journal of Physics: Conference Series **167** (2009) 012047 doi:10.1088/1742-6596/167/1/012047

Cd-Te-In oxide thin films as possible transparent buffer layer in CdTe based solar cells

R Castro-Rodríguez[1], A Martel[2], J Mendez-Gamboa[2], J M Camacho[1] and J L Peña[1]

[1]Applied Physics Department, CINVESTAV-IPN Merida, C.P. 97310, Merida, Yucatan México. [2]Facultad de Ingeniería, Universidad Autónoma de Yucatán. AP 150 Cordemex, 97310 Mérida, Yucatán, México

E-mail: romano@mda.cinvestav.mx

Abstract. Cd-Te-In-oxide thin films were grown by Pulsed Laser Deposition (PLD) technique using CdTe powder embedded in a matrix of indium metallic as target. The films were deposited at different oxygen pressures (P_{O2}) from 15 to 50 mTorr at substrate temperature of 420°C. Sheet resistance (R_{SHEET}) and transmission spectrum were measured as a function of P_{O2}. From measurements of optical transmission, the Photonic Flux Density (PFD) spectrum were obtained and the integral of these PFD for each film were evaluated between energy range of 1.5 eV and 2.4 eV for obtain the amount of photons that can be transferred across the film in this range of solar energy spectrum. These values were evaluated over the R_{SHEET} to be used as a figure of merit. The best choice in our conditions was the films with P_{O2} =28.5 mTorr, where the figure of merit reaches the maximum value.

Keywords: CdTe-In oxide ICTO; PLD; Photonic Flux Density, Figure of merit, Solar cells

PACS: 81.15.Fg, 77.84.Bw, 68.55. a, 78.20.Ci,

1. Introduction

CdTe exhibits very good characteristics as a thin film photovoltaic material: It has a band gap of 1.44 eV [1], close to the theoretical maximum of photovoltaic energy conversion, it's gap is direct that means that only a few microns of the material are needed for the absorption of 90% of solar light. It exhibits a very simple phase diagram and it grows stoichiometric at a substrate temperature greater than 300 °C [2-6]. The typical front contact of CdTe based solar cell is provide by depositing a transparent conductive oxide (TCO) onto glass substrate. The TCO layer has a high optical transparency in the visible and near-infrared regions and high *n*-

type conductivity. This is followed by the deposition of a CdS window layer, the CdTe absorber layer, and finally the back contact. The most efficient CdS/CdTe solar cells needed the inclusion of a buffer layer between the TCO and CdS layers, which improves the quality of the device interface. The role of the buffer layer is both to hinder the any atoms diffusion from TCO and to separate CdS from TCO in order to limit the effect of eventual pinholes that could be present in CdS, since this layer is very thin (80-100 nm). Those TCO must have specific electrical and optical characteristic in order to enhance the transmission of the solar light on the absorbent material of CdTe film [7-9].

The External Quantum Efficiency (EQE) of a solar cell is a parameter that gives information on the photo-current that a cell can produces under a particular wavelength radiation. If the EQE is integrated over the whole solar electromagnetic spectrum, one can evaluate the photo-current that a cell would produce when exposed to white light. In this order, it is important to know the Photonic Flux Density (PFD) at a particular wavelength for one specific TCO buffer layer, because the whole cell EQE depends upon the integral photonic flux, computed by integrating PFD between the wavelengths corresponding to the window and absorber materials, in the case of CdTe based solar cells the wavelengths corresponds at energy of 2.4 eV and 1.5 eV to CdS window layer and CdTe absorber material respectively.

A figure of merit for a TCO may be defined as the ratio of the electrical conductivity to the optical absorption coefficient of the film. Common to all TCO applications is the need for optimizing the electrical and optical coating parameters. Depending on the type of device requiring a transparent electrode, the optical transmission and the electrical conduction of the electrodes should exceed certain minimum values. Ideally both parameters should be as large as possible but their inter-relationship usually excludes the simultaneous achievement of both these criteria [10]. In order to compare the performance of various TCOs, the most widely used figure of merit of Haacke [11], where the optical transmission is take average around of 500 nm (near the solar spectrum maximum).

On the other hand, the interest in the study of ternary oxides Cd-Te-O is due to their potential applications such as TCO buffer layer. Most of the Cd-Te-O compounds have ordered a perovskite structure and possess substantial characteristics similar to indium tin oxide (ITO) and tin oxide (SnO_2), which is as follows: (i) all positive ions have the same closed shell configuration $(n-1)d^{10}$ ns^0 $(n\geq4)$ (ii) a large band-gap between 2.8 and 3.3 eV, and (iii) the ability of ionic substitution to generates conductive carriers. Therefore, these oxides are considerable promise as a host material for TCO and also as buffer layer in CdTe based solar cell.

In this paper Cd-Te-In oxide Films (ICTO) deposited by PLD on a glass substrate at different pressures of reactive oxygen (P_{O2}) are characterized by measurements of optical transmission and sheet resistance. Although the figure of merit of Haacke has been used by many authors [12-14] for TCO characterization in solar cell development and the 500 nm region is an important one, the use of a narrow band is not representative of whole ability of the film to transmit photons. So, instead of transmittance and narrow band we propose the use of full AM 1.5 spectrum photonic flux in order to analyze from the Photonic Flux Density (PFD) spectrum the amount of photons that can be transferred across the film in the range of solar energy spectrum of 1.5 eV and 2.4 eV. Another figure of merit is defined taking the integral of PFD (IPDF) evaluated over the sheet resistance.

2. Experimental Details.

Thin films (160-200 nm thick) were deposited using a typical Pulse Laser Deposition (PLD) technique. Cd-Te-In oxides (CTOs) films were deposited on Corning 7059 glass substrates with a surface area of 38x26 mm^2 with a laser irradiation Nd:YAG in Q-switch mode, which offers energy of ~300 mJ/pulse with a pulse width of 10 ns at repetition rate of 5 Hz. The substrate temperature was measured using a thermocouple in contact with the substrate and the distance target-substrate was 40 mm. The laser beam was reduced to a spot with a diameter of ~1.3 mm using a spherical quartz lens and focused on target to give an energy density of about 2 J/cm^2. During laser irradiation, the target was rotated at 15 rpm. The number total of shots was 4x10^3. A target with 0.5 mm thick and 15 mm of diameter was obtained from high purity CdTe powder (99.999% Cerac) embedded in a matrix of metallic indium (99.999% Cerac), with a nominal ratio 1:1 wt. CdTe and In respectively. The chamber deposition was initially evacuated to 10^{-6} Torr and deposition of films was performed at pressure of reactive oxygen gas (P_{O2}) of 15 at 50 mTorr at T_S=420 °C. Before deposition, the substrates were cleaned in an ultrasonic bath for 10 min with a detergent solution, then rinsed thoroughly with distilled water, and finally with acetone and methanol. After the deposition, substrate was cooled to room temperature to 5 °C-min^{-1} in the partial P_{O2} that was used during the grown.

The Sheet Resistance was measured using a homemade four-contact Van Der Paw resistivity measurement system and a Veeco Dektak 8 perfilometer was used to measurements the thickness of the films. Optical transmittance measurements were performed with a Jobin-Yvon/Spex H20-VIS Spectrophotometer in the 300 to 800 nm wavelength.

3. Results and discussion

Figure 1 shows a graph of the film thickness as a function of P_{O2}. In the figure we can observed that for all pressures, except 35 mTorr, have similar thickness (mean value of 210 nm). The film grown at 35 mTorr has a thickness of 550 nm; which is too big for TCO inclusive; therefore it is not included in optical characterization.

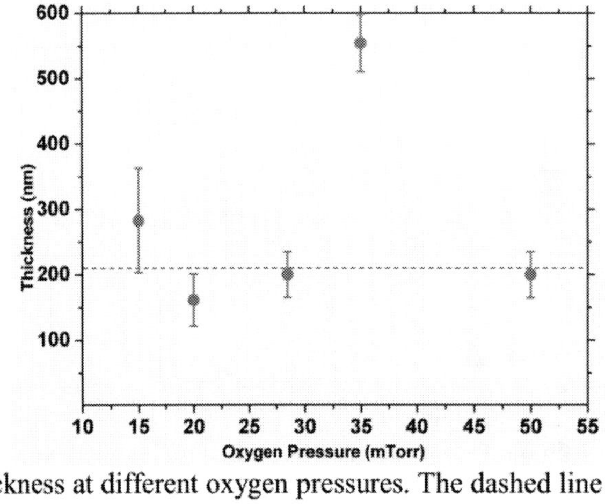

Figure 1. Film thickness at different oxygen pressures. The dashed line is the mean value.

Figure 2 shows a graph of the R_{SHEET} as a function of P_{O2} at room temperature. We can see the lowest R_{SHEET} was obtained P_{O2}=28.5 mTorr. The R_{SHEET} decreases with P_{O2} initially, reaches a minimum ~5.5x10^2 Ω/sq and increases at higher pressure (P_{O2}>28.5 mTorr), reaching ~10^5 Ω/sq at 50 mTorr. The Hall coefficient measured by Van der Pauw electrode configuration of the film at 28.5 mTorr was $-7x10^{-3}$ cm^3-C^{-1}, suggesting that the conduction is *n*-type. Combining the Hall coefficient and conductivity measurements led to a carrier density of ~ $8.9x10^{20}$ cm^{-3} and modest mobility of. $\mu = 9.88$ cm^2-$V^{-1}s^{-1}$.

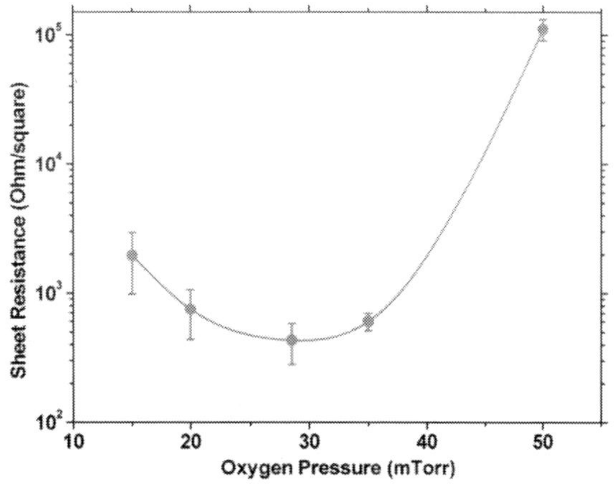

Figure 2. Film sheet resistance versus oxygen pressure.

The optical transmission of the ICTOs films over a wavelength range of 350-800 nm plotted against P_{O2} is shown in Figure 3. At oxygen pressures between 20 and 50 mTorr and for wavelength between 600 and 700 nm, transmission is around 0.80. We can observe that 15, 20 and 50 mTorr curves have similar shape, but different mean values; this behaviour corresponds to thickness variation and similar bang gap. The curve corresponding to 28.5 mTorr shows low transmittance at wavelength below 500 nm, this is maybe due at the presence of new phase formation between CdTe oxide and indium oxides.

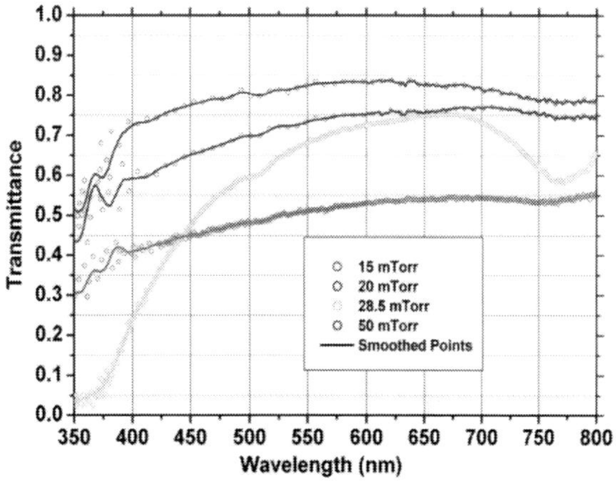

Figure 3. Transmittance spectra of ICTO films at different oxygen partial pressure.

Figure 4 shows the spectral distribution of the PFD of the ICTOs samples defined as:

$$\mathrm{PFD}(h\nu) = \frac{I_S T}{h\nu} \qquad (1)$$

where I_S is the solar AM 1.5 spectra, T is the transmittance spectrum and $h\nu$ the photon energy. This figure represent the amount of AM 1.5 photons transmitted across the film. They are the useful photons for any photovoltaic device. According to these results, the best film is obtained at oxygen pressure of 20 mTorr and the worst one at 15 mTorr. The Integral photonic flux density (IPFD) is also defined as:

$$\mathrm{IPFD} = \int_{Eg(Win)}^{Eg(Abs)} \mathrm{PFD}(h\nu)\,d(h\nu) \qquad (2)$$

where the energy range of the solar spectrum was selected between $Eg_{(Win)}$=2.4 eV and $Eg_{(Abs)}$=1.5 eV corresponding at the band-gap of the window (CdS) layer and absorber material (CdTe) in CdTe based solar cell. The IPFD represents the amount of photons which can be converted into photocurrent by the absorber.

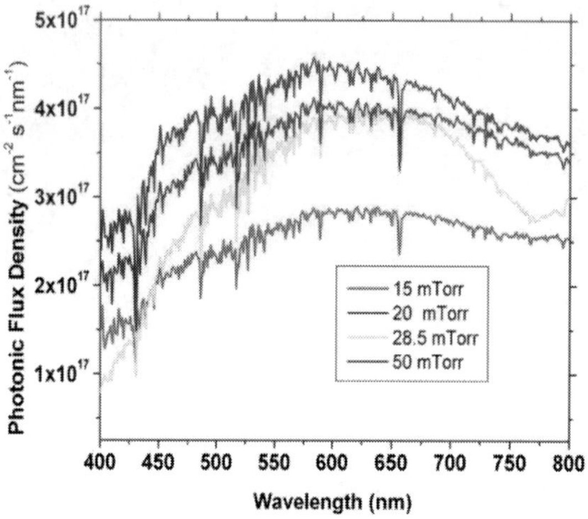

Figure 4. Spectral distribution of the photonic flux density at different oxygen partial pressure.

The above behaviour can be better appreciated in Figure 5, where the IPFD is shown. We can see that the range of amount of photon that can be converted in photocurrent is between $2x10^{17}$ $cm^{-2}s^{-1}$ and $\sim 4x10^{17}$ $cm^{-2}s^{-1}$.

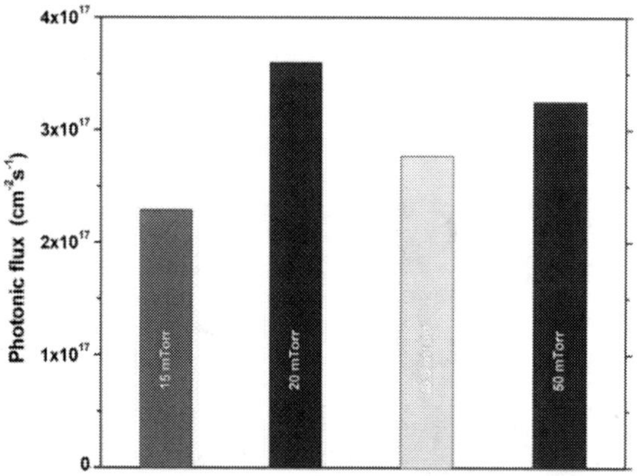

Figure 5. Integral photonic flux of ICTO films at different oxygen partial pressure.

Figure 6 shows the value of our figure of merit, calculated using a new definition:

$$\Theta_{PH} = \frac{IPFD}{R_{SHEET}} \qquad (3)$$

The numerator of the expression (3) determines the photocurrent, whereas the denominator corresponds to the losses by Joule effect. This equation not includes any exponent because the integral photonic flux density has defined physical units. Logarithmic plot beginning at 10^{10} photon-cm^{-2}s^{-1} Ω^{-1} is used for contrast. According to Figure 6, the best choice in our conditions is the P_{O2}=28.5 mTorr. The figure of merit indicates that the films obtained at 20 mTorr and 28.5 mTorr are similar, but the film obtained at 28.5 mTorr is the best one. This figure of merit defined as the ratio of the electrical conductivity to the IPFD coefficient of the film is not contradictory with the figure of merit of Haacke, we think that our figure of merit could play as other roll in TCO to solar cell similar as the Haacke, since the cell photocurrent is determined by the numerator of expression (3), while the Joule losses are proportional to the denominator.

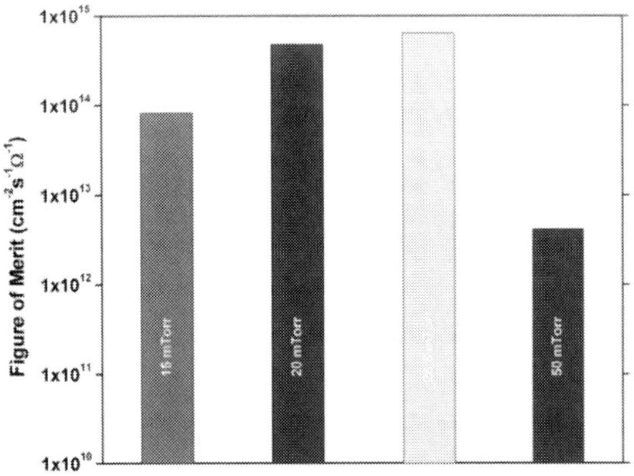

Figure 6. Logarithmic plot of ICTO films figure of merit at different oxygen partial pressure.

4. Conclusions.

We investigated the photovoltaic properties of Cd-Te-In oxide thin films obtained by PLD. From measurements of optical transmission, the Photonic Flux Density (PDF) spectrum were obtained for firms grown between oxygen pressure of 15-50 mTorr, and the integral of PDF for each film were evaluated between energy range of 1.5 eV and 2.4 eV for obtain the amount of photons that can be transferred across the CdS/CdTe solar cell. We found that the best films with potential use as buffer layer in CdTe based solar cells were the films grown at substrate temperature of 420 °C, distance target-substrate of 4.0 cm and oxygen pressure of 28.5 mTorr. We introduced a figure of merit defined as the relation of the integral photonic flux over the sheet resistance and was used to evaluate the ratio of the optical absorption to electrical conductivity of the ICTOs films.

Acknowledgements

This work has been supported by Project No. 59996 CONACyT/Mexico. The authors acknowledge Oswaldo Gómez, Mario Herrera and Roberto. Sánchez for technical support and L. Pinelo for secretarial assistance. A. Martel and J. Méndez thanks the Universidad Autónoma de Yucatán.

References

[1] Bube R H 1998 *Photovoltaic Materials* (London: Imperial College Press) 136
[2] Khairnar U.P, Bhavsar D S, Vaidya R U, Bhavsar G P 2003 *Mater. Chem. Phys.* **80** 421
[3] Lalitha S, Sathyamoorthy R, Senthilarasu S, Subbarayan A, Natarajan K 2004 *Sol. Energy Mater. Sol. Cells* **82** 187
[4] Paulson P D, Mathew X 2004 *Sol. Energy Mater Sol. Cells* **82**, 279

[5] Mathew X, Mathews N R, Sebastian P J, Flores C O 2004 *Sol. Energy Mater. Sol. Cells* **81** 397
[6] Hernández G P, Mathew X, Enríquez J P, Morales B E, Lira M M, Toledo J A, Juárez A S, Campos J 2004 *J. Mater. Sci.* **39** 1515
[7] Rao N, Mvan den Bleek C, Schoonman J 1993 *Solid State Ionics* **59** 263
[8] Ishihara T, Shiokawa K, Eguchi K, Arai H 1998 *Sens. Actuators* **19** 259
[9] Romeo N, Bosio A, Tedeshi R and Canevari V 1998 *2nd World conference and exhibition on photovoltaic solar energy conversion* 446-447
[10] Dawar A L, Joshi J C 1984 *J Mat Science* **19** 1-23
[11] Haacke G 1976 *J. Appl. Phys.* **47** 4086–4089
[12] Zhao Z, Komin V, Viswanathan V, Morel D L and Ferekides C S 2000 *Proc. 28th IEEE PVSC (*IEEE) 662-665
[13] Soliman M, Hussein M M, El - Atawy S and El-Gamal M 2001 *Renewable Energy* **23** 463-470
[14] Li X, Miao W, Zhang Q, Huang L, Zhang Z and Hua Z 2005 *Semicond. Sci. Technol.* **20** 823-828

Effect of Ga in the photocatalytic properties of TiO_2

Ma. E. Manríquez[1], E. Sánchez-Mora[2]*, J. G. Hernández-Cortez[3], A. Moreno-Rodriguez[4] and T. Hernandez García[5]

1 Instituto de Ciencias, BUAP, 4 Sur 104, Puebla, Puebla 72000, México.

2 Instituto de Física, BUAP. AP J-48,72570 Puebla Pue., México

3 Instituto Mexicano del Petróleo, Eje Central L. Cárdenas 152, A.P. 14-805, 07730 México D.F., México.

4 Fac. de Ciencias Químicas, BUAP. 4 Sur 104, Puebla, Puebla 72000, México.

5 Fac. de Ciencias Químicas, UANL, Av. Pedro de Alba s/n Cd. Universitaria C. P. 66400, San Nicolás de los Garza, N. L., México

E-mail: esanchez@sirio.ifuap.buap.mx

Abstract In this work we prepared a series of $GaZn/TiO_2$ (5 wt%, 10 wt%, and 15 wt% of Ga) photocatalysts by the sol-gel method. The X-ray diffraction results showed the presence of TiO_2 in anatase phase. Also, in the $GaZn/TiO_2$ system a small peak assigned to Ga_2O_3 is present. The specific surface area of the Zn/TiO_2 is 97 m^2/g and the specific surface area increases to 132 m^2/g when the amount of Ga increases. The energy band gap values obtained by UV-VIS diffuse reflectance spectroscopy are 2.9, 3.12 and 3.15 eV for $5GaZn/TiO_2$, $10GaZn/TiO_2$ and $15GaZn/TiO_2$ respectively. The results of the photocatalytic activity show that the total destruction of phenol is more efficient in $GaZn/TiO_2$ than in the Zn/TiO_2 photocatalysts.

1. Introduction.

Semiconductors, such as TiO_2, ZnO, ZnS, and SnO_2, have been applied as photocatalysts [1]. Among them, TiO_2 is reported as the most promising one because of its high efficiency, stability, and low cost. In recent years, attention has been diverted to metal-doped TiO_2, which is believed to decrease the band-gap energy. The beneficial effects of the Zn-doping of TiO_2 catalysts have already been reported in wastewater treatment [2,3] and dye sensitized solar cells [4]. We have selected the Zn–TiO_2 because: (a) it has a large contact area to deposit gallium on Zn–TiO_2, (b) it is resistant to the heat treatment required during the solid state synthesis and (c) it is not attacked by highly oxidative radicals produced during the photocatalytic process.

The Ga_2O_3 is a semiconductor with a wide band gap (4.8 eV) and its electronic and luminescent properties have been reported [5,6]. The Ga_2O_3-TiO_2 composite oxide was also reported to be an interesting material having numerous applications in the areas of photocatalysis, electrocatalysis, and materials science [7,8]. In this work, $ZnTiO_2$ (anatase) samples are modified with a Ga^{+3} in 5% 10, and 15% Ga respectively by sol-gel method and tested for a probe reaction, i.e. phenol photo-oxidation out in the aqueous system. From the residual photocatalysts have been determined their crystalline phases by X-ray diffraction (XRD), specific surface areas by BET and semiconductor properties by diffuse reflectance spectra.

2. Experimental

2.1. Synthesis

The $GaZn/TiO_2$ catalysts were prepared by sol-gel method using titanium butoxide (Aldrich 99.99%), ethanol (99.9% Proquim) and HNO_3 (0.1M) was added under continuous stirring to give pH 3. The reaction mixture was heat at 70 °C and appropriate amounts of a solution of $Zn(NO_3)_2$ (Aldrich 99.99%), to obtain the Zn/TiO_2 with 5%wt. Appropriate amounts of $Ga(NO_3)_3$ were added to obtain $GaZn/TiO_2$ powders at 5, 10 and 15% wt of Ga ($5GaZn/TiO_2$, $10GaZn/TiO_2$ and $15GaZn/TiO_2$ samples). Then, samples were dried in vacuum for 24 h at 70 °C. Finally they were annealed at 400°C during 4 h in air. A TiO_2 reference sample was prepared in the same way.

2.2. Characterization

Once the powders were obtained, they were characterized by means of the UV-Vis absorption spectra with a Cary-III spectrometer equipped with a diode array as detector. FTIR characterization was made using a Brucker spectrometer (Vertex 70). X-ray diffraction patterns were recorded by a diffractometer Bruker Advance D-8. Textural properties were determined from the nitrogen adsorption isotherms obtained with a Quantachrome Autosorb-3b.

2.3. Photocatalytic Activity

The photocatalytic activity studies of the samples were carried out as follows: 500 mg of catalyst was added to an aqueous solution of phenol (280 ppm, 200 mL) in air flux (0.5 $L\,min^{-1}$), the reactor used in all experiments was a 500 mL Pyrex glass vessel and the temperature of the system was 30 °C (± 1°C). All experiments were carried out under similar conditions of solar irradiation, i.e., on relatively cloudless days to minimize differences in the degradation due to widely different experimental conditions of photon flux. The experiments were typically initiated at 11:00 a.m. at 2:00 p.m. with a total period of solar exposure (reaction) of 8 hours. The reaction rate was followed by taking samples every 45 min; and analyzing them in a Cary-III spectrometer. The concentration of phenol was calculated from the absorption band at 270 nm by means of a phenol calibration curve.

3. Results and discussion

Fig. 1 shows the nitrogen adsorption-desorption isotherms for the $5GaZnTiO_2$, $10GaZnTiO_2$, and $15GaZnTiO_2$ samples. This figure shows that the isotherms are type IV for the catalysts according to the BDDT (Brunauer, Deming, Deming) classification. Their hysteresis loops are of type H2 and H1 respectively according to the IUPAC classification [9]. The Brunauer-Emmett-Teller (BET) surface area calculations gave a surface area the 133 m^2/g for $15GaZnTiO_2$, 132 m^2/g, and 113 m^2/g for $10GaZnTiO_2$ and $5GaZnTiO_2$ respectively (Table 1) and they are higher than the Zn/TiO_2 sample. The doped Ga exists mostly in the form of Ga_2O_3 on the surface of $ZnTiO_2$ by surface doping, although substitutional doping may also be possible.

X-ray diffraction (XRD) patterns of the $ZnTiO_2$ doped with gallium ions are shown in Fig. 2. These spectra show peaks at the 2Θ values: 25.4°, 37.8°, 48.2°, 54.0°, and 55.2° which correspond to anatase (101), (004), (200), (105), and (211) crystal planes, respectively. In the XRD pattern of $GaZn/TiO_2$ a small peak at 30.65° is assigned to Ga_2O_3 [10] due to the incorporation of Ga ions in the Zn/TiO_2 lattice during the synthesis process. This peak increases as the concentration of Ga ions is increased.

The Figure 3 shows the FTIR spectra of several samples. All the samples show a strong absorption band at 1100 cm^{-1}, which is assigned to internal vibration of TiO_4 tetrahedra and the peak at 550 cm^{-1} assigned to double five-ring vibration. However, we did not find the absorption bands of stretching vibrations of bridged hydroxyls related to tetrahedrally coordinated framework gallium at *ca.* 3615 cm^{-}

[1] to 3625 cm^{-1} for any of the gallosilicate structures [11], which suggests that the gallium species did not incorporate into the framework tetrahedral sites of $ZnTiO_2$.

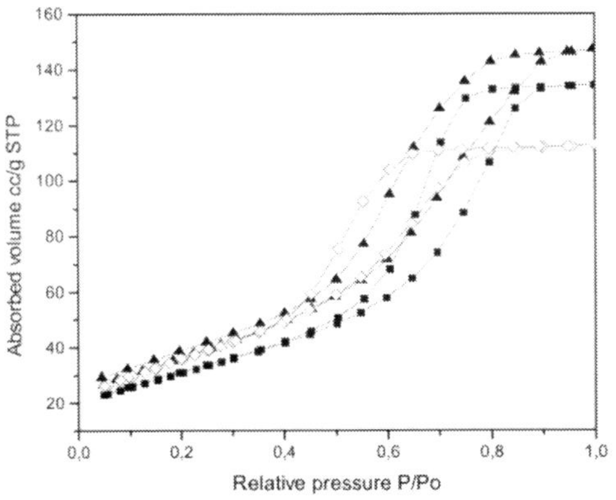

Figure 1. Nitrogen adsorption isotherms for the (■) 5GaZn/TiO_2 (▲) 10GaZn/TiO_2 and (◊) 15GaZn/TiO_2 samples.

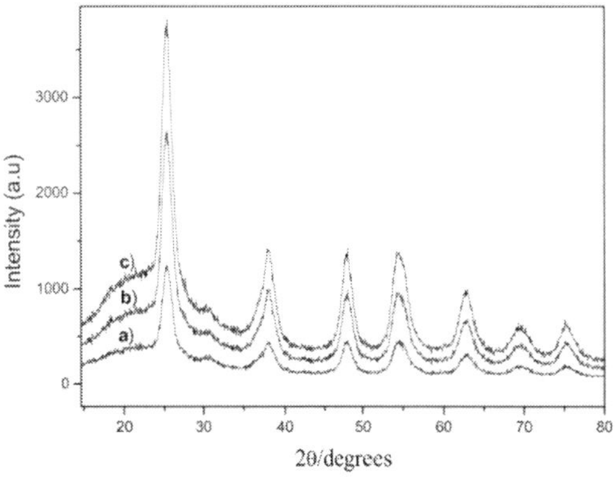

Figure 2. X-ray patterns for the (a) 5GaZn/TiO_2 (b) 10GaZn/TiO_2 and (c) 15GaZn/TiO_2 samples.

Table 1. Specific surface area (SA), pore volume (**Vp**) and average pore size (**Φp**) of GaZn/TiO_2 system.

Samples	SA (m^2/g)	Vp (cc/g)	Φp (Å)
5GaZnTiO₂	113	0.208	73.53
10GaZnTiO₂	132	0.227	68.28
15GaZnTiO₂	133	0.175	52.80
5Zn/TiO₂	105.2	0.217	75.22

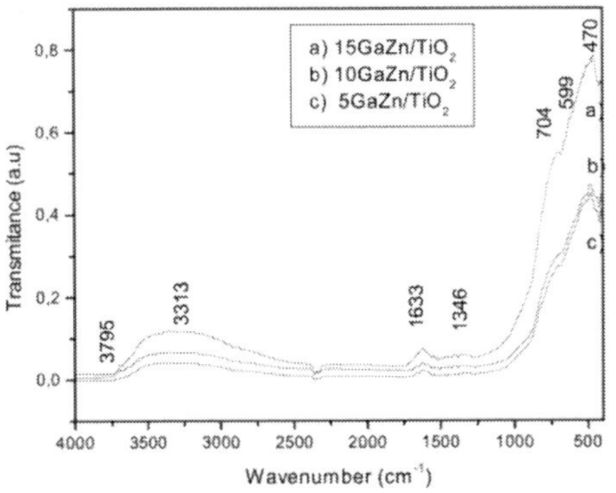

Figure 3. FTIR spectra for the (a) 5GaZn/TiO_2 (b) 10GaZn/TiO_2 and (c) 15GaZn/TiO_2 samples.

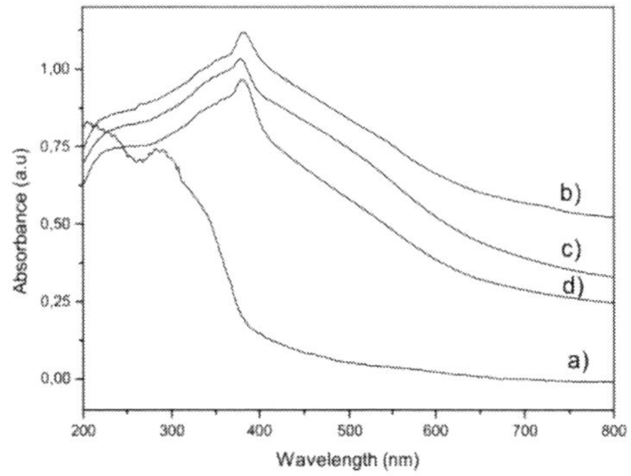

Figure 4 UV-Vis spectra for the (a) Zn/TiO_2, (b) 5GaZn/TiO_2 (c) 10GaZn/TiO_2 and (d) 15GaZn/TiO_2 samples.

An additional shoulder emerges at *ca.* 3795 cm^{-1} for Ga-modified $ZnTiO_2$, it is assigned to the vibration of hydroxyls located at non framework gallium species [12] and it increases in intensity when of gallium content increases.
When the loading amount is low, the hydrated Ga^{3+} cations are closely restricted to the outer surface of the $ZnTiO_2$ crystals, and they rarely have the chance to gain access to the position inside the $ZnTiO_2$ pores [13]. As the loading amount increases, most exchangeable outside sites are occupied and the Ga^{3+} cations have more chance to access to the pores and locate near the mouth of pore. It may be the reason that the activity increased with the increase of loading amount only at the gallium.

Table 2. Band gap values and Results of the Langmuir–Hinshelwood model for phenol decomposition on sol–gel catalysts of $GaZn/TiO_2$

Samples	**Eg (eV)**	**k_1 x10^6 (min-^{1})**	**k_2 (M^{-1})**	**k_1k_2 (min^{-1} M^{-1})**
$5Zn/TiO_2$	2.9	5.59	2450	0.014
$5GaZnTiO_2$	3.12	6.93	3039	0.021
$10GaZnTiO_2$	3.15	9.4	4040	0.039
$15GaZnTiO_2$	3.30	11.92	5162	0.062

The absorption spectra of the $15GaZn/TiO_2$, $10GaZn/TiO_2$ and $5GaZn/TiO_2$ and Zn/TiO_2 samples (Fig. 4) shows an increment in the absorption in the interval from 700 to 400 nm which is due to the absorption of the Ga_2O_3 [14]; the interval from 400 to 350 nm and is due to the TiO_2 fundamental absorption.
The band gap energy (Eg) of the samples were calculated by the equation $\alpha(h\upsilon)=A(h\upsilon- Eg)^{m/2}$, where α is the absorption coefficient, $h\upsilon$ is the photon energy, A is a constant and $m = 1$ represents a direct transition between bands [15]. The band gap energies were 2.9, 3.12 and 3.15 eV for $15GaZn/TiO_2$, $10GaZn/TiO_2$ and $5GaZn/TiO_2$ respectively (Table 2).
The importance of the measured of the band energy resides in that; the rate that electron-hole are generated is determined first by irradiation conditions such as photon energies and fluxes, and secondly by optical absorption characteristics of the photocatalytic particles such as their band gap.
The photocatalytic activity of the samples was studied using, as a test reaction, the decomposition of the phenol. The results obtained by UV-Vis spectra of the phenol as a function of time of sun exposure (reaction) with $15GaZnTiO_2$ photocatalysts is shown in Fig. 5. These spectra show a significant decrease in phenol absorbance (85% in 270 min). Also, no apparent new absorbance peaks, indicating near complete mineralization of phenol. The behavior of the other catalysts is similar to the $15GaZnTiO_2$ sample, but percentage of phenol decomposition is lower to the $15GaZnTiO_2$ catalyst at the same time (270 min) and the concentration of phenol diminishes slowly when the reaction takes place without catalyst (Fig. 6).
For the correct evaluation of data we applied the Langmuir–Hinshelwood model using the integrated equation for heterogeneous photocatalytic reactions in a batch reactor [15,16].

$$\frac{\ln C_0/C}{C_0-C} = -k_2 + \frac{mAk_1k_2t}{V(C-C)}$$

where, V is the liquid volume, C the substrate concentration, t the time, m the mass of catalyst, A the adsorption sites per g of catalyst, k_1 the apparent rate constant and k_2 the apparent adsorption constant. The values of k_1, k_2, k_1k_2 (apparent reactivity) for the catalysts obtained at different temperatures in the decomposition of phenol are tabulated in Table 2. These data were estimated from a graph of (ln (C_0/C))/ $(C_0$-C) against t/(C_0-C). The activity for the catalyst $15GaZnTiO_2$ shows that degradation is favored and according to the high k_2 value we can see that the adsorption capacity of this catalyst is favored due to its higher degree of hydroxylation, hence the OH groups behave as adsorption centers observed from FTIR spectra (see Figure 3). When the Ga^{+3} ion doping is increased in the $ZnTiO_2$ the photocatalytic activities is increased, since the oxide metal incorporation influences the photo reactivity of catalyst by acting as holes traps and by altering the e^-/h^+ pair recombination [17-18].

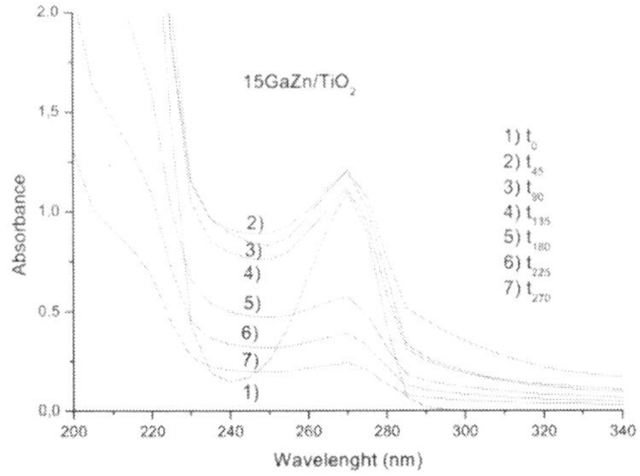

Figure 5 UV-Vis spectra of the phenol as a function of time of sun exposure (reaction) with $15GaZn/TiO_2$ catalyst.

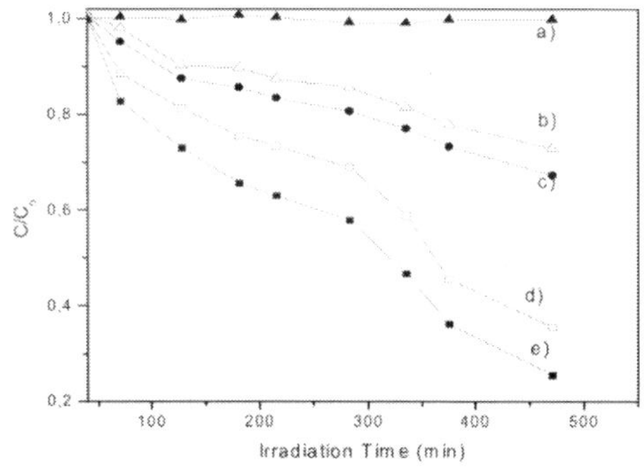

Figure 6. Photocatalytic degradation of phenol a) without catalysts b) $ZnTiO_2$, c) $5GaZnTiO_2$ d) $10GaZnTiO_2$, and e) $15GaZnTiO_2$.

4. Conclusion

Photocatalytic processes by a $GaZn/TiO_2$ catalyst are quantitatively efficient in the decomposition of the water-soluble phenol fraction. The total destruction of water-soluble compounds originating from phenol indicates that the photocatalysis can be employed for water treatment. The process proposed in this investigation uses non-expensive materials for the construction of the catalysts.

5. References.

[1] M. R. Hoffmann, M. S. T.; Choi, W.; Bahnemann, D. W. Chem. Rev., **95** (1995), 69.
[2] A. Kumar,. A. K. Jain, J. Photochem. Photobio. A, **156** (2003), 207.
[3] Z. Zhang, C. Wang, R.,Zakaria, J. Y. Ying, J. Phys. Chem. B, **102** (1998) 10871.
[4] G. Colon, M. Maicu, M.C. Hidalgo, J.A. Navio, Appl. Catal. B: Environmental, **67** (2006) 41.
[5] M. R. Lorenz, J. F. Woods and R. J. Gambino, J. Phys. Chem. Solids, **28** (1967) 403.
[6] T. Harwig, F. Kellendonk and S. Slappendel, J. Phys. Chem. Solids, **39** (1978) 675.
[7] I. A. Rozdin, S. S. Plotkin, V. E. Plyushchev, N. I. Sorokin, Neorg. Mater. **11** (1975) 178.
[8] V. N. Antsiferov, I. V. Filimonova, A. M. Makarov, Russ. J. Appl. Chem., **72** (1999) 1019.
[9] K. Kudo, K. Sayama, A. Tanaka, K. Asakura, K. Domen, K. Maruya, T. Onisi, J. Catal., **120** (1989) 337.
[10] B. M. Reddy, I. Ganesh, E. P. Reddy, A. Fernández, P. G. Smirniotis, J. Phys. Chem. B, **105** (2001) 6227.
[11] R. Fricke, H. Kosslick, G. Lischke and M. Richter, Chem. Rev., **100** (2000) 2303.
[12] L.F. Joly, H. Ajot, E. Merlen, F. Raatz and F. Alario, Appl. Catal. A, **79** (1991) 249.
[13] T. Takata, K. Shinohara, A. Tanaka, M. Hara, JN Kondo and K. Domen, J. Photochem. Photobiol. A, **106** (1997) 45.
[14] T. Chen and K. Tang, Appl. Phys. Lett., **90** (2007) 053104
[15] T. Lopez, R. Gomez, E. Sanchez, F. Tzompantzi and L. Vera, J. Sol-gel Sci. Tech., **22** (2001) 99.
[16] E. Pelizzetti and N. Serpone, Photocatalysis, Fundamentals andApplications, Wiley Interscience, 1989.
[17] J. G. Yu, X. J. Zhao, Q. N. Zhao, Thin Solid Films, **379** (2000) 7.
[18] W. Choi, A Termin, M.R Hoffmann, J. Phys. Chem., **98** (1994) 13669.

Journal of Physics: Conference Series **167** (2009) 012049 doi:10.1088/1742-6596/167/1/012049

Characterization of GaN films grown on GaAs by AP-MOVPE

H Vilchis[1], **V M Sanchez-R and A Escobosa**
Department of Electrical Engineering, CINVESTAV-IPN, Av. Instituto Politécnico Nacional 2508 Col. San Pedro Zacatenco C.P. 07360 México D.F.

E-mail: heber_vil@hotmail.com

Abstract. In this paper we present the results of the synthesis of GaN in an AP-MOVPE system heated by infrared lamps starting from gallium nitride films obtained by nitridation of gallium arsenide. Although dependence of the characteristics of the different parameters of the deposition process on the properties of the layers has been widely studied, the influence of the nature and design of the heating source has been only scarcely reported. We show that the ratio between the two phases depends on the characteristics of the heating source, as well as on other growth parameters. Our results show a compromise between the characteristics of the photoluminescence spectra, the surface morphology and the cubic phase to hexagonal phase ratio. The growth conditions can be adjusted for optimal performance.

1. Introduction

Since the first epitaxial deposit of gallium nitride was realized by H. P. Maruska in 1969 by chemical vapour epitaxy technique [1], the interest on this material has been increasing steadily due to its chemical, electrical and optical properties. Its wide band gap makes this material as one of the most promising semiconductors for applications in optoelectronic devices in the blue and ultraviolet regions, in high temperature and in high power electronic devices. Gallium nitride has two crystalline structures; its hexagonal phase (h-GaN) is stable and its metastable cubic phase (c-GaN). The only difference between both structures is the bond angle to the second neighbor in the cubic structure, which is rotated 60º yielding a different stack of atoms [2]. Both phases show different chemical and physical properties. c-GaN has several advantages to h-GaN like lower phonon scattering lower carrier density, easiness of cleavage and better compatibility with conventional zincblende substrates [3, 4].

Since c-GaN is a metastable phase, an equilibrium process will result in h-GaN. The growth of the cubic phase must be forced using thermodynamic parameters that are far from equilibrium. Techniques such as Molecular Beam Epitaxy (MBE) or Vapour Phase Epitaxy (VPE) are normally carried out on a super saturation regime forcing the species to precipitate and to form the desired compounds [5, 6]; if the substrate presents a cubic phase, the atoms will tend to be arranged in the same structure [7]. Gallium arsenide (GaAs) is the substrate most used because it has a zincblende cubic phase helping the nucleation of c-GaN.

[1] To whom any correspondence should be addressed.

Due to the large lattice mismatch between c-GaN and GaAs, around 20%, and the incorporation of h-GaN, that cannot be avoided [8]. The films have a high density of defects, like dislocations, stacking faults boundary layers degrading their optical and electrical characteristics. Several variants have been reported to improve the quality of the films and to reduce the incorporation of h-GaN, like use RF-radical nitrogen source, ultra-low pressure growth or low temperature deposition or nitridation process before the growth has also been reported [5, 9]. In this paper we present the results of GaN films grown on GaAs considering the distance of the infrared heat source to the suceptor and the V/III molar ratio in order to reduce the incorporation of h-GaN and improving c-GaN.

2. Experimental procedure

Gallium nitride films were obtained at a Metal-Organic Vapour Phase Epitaxy (MOVPE) system shown in figure 1. The process was realized at atmospheric pressure in Mexico City (586 Torr).

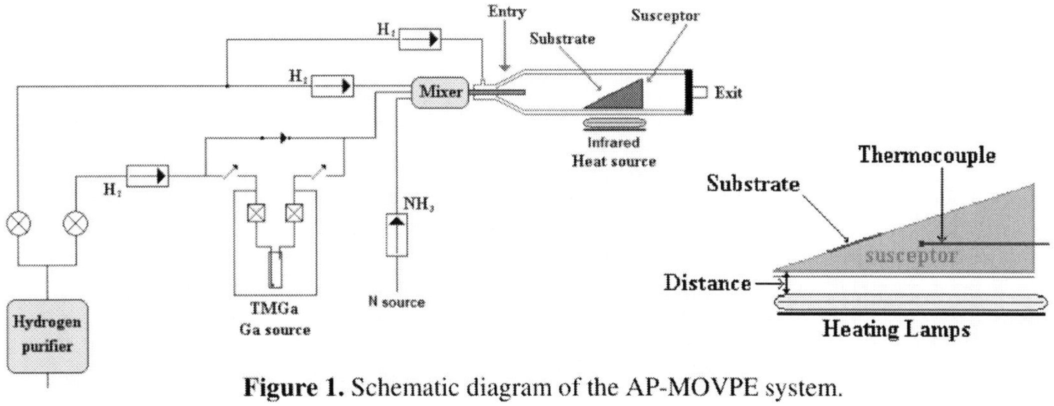

Figure 1. Schematic diagram of the AP-MOVPE system.

The experiments were carried out in a horizontal silica reactor; Trimethylgallium (TMG) and ammonia were the gallium and nitrogen sources. Pd purified hydrogen was used as carrier gas. A SiC coated graphite susceptor could accommodate one 2" wafer but in our experiments we used 1 cm^2 substrates. Six 1.6 kW lamps were used to heat the sample and they were placed directly under the sample holder as show figure 1. The temperature control was measured by a thermocouple inserted in the centre of the susceptor; the actual substrate temperature was previously determined to be 20ºC below the control value, this difference was measured inserting a thermocouple inside the reaction chamber directly over the susceptor. The current needed to maintain a temperature of 850ºC ranged from 41 to 45 A as the distance form the susceptor to the lamps was varied from 8.0 to 12 mm.

The growth process consisted of four steps as follows:

- *Cleaning of the substrate.* The 1 cm^2 GaAs (001) substrate was chemically etched in a solution of H_2SO_4:H_2O_2:H_2O=5:1:1 at 60ºC, immersed in HCl:H_2O=1:1 and rinsed with DI water. Finally the substrate was introduced in the reactor and annealed in H_2 at 520 ºC for 30 minutes to remove the surface oxide.
- *Nitridation process.* GaAs surface was exposed to NH_3 (300 sccm) for 90 minutes at 900 ºC, GaN film was formed over substrate as result of this process. The hydrogen flow was 7 slm.
- *Annealing.* The GaN buffer layer was annealed at 580 ºC for 30 minutes, to improve the crystalline quality of the film.
- *Epitaxial growth.* GaN film was grown at 850 ºC using 500 sccm NH_3 and TMGa for 120 minutes. For the first group of experiments the TMGa flow was kept constant at 0.75 sccm and the distance from the heat source (quartz infrared lamps) to the susceptor was varied between of 8 and 12 mm. For a second set of experiments the distance from the heat source to

the susceptor was kept constant at 10 mm and TMGa flow was varied to obtain different V/III molar ratios according to table 1.

Table 1. Flow rates for Ga source and corresponding molar ratio.

TMGa Flow (sccm)	V/III Molar ratio
0.75	3200
1.0	2400
1.5	1600
2.0	1200
3.0	800

3. Results and discussion

GaN epitaxial films were characterized by different techniques. X-Ray diffraction (XRD) was employed to determine the crystalline structure, atomic force microscopy (AFM) was used to observe the surface morphology, phonon dispersion was characterized by Raman spectroscopy and photoluminescence (PL) was used to determine the emission spectra of the layers. The crystalline structure of the films is influenced by changes of the thermal distribution near substrate due to modifications of the distance from heat source to the susceptor, as shown in figure 2.

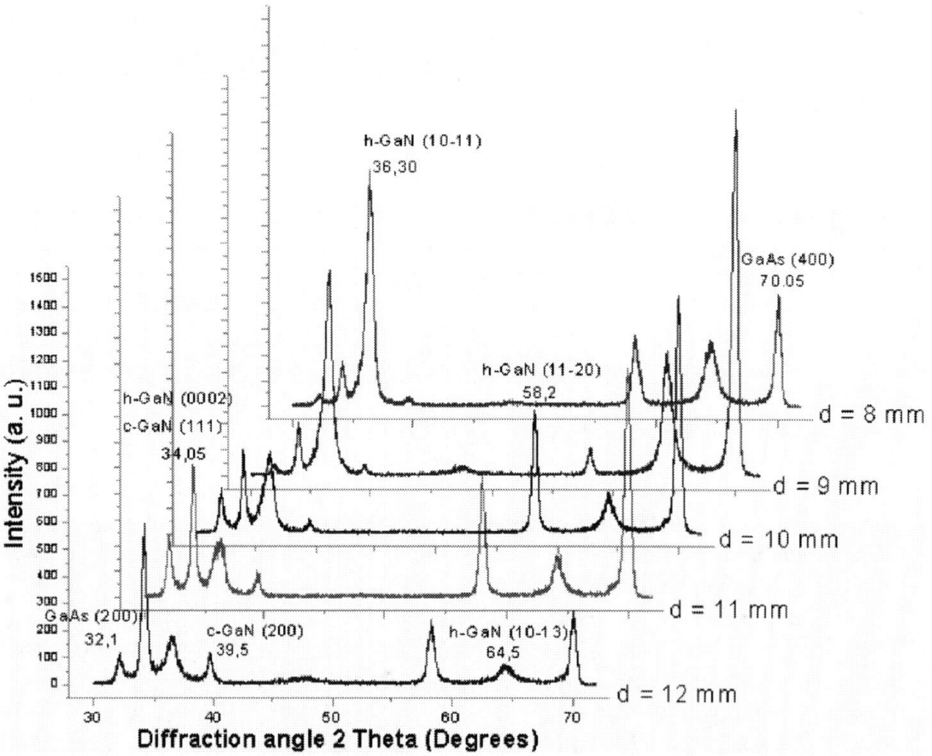

Figure 2. Change of the phase in GaN films due to the influence of the infrared heat source.

Seven peaks are identified in the samples; two of them correspond to GaAs (002) at 32.1° and (004) at 70.05°. The other peaks correspond to GaN, showing the cubic and hexagonal phases with different orientations. For c-GaN they are: (111) at 34.05° and (002) at 39.5°. For h-GaN we can see three peaks, at 36.30° for (10-11) GaN, 58.2° for (11-20) GaN and 64.5° for (10-13) GaN. When the distance is short the hexagonal phase is predominant, however when the distance increases the cubic phase is favoured. Fig. 3 shows the XRD spectrum of sample Nr 59 where the cubic phase is dominant. Increasing the distance of the heating source from the susceptor, results in a less efficient heat concentration, resulting in a less abrupt profile and increase of the temperature of the gas stream inside the reactor. This favors reactions previous to the deposition, affecting the composition of the films leading to an increase of the cubic phase.

The topography of the surface was determined by AFM. Fig. 4 shows a typical microgram of the layers. The roughness of the films was between 321 nm for films with a predominately hexagonal phase and 1046 nm observed when the cubic phase was dominant. It is possible notice the size of the grain and their boundaries.

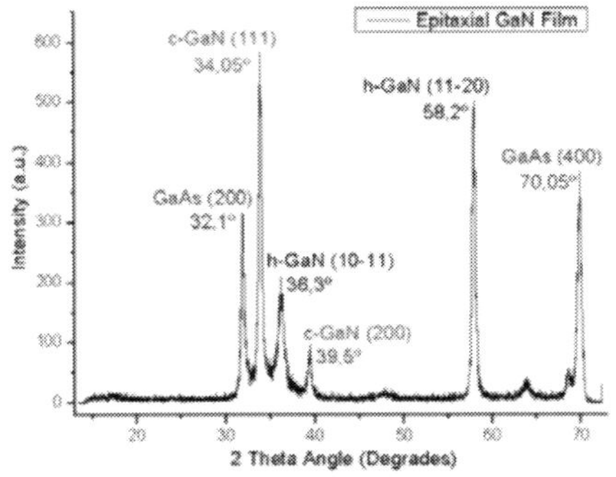

Figure 3. XRD spectrum for epitaxial GaN film with cubic phase favored.

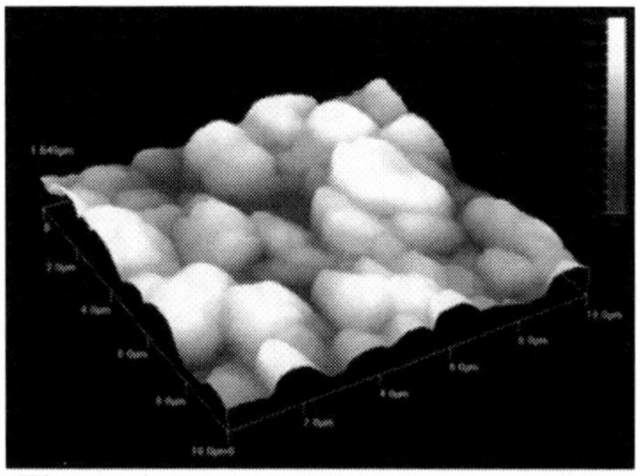

Figure 4. Typical AFM surface image for epitaxial GaN films.

Raman characterization showed transitions of both phases, but the signal from h-GaN was more intense even in predominantly cubic layers. Fig 5 shows the Raman Spectrum of sample Nr 59. The peaks at 563 and 576 cm^{-1} correspond to the E1(TO) and E2 transitions of h-GaN, while the shifts at 548 and 737 are related to TO and LO transitions of the cubic phase [3, 10]. According to the spectrum in figure 3 this film is predominantly cubic but the Raman peaks of the hexagonal phase are more intense and better outlined, this indicates a better crystalline quality of h-GaN.

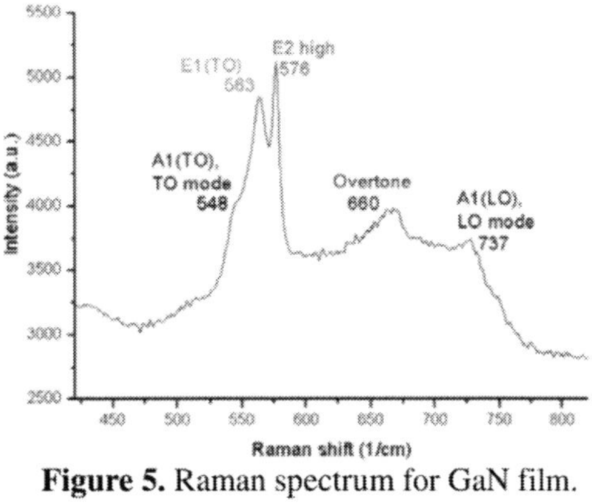

Figure 5. Raman spectrum for GaN film.

Photoluminescence measurements were made using a 325 nm He-Cd laser. A typical spectrum is shown in fig. 6. The hexagonal and cubic phases can be identified by the emission at 361 and 386 nm respectively. Between 450 and 600 nm a broad emission caused by several defects can be seen. This is very common in GaN and it is known as the yellow band [9]. The larger intensity of the band to band peak and the reduction of the yellow band emission are related to a shift of the former towards longer wavelengths as shown in fig. 7. Further studies are needed to determine if this effect is caused by a reduction of the band gap or an increase of doping impurities in the material.

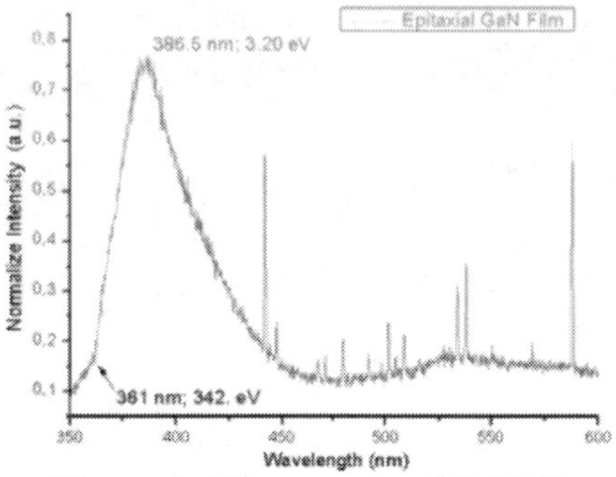

Figure 6. PL spectrum of GaN films.

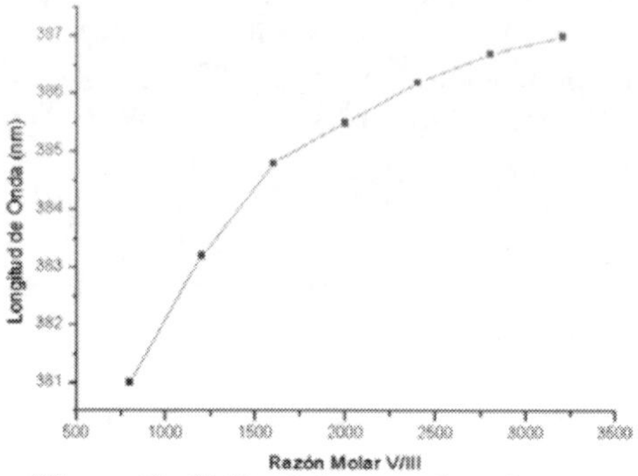

Figure 7. Shift of the wavelength emission due to the V/III molar ratio.

4. Conclusions

The GaN films grown on GaAs at AP-MOVPE system are a mixture of cubic and hexagonal phase. Varying the distance of the infrared heat source was possible to increase the portion of the cubic phase in the epitaxial film. An abrupt temperature profile favours the hexagonal phase while a gradual temperature profile favours the cubic phase.

The optical characteristics mainly of c-GaN layers could be improved adjusting of V/III molar ratio. The band-band emission of c-GaN was better than the band-band emission for h-GaN. Raman spectra indicate a larger density of crystalline effects in the cubic phase.

Acknowledgments

Authors acknowledge the technical support of D. Ramirez in the growth of GaN films, and the help of M. Avendaño, A. Avila and M. Galvan for the photoluminescence, atomic force microscopic and Raman measurements to respectively. This research was partially supported by CONACYT.

References

[1] Maruska H P and Tietjen J J 1969 *Appl. Phys. Lett.* **15** 327
[2] Morkoc H 1999 *Nitride semiconductors and devices* (Berlin: Springer)
[3] Giehler M, Ramsteiner M, Brandt O, Yang H and Ploog K H 1995 *Appl. Phys. Lett.* **67** 733
[4] Okumura H, Ohta K, Feuillet G, Balakrishnan K, Chichibu S, Hamaguchi H, Hacke P, Yoshida S 1997 *J. Crystal Growth* **178** 113
[5] Kikuchi A, Hoshi H and Kishino K 1994 *Jpn. J. Appl. Phys.* **33** 688
[6] Nagayama A, Katayama R, Nakadan N, Miwa K, Yaguchi H, Wu J, Onabe K and Shiraki Y 1999 *Phys. Stat. Sol. A* **176** 513
[7] Garcia S G , *Modelos de crecimiento de GaN por MOCVD y MBE en baja temperatura* 2003 Ph. Tesis (CINVESTAV)
[8] Sun X L, Yang H, Wang Y T, Zheng L X, Xu D P, Zhao D G, Li S F and Wang Z G 1999 *J. Crystal Growth* **212** 397
[9] Escobosa A, Sánchez-R V M, Avendaño M A and Navarro G 2005 *Phys. Stat. Sol. b* **242** 1883
[10] Azhuata T, Sota T, Suzuki K and Nakamura S 1995 *J. of Physics: Condensed Matter* **7** L129

Investigation of $AgInS_2$ thin films grown by co-evaporation

Arredondo C. A[1]**., Clavijo J** [2]**. and Gordillo G.** [1]
[1] Departamento de Física, Universidad Nacional de Colombia, Bogotá.
[2] Departamento de Química, Universidad Nacional de Colombia, Bogotá.Cr.30 N° 45-03, Colombia

E-mail: caarredondoo@unal.edu.co

Abstract. $AgInS_2$ thin films were grown on soda-lime glass substrates by co-evaporation of the precursors in a two-step process. X-ray diffraction (XRD) measurements indicated that these compounds grow in different phases and with different crystalline structure depending upon the deposition conditions. However, through a parameter study, conditions were found to grow thin films containing only the $AgInS_2$ phase with chalcopyrite type structure. In samples containing a mixture of several phases, the contribution in percentage terms of each phase to the whole compound was estimated with the help of the PowderCell simulation package. It was also found that the $AgInS_2$ films present p-type conductivity, a high absorption coefficient (greater than 10^4 cm^{-1}) and an energy band gap Eg of about 1.95 eV, indicating that this compound has good properties to perform as absorbent layer in thin film tandem solar cells. The effect of the deposition conditions on the optical and morphological properties was also investigated through spectral transmitance and atomic force microscopy (AFM) measurements.

1. Introduction

Chalcopyrite type compounds of the groups I – III – VI_2 are direct gap semiconductors, with suitable properties for solar cells manufacturing. The highest conversion efficiency reported until now for simple junction thin film solar cells was obtained with a device fabricated using a layer based on compounds of the Cu-III-VI_2 group as absorber [1]; solar modules based on $Cu(InGa)Se_2$ (CIGS) thin films offer additionally good stability and competitive projected production cost . On the other hand, Ag-III-VI_2 based compounds are characterized by having energy band gap values ranging from 1.87 to 2.03 eV which makes these compounds suitable as absorber layers in two junction tandem solar cells [2].

Different methods have been used for preparing thin films based on Ag-III-VI_2 compounds; thermal evaporation [3], electrodeposition [4] and solution growing [5] are some of them. In this work, thin films of $AgInS_2$ have been grown by co-evaporation of the precursors in a two steps process; for that, the metallic precursors were evaporated from a tungsten boat in presence of elemental sulfur, evaporated from a tantalum effusion cell. The influence of the deposition conditions on the optical, structural and morphological properties were investigated by spectral transmittance, XRD (x-ray diffraction) and AFM (atomic force microscopy) measurements.

---

[1] Tel./fax:571-3165000 Ext. 13017 E-mail adress: ggordillog@unal.edu.co, *caarredondoo@unal.edu.co*

2. Experimental

The $AgInS_2$ thin films were prepared by co-evaporation of the precursor species on a glass substrate, using a system constituted by an evaporation chamber connected to a vacuum system which allows working at pressures of about $2x10^{-5}$ Torr, two tungsten boats (used to evaporate Ag and In respectively), a tantalum effusion cell to evaporate sulfur and a thickness monitor (Maxtec TM-400) with a quartz crystal as sensor, used to measure the evaporated elements flux. The substrate temperature was controlled with a programmable PID controller (Eurotherm 900C). The chemical composition of the $AgInS_2$ films is controlled by controlling the substrate temperature and the evaporated mass of the precursor species.

The deposition of the $AgInS_2$ films was accomplished in two stages. In the first stage an In_xS_y layer is grown by simultaneous evaporation of In and S, keeping the substrate temperature at 300° C. During this stage the flux of In is kept around 2.5 Å/s and the evaporation temperature of S is kept around 135°C. In the second stage the $AgInS_2$ compound is formed by evaporating Ag in a sulfur environment, on the In_xS_y layer, keeping the substrate temperature between 400 and 600°C. During this stage the flux of Ag was kept at about 3 Å/s. The $AgInS_2$ films were submitted to a post-deposition thermal annealing in S ambient during about 1 hour, in order to improve the stability of the chemical composition.

The transmittance measurements were carried out with a VIS – IR Oriel spectrophotometer and the conductivity type was determined through thermoelectric power measurements. The XRD measurements were performed using the Cu Kα radiation of a Shimadzu-6000 diffractometer and the AFM images were obtained with a PSI AP-0100 Scanning Probe Microscope. The film thickness was measured with a Veeco Dektak 150 surface profiler. Thermoelectric power measurements revealed that all the studied $AgInS_2$ films presented p type conductivity.

3. Results and discussion

3.1 Optical characterization

The $AgInS_2$ thin films were characterized through spectral transmittance measurements, in order to study the effect of the synthesis parameters on the optical constants (refractive index n, absorption coefficient α and optical gap Eg). In figure 1 are depicted typical transmission spectra of $AgInS_2$ films, deposited varying the mass of Ag to mass of In ratio as well as the substrate temperature. The studied $AgInS_2$ films were grown varying the deposition parameters as follows:

i) The mass of Ag to mass of In ratio was varied between 0.7 and 1.13, keeping constant the substrate temperature at 300°C and 500°C during the 1st and 2nd stages, respectively, and the masses of In and S at 0.15 and 3 grams, respectively.

ii) The substrate temperature (2nd stage) was varied between 400 and 600°C, keeping constant the amounts of Ag, In and S at 0.15, 0.17 and 3 grams, respectively.

It is observed in figure 1 that the Ag or In rich $AgInS_2$ films, present low transmittance values. This behaviour is seemingly caused by absorption of photons through states within the energy band gap associated to donor states generated by S-vacancies and by anti-site defects. Transmittances around 70% were obtained with $AgInS_2$ films deposited under the following parameter set: M_{Ag}/M_{In}=0.9, substrate temperature of 300°C and 500°C during the 1st and 2nd stages respectively, M_S=3 gr. and evaporation rate of In and Ag of about 2.5 and 3.0 Å/s, respectively.

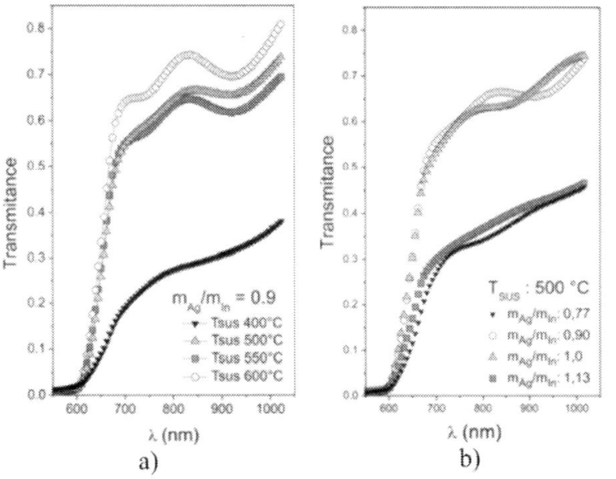

Figure 1: Spectral transmittance measurements of $AgInS_2$ thin films showing the effect of a) substrate temperature and b) mass of Ag to mass of In ratio.

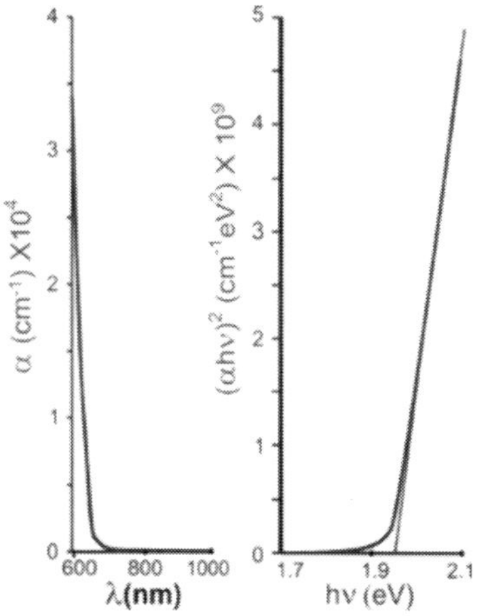

Figure 2. Curves of α vs. λ and $(\alpha h\nu)^2$ vs. hν, corresponding to typical $AgInS_2$ thin films.

On the other hand, the results of figure 1 show that the substrate temperature (2nd stage) also affects the transmittance of the $AgInS_2$ films. At temperatures lower than 500°C, the transmittance decreases significantly, probably because, under this conditions, a high density of native defects are formed in the $AgInS_2$ films, as a consequence of incomplete chemical reaction of the precursors.

In figure 2 are depicted curves of α vs. λ and (αhν)2 vs. hν, which were determined using the transmission spectrum and calculations based on a procedure described in detail in reference [6]. The results of figure 2 show that the $AgInS_2$ films have an absorption coefficient α greater than 10^4 cm^{-1} and an optical gap Eg of 1.93 eV , indicating that the $AgInS_2$ thin films, we have deposited in this work, have suitable properties for using them as absorber layers in two junction tandem solar cells. Theoretical calculations of the conversion efficiency, carried out for two solar cells in tandem operation, revealed that a limit efficiency of 45% can be achieved if the energy gap of the top and button cells are 1.75ev and 1.15 eV respectively [7]. This calculation also predict a limit efficiency of around 40% for a tandem cell fabricated using $AgInS_2$ (Eg=1.9eV) as top cell and $AgInSe_2$ (Eg=1.2eV) as bottom cell.

3.2 Structural characterization

The effect of the substrate temperature and of the Ag mass to In mass ratio, on the structural properties of the $AgInS_2$ films was also studied through XRD measurements. Figure 3 shows typical XRD spectra of $AgInS_2$ films deposited varying the substrate temperature (2nd stage) between 400 and 600ºC and the Ag mass to In mass ratio between 0.77 and 1.13. The results were analyzed using the data reported in the JCPDS data base for these types of compounds and performing theoretical simulation of the XRD pattern, with the help of the PowderCell package. This procedure allowed us to identify the phases present in the samples with a good degree of confidence. It was found that both, the synthesis temperature and M_{Ag}/M_{In} ratio, affect the phase and structure in which the $AgInS_2$ films grow. The diffractograms of figure 2 show that only the samples deposited at substrate temperatures around 500ºC, using an Ag mass to In mass ratio of about 0.9, grow in the $AgInS_2$ phase with tetragonal structure. It was found that, in samples deposited at temperatures greater than 550°C keeping constant the M_{Ag}/M_{In} ratio at around 0.9, the $AgInS_2$ compound grows with a mixture of both, orthorhombic and tetragonal structure. The presence of the InS_2 phase was also identified in this type of samples. The $AgInS_2$ compound grows with both orthorhombic and tetragonal structures when deposited at temperatures below

500°C. When deposited around 500°C using a M_{Ag}/M_{In} ratio different from 0.9, a mixture of $AgInS_2$ orthorhombic and tetragonal together with InS_2 phases were identified.
The lattice parameters determined from the X-ray diffraction spectra, for $AgInS_2$ films grown with chalcopyrite type tetragonal structure, are: a = 5.8980Å and b = 11.1935Å.

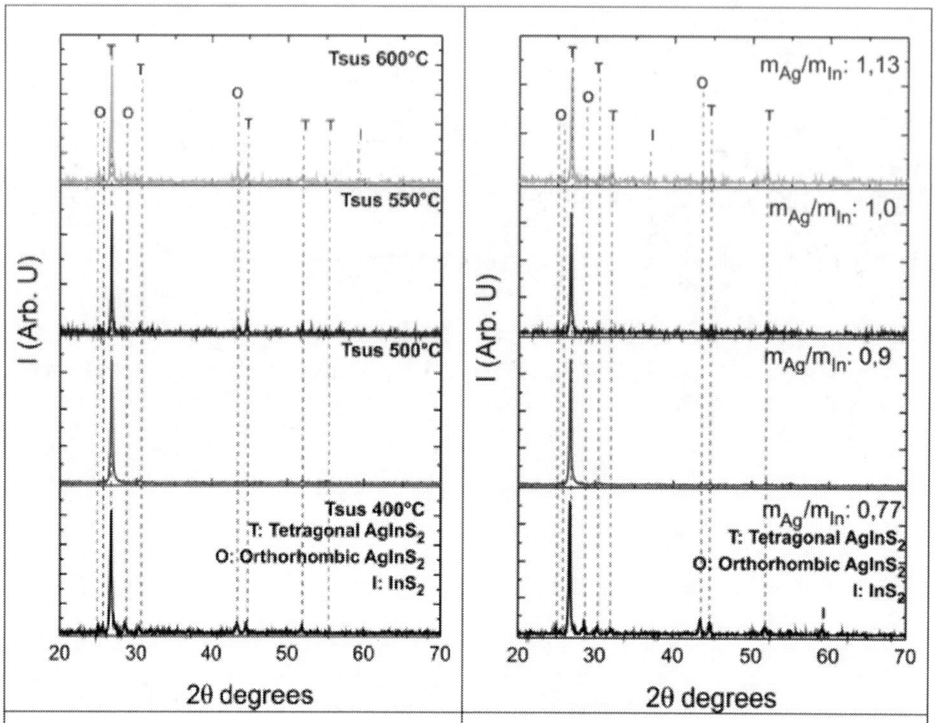

a) b)

Figure 3: Typical XRD spectra of $AgInS_2$ films deposited varying: a) the synthesis temperature between 400 and 600°C and b) the mass of Ag to mass of In ratio between 0.77 and 1.13.

3.3 Morphological characterization

In Figure 4 are shown AFM images taken on typical $AgInS_2$ thin films, deposited varying the M_{Ag}/M_{In} ratio (keeping the substrate temperature (2nd phase) in 500°C) and the substrate temperature (keeping the M_{Ag}/M_{In} ratio in 0.9). In Figure 4, curves are also depicted showing the effect of the substrate temperature and of the M_{Ag}/M_{In} ratio on the grain size of the $AgInS_2$ films.

The results of fig. 4 show that the substrate temperature and the M_{Ag}/M_{In} ratio significantly affecting the films morphology. In general, the grain size decreases by increasing both the substrate temperature and the M_{Ag}/M_{In} ratio.

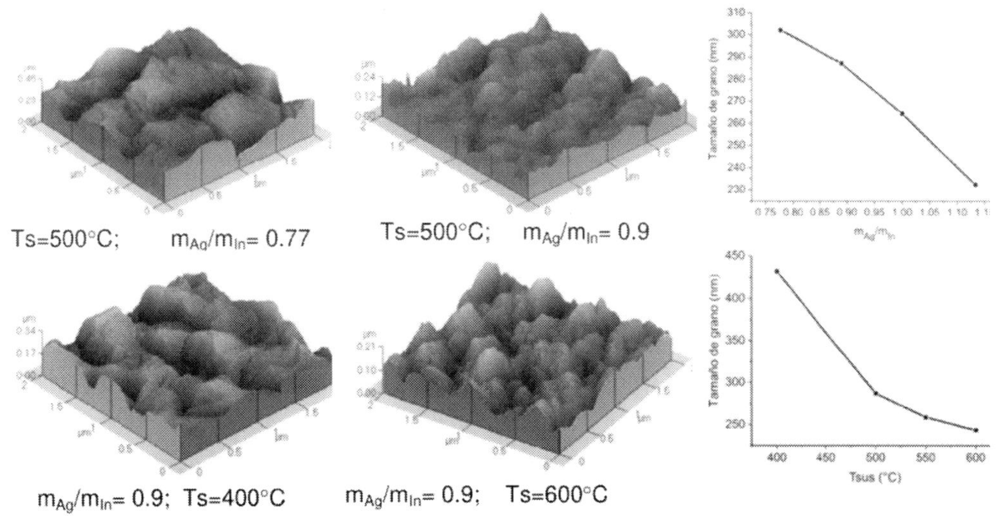

Figure 4: AFM images corresponding to $AgInS_2$ thin films showing the effect of the M_{Ag}/M_{In} ratio and the substrate temperature on the morphology. The variation of the grain size as a function of the M_{Ag}/M_{In} ratio and of the substrate temperature is also shown.

Conclusions

P type $AgInS_2$ thin films with adequate properties to be used as absorber layers in two junction tandem solar cells were grown using a novel procedure based on the co-evaporation of the precursor in a two step process. It was found that the deposition conditions affect the phase and structure in which the $AgInS_2$ films grow. However, chalcopyrite type $AgInS_2$ films with tetragonal structure were grown. The results also revealed that the substrate temperature and the M_{Ag}/M_{In} ratio significantly affect the films' morphology. In general, the grain size decreases by increasing both, the substrate temperature and the M_{Ag}/M_{In} ratio.

Acknowledgements: This work was supported by the Universidad Nacional de Colombia and Colciencias

References

[1] M.A. Contreras, K. Ramanathan, J. AbuShama, F. Hasoon, D.L. Young, B. Egaas, R. Nuofi. *Prog. In Photovoltaics Res. Appl*, 2005, **13**, 209
[2] Z. Aissa, T. Ben Nasrallah, M. Amlouk, J.C. Bernede, S. Belgacem. *Solar Energy Materials & Solar Cells*, 2006, **90**, 1137
[3] E. Yoshinory and N. Hamakawa. *Appl. Phys*, 1995, **34**, 3260
[4] Y. Ueno et al. *Thin Solid Films*, 1990, **91**, 189
[5] R. P. Sharma. *Indian J pure Appl. Phys*, 1995, **33**, 711
[6] M. Gracia, F. Rojas. and G. Gordillo, *20th European PV Solar Energy Conf*, Barcelona, 2005; 1874
[7] F. Meillaud et al. *Solar Energy Materials & Solar Cells*, 2006, **90,** 2952

Journal of Physics: Conference Series **167** (2009) 012051 doi:10.1088/1742-6596/167/1/012051

Study of the influence of annealing parameters on the optical and compositional properties of ZnS, ZnS:Mn and ZnS:Sm grown by Sol Gel

E. Gómez-Barojas[1,4], E. Sánchez-Mora[2], T. Mendoza-Dorantes[3], C. Castillo-Abriz[3] and R. Silva-González[2]

[1]CIDS-IC, BUAP. Ed. 103C, C.U. Col. San Manuel. Puebla, Pue. México 72570

[2]Instituto de Física, BUAP. Apdo. Postal J-48 Puebla, Pue. México. 72570.

[3]FIQ, BUAP. Puebla, Pue. México. 72570.

E-mail: egomez@sirio.ifuap.buap.mx

Abstract. Samples with 3 layers of ZnS were prepared on glass substrate by the Sol Gel method. The annealing parameters were varied as follows: each of the 3 constituent layers of sample 1 was annealed at 400 °C for 12 hrs. For sample 2, the 1rst and 2nd layers were annealed at 300 °C for 2hrs each and the 3rd one at 400 °C for 12 hrs. For sample 3, the 1rst and 2nd layers were annealed at 200 °C for 2hrs each and the 3rd at 400 °C for 12 hrs. Furthermore, 2 samples: one with 3 layers of ZnS:Mn and the other with 3 layers of ZnS:Sm were prepared and annealed similarly to sample 2. AES depth profiles show a transversal diffusion of S in sample 1, a sulfur deficiency in sample 2, and a non distributed excess of S in sample 3. The absorption spectra of the 3 ZnS samples are very similar but the absorption edges shift to smaller energy values for the Mn and Sm doped ZnS samples. Sample 1 presents a PL spectrum with 2 bands at 441 and 537 nm while sample 2 presents a similar spectrum but the bands are 16 and 11 nm blue shifted.

1. Introduction

The Zinc Sulfide (ZnS) is a II-VI semiconductor with a wide direct band gap in the near UV region. Its energy band gap is smaller in cubic ZnS (3.68 eV) [1] than in hexagonal ZnS (3.7 eV) [2]. It is an attractive material to be used as an antirreflecting coating in heterojunction solar cells, and as a phosphor in electroluminescent devices [3].

The object of this work was to prepare samples with three layers of ZnS on glass substrates by the sol-gel method and study the influence of the thermal treatments parameters on the composition and optical properties of the samples. These parameters were varied as follows: Each of the 3 layers of sample 1 was annealed at 400 °C for 12 hrs; we label this annealing condition as (TT400°C). Then, the 1rst and 2nd layers of sample 2 were annealed at 300 °C for 2hrs each and the 3rd at 400 °C for 12 hrs (TT300°C) and for sample 3 the 1rst and 2nd films were annealed at 200 °C for 2hrs and the 3rd at 400

[4] To whom any correspondence should be addressed

°C for 12 hrs (TT200°C). All annealing processes were performed under N_2 flux. Besides, two more samples were prepared: sample 4 with 3 layers of ZnS: Mn^{2+}/glass and sample 5 with 3 layers of ZnS:Sm^{3+}/glass by the same method using the (TT300°C) annealing condition.

2. Experimental

2.1. Synthesis

The ZnS layers were deposited on glass substrates by the Sol-Gel method. The ZnS base deposition solution contained 21.6 ml of metoxyethanol, 2.7970 g of 0.2 M zinc acetate dihydrate, 0.5749 g of 0.4 M thiourea and 0.4 ml of 3.6×10^{-4} M ethanolamine. The solution was heated to 60 °C with constant agitation for 30 min and let aged for 2 days. Then, each of the constituent layers of a sample was deposited by dipping the glass substrate into the deposition solution for 5 min, slowly withdrawing it and let it dried in air. After each deposition an annealing process in N_2 was carried out.

Three ZnS samples were prepared with similar composition solution. However, sample 1 was annealed with the TT400°C condition. Sample 2 with the TT300°C condition and sample 3 with the TT200°C annealing condition. To prepare the ZnS:Mn and ZnS:Sm samples, manganese chloride tetrahydrate (20 wt %) and samarium nitrate hexahydrate (20 wt %) were added to base deposition solutions as sources of Mn and Sm ions, respectively and the layers were annealed with the TT300°C condition. The samples were characterized by the techniques: Auger electron spectroscopy (JEOL, JAPM-7800) in combination with a Sputtering Ar ions system; UV-Visible Spectroscopy (Varian, Cary-100), and Photoluminescence spectroscopy (ARC-275 spectrometer, an R636-10 photomultiplier and UV light-excitation source).

3. Results

3.1. AES

The spectra were recorded at Ep/Ip=5 keV/8×10^{-8} A and a base pressure of 1.7×10^{-7} Pa. Fig. 1 shows Auger depth profiles, relative intensity vs sputtering time, of the ZnS samples annealed with the conditions: (a) TT400°C, (b) TT300°C and (c) TT200°C. In (a) the relative intensity of S is greater than that of Zn, (I(S) > I(Zn)), the shape of the S profile is not plane but curved, probably due to a transversal diffusion of S toward the sample surface. In b) the I(Zn) > I(S) through depth, except for the depression of the S profile which is very likely due to the different annealing temperature of the 3rd layer, it looks more homogeneous than that in (a). In c) the ZnS layer elemental composition is not homogeneous, since the S profile consists of 3 large peaks while the Zn profile is of very low intensity. The S profile in Fig. 1(a) could be due to the formation of SO_x species in gas phase that migrate to the surface sample to be desorbed [4, 5]. A similar S transversal diffusion effect has been also observed in CdS/CdTe heterostructures [6]. It was not possible to perform this analysis in the doped ZnS samples because they present severe charging effects.

3.2. UV-Vis

3.2.1. ZnS samples

Fig. 2) shows UV-Vis absorption and transmission spectra. The lower part of the absorption spectrum in Fig. 2a is slightly shifted to larger wavelengths as the annealing temperature is increased. The spectrum corresponding to the sample with the (TT300°C) annealing condition shows some oscillations in the lower part (see the inset) due to the combination of annealing temperatures used. In 2 (b) transmittance spectra are shown. The sample with TT300°C annealing condition shows well defined interference fringes. The fringes are associated with the uniformity of a film thickness, when the thickness is not uniform or it is slightly tapered the interference effects are destroyed and the transmission is a smooth curve as the spectrum of the sample with TT400°C annealing condition.

Since UV-Vis spectra were recorded from each constituent layer, the thickness, t, refraction index, n; absorption coefficient, α; and energy band gap, Eg were calculated using the transmittance spectra and the Swanepoel method [7]. The estimated data are listed in Table 1.

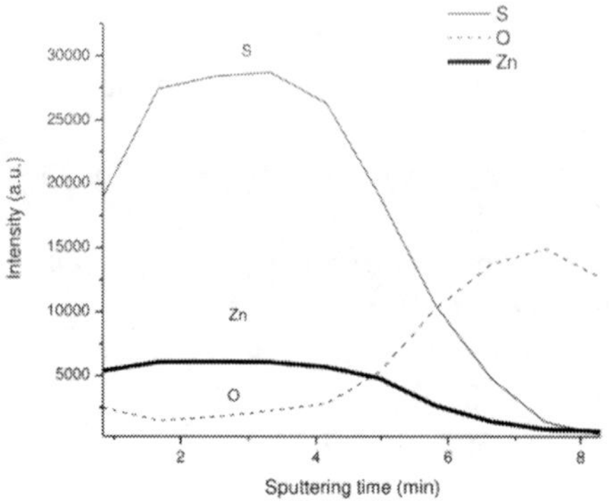

Fig. 1. (a) Auger depth profile of the ZnS sample with TT400°C annealing condition.

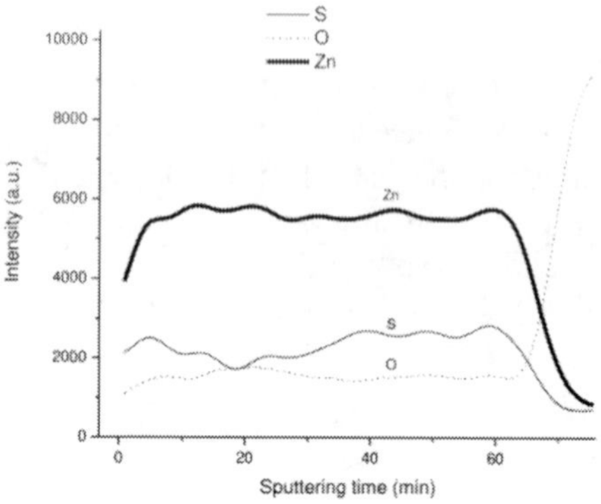

Fig. 1. (b) Auger depth profile of the ZnS sample with TT300°C annealing condition.

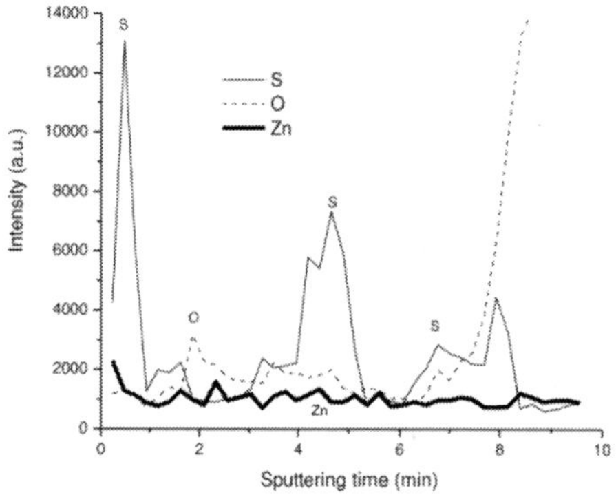

Fig. 1. (c) Auger depth profile of the ZnS sample with TT200°C annealing condition

Table 1. Calculated optical parameters of the ZnS samples with different annealing conditions.

Sample, annealing condition	Eg (eV)	n	*t* (nm)	α^2(cm^{-1})
ZnS, TT400°C	3.65	$1.83_{\lambda=497\ nm}$	425.4	$1.3\text{x}10^8$
ZnS, TT300°C	3.67	$1.67_{\lambda=540\ nm}$	572.7	$5.5\text{x}10^7$
ZnS, TT200°C	3.70	$1.52_{\lambda=561\ nm}$	502.6	$1.7\text{x}10^7$

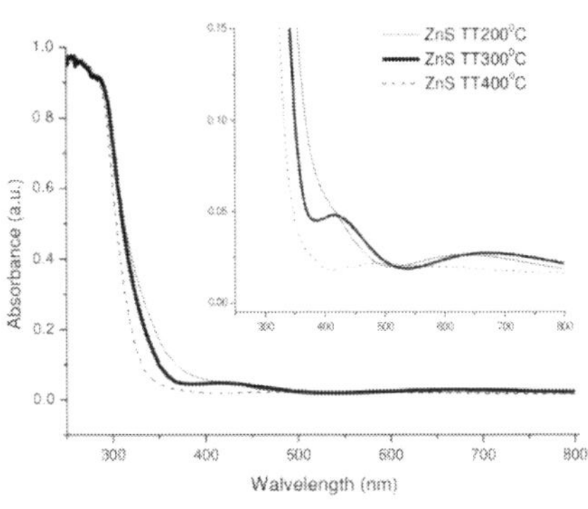

Fig. 2 (a). Absorbance spectra of the ZnS samples with TT200°C, TT300°C and TT400 °C annealing conditions:

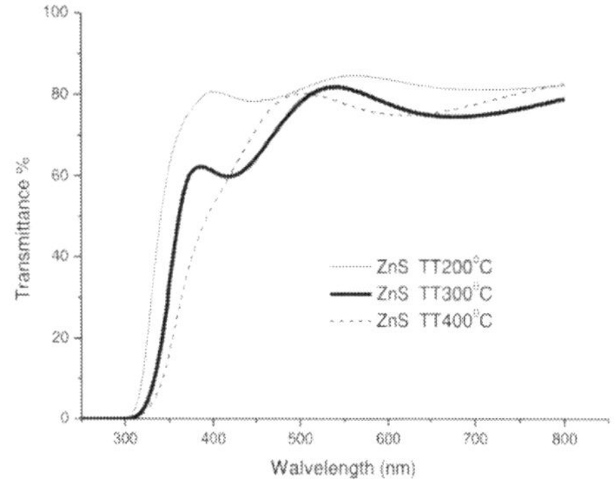

Fig. 2 (b) Transmittance spectra of the ZnS samples with TT200°C, TT300°C and TT400 °C. annealing conditions

3.2.2. Samples with 3 layers of ZnS:Mn and ZnS:Sm

Fig. 3 shows UV-Vis spectra of the ZnS:Mn and ZnS:Sm samples (a) absorbance and (b) transmittance spectra. Correspondent ZnS spectra are included for comparison. In (a) the absorption threshold of the doped ZnS samples are shifted to smaller energy values (3.26 eV for ZnS:Mn and 3.06 eV for ZnS:Sm) with respect to the one of the undoped ZnS. In b) the transmittance percentages of ZnS:Mn and ZnS:Sm are smaller than the one of the undoped ZnS.

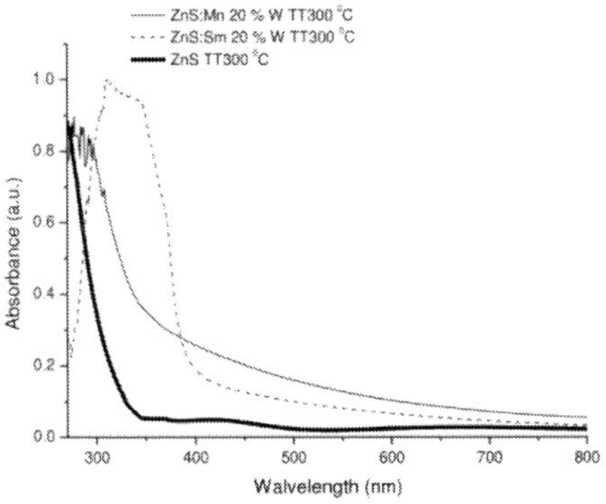

Fig. 3. (a) Absorbance vs wavelength spectra of the ZnS:Mn, ZnS:Sm and ZnS samples:

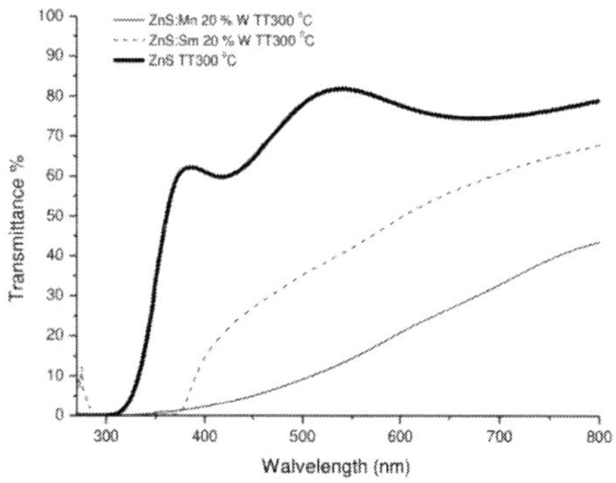

Fig. 3. (b) Transmittance vs wavelength spectra of the ZnS:Mn, ZnS:Sm and ZnS samples.

3.3. Photoluminescence

Fig. 4 shows luminescence spectra of the ZnS samples whose annealing conditions were: a) TT400°C and b) TT300°C. In a) the emission band with maximum at 441 nm (2.81 eV) in the blue range is assigned to Zn vacancies and the emission band at 537 nm (2.31 eV) is assigned to vacancies located near the ZnS valence band [8,9] that act as acceptors in the semiconductor terminology. In b), the PL spectrum shows 2 broad shoulders at 425 nm (2.91 eV), and 526 nm (2.36 eV). These bands are similar to the ones in a) but blue shifted due to the change in the thermal treatment condition. The doped ZnS samples presented similar spectra to the above (not shown)

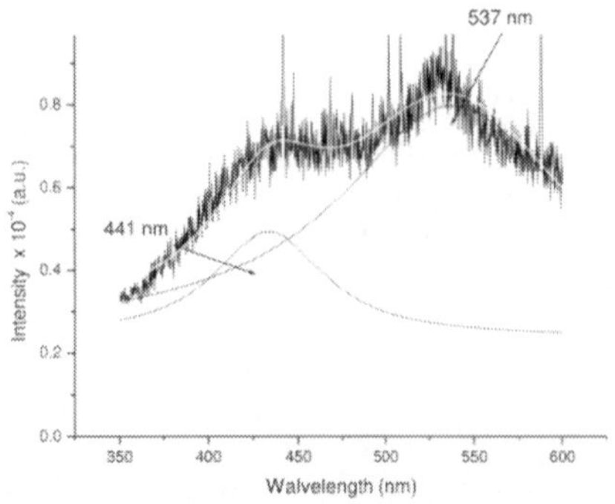

Fig. 4. (a) Photoluminescence spectrum of the ZnS sample with TT400°C annealing condition.

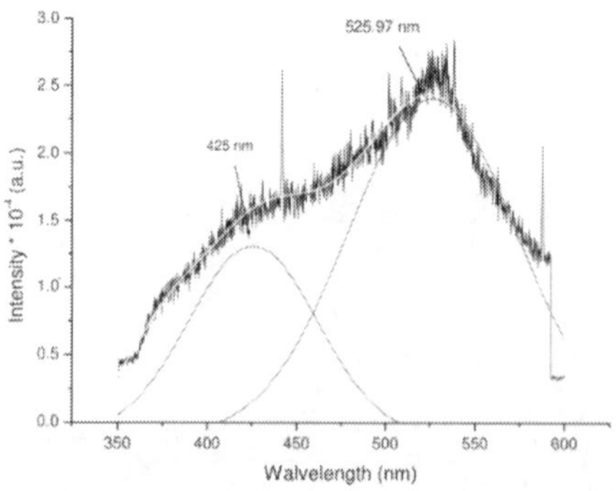

Fig. 4. (b) Photoluminescence spectrum of the ZnS sample with TT300°C annealing condition.

4. Conclusion

From UV-Vis characterization results, it is observed that having used a new solution with 2 days of age for each constituent ZnS layer a large and equal amplitude of the absorption edges of the three samples were obtained. The absorbance borders of the doped ZnS are shifted to larger wave lengths with respect to that of ZnS. The ZnS sample annealed at the TT300°C condition presents better features for example: good absorbance amplitude and a large transmittance percentage. The AES depth profiles show that the S transversal diffusion can be avoided by performing the annealing at TT300°C condition. The ZnS sample with TT300°C annealing condition presents greater PL intensity and and the bands are 16 and 11 nm blue shifted with respect to the one with TT400°C annealing condition. The ZnS:Mn and ZnS:Sm samples require further investigation since no great change in the PL spectra was observed.

5. Acknowledgment

Work partially supported by VIEP, BUAP trough projects Nos. 11/EXC/07/G.

6. References

[1] Landolt, H.; Bomstein, R. Numerical data and functional relationships in science and technology, New Series. Springer-Verlag.; Berlin, 1999
[2] Q. Xion. Nano Letters. 4, 9 (2004) 1663.
[3] B. Su and K. L. Choy. J. Mater. Chem. 10 (2000) 949
[4] E. Gómez-Barojas, R. Silva-González, J. Pantoja- Enriquez. Solar Energy Materials & Solar Cells 90 (2006) 2235
[5] X. Q. Wang, D. Xu, S. G. Li, J. Huang, and W. Du. Crys. Res. Technol. **41**,1,55 (2006) 55
[6] X. Q. Wang, D. Xu, X. F. Cheng, M. K. Lu, D. R. Yuan, J. Huang, G. W. Lu, H. X. Ning, X. L. Duan, Z. M. Wang, G. T. Lu, S. G. Li, Y. Chen, and Y. Q. Zhou. Mater. Res. Bull. **37** (2002) 1859.
[7] Monjoy Sreemany, Suchitra Sen, J. Matter. Chem. and Phys. 83 (2004)169
[8] W. Chen, R. Sammynaiken, Yining Huang, J. Olle Malm, R. Wallenberg, J. Olor Bovin., V. Zwiller and N. A. Kotov. J. Appl. Phys. 89, 2 (2001) 1120
[9] H. C. Warad, S. C. Ghosh, B. Hetmtanon, C. Thanachayanont and J. Dutta, J. Sci and Tech. of Adv. Materials. 6 (2005) 296

Thermal evolution of CaO-doped HfO_2 films and powders

S A Barolin[1], M C Caracoche[2], J A Martínez[2], P C Rivas[3], M A Taylor[2], A F Pasquevich[2] and O A de Sanctis[1,4]

1Lab. Materiales Cerámicos, FCEIyA, Universidad Nacional de Rosario, IFIR-CONICET
2Departamento de Física, FCE, Universidad Nacional de La Plata, Argentina – IFLP-CONICET
3Facultad de Ciencias Agronómicas y Forestales, Universidad Nacional de La Plata, Argentina - IFLP

oski@fceia.unr.edu.ar

Abstract. Solid solutions of ZrO2 and HfO2 are potential electrolyte materials for intermediate-temperature SOFC because both are oxygen-ion conductors. The main challenge for these compounds is to reduce the relatively high value of the activation energies vacancies diffusion, which is influenced by several factors. In this work the thermal evolution of CaO-HfO_2 materials have been investigated. (CaO)y-Hf(1-y)O(2-y) (y = 0.06, 0.14 y 0.2) coatings and powders were synthesized by chemical solution deposition (CSD). Films were deposited onto alumina substrates by Dip Coating technique, the burning of organic waste was carried out at 500 °C under normal atmosphere and then the films were thermally treated at intervals of temperature rising to a maximum temperature of 1250 °C. By means Glazing Incidence X-ray Diffraction (φ-2θ configuration) the phases were studied in the annealed films. On the other hand, the thermal evolution and crystallization process of powders were analyzed in-situ by HT-XRD. The phenomena crystallization occurred in films and powders were analyzed. The activation energies of diffusion of oxygen vacancies of HfO2-14 mole% CaO and HfO2-20 mole% CaO films were measured from the thermal evolution of the relaxation constant measured by Perturbed Angular Correlation Technique.

1. Introduction

The solid solutions based HfO_2 have taken very technological importance in recent years, not only for their applications as solid electrolyte in fuel cells and gas sensors, but also for its use as picture-Detectors high-energy particles (scintillation) [1]. Furthermore, due to the miniaturization of devices, the demand for functional materials in restricted dimensions, such as thin films, is imperative. The required properties of solid solutions based on HfO_2 for any of their applications, are dependent on the structures and the stability of cubic phases metaestables or tetragonal. The objective of this work is to study the evolution of phases in films and powders in the system CaO-HfO_2 prepared from the precursor solution.

[4] Author to whom correspondence should be addressed

2. Experimental procedure

2.1. Samples preparation

CaO-HfO_2 precursor solutions were prepared by using Calcium acetate and Hafnium buthoxide (Hf(OC4H9)4), as source materials, with buthanol as solvent. Acetoin (3-hydroxy-2-butanone, CH3COCH(OH)CH3) was used as chelating agent. Commercial starting materials were used in their as-received state from Aldrich. Acetoin was added to Hafnium buthoxide - buthanol solution (the molar ratio of the chelating agent to the alkoxide was R = 4). The compositions were $Hf_{1-x}\ Ca_x\ O_{2-x}$ with x = 0,06; 0,14; 0,20. The solution was suitable for the deposition of uniform and crack-free layers by Dipping-coating, with a thickness of ~150 nm per layer deposited onto alumina substrates. After deposition, each layer was dried at 200 °C for 3 min and pyrolyzed at 500 °C for 6 min. Additional layers were dip-coated to build up the desired thickness. Several samples of multilayer coating was heat-treated by quickly placing it into a preheated furnace at temperatures ranged between 500 and 1250 °C for 60 minutes. To derive gel powders the parent solutions were kept overnight in an oven maintained at 60◦C under 100 Pa of pressure. After solvent evaporation the residue was heating at 200 °C. The dried gel was crushed in a agate mortar to get fine amorphous powders.

2.2. X-ray diffraction

The structure of the films was measured using an X-ray diffractometer with Cu K_α radiation (Philips X-Pert Pro) and a graphite monochromator (the step size being of 2θ = 0.02° and a 1 second time per step). For annealed films, a grazing incidence configuration (GIXRD) was used for measuring at room temperature. High temperature in-situ X-ray diffraction (HT-XRD) patterns of powders were obtained in air atmosphere making use of an Anton-Paar HTK10 high-temperature chamber. The spectra were taken at nominal temperatures (temperature of heating filament) equal to the temperatures used in annealing treatment of films. In order to overcome partly uncertainty in the determination of temperatures, a second S-thermocouple was placed onto the sample to measure directly the temperature of the powders (TR). Diffraction spectra were exploited to calculate average crystallite sizes of the solid solution from the magnitude of the half-width of the main line for c-HfO_2 (2θ = 30°), instrumental broadening of 0.130°), using the Scherrer equation.

2.3. Thermal analyses

By means of a DTG-60H Shimadzu analyzer, simultaneous Differential Thermal Analisys (DTA) and Thermogravimetric Analysis (TGA) measurements were carried out on the as-obtained powder gels from room temperature to 800◦C in normal atmosphere. Alumina crucibles and α-alumina reference were used.

2.4. Oxygen vacancies movement

The activation energy of oxygen vacancies diffusion was measured by PAC experiment. The PAC method is a nuclear technique by which the hyperfine interaction between a radioactive nucleus acting as a probe and the internal fields of the solid in the nearest neighbourhood can be measured [2]. PAC experiment, gives the desired information about the electrical field gradient (EFG) produced by the surrounding ions at the probe nucleus via the determination of the four quantities, i.e., relative fraction, quadrupole frequency, asymmetry parameter, and frequency spread, associated with each unequivalent Hafnium lattice site. In fact, atomic or defect movements cause a further perturbation of the angular correlation which may be reflected in the PAC spin rotation curves by a damping in the amplitude obeying to a factor of the form $e^{-\lambda t}$ where λ is called the relaxation constant. In stabilized hafnias or zirconias, this situation appears with the onset of the thermally activated vacancy-hopping process. The jump frequency, v obeys an Arrhenius law and the correlation time $\tau = 1/\nu$ can then be associated with the activation energy of the jump through the equation: $\tau = \tau_\infty\ e^{Eac/kT}$ (1). It has already been reported that an exponential decrease of the relaxation constant with the reciprocal absolute temperature implies a slow diffusion movement for which the correlation time between two successive atomic configurations is equal to λ^{-1}. By contrast, an exponential increase of λ with the reciprocal absolute temperature, generally found at higher temperatures, is typical of a fast relaxation regime, for which the correlation time is proportional to λ. Hence, considering equation (1), the activation energies for movement of the vacancies can be drawn from the slopes of the $\lambda\ (T^{-1})$ function [3].

3. Results and discussions.

3.1. Thermal evolution of as-prepared materials.

Figure 1 show DTA – TGA curves for CaO 20 mol% – HfO_2 powder (no substantial differences for other compositions). The TGA curve shows a weight loss assigned to the decomposition and burning of the residual organic groups and no weight loss at temperatures > 520°C is observed, so the thermal decomposition of the gel is completed at that temperature. The DTA curves display intense exothermic peaks, one at centered at 350 °C and another around 490 °C (488, 492 and 498 °C for 6, 14 and 20 mole% of CaO, respectively) that are attributed to burning of the residual organic groups and to removal of hydroxyls, with a simultaneous onset of the crystallization of the cubic/tetragonal phase, respectively.

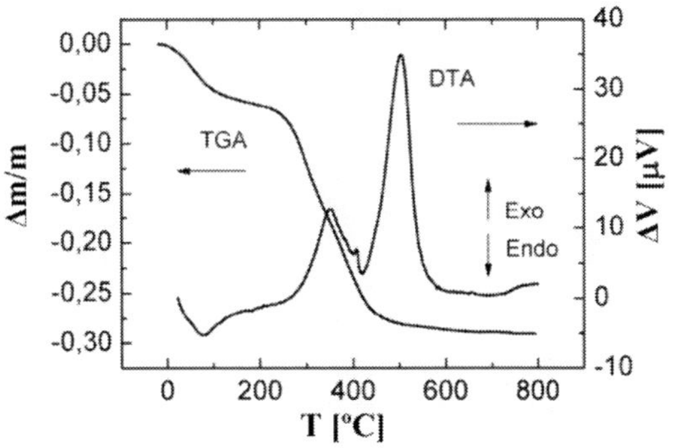

Figure 1. Thermogravimetric and Differential thermal analysis of powder CaO 20 mol% – HfO_2

3.2. Crystalline phases and metastable phase stabilization.

Figure 2 shows in-situ HT-XRD spectra of the CaO– HfO_2 powders that were taken at three temperatures for each composition: i) at 500 °C as nominal temperature, ii) at the highest temperature that the spectrum only exhibited diffraction peaks corresponding to cubic hafnia and iii) at the temperature that the monoclinic phase appears. The amorphous phase unambiguously transforms to cubic phase in the 14 and 20 mole% of CaO doped- HfO_2 powders. On the other hand, the metastable crystalline phase growth in the 6 mole% of CaO doped- HfO_2 powder could not be clearly identified as tetragonal phase or cubic phase due to its early destabilization by irruption of monoclinic phase at low temperature. The X-ray diffraction result follows the same trend that showed the DTA analysis, indicating that for increasing CaO content the crystallization temperature increases.

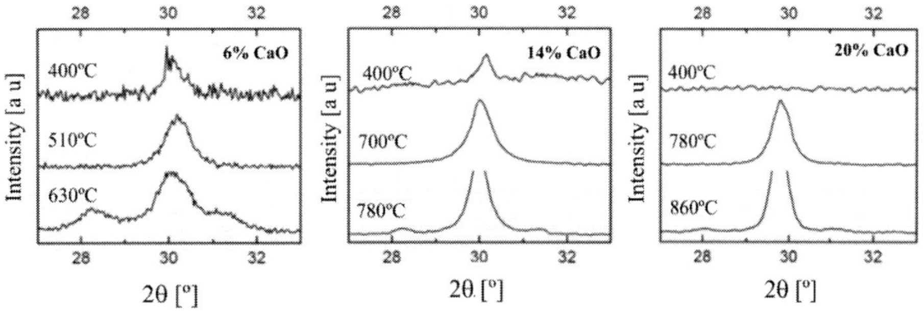

Figure 2. HT-XRD pattern of CaO - HfO_2 powders

Figure 3 shows ex-situ GI-XRD spectra of the CaO–HfO_2 films that were taken at room temperature on samples of annealed films at: i) 500 °C, ii) the highest temperature that the spectrum only exhibited diffraction peaks corresponding to cubic hafnia and iii) temperature that the monoclinic phase appears. The appearance of the monoclinic phase in the films occurred at higher temperatures than the ones in the powders; at 800 °C, 1000 C° and 1100 °C for the 6, 14 and 20 mole% of CaO doped- HfO_2 film, respectively.

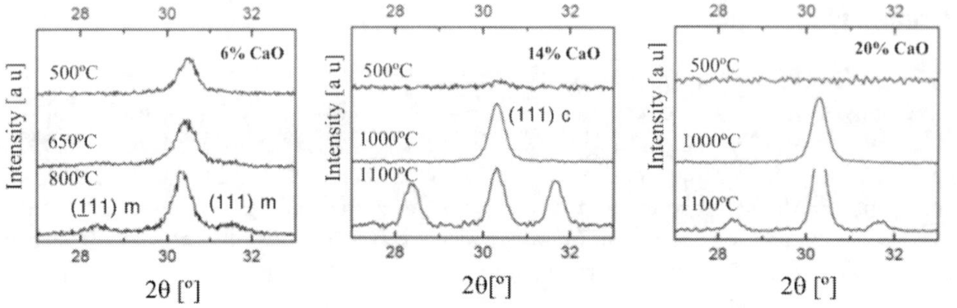

Figure 3. GI-XRD pattern of annealed for CaO - HfO_2 films

Table 1 shows the thermal evolution of fractions of cubic phase and monoclinic phase in annealed films and thermal treated powders of composition xCaO-(1-x)HfO_2 with x = 0.06, 0.14 and 0.2.

Table 1. Thermal evolution phase fractions for CaO – HfO_2 materials

Comp.		$Hf_{0,94}Ca_{0,06}O_{1,94}$				$Hf_{0,86}Ca_{0,14}O_{1,86}$				$Hf_{0,20}Ca_{0,20}O_{1,80}$			
Temp. [°C]		Film		Powder		Film		Powder		Film		Powder	
T_N	T_R	c	m	c	m	c	m	c	m	c	m	c	m
500	400	100	0	100	0	100	0	100	0	¿?	¿?	¿?	¿?
650	510	94	06	100	0	100	0	100	0	100	0	100	0
800	630	79	21	72	28	100	0	100	0	100	0	100	0
900	700	64	36	40	60	100	0	100	0	100	0	100	0
1000	780	39	61	22	78	> 99	< 01	95	05	100	0	100	0
1100	860	0	100	17	83	36	64	71	29	79	21	98	02
1250	1130	0	100	21	79	0	100	61	39	75	25	87	13

The metastable tetragonal/cubic phases become unstable at increasing temperature of thermal treatment if the metastable–monoclinic transformation is not prevented. The Calcium cations are able to be substituted for Hf^{+4} ions in the cation lattice and consequently create vacancies in the oxygen sublattice in order to satisfy electroneutrality requirements. The oxygen vacancies increase the cubic phase stability due to the transformation of the Hf^{+4} surroundings from 8-folded oxygen coordination to 7-folded oxygen coordination. The question is to know the minimum dopant concentration for full cubic stabilization. Table 1 shows that the stability of the cubic phase increases for increasing CaO content for both films and powders. However, in the powders (TR) the irruption of monoclinic phase occurred at lower temperatures than in the films (TN). For thermal treatment at temperatures of 1000 °C, the cubic maintains as the unique phase in the 14 and 20 mole% of CaO doped-films.

This different behavior between films and powders has been already observed in the system Y_2O_3 doped-ZrO_2 [4]. The highly defective nature of the structure in the film coating, produced by presence of residual lattice strains and asymmetric grains, were been aiding the cubic phase stabilization.

The thermal evolution of the grain size vs. 1/kT(K) of the cubic phase in the HfO2 - 20 mole% of CaO film and powder, make possible to infer that the mechanisms operated during the growth were dissimilar in each one of such materials (Figure 4). The change of slope at around 700 °C would indicate the starting of a recrystallization process during the thermal evolution of the powder, which would be indicating that in the powder the nucleation developed mainly by homogenous nucleation. Contrarily, the constant slope in the all temperature range and the low value of activation energy of the growing grain, it would evidence a heterogeneous nucleation in the film.

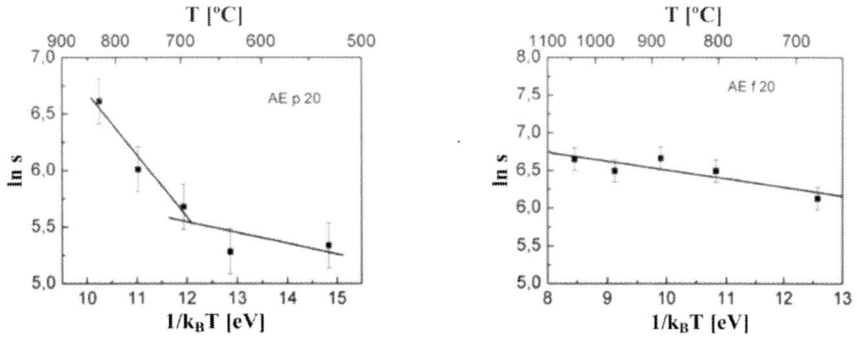

Figure 4. Thermal evolution of the size grain as function of 1/kT(K) of HfO_2 - 20 mole% of CaO

3.3. Oxygen vacancies diffusion.
Concerning the movement of oxygen vacancies, the $\lambda(T^{-1})$ function is plotted in figure 5. For both CaO 14 mole% doped- HfO_2 and 20 mole% CaO 20 mole% doped- HfO_2 films a slow-fast two regime behaviour can be observed. The movements at temperatures lower than 800°C correspond, according to the dynamic model assumed, to a slow diffusion ($\tau \propto \lambda^{-1}$) that is related to phenomena that are not linked to the oxygen vacancies diffusion. For temperatures higher than 800°C, in turn, the fast relaxation dynamical effect due to oxygen vacancies movement can be considered Arrhenius-like and an activation energies of 0.54 ± 0.04 eV and 0.70 ± 0.05 eV could be drawn from the $\lambda(T^{-1})$ slopes corresponding to HfO_2-14 mole% CaO film and HfO_2-20 mole% CaO film, respectively. Unexpectedly, the film that exhibits higher activation energy is the one where the CaO doping is higher, in which consequently the concentration of oxygen vacancies is higher also. This fact suggests that high levels of CaO content in the structure of the CaO doped- HfO_2 films performs a situation that impedes partly the oxygen movement. High CaO content would produce an increase of the $\left[V_O^{\bullet\bullet}Ca''\right]$ association or the growth of order nanophases formed by defective-oxygen compounds that were unappreciable in XRD [5].

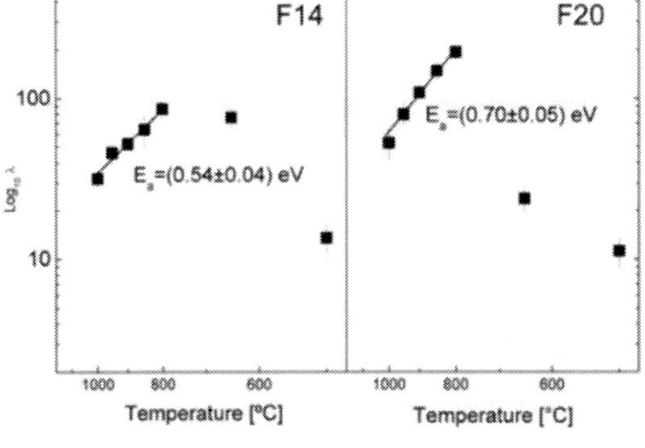

Figura 5. Activation energies deduced from the relaxation constants λ experienced by probes in the cubic solid solution.

4. Conclusions

Powders and coatings of metastable cubic hafnia have been successfully prepared by chemical solution deposition technique. The burning of residual organic groups, hydroxyls removal and onset of crystallization performed at temperatures lower than 500 °C.

The maximum temperature up to the stabilization is retained increases when CaO content increases in both powder and film. However, the irruption of monoclinic phase occurred at higher temperatures in the films. Films with 14 mole% doped- HfO_2 and 20 mole% CaO 20 mole% retained the cubic phase after annealing treatment at 1000 °C for 1 hour.

The determination of activation energies for oxygen vacancies movements allows to find that very high CaO content promotes the phenomena that impede partly the oxygen movement in the CaO – HfO_2 films.

References

[1] M Kirm, J Aarik, M Jürgens and I Sildos 2005 *Nuclear Instruments and Methods in Physics Research* A **537** 251-255

[2] P C Rivas, M C Caracoche, J A Martínez, A M Rodríguez, R Caruso, N Pellegri and O de Sanctis 1997 *J. Mater. Res.* **12** 493

[3] J A Gardner, H Jaeger, H T Su, W H Warner and J C Haygarth 1988 Physica B**150**, 223

[4] R Caruso, E Benavidez, O de Sanctis, M C Caracoche, P C Rivas, M Cervera, A Caneiro and A Serquis 1997 Phase Structure and Thermal Evolution in Coating Films and Powders Obtained by Sol-gel Process. Part. II: ZrO_2-2.5 % mol Y_2O_3 *J. Mater. Res.* **12** [10] 2594-601

[5] M C Caracoche, J A Martínez, P C Rivas, M A Taylor, A F Pasquevich, S Barolin, O A de Sanctis 2008 Nanostructures in calcia stabilized hafnia thin films observed by PAC as a function of temperature *Hyperfine Interaction* 0304-3843

Wettability, optical properties and molecular structure of plasma polymerized diethylene glycol dimethyl ether

T C A M Azevedo[1], M A Algatti[1,3], R P Mota[1], R Y Honda[1], M E Kayama[1], K G Kostov[1], R S Fernandes[1], N C Cruz[2] and E C Rangel[2]

[1] FEG-DFQ-UNESP, Av. Ariberto Pereira da Cunha 333, 12516-410 – Guaratinguetá, SP, Brazil

[2] UNESP, Avenida Três de Março, 511, 18087-180 Sorocaba, SP, Brazil

E-mail: algatti@feg.unesp.br

Abstract. Modern industry has frequently employed ethylene glycol ethers as monomers in plasma polymerization process to produce different types of coatings. In this work we used a stainless steel plasma reactor to grow thin polymeric films from low pressure RF excited plasma of diethylene glycol dimethyl ether. Plasmas were generated at 5W RF power in the range of 16 Pa to 60 Pa. The molecular structure of plasma polymerized films and their optical properties were analyzed by Fourier Transform Infrared Spectroscopy (FTIR) and Ultraviolet-Visible Spectroscopy, respectively. The IR spectra show C-H stretching at 3000-2900 cm^{-1}, C=O stretching at 1730-1650 cm^{-1}, C-H bending at 1440-1380 cm^{-1}, C-O and C-O-C stretching at 1200-1000 cm^{-1}. The refraction index was around 1.5 and the optical gap calculated from absorption coefficient presented value near 3.8 eV. Water contact angle of the films ranged from 40° to 35° with corresponding surface energy from 66 to 73×10^{-7} J. Because of its favorable optical and hydrophilic characteristics these films can be used in ophthalmic industries as glass lenses coatings.

1. Introduction

Material processing using low-temperature low-pressure RF excited plasmas is essential in many current scientific and technological issues encompassing microelectronics, optical and biomaterial industries [1-13]. The main reason is that within such kind of plasmas electrons may attain energy in excess of several eV in comparison with the heavy particles present in the discharge. This unusual thermal non-equilibrium situation is very efficient in molecular fragmentation through electron impact. This feature of such plasmas gives rise to a very reactive chemistry in a relative cold environment whose kinetics is not easily controlled [1-9]. Therefore it is essential to perform different plasma diagnostics in order to probe the trends of chemical species as well as the electrons behavior for different plasma parameters, such as RF power coupled to the plasma chamber, gas pressure, gas flow and so on [2,3,7].

In the field of biomaterial science and technology, plasma polymerized polyethylene glycol dimethyl ether has been attracting the attention of the scientific community due to its non-fouling properties [14-19]. If appropriate plasma parameters are set, polymer films can be synthesized keeping

[3] To whom the correspondence should be addressed

quite similar molecular structure to that of polyethylene oxide-like (PEO-like) with the advantage that these films are not soluble in water. The aqueous solubility of PEO makes it inappropriate for many biomaterial applications. In order to retain the monomer structure within the plasma deposited films and consequently its functionality, many different issues have been addressed in recent literature as for instance: film deposition under low mean RF power level by controlling the power supply on/off ratio [19,20], decreasing of monomer residence time and consequently the reduction of its interaction with the plasma environment [20], cooling of substrate with liquid nitrogen [21], energy reduction of the ions reaching the substratum [4], and so on.

In order to determine the appropriate experimental parameters that would result in customized hydrophilic transparent film in plasma polymerization of PEO-like coatings this paper deals with plasma polymerization of diglyme at relatively low RF power levels using silicon and quartz plates as substrates. Molecular structure, optical properties and the hydrophilic/hydrophobic characteristics of plasma polymerized diglyme at 5 W of RF power and for different values of pressure within plasma chamber, were characterized using FTIR, ultraviolet spectroscopy, and contact angle measurements respectively. The optical characteristics of plasma polymerized diglyme were obtained by measurements of refraction index, absorption coefficient and optical gap

2. Experimental setup

Diglyme ($CH_3O(CH_2CH_2O)_2CH_3$) plasmas were excited by a RF power supply operating in 13.56 MHz whose output power could be adjusted from 0 to 300 W (Tokyo HY-Power model RF-300™). The RF power was coupled to the plasma reactor through an appropriate matching network (Tokyo HY-Power model MB-300™) that allows minimizing the reflected RF power. All experiments were run with the RF output kept fixed on 5 W in diglyme atmospheres in the pressure range from 16 Pa (120 mTorr) to 60 Pa (450 mTorr) within a stainless steel cylindrical, 210 mm of internal diameter and 225 mm long, parallel plate electrodes plasma reactor. This reactor is provided with eight lateral entrances, positioned at the mid plane between the electrodes, which can be used for setting optical, electrical and mass diagnostics and the low (mechanical pump) and high (turbo-molecular pump) vacuum systems. The vacuum inside the plasma chamber is monitored by pirani™ (thermocouple) and penning™ (inverse magnetrom) gauges. The turbo-molecular pump is connected to the chamber through a gate valve and is used for cleaning purposes. The pressure is pumped down to 1.33×10^{-4} Pa (10^{-6} Torr), and the chamber is purged with argon several times before each experiment. The inner side of the plasma chamber was polished up to the optical quality (roughness of 0.5 microns or less) in order to minimize the retention of impurities and facilitate the cleaning process. The plasma chamber walls were heated with a temperature-controlled belt in order to minimize the monomer's condensation as well as the humidity. Diglyme from a stainless steel bottle was fed into the plasma chamber through a needle valve. The mass spectrometry was realized using a mass spectrometer and energy analyzer (Hiden Analytical model EQP-300™), operating in the mass and energy range from 1 to 300 amu and from 0 to 100 eV, respectively. The background mass spectra with the plasma turned off were carried out at the same pressure conditions of each diglyme plasma polymerization processes. These spectra were used to control the number of chemical species resulting from the fragmentation within the mass spectrometer and also to detect the presence of contaminants resulting from the plasma chamber cleaning process. The block diagram and the photograph of the experimental setup are presented in figures 1 and 2, respectively.

Diglyme films were deposited on silicon and quartz substrates for one hour for all pressure conditions. Film thickness was determined through step meter technique using an Alpha Step Tencor-100™. Typical polymer film thickness varied from 100 nm to 140 nm. Diglyme films for IR and ultraviolet-visible (UV-Vis herein) analysis were deposited on glass plates previously covered with aluminium films and quartz plates, respectively. FTIR and UV-Vis spectra were obtained using Jasco FTIR-100™ and Perkin Elmer Lambda-9™ spectrometers, respectively. The hydrophobic/hydrophilic characteristics of plasma-polymerized Diglyme were investigated through contact-angle measurements using water as probe liquid and the Ramé-Hart goniometer model 100™.

Journal of Physics: Conference Series **167** (2009) 012053 doi:10.1088/1742-6596/167/1/012053

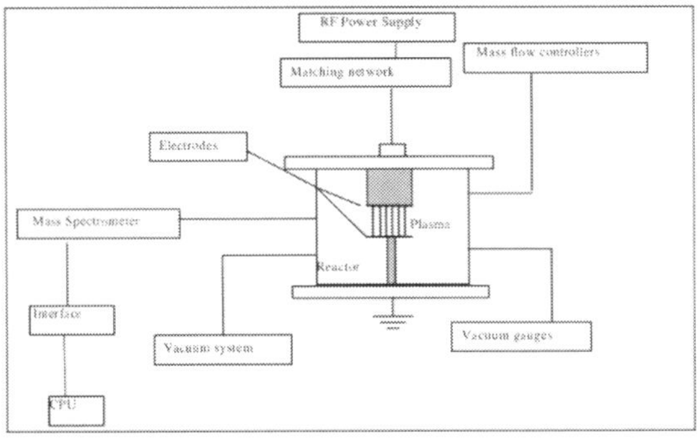

Figure 1. Block diagram of the experimental setup used in plasma polymerization of diglyme films.

Figure 2. Photograph of the experimental arrangement.

3. Results and Discussion

Figure 3 shows typical infrared spectra of diglyme plasma polymers. The absorption bands at 2960 cm^{-1} and 2900 cm^{-1} are associated to C-H stretching modes in CH_3 and CH_2 groups, respectively. The absorption band between 1730 cm^{-1} and 1650 cm^{-1} is due to C=O stretching. The absorptions observed in the wavenumber range from 1440 cm^{-1} to 1380 cm^{-1} are due to C-H bending modes. The strong band in the range from 1200 cm^{-1} to 1000 cm^{-1} is associated with C-O and C-O-C stretching modes. The IR spectra of plasma polymerized diglyme films demonstrate the preservation of some functional groups of diglyme monomer's structure, i.e. $CH_3O(CH_2CH_2O)_2CH_3$. The absorption peak due to O-H stretching near 3500-3200 cm^{-1} indicates the presence of the polar radical OH incorporated into the polymer structure at the pressure of 59 Pa (440 mTorr). This behavior may be understood if one takes into account the spectrum of the monomer Diglyme shown in figure 4. It can be seen a strong absorption peak due to O-H stretching from the water molecules incorporated to the monomer.

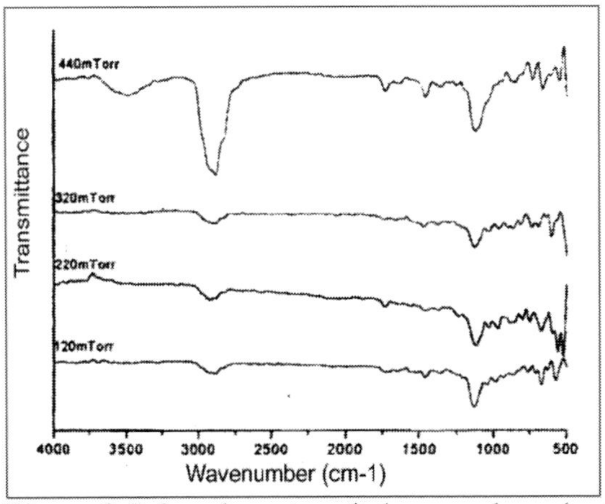

Figure 3. Infrared spectra of plasma polymerized diglyme films for different values of gas pressure.

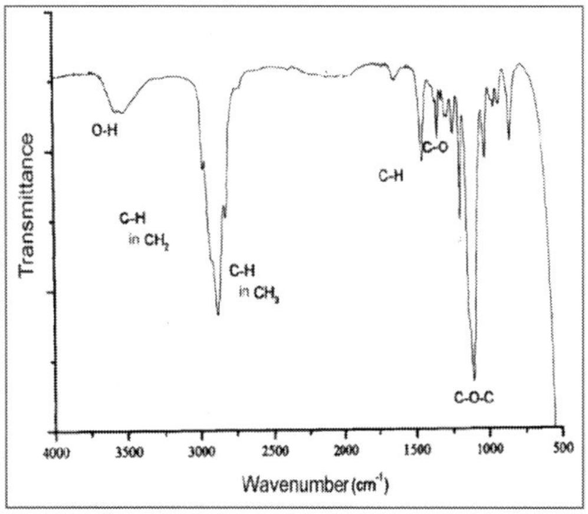

Figure 4. Infrared spectra of diglyme monomer.

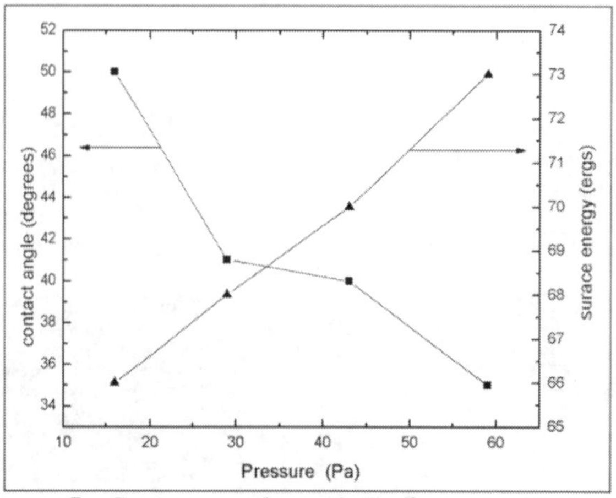

Figure 5. Contact angle and Surface energy versus pressure for plasma polymerized diglyme films.

The comparison between IR spectra in figures 3 and 4 allows us to conclude that for RF power kept fixed on 5 W plasma discharges were favorable for the retention of main monomer's functional groups on plasma polymerized Dyglime films. These results show that the PEO-like characteristics of Diglyme films were preserved for the pressure range investigated in the experiment.

The results presented in figure 5 show that the water contact angle decreases monotonically with the increasing of pressure with the corresponding increasing of surface energy. This is in close agreement with the IR spectra that show the incorporation of O-H radical into the plasma polymerized diglyme film with the increasing of pressure. It also should be stressed that the increasing of pressure enhances the PEO-like characteristics of the films due the reduction of the electron mean free path within the discharge which favors the electronic elastic scattering [2,3].

The refractive index and the optical-absorption coefficient have been determined for plasma polymerized diglyme films deposited at different values of gas pressure. The calculations were performed from the data of the UV-Vis transmission spectra using the procedure outlined in [22,23]. The refractive index of about 1.5, calculated for photon energy of 2.0 eV, remained approximately constant for plasma polymerized diglyme films deposited at different values of gas pressure inside the plasma chamber. These results show that these films may be used as anti-reflecting protective coatings in ophthalmic industries. The optical gap of the plasma polymerized Diglyme films at a fixed relative low value of RF power (5 W) for pressure ranging from 16 Pa to 60 Pa remained approximately constant around 3.8 eV.

4. Conclusions

Plasma polymerized diglyme films deposited at 5 W RF power and monomer pressure varying from 16 Pa (120 mTorr) to 60 Pa (450 mTorr) are transparent to visible light and therefore can be used in ophthalmic industries as protective anti-reflecting coatings. The IR analysis of polymeric films revealed that the operation of the plasma chamber at relative low RF power level was successful in producing plasma polymerized diglyme whose structure resembles the monomer's one. The increasing of the gas pressure inside the plasma chamber favours the fragmentation of the monomer by electron impact and the recombination processes keeping many monomers' functional groups on the polymeric films structure. These deposition conditions enhanced the PEO-like characteristics of plasma polymerized diglyme films keeping the refractive index and the optical gap approximately constant around 3.8 eV. Therefore one can conclude that the careful setup of plasma diagnostics such as mass spectrometry and optical spectroscopy result in thin plasma polymerized diglyme films with customized optical properties.

Acknowledgments
The authors would like to thank FAPESP, FUNDUNESP and CNPq for financial support. José Benedito Galhardo is acknowledged for technical assistance during the plasma chamber assembling.

References

[1] d'Agostino R, Favia P, Oehr C and Wertheimer M R 2005 Low-temperature plasma processing of materials: past, present and future *Plasma Process. Polym.* **2** 7-15
[2] d'Agostino R, Favia P, Kawai Y, Ikegami H, Sato N and Arefi-Konsari F 2008 (eds) *Advanced Plasma Technology* (Weinheim, Germany: Wiley-VCH)
[3] Hippler R, Kersten H, Schmidt M and Shoenbach K H 2008 (eds) *Low Temperature Plasmas: Fundamentals, Technologies and Techniques* (Weinheim, Germany: Wiley-VCH)
[4] Biederman H 2004 (ed) *Plasma Polymer Films* (London: Imperial College Press)
[5] Bogaerts A, Neyts E, Gijbels R, van der Mullen J 2002 *Spectrochim. Acta* B **57**, 609
[6] Chu P K, Chen J Y, Wang L P, Huang N 2002 *Mat. Sci. Eng. Rep.* **36**, 143-206
[7] Shul R J, Pearton S J 2000 (eds.) *Handbook of Advanced Plasma Processing Techniques*, (Springer, Berlin)
[8] Inagaki N 1996 *Plasma Surface Modification and Plasma Polymerization*, (Technomic Publishing Company, Inc., USA)
[9] Yasuda H 1985 *Plasma Polymerization*, (New York: Academic Press, Inc., USA)
[10] Roth J R 1995 *Industrial Plasma Engineering: Principles*, v. 1, (UK: IOP Publishing)
[11] Roth J R 2001 *Industrial Plasma Engineering: Applications*, v. 2, (UK: IOP Publishing)
[12] d'Agostino R 1990 (ed.) *Plasma Deposition, Treatment and Etching of Polymers*, (New York: Academic Press)
[13] d'Agostino R, Favia P, Fracassi F 1997 (eds.) *Plasma Processing of Polymers, NATO ASI series E: Applied Sciences vol. 346,* (Dordrecht, The Netherlands: Kluwer Academic Publishers)
[14] Alcantar N A, Aydil E S, Israelachvili J N 2000 *J. Biomed. Mat. Res.* **51**, 343
[15] Favia P, d'Agostino R 1998 *Surf. Coat. Technol.* **98**, 1102
[16] Hendricks S K, Kwok K, Shen M, Horbett T A, Ratner B D, Bryers J D 2000 *J. Biomed. Mat. Res.* **50**, 160
[17] Sardella E, Gristina R, Senesi G S, d'Agostino R and Favia P 2004 Homogeneous and micropatherned plasma-deposited PEO-like coatings for biomedical surfaces *Plasma Process. Polym.* **1** 63-72
[18] Sardella E, Favia P, Gristina R, Nardulli M and d'Agostino R 2006 Plasma-aided micro- and nanopatterning processes for biomedical applications *Plasma Process. Polym.* **3** 456-69
[19] Cicala G, Creatore M, Favia P, Lamendola R, d'Agostino R 1999 *Appl. Phys. Lett.* **75**, 37
[20] Oehr C, Müller M, Elkin B, Hegemann D, Vohrer U 1999 *Surf. Coat. Technol.* **116-119**, 25
[21] Biederman H, Slavínská D 2000 *Surf. Coat. Technol.* **125**, 371
[22] Durrant E, Mota R P, Moraes M B 1992 *J. Appl. Phys.* **71**, 448
[23] Cisneros J L, Rego G B, Tomyiama M, Bilac S, Gonçalves J M, Rodriguez A E, Argüello Z P 1983 *Thin Solid Films* **100**, 155

Amorphous hydrogenated carbon films treated by SF_6 plasma

N M S Marins[1], R P Mota[1], D C R Santos[1], R Y Honda[1], M E Kayama[1], K G Kostov[1], M A Algatti[1], N C Cruz[2], E C Rangel[2]

[1] Laboratório de Plasma, Faculdade de Engenharia, UNESP, Av. Dr. Ariberto Pereira da Cunha-333, 12516-410, Guaratinguetá, SP, Brazil.
[2] Laboratório de Plasmas Tecnológicos, Unidade Diferenciada Sorocaba/Iperó, UNESP, Av. Três de Março-511, 18085-180, Sorocaba, SP, Brazil.

E-mail: nazir@feg.unesp.br

Abstract. This work was performed to verify the chemical structure, mechanical and hydrophilic properties of amorphous hydrogenated carbon films prepared by plasma enhanced chemical vapor deposition, using acetylene/argon mixture as monomer. Films were prepared in a cylindrical quartz reactor, fed by 13.56 MHz radiofrequency. The films were grown during 5 min, for power varying from 25 to 125 W at a fixed pressure of 9.5 Pa. After deposition, all samples were treated by SF_6 plasma with the aim of changing their hydrophilic character. Film chemical structure investigated by Raman spectroscopy, revealed the increase of sp^3 hybridized carbon bonds as the plasma power increases. Hardness measurements performed by the nanoindentation technique showed an improvement from 5 GPa to 14 GPa following the increase discharge power. The untreated films presented a hydrophilic character, which slightly diminished after SF_6 plasma treatment.

1. Introduction

Amorphous hydrogenated carbon (a-C:H) films have attracted great interest due to their promising properties, such as high hardness, wear resistance, low friction coefficient, chemical inertness and biocompatibility [1-6]. They are usually applied as protective coating for optical windows, hard disk and fibers [7], magnetic media [8], machine parts [9], biomedical coating and biosensors [10]. These films are usually synthesized by glow discharge techniques as Plasma Enhanced Chemical Vapor Deposition (PECVD), and their properties depend on the conditions of the process such as discharge power, excitation frequency, deposition time, monomer kind and pressure, among others. The combination of these parameters determines the chemical structure of the films, whose sp^3/sp^2 proportion rate is responsible for their mechanical, chemical and tribological properties [11,12]. Plasma treatments are interesting because they alter the surface properties and preserve the bulk ones. Fluorine-containing gases are widely used for etching processes and plasma treatments to modify the conventional polymer surfaces [13,14], for instance, to produce hydrophobic polymers used in packing and non-sticking applications [15,16]. This work aimed to verify the hardness and wettability of a-C:H films prepared from acetylene/argon mixture plasma at different radiofrequency (RF) power. Following the deposition, the films were submitted to SF_6 plasma with the purpose of modifying their hydrophilic character.

2. Materials and Methods

Films were prepared in a cylindrical quartz reactor, fitted with two parallel plate electrodes. In this equipment, the upper electrode is connected to 13.56 MHz RF power supply and the lower one is grounded. Plasmas were established in a mixture of acetylene and argon, respectively in the proportion of 30% and 70% related to the total gas pressure fixed at 9,5 Pa. Deposition time was 5 min and the excitation power varied from 25 to 125 W. Films were deposited onto glass and silicon substrates, placed on the lower electrode. The substrates temperature was kept fixed at 25 ^{0}C during the deposition process. After the deposition, the samples were treated with SF_6 radiofrequency plasmas (13.56 MHz) for 5 min. This treatment was performed in a stainless steel cylindrical vessel, fitted with two plate electrodes, using 70 W of power and 13.3 Pa of pressure. Film chemical structure and composition were investigated by Raman spectroscopy in a Renishaw Micro-Raman Spectrometer, with argon-ion laser excitation wavelength at 514 nm. Thickness measurements were carried out by Alpha-Step Tencor profilemeter, and hardness was evaluated by a Nanoindenter Hysitron Triboindenter. Film wettability was evaluated by contact angle technique, using a Ramé-Hart 100-00 Goniometer. Such analysis was performed before and after the SF_6 plasma treatment.

3. Results and discussions

Figure 1 shows Raman spectra of the a-C:H films deposited at discharge power ranging from 25 W to 125 W. As can be seen, the two bands at approximately 1350 cm^{-1} and 1570 cm^{-1} fitted with Gaussian curves correspond to D and G peaks, respectively. The D peak is associated with sp^3 hybridized carbon bond and G peak is related to sp^2 state. This figure indicates that a-C:H films is predominantly formed by sp^2 bonding. Both peak positions shift towards lower wavenumbers as the power increased, which suggests an increase in the sp^3 bonding [17]. In this sense, we calculated the ratio of D and G peak areas, I_D /I_G, which is depicted in figure 2. As can be seen in this figure, the I_D /I_G intensity ratio improved from 0.25 to 0.55 as the power increased. This finding confirms the rise in the sp^3 states and can be explained by processes involved in plasma polymerization. It is known that higher discharge power induces a greater molecular fragmentation of the precursor mixture (acetylene and argon), and the film grows through diverse recombination between dangling bonds and free radicals. The probability of C-H bond breakage is higher than C-C ones and it leads to loss of hydrogen atoms and consequently film carbonization. Besides, during the deposition, the film is also bombarded by all species presented in the discharge, promoting recombination via chain crosslinking. All these processes are intensified by heavy species as argon ions, which is abundant into the discharge [17].

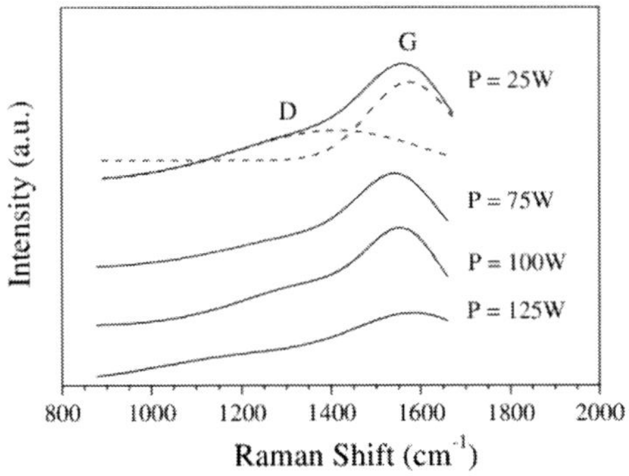

Figure 1. Raman spectra of samples deposited at different applied powers, prior to SF_6 treatment.

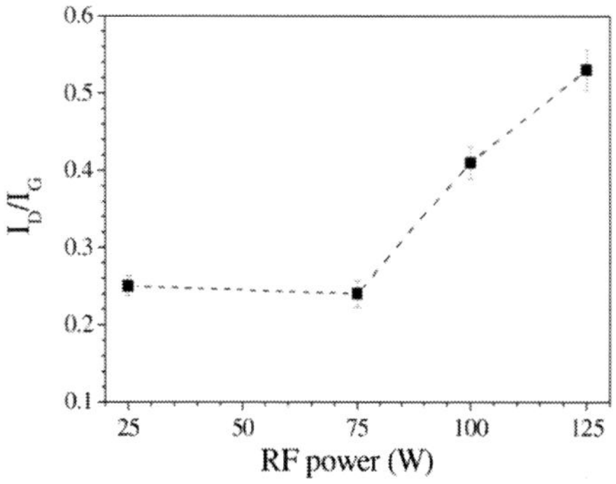

Figure 2. I_D/I_G ratio of the films as a function of the deposition power.

In figure 3 the hardness of the films is plotted as a function of the discharge power. The hardness of the a-C:H film deposited at 75 W is about the same (within the error bar) as the hardness of the films deposited at 25 W (5 GPa), and attains larger values at 100 W (14 GPa) and 125 W (12 GPa). This is related to I_D/I_G ratio of the films, which is the same at lower powers, however increases for power above 75 W. As explained before, I_D/I_G rate is related to sp^3 groups, which control the mechanical properties, therefore the increase of power promoted the hardness improvement. It is believe that long deposition time as well as high discharge power cause increase of the substrate temperature and consequently, the structure hardness is reduced. It is important to note that a-C:H films present hardness values ranging from 9 to 16 GPa [18] and the hardness of the polymer films deposited from acetylene-argon mixtures achieves 14 GPa.

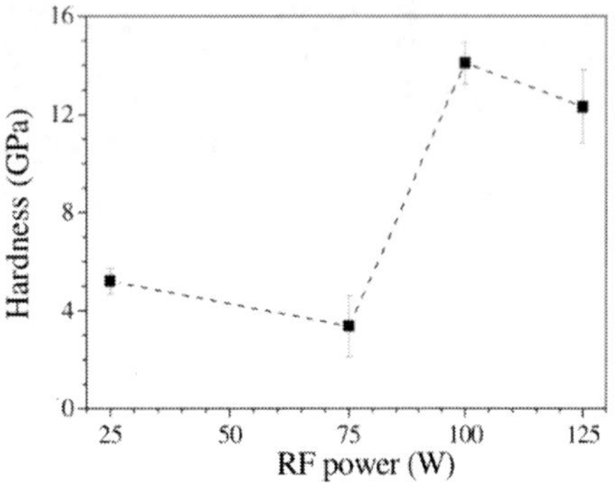

Figure 3. Film hardness as a function of the discharge power.

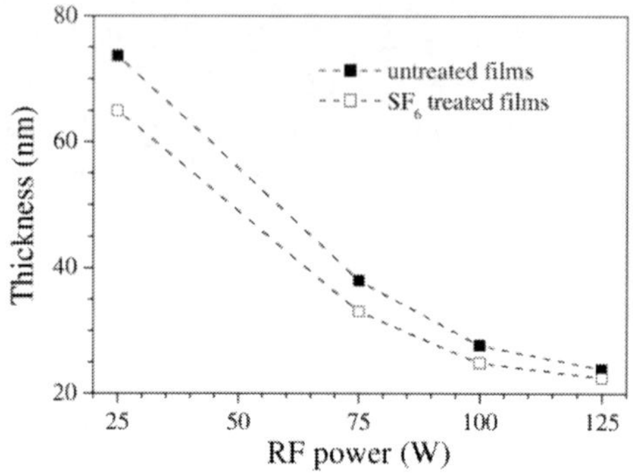

Figure 4. Film thickness as a function of the RF power.

Figure 4 shows the thickness of the a-C:H films treated and untreated by SF_6 plasma, as a function of the discharge power. The thickness of untreated sample diminished from 74 to 24 nm with increasing the power, probably due to sputtering processes caused by argon ions. During the collisions, these heavy ions can transfer enough energy to remove a portion of the film surface increasing the sputtering effect. After SF_6 treatment, the thickness present the same tendency as the power increases, however their values are slightly lower than the untreated sample thickness. In this case, the highly reactive fluoride atoms are responsible for etching process. Ablation processes due to creation of volatile groups which are taken to the plasma also contribute for thickness reduction. The etching effect is more intense for soft films, as can be observed in the case of the sample deposited at 25 W.

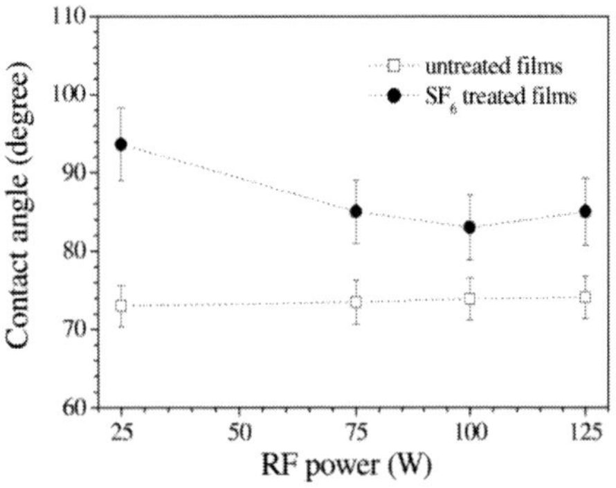

Figure 5. Contact angle of the untreated and treated films as a function of the RF power

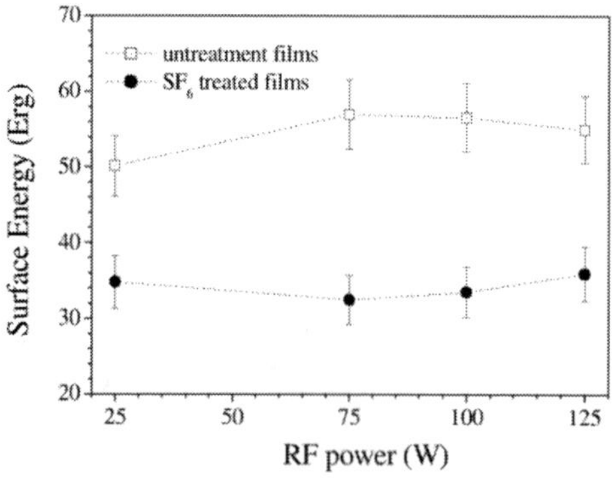

Figure 6. Surface Energy of the untreated and treated films as a function of the power.

Figure 5 shows the contact angle measurements of the untreated and SF_6 treated films. As can be seen, the contact angle measurements of the as-deposited films kept the same value around 73° independently on the deposition power. This result is consistent with surface energy measurements which present a constant value of 58 Erg, as can be seen in the figure 6. Although the film chemical structure was altered by different RF power, the hydrophilic character of the untreated films was preserved. After SF_6 treatment, there was an increase of the contact angles, mainly for the sample deposited at the lower power (94°). Probably, the fluorine atoms removed hydrogen atoms from the polymeric chain and recombine with dangling bonds. Besides, the fluorine atoms can be linked to free radicals trapped in the film structure. These processes promoted the loss of some polar groups and the film acquired a slightly hydrophobic character. The quantity of free radicals of plasma polymerized films depends on the chain mobility. Thus, the samples deposited at higher power may not have enough mobility to allow recombination with free radicals, therefore their contact angle was around 84°. In figure 6 it can be seen that the surface energy of the films decreased from approximately 54 Erg to 36 Erg after the treatment, but they are practically the same for the films deposited at different RF powers, in spite of the structural differences among them.

4. Conclusions

The plasma polymerized films deposited from acetylene-argon mixture presented a structure predominantly containing sp^2 hybridized bonds, but the sp^3 groups tend to rise as the discharge power increases from 75 to 125 W. This alteration of the chemical structure promoted the hardness improvement from 5 GPa to 14 GPa, which characterize these polymers as a-C:H films. The thickness of the untreated films decreased with the deposition time, probably due to sputtering and substrate heating. The SF_6 treatment also diminished the thickness, probably due to etching effect, and the films become less hydrophilic than the untreated film.

Acknowlegments

The authors thank the CAPES, CNPQ, FAPESP and FUNDUNESP for financial support and the National Institute for Space Research (INPE) for technical assistance.

References

[1] Robertson J 2002 *Mat. Sci. Eng. R* **37** 129
[2] Robertson J 2001 *Thin Solid Films* **383** 81
[3] Tainen V 2001 *Diam. Relat. Mater.* **10** 153
[4] Field S K, Jarrat M and Teer D G, 2004 *Tribology International* **37** 949
[5] Liu F-X, Yao K-L and Liu Z-L 2007 *Diam. Relat. Mater.* **16** 1746
[6] Zaharia T, Sullivan J L, Saied S O, Bosch R C M and Bijker M D 2007 *Diam. Rel. Mat.* **16** 623
[7] Dowling D P, Donnelly K, O'Brien T P, O'Leary A, Kelly T C and Neuberger W 1996 *Diam. Relat. Mater.* **5** 492
[8] Schlatter M 2002 *Diam. Relat. Mater.* **11** 178
[9] P. Beer, M. A. Djouadi, R. Marchal, A. Sokolowska, M. Lambertin, A. Czyzniewski and W. Precht 1999 Vacuum **53** 363
[10] Shi B, Ajayi O, Fenske G, Erdemir A and Liang H 2003 Wear **225** 1015
[11] Bull S J 1995 *Diam. Relat. Mater* **4** 5-6 827
[12] Chu P K and Li L 2006 *Mat. Chem. And Phys.* **96** 243
[13] Jiang M and Ning Z 2006 *Surf. Coat. Tech.* **200** 3682
[14] Katoh M, Ohte T, Kojima A and Ohtani S 1994 *J. Photopolymer Sci. and Tech.* **7** 2 319
[15] Prat R, Koh Y J, Babukutty Y, Kogoma M, Okazaki S and Kodama M 2000 *Polymer* **41** 7355
[16] Korotkov R Y, Goff T and Ricou P 2007 *Surf. Coat. Tech.* **201** 7207
[17] Rangel E. C., Cruz N. C., Kayama M. E., Rangel R. C. C., Marins N. and Durrant S. F. 2004 *Plasma and polymers* **9** 1
[18] Liang J H, Chen M H, Tsai W F, Lee S C, Ai C F 2007 *Nucl. Instrum. Meth. B* **257** 696

Journal of Physics: Conference Series **167** (2009) 012055 doi:10.1088/1742-6596/167/1/012055

Plasma-polymerized hexamethyldisilazane treated by nitrogen plasma immersion ion implantation technique

R Y Honda[1], R P Mota[1], R G S Batocki[1], D C R Santos[1], T Nicoleti[1], K G Kostov[1], M E Kayama[1], M A Algatti[1], N C Cruz[2], L Ruggiero[3]

[1] Laboratório de Plasma, Faculdade de Engenharia, UNESP, Av. Dr Ariberto Pereira da Cunha-333, 12516-410, Guaratinguetá, SP, Brazil.

[2] Laboratório de Plasmas Tecnológicos, Unidade Diferenciada, UNESP, Av. Três de Março-511, 18085-180, Sorocaba, SP, Brazil.

[3] Faculdade de Ciências, UNESP, Av. Luis E. Carrijo Coube 14-1, 17033-360, Bauru, SP, Brazil.

E-mail: honda@feg.unesp.br

Abstract. This paper describes the effect of nitrogen Plasma Immersion Ion Implantation (PIII) on chemical structure, refraction index and surface hardness of plasma-polymerized hexamethyldisilazane (PPHMDSN) thin films. Firstly, polymeric films were deposited at 13.56 MHz radiofrequency (RF) Plasma Enhanced Chemical Vapour Deposition (PECVD) and then, were treated by nitrogen PIII from 15 to 60 min. Fourier Transformed Infrared (FTIR) spectroscopy was employed to analyse the molecular structure of the samples, and it revealed that vibrations modes at 3350 cm^{-1}, 2960 cm^{-1}, 1650 cm^{-1}, 1250 cm^{-1} and 1050 cm^{-1} were altered by nitrogen PIII. Visible-ultraviolet (vis-UV) spectroscopy was used to evaluate film refractive index and the results showed a slight increase from 1.6 to 1.8 following the implantation time. Nanoindentation revealed a surface hardness rise from 0.5 to 2.3 GPa as PIII treatment time increased. These results indicate nitrogen PIII is very promising in improving optical and mechanical properties of PPHMDSN films.

1. Introduction

PIII is a highly efficient technique for surface modification of metal, semiconductor and polymeric materials [1,2,3,4,5] and, it has been applied in different areas, such as aeronautic, electronic, optical and biomedical. In this implantation process, the samples are immersed in gas discharge and are polarized with high negative voltage pulses. Plasma ions are accelerated toward the samples and implanted into their surfaces. This bombardment promotes bond breakage and recombination processes on chemical structure of the target surface and consequently, it causes modifications in its chemical, optical, electrical, biological, mechanical and tribological properties [6,7,8]. However, the degree of these modifications is strongly dependent on plasma parameters (frequency, discharge power, gas pressure, gas specie), pulse characteristics (amplitude, repetition rate and work time) and target material [8]. In this work, plasma-polymerized HMDSN were used as target, because they are interesting to many applications, such as optical, protective and biocompatible coatings [9,10,11,12]. Refractive index and surface hardness of HMDSN films treated by nitrogen ions were correlated to molecular chemical modification induced by PIII technique.

2. Materials and Methods

PPHMDSN films, with 295 nm average thickness, were prepared by PECVD in a cylindrical stainless steel reactor fitted with two parallel plate electrodes. RF power supply (13.56 MHz, 50 W) was connected to upper electrode and the lower one, used as substrate holder, was grounded. HMDSN vapour pressure was 40 mTorr and deposition time was 30 min. PIII was performed in the same reactor, fixed at the lower electrode to a high voltage pulser. Nitrogen plasma of 50 W power and 40 mTorr pressure was established while negative pulses of 30 kV amplitude and 100 Hz repetition rate were applied. PIII treatment time varied from 15 to 60 min. Thickness measurements were carried out by Alpha-Step Tencor profilemeter. Molecular chemical structure of the film surfaces was investigated by Fourier Transformed Infrared (FTIR), using a Perkin Elmer FTIR 1600 spectrometer. Refractive index was analysed by visible-ultraviolet (vis-UV) using a Hitachi U-3501 spectrometer and hardness was evaluated by a nanoindenter Hysitron Triboindenter.

3. Results and discussions

Figure 1 shows the IR spectra of PPHMDSN films submitted to different time of nitrogen PIII. As can be noticed, as-deposited film spectrum is characterized by C–H stretching at 2960 cm^{-1} and 2900 cm^{-1}, Si–CH_3 stretching at 1250 cm^{-1}, Si–O stretching at 1050 cm^{-1}, Si–N stretching at 940 cm^{-1}, Si–$(CH_3)_x$ stretching at 840 cm^{-1} and Si–C stretching at 800 cm^{-1}. Although HMDSN monomer contains no oxygen (O), free radicals trapped in PPHMDSN film structure react with atmosphere O during the deposition process or when the samples are removed from the reactor [3,7]. Therefore, O-containing groups usually appear into PP film molecular chain. Observing IR spectra of the implanted films, it can be seen that some of the original PPHMDSN bands were intensified with the rise of nitrogen PIII treatment time. Besides, there was the appearance of three new absorption bands: O–H stretching at 3400 cm^{-1}, Si–H stretching at 2130 cm^{-1} and the band centered around at 1650 cm^{-1}. The later is attributed to unsatured carbon bonds, such as C=O and C=N. It is possible that N–H stretching at 3350 cm^{-1} and N–H bending at 1650 cm^{-1} are overlapped by large bands, because N ions are highly reactive and are introduced into the polymeric structure. Chemical bonds and their related absorption peaks are shown in Table 1.

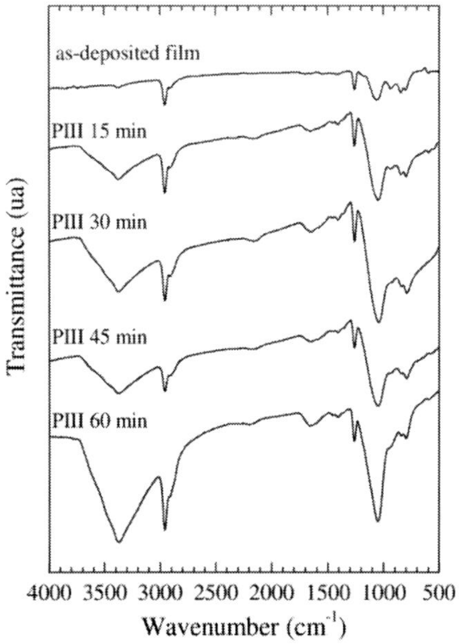

Figure 1. IR spectra of PPHMDSN films treated by nitrogen PIII for different time.

Table 1. PPHMDSN film infrared spectra vibration modes.

Vibration modes	Absorption peak (cm^{-1})
O–H stretching	3400
N–H stretching	3350
C–H stretching in CH_3 groups	2960
C–H stretching in CH_2 groups	2900
Si–H stretching	2130
C=O stretching	1700
N–H bending in NH_2 groups / C=C stretching	1650
C=N /C=O bending	1600
CH_3 bending in Si–$(CH_3)_x$ groups	1230
Si–O stretching	1050
Si–N stretching	940
Si–$(CH_3)_x$ rocking – stretching	850
Si–C stretching	800

These molecular chemical modifications are attributed to three concomitants effects of ion implantation processing: chain breakage, unsaturation and crosslinking. It is known that the impact of energetic ions with a target material causes chemical bond breakages and promotes emission of atomic and molecular species [8]. In polymeric material, hydrogen (H) bonds are preferentially lost because they belong to chain termination, and H atom progressive emission induces the appearance of dangling bonds. These dangling bonds can recombine through unsaturation processes such as double bond between carbon atoms or between carbon atoms and other species present in the polymeric structure. The presence of C=C, C=O and C=N bonds is suggested around 1650 cm^{-1} band, although it is difficult to distinguish them in the spectra. Chain crosslinking can also occur due to dangling bonds, and it is responsible for the chain mobility decrease and, consequently, for the structure stiffness. These factors increase the density of the samples while their thicknesses diminish. Figure 2 shows PPHMDSN film thickness measurements after and before (t=0) PIII treatments.

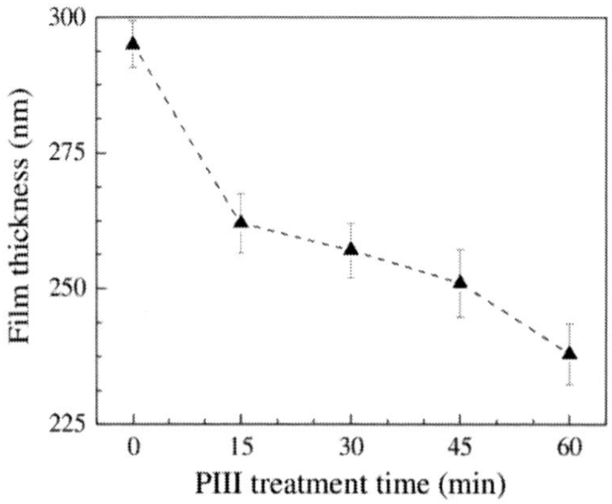

Figure 2. Thickness of PPHMDSN films as a function of implantation time.

As can be seen in figure 2, the thickness of the films diminished from 295 nm to 238 nm as the treatment time increased. This can confirm the density rise of the samples, although the thickness decrease can be also related to sputtering processes frequently observed in ion implantation treatments [8]. The density increase promotes alterations in the optical properties of the samples, which was evaluated through refractive index measurements show in figure 3. In that figure, it can be noticed a slight refractive index enhancement from 1.65 to 1.78. Moreover, the density rise is directly related to hardness improvement. Figure 4 shows the hardness measurements performed at 70 nm depth, as a function of treatment time. As can be seen, the PPHMDSN film hardness rose from 0.5 to 2.3 GPa as nitrogen PIII treatment time increased. It corroborates the unsaturated and crosslinking process enhancement suggested before. In this sense, it is necessary to emphasize that ion implantation usually promotes modifications into the film surface, however IR and vis-UV analyses were performed in the film bulk. Therefore, any alteration in these analyses shows that PIII is an effective technique for surface modifications.

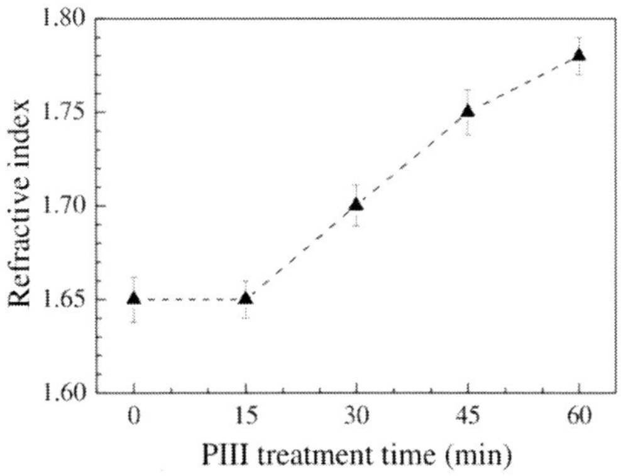

Figure 3. Refractive index of PPHMDSN films measured at a wavelength of 628.34 nm as a function of nitrogen PIII processing time.

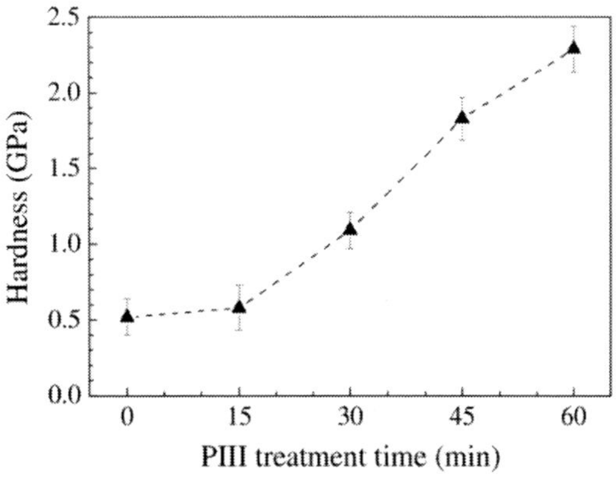

Figure 4. Hardness of PPHMDSN treated by nitrogen PIII, measured at 70 nm depth, as a function of implantation time.

4. Conclusions

The results of IR measurements show the O–containing group and unsatured bond appearance, as well as Si bonds increase, after nitrogen PIII. These molecular chemical modifications induced a rise of the density, refractive index and surface hardness of the PPHMDSN thin films. Nevertheless, the degree of these alterations looks directly dependent on the implantation time. It is attributed to higher amount of insaturation and crosslinking recombination processes at longer treatment time. All analysis performed in this work show nitrogen PIII is efficient to modify the molecular chemical structure of PPHMDSN films and their surface properties.

Acknowledgements

Authors thank Instituto Nacional de Pesquisas Espaciais (INPE) for infrared analyses, Fundação de Amparo a Pesquisa do Estado de São Paulo (FAPESP) and FUNDUNESP for financial support.

References

[1] Poon R W Y, Ho J P Y, Liu X, Chung C Y, Chu P Y, Yeung K W K, Lu W W and Cheung M C 2005 Improvements of anti-corrosion and mechanical properties of NITi orthopedic materials by acetylene, nitrogen and oxygen plasma immersion ion implantation *Nucl. Instrum. Meth. B* **237** 411

[2] Fu R K Y, Cheung I T L, Mei Y F, Shek C H, Siu G G, chu P K, Yang W M, Leng Y X, Huang Y X, Tian X B and Yang S Q 2005 Surface modification of polymeric materials by plasma immersion ion implantation *Nucl. Instrum. Meth. B* **237** 417

[3] Gan B K, Bilek M M M, Kondyurin A, Mizuno K and McKenzie D R 2006 Etching and structural changes in nitrogen plasma immersion ion implanted polystyrene films *Nucl. Instrum. Meth. B* **247** 254

[4] Sui J and Cai W 2006 Mechanical properties and anti-corrosion behaviour of the diamond-like carbon films *Surf. Coat. Tech.* **201** 1323

[5] Yeung K W K, Chan Y L, Lam K O, Liu X M, Wu S L, Liu K Y, Chung C Y, Lu W W, Chan D, Luk K D K, Chu P. K and Cheung K M C 2008 New plasma surface-treated memory alloys: towards a new generation of "smart" orthopaedic materials *Mat. Sci. Eng. C* **28** 454

[6] Burakowski T and Wierzchon T 1999 *Surface Engineering of Metals* (Flórida, CRC Press)

[7] Lawrynowicz Z 1992 *Metallurgy, Heat Treatment, Surface Engineering* 115–117 50

[8] Santos D C R, Rangel R C C, Mota R P, Cruz N C, Schreiner W H and Rangel E C 2004 Modification of plasma polymer films by ion implantation *Mater. Res.* **7-3** 493

[9] Chen C-H, Yang M-R and Wu S-K 2008 Polymerized hexamethyldisilazane coated on equiatomic TiNi shape memory alloy using DC-pulsed plasma assisted chemical vapor deposition *Surf. Coat. Tech.* **202** 2709

[10] Ladwig A, Babayan S, Smith M, Hester M, Highland W, Koch R and Hicks R 2007 Atmospheric plasma deposition of glass coatings on aluminum *Surf. Coat. Tech.* **201** 6460

[11] Wagner N J, Gerberich W W and Heberlein J V R 2006 Thermal plasma chemical vapor deposition of wear-resistant, hard Si-C-N coatings *Surf. Coat. Tech.* **201** 4168

[12] Silva M L P, Tan I H, Nascimento A P, Galeazzo and Jesus D P 2003 Use of plasma polymerized highly hydrophobic hexamethyldisilazane films for sensor development *Sens. Act. B* **91** 362

Journal of Physics: Conference Series **167** (2009) 012056 doi:10.1088/1742-6596/167/1/012056

Polyurethane coating with thin polymer films produced by plasma polymerization of diglyme

M A Ribeiro[1], A S Ramos[1], M I Manfredini[1], H A Alves[1], R Y Honda[2], K G Kostov[2], E. F. Lucena[2], E C T Ramos[1], R P Mota[2, 3], M A Algatti[2], and M E Kayama[2]

[1] UNIVAP, São José dos Campos, SP, Brazil
[2] FEG-DFQ-UNESP, Av. Ariberto Pereira da Cunha 333, 12516-410 – Guaratinguetá, SP, Brazil

E-mail: rmota@feg.unesp.br

Abstract. Aqueous-based polyurethane dispersions have been widely utilized as lubricants in textile, shoes, automotive, biomaterial and many other industries because they are less aggressive to surrounding environment. In this work thin films with different thickness were deposited on biocompatible polyurethane by plasma polymerization process using diethylene glycol dimethyl ether (Diglyme) as monomer. Molecular structure of the films was analyzed by Fourier Transform Infrared spectroscopy. The spectra exhibited absorption bands of O-H (3500-3200cm^{-1}), C-H (3000-2900cm^{-1}), C=O (1730-1650cm^{-1}), C-O and C-O-C bonds at 1200-1600cm^{-1}. The samples wettability was evaluated by measurements of contact angle using different liquids such as water, glycerol, poly-ethane and CMC. The polyurethane surface showed hydrophilic behavior after diglyme plasma-deposition with contact angle dropping from 85^0 to 22^0. Scanning Electron Microscopy revealed that diglyme films covered uniformly the polyurethane surfaces ensuring to it a biocompatible characteristic.

1. Introduction

Medical grade Polyurethanes are one of the most biocompatible and blood-compatible materials known up to present days. These materials are extensively used in the development and manufacturing of medical devices ranging from catheters to artificial organs [1,2]. In spite of its usefulness as biomaterial many drawbacks must be addressed for a tailored application as for instance the hydrophobic characteristic and the hydrolytic instability of many medical-grade polyurethanes. Therefore it is of paramount importance the set up of surface treatments in order to improve the coupling of polyurethanes with the living environment. In this scenario low-pressure plasma material processing may play an important role [3-10]. The key feature of such kind of plasmas is that electrons may attain energy in excess of several eV in comparison with the heavy particles present in the discharge. This unusual non-thermal equilibrium situation is very efficient in molecular fragmentation through electron impact giving rise to a very reactive chemistry in a relative cold environment whose kinetics is not easily controlled [3-5]. Therefore it is essential to perform different plasma diagnostics in order to probe the trends of chemical species as well as the electrons behavior for different plasma

[3] To whom the correspondence should be addressed

parameters, such as RF power coupled to the plasma chamber, gas pressure, gas flow and so on [3-5,9,10].

In the field of biomaterial science and technology, plasma polymerized polyethylene glycol dimethyl ether has been attracting the attention of the scientific community due to its non-fouling properties [11-15]. If appropriate plasma parameters are set, polymer films can be synthesized keeping quite similar molecular structure to that of polyethylene oxide-like (PEO-like) with the advantage that these films are not soluble in water. The aqueous solubility of PEO makes it inappropriate for many biomaterial applications. In order to retain the monomer structure within the plasma deposited films and consequently its functionality, many different issues have been addressed in recent literature as for instance: film deposition under low mean RF power level by controlling the power supply on/off ratio [16,17], decreasing of monomer residence time and consequently the reduction of its interaction with the plasma environment [17], cooling of substrate with liquid nitrogen [18], energy reduction of the ions reaching the substratum [4,5], and so on.

In order to determine the appropriate experimental parameters that would result in customized hydrophilic transparent film in plasma polymerization of PEO-like coatings this paper deals with plasma polymerization of diglyme at relatively low RF power levels using silicon, quartz plates and medical grade polyurethane rectangular pieces of 2 cm^2 as substrates. Molecular structures of plasma polymerized diglyme and polyurethane were characterized using Fourier transform infrared spectroscopy (FTIR here in).

2. Experimental setup

Diglyme ($CH_3O(CH_2CH_2O)_2CH_3$) plasmas were excited by a RF power supply operating in 13.56 MHz whose output power could be adjusted from 0 to 300 W (Tokyo HY-Power model RF-300^{TM}). The RF power was coupled to the plasma reactor through an appropriate matching network (Tokyo HY-Power model MB-300^{TM}) that allows minimizing the reflected RF power. All experiments were run with the RF output kept fixed on 5 W in diglyme atmospheres in the pressure of 20 Pa (150 mTorr) within a stainless steel cylindrical, 210 mm of internal diameter and 225 mm long, parallel plate electrodes plasma reactor. This reactor is provided with eight lateral entrances, positioned at the mid plane between the electrodes, which can be used for setting optical, electrical and mass diagnostics and the low (mechanical pump) and high (turbo-molecular pump) vacuum systems. The vacuum inside the plasma chamber is monitored by pirani^{TM} (thermocouple) and penning^{TM} (inverse magnetrom) gauges. The turbo-molecular pump is connected to the chamber through a gate valve and is used for cleaning purposes. The pressure is pumped down to 1.33 x 10^{-4} Pa (10^{-6} Torr), and the chamber is purged with argon several times before each experiment. The inner side of the plasma chamber was polished up to the optical quality (roughness of 0.5 microns or less) in order to minimize the retention of impurities and facilitate the cleaning process. The plasma chamber walls were heated with a temperature-controlled belt in order to minimize the monomer's condensation as well as the humidity. Diglyme from a stainless steel bottle was fed into the plasma chamber through a needle valve. The mass spectrometry was realized using a mass spectrometer and energy analyzer (Hiden Analytical model EQP-300^{TM}), operating in the mass and energy range from 1 to 300 amu and from 0 to 100 eV, respectively. The background mass spectra with the plasma turned off were carried out at the same pressure condition of diglyme plasma polymerization processes. These spectra were used to control the number of chemical species resulting from the fragmentation within the mass spectrometer and also to detect the presence of contaminants resulting from the plasma chamber cleaning process. The block diagram and the photograph of the experimental setup are presented in figures 1 and 2, respectively.

Diglyme films were deposited on medical grade polyurethane samples, silicon and quartz substrates for one hour. Film thickness was determined through step meter technique using an Alpha Step Tencor-100^{TM}. Typical polymer film thickness was 50 nm. Diglyme and polyurethane structures were studied through infrared spectra obtained with a Perkin Elmer Lambda-9^{TM} spectrometer. The hydrophobic/hydrophilic characteristics of plasma-polymerized Diglyme were investigated through contact-angle measurements using de-ionized water as probe liquid and a Ramé-Hart goniometer model 100™.

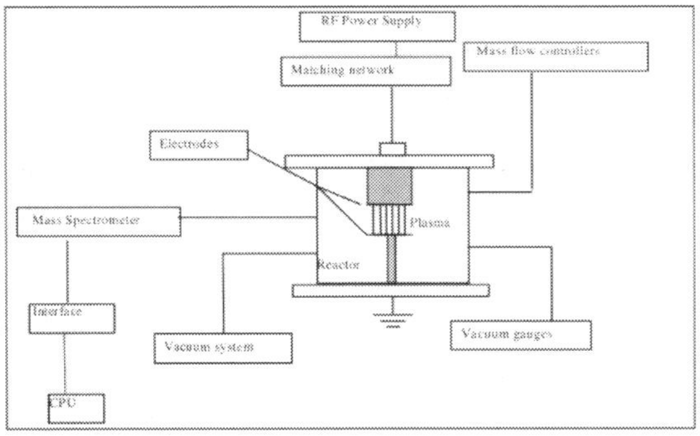

Figure 1. Block diagram of the experimental setup used in plasma polymerization of diglyme films.

Figure 2. Photograph of the experimental arrangement.

3. Results and Discussion

Figure 3 shows typical infrared spectra of diglyme plasma polymers deposited on polyurethane substrate (top spectrum) and polyurethane substrate (bottom spectrum). The absorption band centered in 2897cm^{-1} is associated to C-H stretching modes in CH_3 and CH_2 groups. The absorption band between 1730 cm^{-1} and 1650 cm^{-1} is due to C=O stretching. The absorptions observed in the wavenumber range from 1440 cm^{-1} to 1380 cm^{-1} are due to C-H bending modes. The strong band in the range from 1200 cm^{-1} to 1000 cm^{-1} is associated with C-O and C-O-C stretching modes. The IR spectra of plasma polymerized diglyme films demonstrate the preservation of some functional groups of diglyme monomer's structure, i.e. $CH_3O(CH_2CH_2O)_2CH_3$.

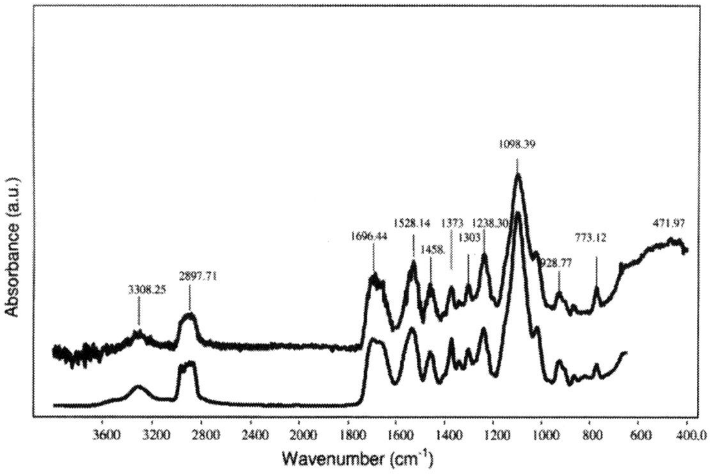

Figure 3. Infrared spectra of plasma polymerized diglyme film deposited on polyurethane surface (top) and polyurethane substrate (bottom).

The accordance between both spectra shows that all IR bands of plasma polymerized diglyme are present in polyurethane spectrum. This means that film thickness was not enough to avoid the appearance of polyurethane bands in diglyme/PUD spectrum. These results allow us to conclude that for the RF power kept fixed on 5 W the plasma discharges were favorable for the retention of the main monomer's functional groups on plasma polymerized Diglyme films. These results show that the PEO-like characteristics of Diglyme films were preserved for the values of RF power and gas pressure used in the experiment.

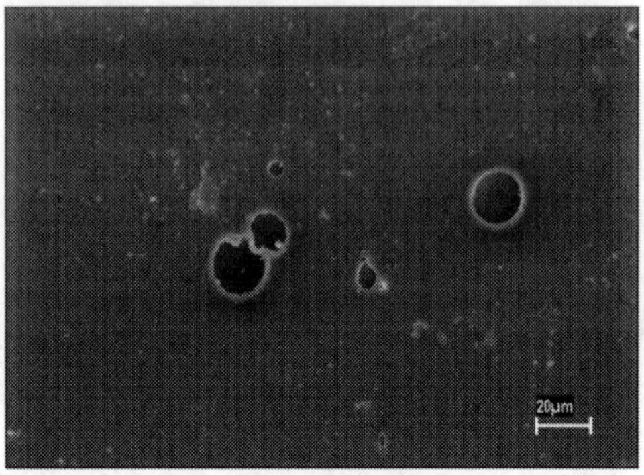

Figure 4. Micrograph 500 times magnified of aqueous-based polyurethane dispersions.

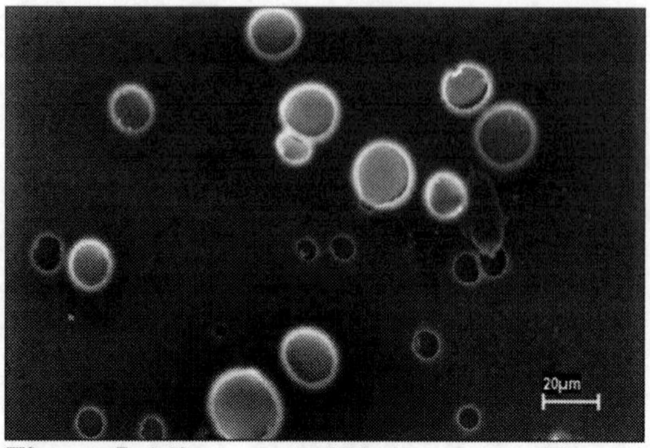

Figure 5. Micrograph 500 times magnified of aqueous-based polyurethane covered with plasma polymerized diglyme films.

Figures 4 and 5 show micrographs of aqueous-based polyurethane dispersions and aqueous-based polyurethane covered with plasma polymerized diglyme films. The pinholes shown in figure 4 are inherently to polyurethane synthesis process. The mean diameter of such pinholes is about tens of microns. It can be seen in figure 5 that plasma polymerized diglyme covers the polyurethane surface as well as the pinholes. The differences on pinholes brightness are due the decreasing of pinhole's depth caused by diglyme deposition. This demonstrates that plasma polymerized diglyme covers uniformly the polyurethane surface. Such features of diglyme films are reinforced if one considers the micrographs shown in figures 6 and 7. It can be seen in figure 7 that diglyme film covers uniformly the smooth glass plate shown in figure 6. Therefore plasma polymerized diglyme may be used as biocompatible protective coating since it is resistant to the aggressive body environment and may be used as a permeable membrane on polyurethane catheter's surface previously embedded with heparin. The contact angle measurements with de-ionized water as a probe liquid revealed a drastic dropping from 85^0 to 22^0 soon after plasma-deposition of diglyme on polyurethane surfaces. This result shows that the experimental condition used in the present work was profitable on the improving of hydrophilic character of polyurethane surface.

Figure 6. Micrograph 5000 times magnified of glass substrate.

Figure 7. Micrograph 5000 times magnified of plasma polymerized diglyme films deposited over glass substrate.

4. Conclusions

Plasma polymerized diglyme films deposited at monomer pressure of 20 Pa (150 mTorr) and 5 W of RF power coupled to the plasma chamber present a PEO-like characteristic that makes it resistant to body environment and may be used as a permeable-selective coatings for controlled releasing of anticoagulant substances like heparin for instance. The IR analysis of polymeric films revealed that the operation of the plasma chamber at relative low RF power level was successful in producing plasma polymerized diglyme whose structure resembles the monomer's one. Therefore one can conclude that the experimental conditions used in this work resulted in thin plasma polymerized diglyme films with customized biocompatible properties.

Acknowledgments

The authors would like to thank FAPESP, FUNDUNESP and CNPq for financial support. José Benedito Galhardo is acknowledged for technical assistance during the plasma chamber assembling.

References

[1] Ratner B D, Hoffman A S, Schoen F J, Lemons J E 2004 (eds) *Biomaterials Science: An Introduction to Materials in Medicine* (Academic Press, New York)

[2] Zdrahala R J, Zdrahala I J 1999 Biomedical Applications of Polyurethanes: A Review of Past Promises, Present Realities, and a Vibrant Future *J. Biomater. Appl.* **14**, 67-90

[3] d'Agostino R, Favia P, Oehr C and Wertheimer M R 2005 Low-temperature plasma processing of materials: past, present and future *Plasma Process. Polym.* **2** 7-15

[4] d'Agostino R, Favia P, Kawai Y, Ikegami H, Sato N and Arefi-Konsari F 2008 (eds) *Advanced Plasma Technology* (Weinheim, Germany: Wiley-VCH)

[5] Hippler R, Kersten H, Schmidt M and Shoenbach K H 2008 (eds) *Low Temperature Plasmas: Fundamentals, Technologies and Techniques* (Weinheim, Germany: Wiley-VCH)

[6] Biederman H 2004 (ed) *Plasma Polymer Films* (London: Imperial College Press)

[7] Bogaerts A, Neyts E, Gijbels R, van der Mullen J 2002 *Spectrochim. Acta* B **57**, 609

[8] Chu P K, Chen J Y, Wang L P, Huang N 2002 *Mat. Sci. Eng. Rep.* **36**, 143-206

[9] Shul R J, Pearton S J 2000 (eds.) *Handbook of Advanced Plasma Processing Techniques*, (Springer, Berlin)

[10] d'Agostino R, Favia P, Fracassi F 1997 (eds.) *Plasma Processing of Polymers, NATO ASI series E: Applied Sciences vol. 346,* (Dordrecht, The Netherlands: Kluwer Academic Publishers)

[11] Alcantar N A, Aydil E S, Israelachvili J N 2000 *J. Biomed. Mat. Res.* **51**, 343

[12] Favia P, d'Agostino R 1998 *Surf. Coat. Technol.* **98**, 1102

[13] Hendricks S K, Kwok K, Shen M, Horbett T A, Ratner B D, Bryers J D 2000 *J. Biomed. Mat. Res.* **50**, 160

[14] Sardella E, Gristina R, Senesi G S, d'Agostino R and Favia P 2004 Homogeneous and micropatherned plasma-deposited PEO-like coatings for biomedical surfaces *Plasma Process. Polym.* **1** 63-72

[15] Sardella E, Favia P, Gristina R, Nardulli M and d'Agostino R 2006 Plasma-aided micro- and nanopatterning processes for biomedical applications *Plasma Process. Polym.* **3** 456-69

[16] Cicala G, Creatore M, Favia P, Lamendola R, d'Agostino R 1999 *Appl. Phys. Lett.* **75**, 37

[17] Oehr C, Müller M, Elkin B, Hegemann D, Vohrer U 1999 *Surf. Coat. Technol.* **116-119**, 25

[18] Biederman H, Slavínská D 2000 *Surf. Coat. Technol.* **125**, 371

Journal of Physics: Conference Series **167** (2009) 012057 doi:10.1088/1742-6596/167/1/012057

DETERMINATION OF THE GELATINIZATION TEMPERATURE OF STARCH PRESENTED IN MAIZE FLOURS

D.F. Coral[1], P. Pineda-Gómez[1,2], A. Rosales-Rivera[1] and M.E. Rodriguez-Garcia[3]

[1]Laboratorio de Magnetismo y Materiales Avanzados, Facultad de Ciencias Exactas y Naturales, Universidad Nacional de Colombia, Manizales, Caldas, Colombia.

[2]Universidad de Caldas, Manizales, Caldas, Colombia

[3]Departamento de Nanotecnología, Centro de Física Aplicada y Tecnología Avanzada, UNAM, Querétaro,Qro., México.

E-mail: `dfcoralc@unal.edu.co`

Abstract. A study of the influence of the moisture and the grain size on the gelatinization temperature (Tp1) of starch from four industrial maize flours as well as an unprocessed maize sample is presented. The gelatinization is a phase transition that can be observed using Differential Scanning Calorimetry (DSC), and it is manifest as an endothermic peak in the DSC thermogram. Results show that for 60, 65, 70 and 75% of moisture, the endothermic peak temperature increases from 70 to 75°C when the moisture increases, and decrease when the grain size increases from 250 to 420μm. A study of the gelatinization enthalpy (ΔHp) is also presented, this is related with the gelatinized starch present in the sample. The enthalpy decrease when the moisture increases and it decrease when grain size is increased.

1. Introduction

A proper understanding of the starch phase transitions is extremely important in food processing. Native starch is not widely used in the food industry due to its poor functional properties, therefore most starches currently incorporated into foods are chemically modified. The use of physical modification techniques such as heat-moisture treatment without destroy its granular structure has been widely applied. The mean thermal transition of the starch is gelatinization, it is used to describe the molecular behavior of starch related with heat and moisture content. In this process the starch changes its semi-crystalline phase to an amorphous phase. In excess of water, the hydrogen bridges are broken allowing water be associated with the free hydroxyl groups. This change, in turn, facilitates its molecular mobility in the amorphous regions and allowing the swelling of the grains [1]. The most important parameter in the gelatinization study is the temperature. In the gelatinization process, To is defined as the initial temperature, Tp1 is the begin of gelatinization or crystal melting and Te is the final temperature. The gelatinization temperature range ΔT = Te-To and the energy necessary to complete the process or gelatinization enthalpy ΔHp are also important. The starch is composed by amylose and amylopectin, both with a semicristaline phase [2, 3]. If there is an excess of water and the temperature increases, crystals of starch are molten cooperatively at Tp1. When the water content is limited, only a few part of crystals are melted by this mechanism, and the rest produce a second transition at high temperature (Tp2) which is in agree with the Flory theory [4]. Results of Differential Scanning Calorimetry (DSC) are presented to determinate gelatinization parameters. The gelatinization process is presented in the DSC as and endothermic peak. The results are used to determinate the width of the endothermic peak (ΔT) and the peak high index (PHI), these parameters are used to determinate the homogeneity and uniformity of the starch gelatinization [5]. This study was made using four different maize flours, and one sample of natural maize with the aim of compare the degree of gelatinized starch in relation with the natural product.

2. Experimental

In order to study the gelatinization transitions in maize starch, four industrial white maize flours produced in Colombia and one sample of native milled maize, were used. Two grain sizes, 250 and 4200 μm, were analized and these were obtained using sieves number 40 and 60. The mass of each sample was 10 mg ± 0.1 with 60, 65, 70 and 75% (w/w) of deionised water. The industrial flours were identified as A, B, C, D and the native milled maize sample was E. For the DSC analysis a Q100 TA Instruments calorimeter was used. The samples were placed in aluminum hermetic sealed pans, and were stabilized at room temperature during 15 minutes, then samples were heated in a ramp at 10°C/min using an empty pan as reference and a nitrogen environment [6].

3. Result and Discusion

Figure 1 presents a typical DSC thermogram for the sample E, natural milled maize with 75% (w/w) of water and grain size 250μm. The endothermic peak shows the gelatinization transition of the starch. The area between the base line and the thermogram represent the gelatinization enthalpy (ΔHp), and it is related with the amount of starch in amorphous phase. The gelatinization temperature (Tp1) was 70.07°C and the gelatinization enthalpy was 2.228J/g, the ΔT for this sample was 19.5°C.

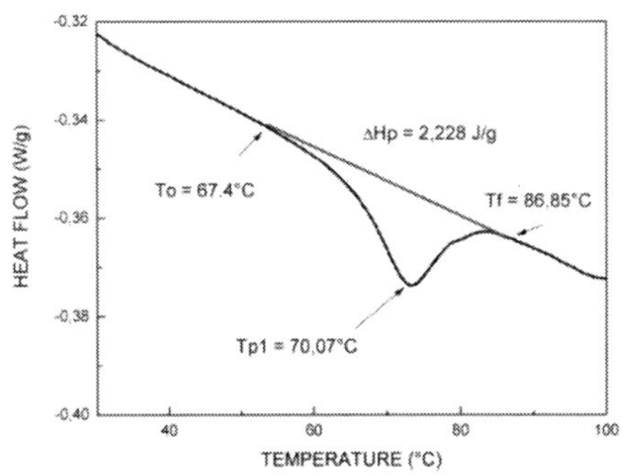

Figure 1. Typical DSC thermogram for the sample E, with 75% (w/w),grain size: 250μm.

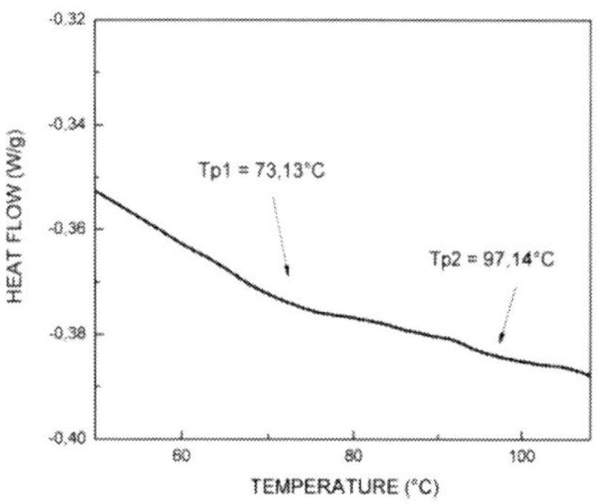

Figure 2. Typical DSC thermogram for the sample B, with 60% (w/w),grain size: 420μm.

Figure 2, presents two peaks behavior for the sample B with 60% (w/w) of moisture and grain size 420μm. A second peak is observed at 97,14°C when the moisture is not enough to complete the transition. All samples were analyzed in the same way, using different moisture content. Results are presented in tables 1 and 2.

Table 1. Summary of gelatinization parameters for the samples with grain size 250μm.

SAMPLE	*MOISTURE* (%)	*Tp*1(°C)	ΔT (°C)	*Tp*2(°C)	ΔHp (J/g)	PHI (%)
A	60	76,31	36,26	—	1,310	9,894
	65	72,16	44,08	—	2,914	33,727
	70	71,64	32,88	—	0,566	3,447
	75	70,10	50,34	—	1,773	14,887
B	60	70,88	28,77	97,14	0,807	2,747
	65	72,80	40,64	—	1,309	16,261
	70	71,59	42,05	—	0,267	2,473
	75	70,81	44,08	—	2,106	17,832
C	60	74,09	31,98	99,7	1,000	9,209
	65	70,33	27,87	97,72	0,910	34,137
	70	71,81	29,69	—	0,415	3,606
	75	71,59	29,35	—	2,228	12,076
D	60	74,41	40,87	—	1,591	15,462
	65	68,05	23,39	—	0,497	27,584
	70	71,56	29,69	—	0,341	3,225
	75	70,64	41,39	—	1,739	10,835
E	60	73,99	48,93	—	22,89	93,856
	65	72,11	29,11	—	1,246	11,010
	70	71,07	23,82	—	2,202	32,099
	75	70,07	19,05	—	2,228	46,320

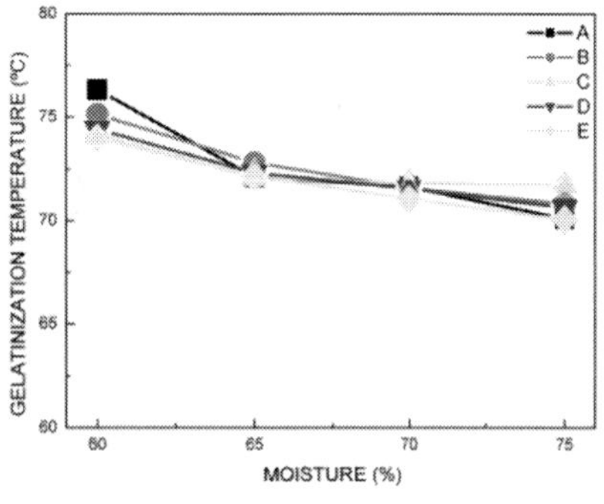

Figure 3. Tp1 vs moisture for samples with grain size 250μm.

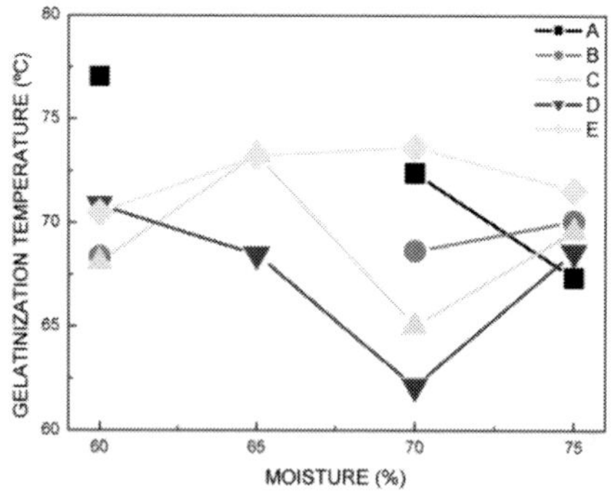

Figure 4. Tp1 vs moisture for samples with grain size 420μm

In figure 3 the relation between the moisture content and Tp1, for samples with 250μm is presented. It shows that Tp1 moves to lower temperatures when the moisture content is incremented. This can be explained because the water favors the beginning of the process [6]. Results in figure 3 indicate that the higher moisture contributes to produce starch gelatinization;

the gelatinzation values obtained for this samples are in agreed with other studies [1]. Water molecules act as plasticizers agents of polymeric molecules [8], it means that for the lower values of moisture, the higher values of Tp1 were obtained. A plot of Tp1 vs moisture content, for samples with 420μm, is presented in the figure 4. In comparation with results for the 250μm, these samples present a lower Tp1, therefore, the same water effect is also observed. For 65 and 70% of moisture the sample C does not exhibit a characteristic gelatinization peak, it could be due to the increases of the grain size that do not allow the entrance of water into the whole grain or grains are already pre-gelatinizated or gelatinizated. The influence of the grain size in the gelatiniztion process is evident in the lost of behavior between the samples with 65 and 70% of mositure, also in the increment of Tp1 for the same sample at the same moisture content. It was found tha the gelatinization temperature is not an intrinsic property, it depends on the process parameters such as water content and particle size.

Table 2. Summary of gelatinization parameters for the grain size 420μm.

SAMPLE	*MOISTURE* (%)	*Tp*1(°C)	ΔT (°C)	*Tp*2(°C)	ΔHp (J/g)	*PHI*(%)
	60	77,03	26,84	110,65	0,726	11,04
	65	71,47	17,63	118,69	0,147	22,920
A	70	72,39	10,78	100,19	0.544	0,750
	75	67,33	23,98	—	1,403	2,120
	60	68,35	25,49	101,21	0,627	2,150
	65	68,28	26,17	82,13	0,244	35,000
B	70	68,63	21,42	—	0,245	0,350
	75	70,08	24,54	—	0,428	0,610
	60	68,10	17,82	88,52	0,248	60,000
	65	—	—	—	—	—
C	70	—	—	—	—	—
	75	69,59	25,6	—	0,477	0,690
	60	70,85	27,07	98,76	0,573	2,760
	65	68,46	29,43	—	0,635	0,930
D	70	62,13	22,75	—	0361	0,580
	75	68,58	23,58	—	0,440	0,640
	60	70,46	17,65	—	2,458	3,610
	65	73,20	15,36	—	0,278	0,380
E	70	73,64	5,47	—	0,171	0,230
	75	71,56	27,65	—	4,403	6,550

The higher Tp1 values mean that most energy is necessary to initiate the gelatinization process, differences in ΔHp indicate that different amounts of amylopectin crystals have been melted [9, 10]. The ΔHp for samples is presented in tables 1 and 2. Differences of ΔHp, in comparation with the natural sample, indicate that industrial samples are pregelatinizated due to previous hydrothermal treatment. Is possible to observe that the ΔT increase when the moisture content decrease, the bigger ΔT value correspond to the lower water content. The PHI is presented in tables 1 and 2. It is calculated dividing the ΔHp *100 with the difference between the Tp1 and To [5]. The high value of PHI was found for the sample E with grain 250μm, due that there is no thermal process that degrades chains of amylose and amylopectin.

The low value of PHI was found for the sample C with grain 420μm. Samples with the lower moisture content present the lower value of PHI. With these results is possible evidence that the gelatinization process is benefited for higher moisture contents, and for lower grain sizes. A second peak behavior, identified as Tp2, in agree with the Flory theory, was observed for samples with the lower moisture content (tables 1 and 2). The Tp2 is related with a second melt of amylopectin crystals, which need more energy to be melted. For grain size 250μm, only samples B and C present this phenomenon, for the grain size 420μm, only the sample E does not present Tp2, because a wide ΔT in for the first peak was presented.

4. Conclusions

A study of the gelatinization parameter for instant maize flours as well as a native corn flour were made using DSC. The Tp1 was found between 70 to 75°C for all studied samples. Tp1 decreases for high values of moisture content and it increases when the grain size increases. The second peak (Tp2) is related to the amount of water present in the system and it appears only for samples with lower water content. Industrial samples show lower enthalpy in comparation with the native sample, which indicates that the industrial process affects the starch molecule causing a pre-gelatinization. The knowledge of the Tp1 and Tp2 lead make a control in the industrial process of flours preparation.

References

[1] Evans, I.D., Haisman, D.R., 1982. *The Effect of Solutes on the Gelatinization Temperature Range of Potato Starch. Starch, 34: 224-231.*

[2] Isela Rojas-Molina, Elsa Gutierrez-Cortez, Alin Palacios-Fonseca, Leticia Baños, J. L. Pons-Hernandez, S. H. Guzman-Maldonado, P. Pineda-Gomez, M. E. Rodrigues-Garcia, M.E., 2007. *Study of the Changes in the Endosperm of Quality Protein Maize During Traditional Nixtamalization Process. Cereal Chemistry,84(4): 304-312.*

[3] Roos, Y.H., 1995. Transition in Foods. Food Science and Technology. *International Series. Academic Press.*

[4] Biliaderis, C.G., 1992. *Structures and Phase Transitions of Starch in Food Systems. Food Technology, 46(6): 98-100, 102-104, 106, 108-109.*

[5] Sandhu, K. S., Singh, N., 2007. *Some Properties of Corn Starches II: Pysicochemical, Gelatinization, Retrogradation, Pasting and Gel textural Properties. Food Chemistry, 101: 1499-1507.*

[6] Lelievre, J., 1976. *Theory of Gelatinization in a Starch-water Solute System . Polymer, 17: 854-858.*

[7] Hoover, R., Manuel, H., 1996.*The Effect of Heat-Moisture Treatment on the Structure and Physicochemical Properties of Normal Maize, Waxy Maize, Dull Waxy Maize and Amylomaize V Starches. Journal of Cereal Science, 23: 153-162.*

[8] Ferry, J.D., 1980.Viscoelastic Properties of Polymers. 3ra edicion. *John Wiley and Sons, Nueva York.*

[9] Liu, Q., Thompson, D. B., 1998. *Effects of Moisture content at Different Gelatinization Heating Temperatures on Retrogradation of Waxy-type Maize Starches. Carbohydrate Research, 314: 221-235.*

[10] McPherson, A.E., Jane, J., 1999. *Comparison of Waxy Potato with Other Root and Tuber Starches. Carbohydrate Polymers, 40, 57-70.*

Characterization of single crystalline ZnTe and ZnSe grown by vapor phase transport.

A.B. Trigubó[1,2], M.H. Aguirre[3], M.C. Di Stefano[1], A.M. Martínez[2], R. D´Elía[2], H. Cánepa[2], E. Heredia[2].

[1] FRBA-UTN, (1179) Buenos Aires, Argentina.
[2] CINSO-CITEFA: (1603) Villa Martelli, Pcia de Buenos Aires, Argentina.
[3] Dpto de Quím Inorg, Fac de Cs Quím, Univ Complutense, (28040) Madrid, España.

atrigubo@citefa.gov.ar

Abstract. Tubular furnaces were designed and built to obtain single crystalline ZnTe and ZnSe ingots using respectively physical and chemical transport methods. Different temperature profiles and growth rates were analyzed in order to optimize the necessary crystalline quality for device development.
Optical and scanning electron micrographs of the corrosion figures produced by chemical etching were used to obtain the dislocation density and the misorientation between adjacent subgrains in ZnTe and ZnSe wafers. Structural quality of the single crystalline material was determined by transmission electronic microscopy. Optical transmittance was measured by infrared transmission spectrometry and the resulting values were compared to commercial samples.

1. Introduction

ZnTe is a useful II-VI compound for optical devices because of its wide energy gap (about 2.3 eV at room temperature) [1]. This semiconductor material has been found to become photorefractive when it is appropriately doped with V or with Mn and V. The combination of photorefractivity and semiconductivity make this material attractive for many applications as optical power shielding and holographic interferometry. Moreover, ZnTe:V have a superior photorefractive performance at wavelengths from 0.6 to 1.3 μm than other photorefractive materials. This semiconductor might also be used for windows due of its important optical transmission.

The structures based on ZnSe for laser emission in the blue-green range of the spectrum have been mainly grown on GaAs due to the high quality of these commercially available substrates. However ZnSe homoepitaxy is a promissory line of investigation with the aim of diminishing the density of structural defects [2]. As ZnSe transmits in a wide range of the infrared spectrum it is also used for lenses and windows.

In this paper we study the growth of single crystalline ZnTe and ZnSe [3] by vapor phase transport. These materials have notable electro-optical properties [4]; however, as the quality of detectors and optical devices are critically dependent on its crystalline properties, our aim was to determine the best growth conditions to obtain a high crystalline quality. The crystal quality was determined by chemical etching and transmission electron microscopy (TEM). The optical transmission under incident IR radiation was evaluated with the purpose of using ZnTe and ZnSe in the manufacturing of lenses and windows.

2. Experimental Procedure.

Electrical and optical properties of semiconductors are determined by their impurities, so it is important not only to diminish them but also to know which ones are present. The elements used in the synthesis of ZnTe and ZnSe can be obtained with a purity of 6N for Zn and Te, and 5N for Se. These elements were purified by dynamic vacuum distillation previous to their employment in the synthesis [5-7].

Tubular furnaces for single crystalline growth of semiconductor ingots of the II-VI family were designed, built and used several times with different temperature profiles to get the best conditions for each single crystalline compound synthesis. Vapor Phase Transport without and with gaseous carrier, molecular iodine in our case, was the chosen technique. It requires a particular temperature profile with the corresponding growth conditions control: A) growth speed in the first case for single crystalline ZnTe growth (figure 1a); and B) ampoule temperature gradient, growth time and amount of iodine in the latter case for single crystalline ZnSe growth (figure 1b).

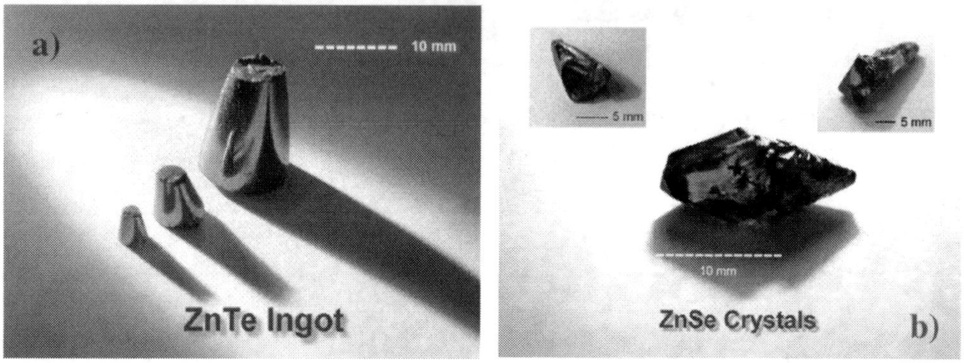

Figure 1: a) ZnTe ingot grown by Physical Vapor Phase Transport. b) ZnSe ingot grown by Chemical Vapor Phase Transport with iodine as gas carrier.

Various reagents were used for the chemical etching of each one of the semiconductor compounds. The micrographic results for each material were compared to determine the relative advantages of the

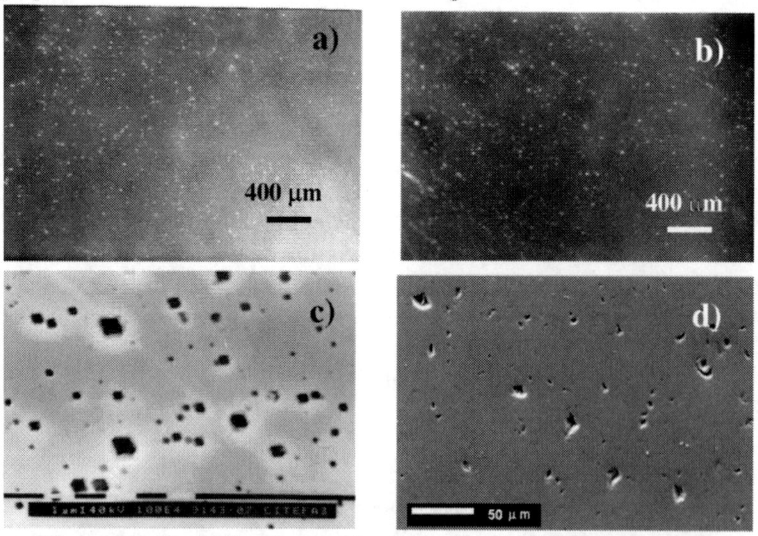

Figure 2: a) OM micrograph of a chemically etched ZnTe wafer. b) OM micrograph of a chemically etched ZnSe wafer. c) SEM micrograph of the same ZnTe wafer of figure 2a. d) SEM micrograph of the same ZnSe wafer of figure 2b.

chemical solutions. Data analysis determined the following chemical selection: $K_2Cr_2O_7$+HCl for ZnTe and $KMnO_4$+H_2SO_4 for ZnSe. Figures 2a and 2b show micrographs obtained by optical microscopy (OM), and figures 2c and 2d correspond to micrographs respectively obtained by scanning electronic microscopy (SEM) of the same chemically etched wafers of ZnTe and ZnSe crystals.

A precision ion polishing system (PIPS), Model Gatan 691 (VAr+< 4 kV) was employed to get thin specimens from ZnTe and ZnSe in planar view geometry. The TEM examination of the samples was performed with a JEOL JEM 2000 FX (W termoionic filament) at 200 kV of operation. A JEOL JEM 3000 F (Field Emission Gun) at 300 kV of operation was used to obtain LRTEM and HRTEM images by means of a Gatan CCD camera.

Transmission spectra of ZnTe and ZnSe crystals, grown at our laboratory and commercial ones, were obtained using a Perkin-Elmer System 2000 Fourier transmission infrared equipment (FTIR).

3. Results and Discussion.

Table 1 shows the dislocation densities and adjacent subgrain misorientation obtained as a statistical average of different measurements in the same wafer with the chosen reagent for the semiconductor compound.

Table 1: dislocation density and adjacent subgrain misorientation.

Semiconductor Compound	Reagent	$\delta_{\perp}(cm^2)$	$\varphi('')$
ZnTe	$K_2Cr_2O_7$+HCl	2.31 10^5	11
ZnSe	$KMnO_4$+H_2SO_4	6.2 .10^5	22

3.1. ZnTe TEM characterization

3.1.1. *Jeol JEM 2000 FX.*

Figure 3a shows the [1 1 0] zone axis of the transmission electron diffraction (TED) pattern of the ZnTe specimen. The TEM images corresponding to **g** = [0 0 -2], [0 0 2] and [-2 2 0] have shown no dislocations in the semiconductor material. As an example, the image for **g** = [0 0 -2] is illustrated in figure 3b.

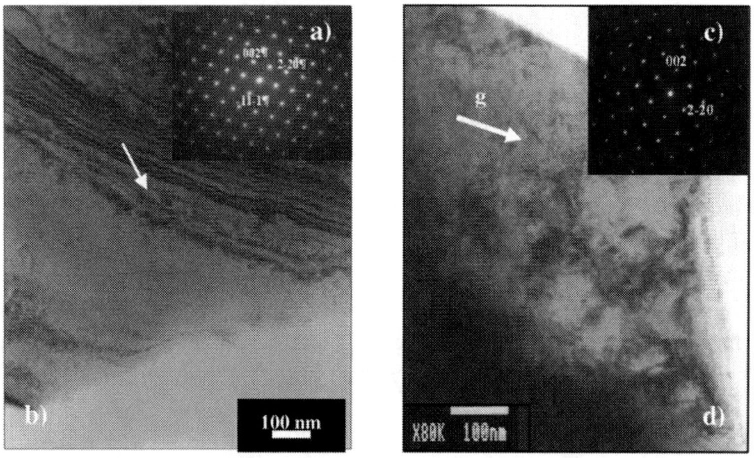

Figure 3: a) [1 1 0] TED pattern. b) LRTEM image with **g** = [0 0 -2]. c) [1 1 0] TED pattern. d) LRTEM image with **g** = [2 -2 0].

3.1.2. Jeol JEM 3000 F

The high resolution TEM images were obtained with the [1 1 1] zone axis (figure 4a). In figure 4b no dislocations are present confirming a low dislocation density. It can also be observed the excellent structural crystalline order between dislocations whose etch pits were revealed by chemical reagents in the previous analysis.

3.2. ZnSe TEM characterization

3.2.1. Jeol JEM 2000 FX

Figure 3c is the transmission electron diffraction diagram with [1 1 0] zone axis. TEM images with **g** = [1 -1 1] and **g** = [2 -2 0] have shown no dislocations in the semiconductor material. Figure 3d is an example of this fact (**g** = [2 -2 0]).

3.2.2. Jeol JEM 3000 F

Dislocations presence have been analyzed by choosing the (1 1 1) zone axis in the TED diagram (figure 4c). High resolution TEM images were obtained with different magnifications. None of them show dislocations confirming the low dislocation density. Figure 4d corresponds to a larger TEM magnification and, as in the case of ZnTe, it can be observed the excellent structural crystalline order between dislocations revealed by chemical etching in the previous stage.

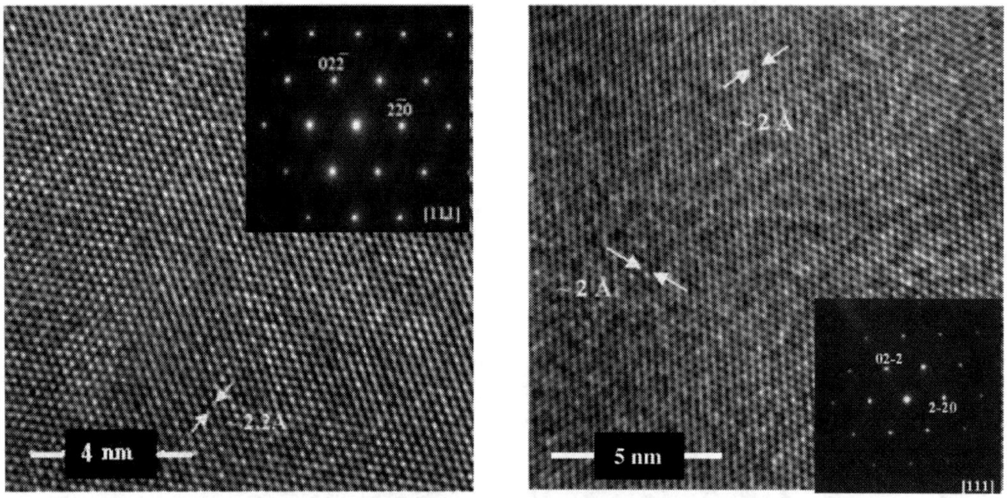

Figure 4: a) [1 1 1] TED pattern. b) HRTEM image with [1 1 1] zone axis. c) [1 1 1] TED pattern. d) HRTEM image with [1 1 1] zone axis.

3.3. ZnTe and ZnSe transmission spectra.

Figure 5a and 5b compare the transmission spectra obtained from one of the samples of ZnTe and ZnSe grown in the Solid State Materials Research Centre (CINSO), and one commercial sample of each type used for optical windows. Both spectra confirm that these materials can be used in manufacturing optical devices and in windows for IR detectors in a broad range of wavelengths, especially in the first and second atmospheric window (3-5 μm and 8-14 μm respectively). It was determined that the polishing improvement of materials produced also an improvement of the transmittance values.

The spectral curves indicated undesirable oscillations in the H_2O and CO_2 absorption wavelengths due to the internal calculation of the FTIR program, since in these absorption bands the measurement error increases due to very small signal values.

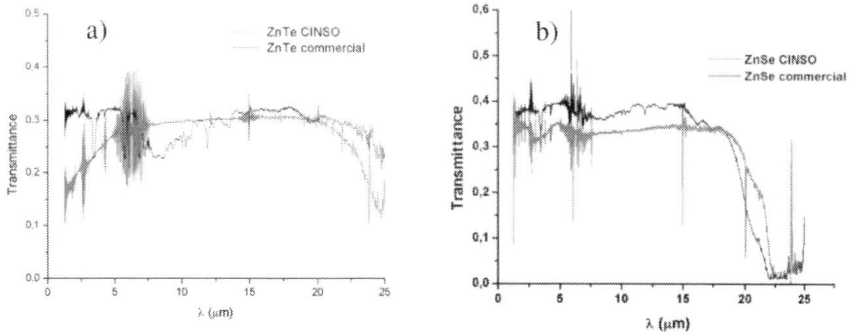

Figure 5: Commercial and CINSO samples transmission spectra. a) ZnTe b) ZnSe

4. Conclusions

Transmission electron microscopy micrographs confirmed the low dislocation density and the excellent crystalline quality of the grown materials between dislocations previously studied by chemical etching.

The semiconductor material grown at CINSO and the commercial wafers have comparable FTIR spectra, suggesting their possible use as IR transmission windows.

Acknowledgements

The authors are grateful to Dr. J. Codnia of the CEILAP-CONICET-CITEFA for the FTIR spectra and to Professor M.Á. Alario Franco of the Inorganic Chemistry Department of the UCM-Spain for providing facilities at Microscopy Centre "Luis Bru".

This work has been partially supported by CONICET and FRBA-UTN through the projects PIP 5468-05 and 25/C080.

References

[1] Kitamura N, Kakehi M and Wada T 1977 *Japan. J. Appl. Phys.* **16** 9 1541
[2] Korostelin Y U, Koslovsky V I, Nasibov A S and Shapkin P V 1999 *J. Crystal Growth* **197** 445
[3] Chen A B and Sher A 1995 *Semiconductors Alloys* (New York: Plenum)
[4] Atroshchenko L V, Gal'chinetskii L P, Galking S N, Ryzhikov V D, Silin V I and Shevtsov N I 1999 *J. Crystal Growth* **197** 471
[5] Heredia E, Trigubó A B, Gilabert U, Marrero J, Ledesma A, Cánepa H and D'Elía R 2000 *Superficies y Vacío* **10** 27
[6] Cánepa H, D'Elía R, Heredia E and Trigubó A B 2003 *Proc Jornadas SAM-Congreso CONAMET- Simposio Materia (Bariloche, Argentina, 17-21 November 2003)*
[7] Heredia E, Trigubó A B, Marrero J and Cánepa H 1993 *Proc. AFA* **5** 438. *(Rosario, Argentina, 12-15 October 1993)*

Thickness effect on electric resistivity on polystyrene and carbon black- based composites

S. Hernández-López[1], E. Vigueras-Santiago[1], M. Mayorga-Rojas[2], D. Reyes-Contreras[2]

[1]Laboratorio de Investigación y Desarrollo de Materiales Avanzados (LIDMA) Facultad de Química, Paseo Colón Esquina con Paseo Tollocan, s/n, CP 50000, Toluca estado de México.

[2]Facultad de Ciencias, Universidad Autónoma del Estado de México. Av. Instituto Literario 100 Ote. C. P. 50000, Toluca, Estado de México, México.

E-mail: eviguerass@uaemex.mx

Abstract. Changes on electrical resistivity were experimentally studied for polystyrene and carbon black-based composites respect to the temperature. 22% w/w carbon black composite films at 30 μm, 2mm y 1cm thick were submitted to thermal heating-cooling cycles from room temperature to 100°C, slightly up to T_g of the composite. For each cycle changes on electrical resistivity constituent a hysteresis loop that depends on the sample thickness. The changes during the heating stage could be explained as a consequence of the thermal expansion and mobility of the polymer chains at T_g, producing a disconnecting of the electrical contacts among carbon black particles and an important increasing (200%) of the electrical resistivity. For each cycle, the hysteresis loop was observed in thicker samples, whereas for 30 ? m thickness sample the hysteresis loop was lost after four cycles.

1. Introduction

Is well know that in conductive polymer composites with carbon black (CB) the conductive to dielectric transition could carried out in two ways: one is modifying the CB particles concentration [1] and the other is by heating until some polymer transition is achieved as melting (T_m) or glass transition (T_g) temperatures in crystalline/semicrystalline and amorphous polymers, respectively [2]. The existence of these resistivity modifiers mechanisms has allowed the use of those type of composites materials in a broad gamma of practical applications [3]. Preparation, optimization and modification of the electrical properties in polymer composites are dependent of the couple polymer - carbon particles. This includes the physical and chemical properties of the polymer matrix; the geometry, size and morphology of the CB particles as essential factors for the final characteristics of the polymer composites [1-3]. The dependence of the resistivity with temperature could be metallic type as semiconductor one [4,5] and some polymer composites show it as the well known positive temperature coefficient (PTC). This PTC response consists in an increasing of the resistivity in some

magnitude orders when the composite is heated until T_m or T_g [2,6]. Significant attention has been paid to this effect in polymer composites based on semicrystalline thermoplastic polymers and epoxy resins or thermoset matrixes [7]. Temperature dependence of electrical conductivity in polymer composites are a function of the heating and cooling rate [7], of the number of heating-cooling cycles [8], the polymer-filler composite pair and their proportions [1-11]. Many are the factors involved in that behavior, however there are a few studies related with the sample thickness [9] as another variant which could be involve in the temperature response. Is worthy note that important information about mechanical and electrical stability, reversibility and electrical reproducibility of the systems trough these studies represent possible application such a thermal switches on temperature control, among others technological applications.

The basic mechanism of electro-thermal behavior is still understood because of the complex nature of the composites. However an accepted way for explaining the resistivity changes as a function of temperature considers the polymer matrix expansion during the heating process where the CB particles separates increasing the distance among them and diminishing the electrical contacts (conductive networks) [10].

In this work the electro-thermal behavior of polymer composites based on atactic (amorphous) polystyrene polymer and Vulcan XC72-CB was studied. We show that the conductive composite thickness plays an important roll on electro-thermal behavior in these kinds of composites during heating-cooling cycles at their T_g. An electro-thermal open hysteresis loop was observed in all samples for the first cycles. At the end of the fourth cycle, the resistivity vs temperature curve was reversible for the thinnest sample and a close hysteresis loop was observed for thicker samples.

2. Experimental

Polystyrene (PS) 10,000 Mw, analytic grade tetrahydrofurane (THF) were purchased from Sigma-Aldrich, Co. Carbon black (CB) Vulcan XC72 was donated by Cabot Co. and they were used without any other treatment. Glass transition (T_g) of 22% PS+CB composite was 98°C as determined by calorimetry in a SDT Q600 Modulus, under nitrogen atmosphere and a heating rate of 10°C/min. Composite was obtained by ultrasonic mixing using THF as solvent [12]. Films of 30 μm thickness were deposited on glass substrates by spin coating using that solution. Then the solvent was evaporated from the remaining solution and cylinders of 2 mm and 1 cm heights and 4.5 mm radius were obtained by compressing moulding at 120 ^{0}C [12]. For films, electrical contacts were painted as parallel lines on the same face. For bulk samples both faces were full covered with conductive silver paste [12]. All samples were submitted to six heating-cooling cycles into a home made oven. During the heating samples were heated from room temperature (25°C) to 100°C (T_g of the composite) at a rate of 1.5°C/min. The cooling stage was from 100 to 25°C at 0.2°/min. Electrical resistance and temperature were measured simultaneously at intervals of one second using digital multimeters coupled to a computer.

3. Results and discussion

The effect of the composite thickness was relevant from the initial resistivity. There was a difference of three magnitude orders at least between the thinnest (30 μm) and the thicker films (Figure 1). The changes on electrical resistivity during the thermal treatments evidenced significantly variation according to the sample thickness too. They also showed a common behaviour which consisted on the different resistivity value between the initial (before the thermal treatment) and the end of the first cycle (Figure 1). This difference was really important for the thin layer (30μm) for about one magnitude order higher. For thick samples that difference was depreciable (Figure 1).

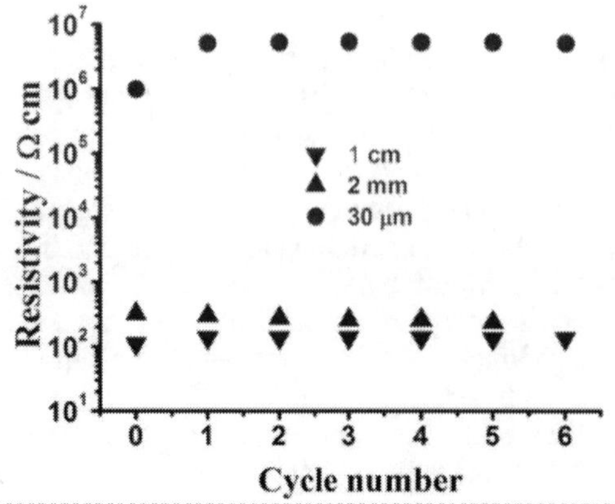

Figure 1. Resistivity values after each cycle number

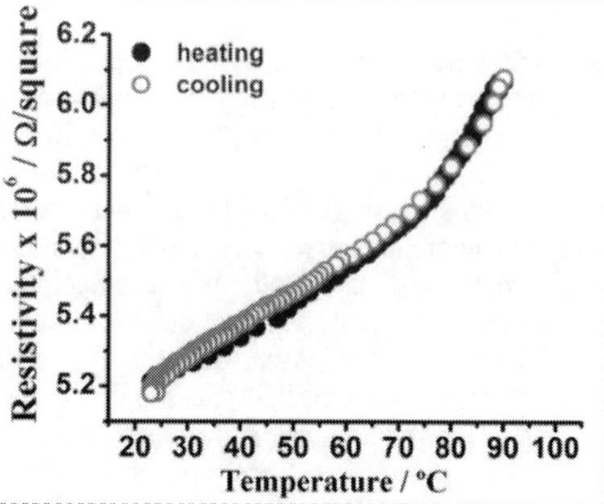

Figure 2. Resistivity behavior for the 30 μm film at fourth thermal cycle

With the increasing of the cycle's number, all samples tended to a stable resistivity value. This asymptotic behaviour was reached after the fourth cycle for 30 μm and 2 mm thickness samples (squares and triangles curves in Figure 1). For 30 μm film the resistivity changes were reversible (Figure 2) through the all trajectory whereas the resistivity for thick samples builds a close hysteresis loop after the fourth cycle (Figures 3 and 4). An explanation of those behaviours is related with a redistribution of the CB particles: during the heating, the thermal polymer expansion and the consequent increasing in the molecular movement lets a redistribution of them, a disconnection-reconnection of the conductive networks allowing a new conductive configuration.

This effect was more evident in the thinnest film due to the electro thermal response follow a reversible behaviour of the resistivity during the heating-cooling treatments meaning that a conductive network in equilibrium is accomplished. For thicker samples this equilibrium configuration was not reached even after 6 cycles following a close hysteresis loop. In this case there are continuous contact configuration changes of the carbon particles during the heating and cooling process, however the average of all of them predominate and render the same final resistivity value.

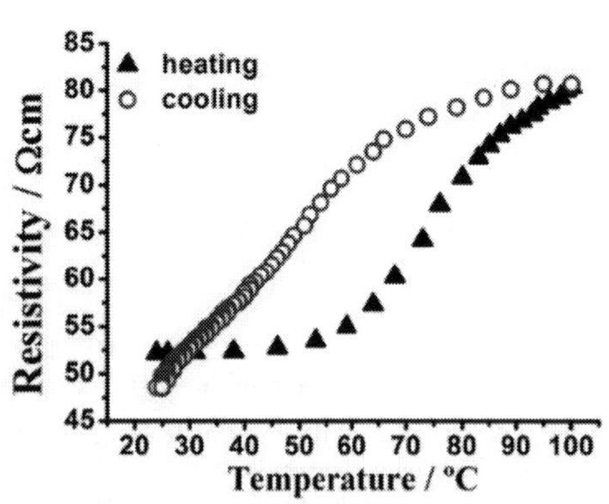

Figure 3. Open hysteresis loop for the resistivity as a function of temperature showed by the 2 mm sample at the third cycle.

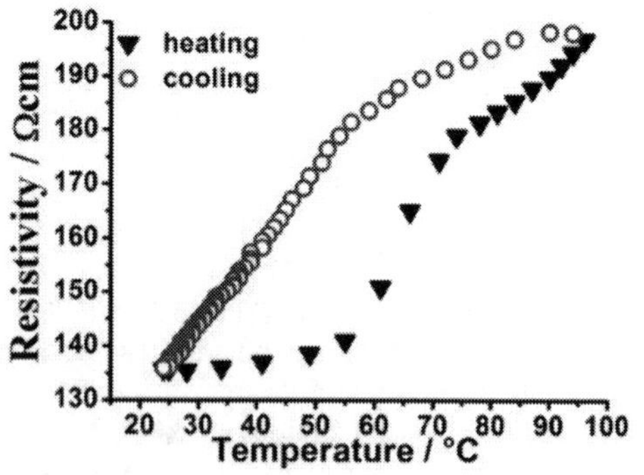

Figure 4. Close hysteresis loop for the 1 cm sample at the fourth cycle.

This redistribution explained in terms of the polymer movement has an important dependence of the thickness. As the thickness diminishes the contact's number among the conductive particles also diminishes in the same direction as the thickness do, according to the description in [13]. As a consequence the redistribution and probability of conductive particles to acquire the same conductive configuration is high for thin films, attributing them the observed reversibility.

4. Conclusions

Electrical resistivity of PS + CB composites reaches a stable final value after several heating-cooling cycles dependently of the sample thickness. For a thin composite film (30 μm) the hysteresis loop was lost after four cycles. For thicker samples the resistivity as temperature dependent showed a close hysteresis loop after being submitted to six treatments. We consider that those behaviours are associated to a change in spatial configuration and original distribution of the CB particles due to a increasing of the polymer chains motion as the temperature reaches its T_g. The number of CB-electrical contacts was clearly modified for the thin film during the first four cycles until it reached equilibrium. However for thicker samples the resistivity changed during the heating and cooling steps but at the end of the cycles that value was practically the same, meaning that total average of CB contacts was no modified.

5. Acknowledgments

This work was supported by CONACYT under the project 25496.

References

[1] Zhang W, Dehgheni —Sanij A A 2007 Carbon Based Conductive Polymer Composites *Matter Sci.* 42 3408

[2] Azulay D, Eylon M, Eshkenazi O, Toker D, Balberg M, Shimoni N, Millo O, Balberg I 2003 Electrical-thermal switching in carbon-black-polymer composites as a local effect *Phys Rev Lett.* 90 236601-1

[3] Donnet J B, Bansal R. L, Wang M.-J (Eds.) 1993 Carbon Black New York Dekker

[4] Swapan K, Bhattacharya A, Basu S, Sadhan K 1980 Effect of temperature on the electrical conductivity of poly(vinyl chloride)-cupper composites *J. Appl. Polym. Sci.* 25 111

[5] Hindermann-Bischoff M, Ehrburger-Dolle F 2000 Electrical conductivity of carbon black-polyethylene composites experimental evidence of the change of cluster connectivity in the PTC effect. *Carbon* 39 375

[6] Grayson Alexander M 1999 Anomalous temperature dependence of the electrical conductivity of carbon-poly(methyl methacrylate) composites *Mater. Res. Bull.* 34 603

[7] Hirano S, Kishimoto A 1998 Effect of heating rate on positive-temperature-coefficient-of-resistivity behavior of conductive composite thin films *Appl. Phys. Lett.* 73 3742

[8] Wan Y, Wen D 2005 Stabbility of thermo-sensitive properties of carbon-black/styrene-butadiene-rubber composites membranes *Smart Mater. Struct.* 14 941

[9] Liu L, Matitsine S M, Rozanov K N 2005 The Thickness Dependence of Resonance Frequency in Anisotropic Composites with Long Conductive Fibers *Electomagnetics* 25 69

[10] Heaney M B 1996 Resistance-expansion-temperature behavior of a disordered conductor-insulating composite *Appl. Phys. Lett.* 69 2602

[11] Choi E S, Brooks J S, Al-Haik M S, Hussaini M Y, Garmestani H, Li D, Dahmen K 2003 Enhancement of thermal and electrical properties of carbon nanotube polymer composites by magnetic field processing *J. Appl. Phys.* 94 6034

[12] San Juan Farfán R, Hernández López S, Martínez Barrera G, Camacho López M A, Vigueras Santiago E 2005 Electrical Characterization of Polystyrene-Carbon Black Composites. *Phys. Stat. Sol. (c)* 2 3762

[13] Carmona F, Ravier J 2002 Electrical properties and mesoestructure of carbon black-filled polymers *Carbon* 40 151

Journal of Physics: Conference Series **167** (2009) 012060 doi:10.1088/1742-6596/167/1/012060

Ab initio study of a $TiO_2/LaAlO_3$ heterostructure

M Weissmann and V Ferrari

Departamento de Física, Comisión Nacional de Energía Atómica, Gral. Paz 1499, 1650 San Martín, Buenos Aires, Argentina.

E-mail: weissman@cnea.gov.ar, ferrari@tandar.cnea.gov.ar

Abstract.
In this work we explore the origin of the ferromagnetism appearing when a TiO_2 film is grown on another non-magnetic oxide as a substrate such as $LaAlO_3$ (001), concentrating on the role played by the oxygen vacancies in this phenomenon. Using Density Functional Theory ab-initio methods, we study the free-standing anatase film as well as the interfaces with either the LaO or AlO_2 planes of $LaAlO_3$. Our results show that the interface LaO/TiO_2 is favored against the AlO_2/TiO_2 one if no oxygen vacancies are present in the interface whereas the contrary happens when there are oxygen vacancies. In both cases, the cohesive energy is of the same order of magnitude but only at AlO_2/TiO_2 we found a magnetic solution.

1. Introduction

Intense search for ferromagnetism above room temperature in dilute magnetic semiconductors and insulators has taken place in recent years. The fabrication of these materials offers exciting possibilities for spintronics and the use of wide gap oxides is appealing for magneto-optical devices. Room-temperature ferromagnetism has been observed in both doped [1] and undoped insulating oxide thin films such as TiO_2, ZnO and HfO_2 [2, 3] grown over other oxides or semiconductors. This magnetic order can be weak and the determination of its intrinsic character may require complementary, independent and careful techniques to be probed [4, 5]. Many authors have attributed an important role to oxygen vacancies [6, 7] and other defects [8] to produce this magnetic ordering, but the effect of the substrate or the interface has not been taken into account in the interpretations.

Considering the open questions in the aforementioned systems, in this work we explore the possibility of observing magnetism in an heterostructure formed by a simple oxide such a TiO_2 anatase and a $LaAlO_3$ (LAO) substrate. Experiments show that large terraces are formed on the substrate surface, and that the growth is epitaxial on the (001) direction [9]. However, it is not clear if the surface is mostly composed of LaO planes or of AlO_2 planes [10, 11]. Spectroscopic experiments show that there are mixed-valence Ti ions in these films, which may be due to the presence of oxygen vacancies [12]. The magnitude of the magnetization has been found to be proportional the oxygen content in some experiments [13]. We therefore concentrate on the effect of oxygen vacancies in the electronic structure and in the interfacial structural relaxation, as the electric fields due to the dipole layer at the interface may lead to important changes in the positions of the atoms with respect to the bulk materials. For this purpose we perform ab-initio calculations for the two interfaces in a TiO_2/LAO heterostructure using computational codes such as SIESTA [14] or Wien2k [15] and also different unit cells.

We first study separately the component systems: TiO_2 and LAO in bulk and as free standing slabs, representing the deposited anatase film and the substrate, respectively. As the ab-initio codes allow only the study of three dimensional periodic systems, the films are represented by repeated slabs with enough free space between them. For the heterostructure, two different types of unit cell were used: trilayer slabs LAO/TiO_2/LAO and superlattices ...TiO_2/LAO/TiO_2/LAO...

The calculations are performed within the density functional theory (DFT) [16] using the full potential augmented plane waves method, as implemented in the Wien2k code [15]. We use the local density approximation (LDA) [17] for the exchange and correlation, and small muffin tin radii to allow for lattice relaxation. The muffin tin radii used were R_{Ti}^{MT} = 1.7 bohr, R_{O}^{MT} = 1.4 bohr, R_{La}^{MT} = 2.5 bohr, R_{Al}^{MT} = 1.7 bohr. The number of plane waves in the interstitial region is mostly restricted to RKMax=6 although in many cases it was increased to RKMax=7 and no qualitative difference was observed. The effect of increasing the number of k-points was also checked.

2. TiO_2 anatase and LAO: Bulk and surface properties

TiO_2 anatase and LAO are both conventional band insulators, with in-plane bulk lattice parameter a=3.79 Å [18] (Fig. 1). The small mismatch between the in-plane lattice constants allows for a good epitaxial experimental outcome [9].

TiO_2 anatase presents a 3.2 eV band gap between filled oxygen $2p$ bands and unfilled Ti $3d$ conduction bands. The calculated band gap is 2.2 eV as shown in Fig. 2(a). When oxygen vacancies are formed in the bulk anatase structure the calculations show that the system becomes metallic and a vacancy level appears inside the gap, very close to the conduction band. The small band gap and the vacancy levels very close to the conduction band are well known features of the LDA approximation.

LAO presents a 5.6 eV gap between filled O-$2p$ bands hybridized with Al-p states and unfilled conduction bands composed mostly of La-$4f$ states. The calculated gap is 4 eV, as shown in Fig. 2(b).

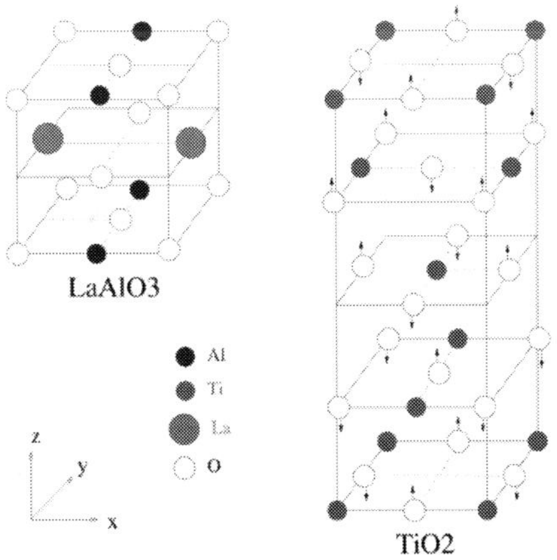

Figure 1. Bulk unit cells of the two component systems. The arrows on the oxygen atoms in the anatase structure indicate that they are off-plane.

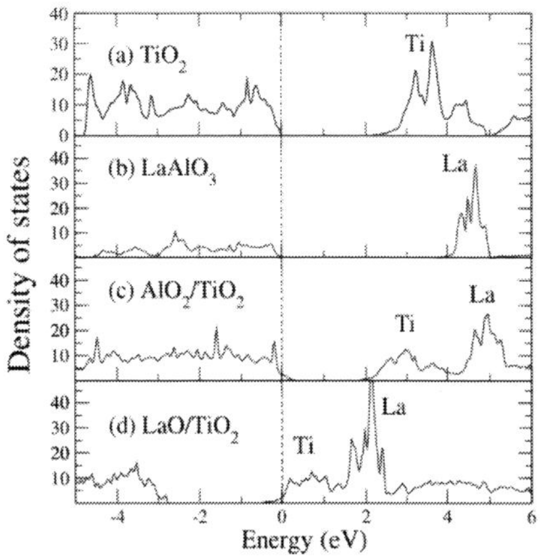

Figure 2. Density of States of: (a) bulk TiO_2 anatase, (b) bulk LAO (c) and (d) Superlattices either with AlO_2 or LaO planes in the interfaces (relaxed structures).

To model the properties of the free surface of an anatase film, we use a slab with 5 or 9 TiO_2 layers. A DFT calculation using the SIESTA code [14] allows the unit cell to change its shape under relaxation, which is not possible with the Wien2k code. This results in a strong deformation of the (001) surface, that goes from a square into a rectangular shape, and also makes the slab thinner. The free standing anatase film would therefore deform in that way, but as we are studying its epitaxial growth over LAO we keep the in plane lattice constant fixed and the cell shape square. Relaxation is only considered in the z direction, perpendicular to the slab. With this constraint we create oxygen vacancies in the surface of the slab and relax the atomic positions using the Wien2k code. The vacancies seem to be crucial for the appearance of magnetism: on one hand, magnetic solutions appear if there are oxygen vacancies at the surface and on the other hand the value of the magnetic moment increases with the number of vacancies. In our calculations, one vacancy per surface cell (50%) gives after relaxation a magnetic moment of 0.3 μ_B, localized at the Ti atom in that surface, while two vacancies per surface cell give 2 μ_B. In Fig. 3 we show the charge and spin density maps near the surface for the anatase slab with two vacancies per surface (all the surface oxygen atoms removed). The system is clearly magnetic, and the magnetic moment is mostly localized in the Ti superficial atom.

The other building block of the heterostructures is the slab of LAO. Along the (001) direction this slab alternates layers of AlO_2 (negatively charged) and LaO (positively charged). Therefore, an integer number of unit cells in the z direction would produce a large dipole and it is more convenient to use an odd number of layers. For the superlattices we consider 5 layer slabs with either LaO or AlO_2 termination. In both cases the system becomes metallic within the LDA approximation and the interesting change with respect to the bulk density of states, shown in Fig. 2(b), is that the LaO termination shifts the unoccupied surface La states substantially towards the Fermi energy.

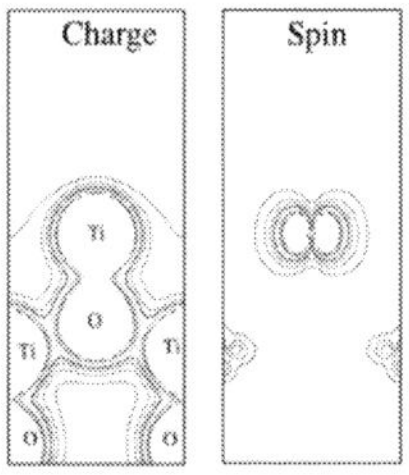

Figure 3. Charge and spin density of the surface of a TiO_2 slab with no oxygen atoms at the surface.

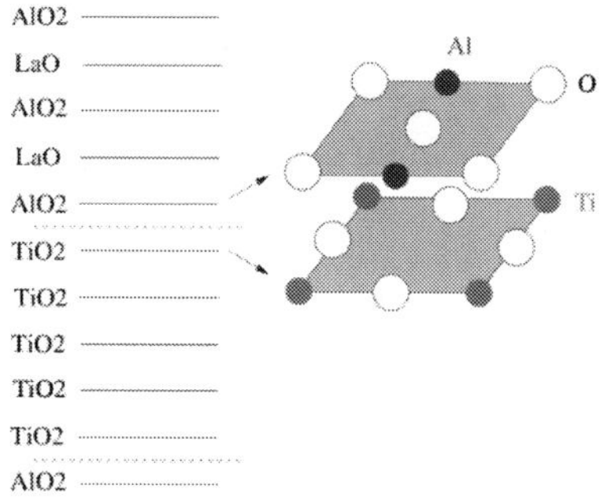

Figure 4. Schematic diagram for the unit cell of one of the superlattices studied.

3. TiO_2/LAO superlattices

We perform calculations on superlattices like the one shown schematically in Fig. 4. We have considered the 2 cases: the TiO_2 plane facing the AlO_2 or the LaO termination. In both situations we assume that the oxygen atoms of the LAO surface face the Ti atoms of the anatase surface. The lateral lattice parameter is kept fixed at the experimental value for LAO but the distance between the two slabs composing the heterostructure is obtained by minimizing the total energy. The fact that the total energy has a minimum when changing this distance verifies that there is bonding between the two materials and also gives the optimal value for that distance. This fixes the size of the superlattice unit cell along the z axis for each case . Afterwards, all the atoms are allowed to relax until the force on each one is less than 2 mev/Å.

We find that there is a competition between the tendency of the anatase film to shrink (as in the free standing slab) and the tendency to bond to LAO. In particular, if an oxygen from the anatase surface faces an Al atom the bond is strong.

Fig. 2(c) and (d) show the density of states of the two superlattices without vacancies and after relaxation. Both interfaces are metallic without vacancies. This metallicity is not altered by the formation of oxygen vacancies or by structural relaxation.

Due to the anatase structure there are two types of AlO_2/TiO_2 interfaces. This can be inferred from Fig. 1 and Fig. 4, but it is shown more clearly in Fig. 5. To simulate growth conditions in a typical experiment [19], we introduced oxygen vacancies at both layers of the interface and found that it is more favorable to perform them in the TiO_2 layer rather than in the AlO_2 plane.

We find magnetic solutions in the unrelaxed geometry, with and without vacancies, but the magnetic moments decrease with structural relaxation. Without vacancies the moments are located in some of the oxygen atoms and disappear with relaxation. With vacancies they are located in the Ti superficial atoms and do not disappear.

Fig. 6 shows the charge and spin densities of the superlattice, when there are no oxygen atoms in the TiO_2 side of the interfaces. Although both interfaces have the same number of missing oxygen atoms, only one of them is magnetic, depending on the relative orientation of the second neighbor layers.

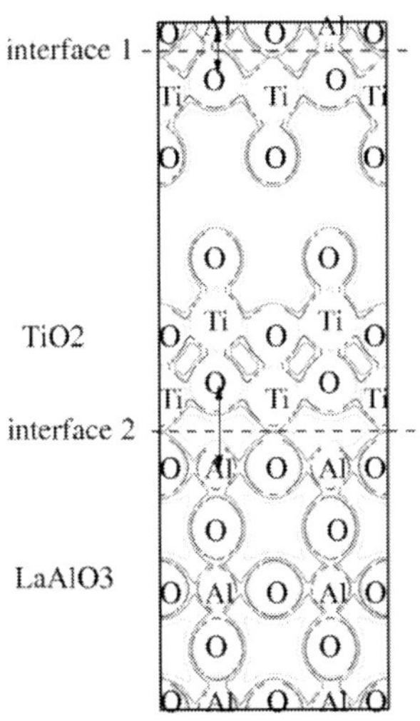

Figure 5. Charge density for the superlattice with AlO_2/TiO_2 interface without vacancies. Distance from the surface Al to the neighbor anatase oxygen is shorter in interface 1.

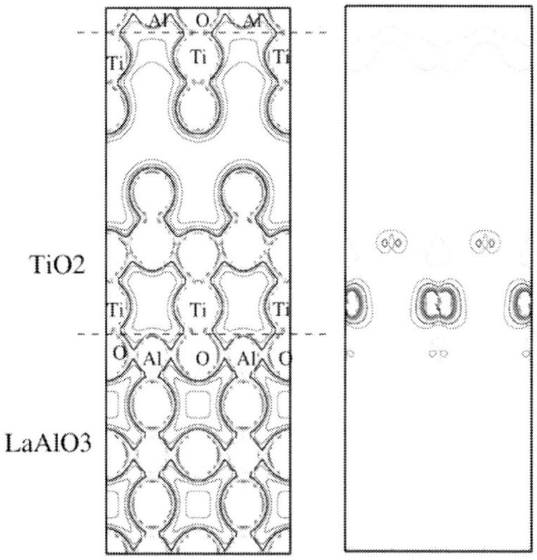

Figure 6. Charge (left) and spin (right) density maps for the system of Fig. 5 with 2 vacancies. Only one interfacial Ti ion is magnetic.

We also study the LaO/TiO_2 heterointerface and perform oxygen vacancies to simulate experimental growth conditions. Introducing vacancies in this case is less favourable compared to the other interface and magnetic solutions are only found with many oxygen vacancies present, as shown in Table 1.

The cohesive energy of a superlattice, given in Table 1, is obtained subtracting from its total energy the energy of the relaxed slabs of TiO_2 and LAO that compose it. The supercell with LaO/TiO_2 interface has more cohesive energy if no vacancies are present while the one

Journal of Physics: Conference Series **167** (2009) 012060 doi:10.1088/1742-6596/167/1/012060

with AlO_2/TiO_2 interface is preferred if there are oxygen vacancies. This last case presents a magnetic moment, localized at the superficial Ti atom.

Table 1. Cohesive energies and total magnetic moments per unit cell for the two types of superlattices considered. The number of vacancies (vac) is the number of missing oxygen atoms from the anatase side of each interface.

case	Ti/Al		Ti/La	
	E (eV)	μ (μ_B)	E(eV)	μ (μ_B)
no vac	4.89	0.01	6.86	0
1 vac	6.65	0.26	6.01	0
2 vac	6.62	1.28	5.94	0.57

4. Conclusions

The calculated cohesive energies allow us to draw some conclusions referring to experiments where the TiO_2 film grows over a LAO substrate with terrace-like structure where there are both LaO and AlO_2 layers. When the film is in contact with the LaO layer, the magnetic atom will not be at the interface with the substrate but rather at the free surface of the anatase film. In that case, the role of the substrate would be just epitaxial, fixing the structure of the anatase film grown on it. If in turn the film is in contact with the AlO_2 layer, it is likely that oxygen vacancies occurring during the growth will be located at that interface and the Ti at the interface would be magnetic.

References

[1] Chambers S A 2006 *Surface Science Reports* **61** 345–381
[2] Yoon S D, Chen Y, Yang A, Goodrich T L, Zuo X, Ziemer K, Vittoria C and Harris V G 2007 *Journal of Magnetism and Magnetic Materials* **309** 171–175
[3] Sudakar C, Kharel P, Suryanarayanan R, Naik J T V, Naik R and Lawes G 2008 *Journal of Magnetism and Magnetic Materials* **320** L31–L36
[4] Kaspar T C, Droubay T, Heald S M, Engelhard M H, Nachimuthu P and Chambers S A 2008 *Phys. Rev. B* **77** 201303
[5] Golmar F, Navarro A M M, Torres C E R, Sánchez F H, Saccone F D, dos Santos Claro P C, Benítez G A and Schilardi P L 2008 *Appl. Phys. Lett.* **92** 262503
[6] Park M S, Rhim S H and Freeman A J 2006 *Phys. Rev. B* **74** 205416
[7] Hong N H, Sakai J, Poirot N and Brizé V 2006 *Phys. Rev. B* **73** 132404
[8] Xu Q *et al.* 2008 *Appl. Phys. Lett.* **92** 082508
[9] Lotnyk A, Senz S and Hesse D 2007 *Thin Solid Films* **515**
[10] Francis R J, Moss S C and Jacobson A J 2001 *Phys. Rev. B* **64** 235425
[11] Lanier C H, Rondinelli J M, Deng B, Kilaas R, Poeppelmeier K R and Marks L D 2007 *Phys. Rev. Lett.* **98** 086102
[12] Sasahara A, Droubay T C, Chambers S A, Uetsuka H and Onishi H 2005 *Nanotechnology* **16** S18–S21
[13] Pan D, Xu G, Lv L, Yong Y, Wang X, Wan J, Wang G and Sui Y 2006 *Appl. Phys. Lett.* **89** 082510
[14] Soler J M, Artacho E, Gale J D, García A, Junquera J, Ordejón P and Sánchez-Portal D 2002 *J. Phys.: Condens. Matter* **14** 2745–2779
[15] Blaha P, Schwarz K, Madsen G, Kvaniscka D and Luitz J 2001 *WIEN2k: An Augmented Plane Wave plus Local Orbitals Program for Calculating Crystal Properties* TU Wien, Austria http://www.wien2k.at
[16] Hohenberg P and Kohn W 1964 *Phys. Rev.* **136** B864–B871
[17] Perdew J P and Zunger A 1981 *Phys. Rev. B* **23** 5048–5079
[18] Kitazawa S, Choi Y, Yamamoto S and Yamaki T 2006 *Thin Solid Films* **515** 1901–1904
[19] Brinkman A, Huijben M, van Zalk M, Huijben J, Zeitler U, Maan J C, van der Wiel W G, Rijnders G, Blank D H A and Hilgenkamp H 2007 *Nature Materials* **6** 493

Journal of Physics: Conference Series **167** (2009) 012061 doi:10.1088/1742-6596/167/1/012061

Molecular architectures based on π-conjugated block copolymers for global quantum computation

C A Mujica Martinez[1,3], **J C Arce**[1], **J H Reina**[2] **and M Thorwart**[3,4]

[1]Universidad del Valle, Departamento de Química, A. A. 25360, Cali, Colombia
[2]Universidad del Valle, Departamento de Física, A. A. 25360, Cali, Colombia
[3]Institut für Theoretische Physik IV, Heinrich-Heine-Universität Düsseldorf, 40225 Düsseldorf, Germany
[4]Freiburg Institute for Advanced Studies (FRIAS), Universität Freiburg, 79104 Freiburg, Germany

E-mail: `camujica@univalle.edu.co`, `j.reina-estupinan@physics.ox.ac.uk` `jularce@univalle.edu.co`

Abstract. We propose a molecular setup for the physical implementation of a barrier *global* quantum computation scheme based on the electron-doped π-conjugated copolymer architecture of nine blocks PPP-PDA-PPP-PA-(CCH-acene)-PA-PPP-PDA-PPP (where each block is an oligomer). The physical carriers of information are electrons coupled through the Coulomb interaction, and the building block of the computing architecture is composed by three adjacent qubit systems in a quasi-linear arrangement, each of them allowing qubit storage, but with the central qubit exhibiting a third accessible state of electronic energy far away from that of the qubits' transition energy. The third state is reached from one of the computational states by means of an on-resonance coherent laser field, and acts as a barrier mechanism for the direct control of qubit entanglement. Initial estimations of the spontaneous emission decay rates associated to the energy level structure allow us to compute a damping rate of order 10^{-7} s, which suggest a not so strong coupling to the environment. Our results offer an all-optical, scalable, proposal for global quantum computing based on semiconducting π-conjugated polymers.

1. Introduction

The quantum processing of information has caused great impact from the standpoint of its potential uses, to the point of being considered a 'second quantum revolution' [1]. One of the advantages of these technologies is their capability of encrypting information with extreme security, avoiding the interception of communications between two remote points [2]. Another advantage is the possibility of achieving quantum computation, which exploits the interference principle and the existence of non-local correlations between interacting quantum objects, which have become fundamental physical resources in the development of quantum information protocols. This would allow a quantum computer to carry out tasks that are intractable for a classical computer (e.g. the factorization of a large prime number [3] or the evaluation of a whole domain of a function in a single computational step [2, 4]); nevertheless, the experimental challenge implied by the construction of this kind of devices stems from the difficulty of *locally* controlling the quantum dynamics of the quantum bits (qubits) and the decoherence caused by the interactions with the environment [5, 6, 7].

A qubit is a quantum system in which the Boolean states (classical bits) 0 and 1 are represented by a pair of orthonormal quantum states identified as $|0\rangle$ and $|1\rangle$; the two states form a computational base, and any other state of the qubit can be written as a superposition $\alpha|0\rangle + \beta|1\rangle$, with α and β complex numbers in general, such that $|\alpha|^2 + |\beta|^2 = 1$. Physically, a qubit can be a microscopic system like an atom [8, 9], a nuclear spin [10], a polarized photon [11], or more complex structures like quantum dots [12, 13, 14, 15], molecular arrays [16, 17], and photosynthetic biomolecular systems [18], among many others.

Quantum dots are advantageous due to the existing industry for their nanofabrication and the ease of incorporating them into current opto-electronic devices [19]; besides, quantum computation schemes using quantum dots have been reported [12, 13, 14, 15, 20, 21, 22]. These quantum dots can be constructed from inorganic or organic semiconductors, the latter being of special interest due to previous studies of such [23], where the quantum dots are really organic heterostructures (block copolymers). They are easier to construct than the inorganic systems, since they do not require expensive pieces of equipment as required for molecular beam epitaxy or metal-organic chemical vapour deposition. In principle, it is possible to construct an unlimited variety of organic heterostructures, since the interface between the materials is a chemical carbon-carbon bond, in contrast to inorganic ones, where it is required that the materials exhibit similar lattice constants to avoid interfacial stress [24], which notably limits the variety of heterostructures that can be synthesised.

Nowadays, many theoretical schemes for quantum computation in the solid state can be found. Universal quantum computation requires controlling the dynamics of individual qubits and the interactions between them, the latter being the main cause of decoherence (loss of coherence between quantum superpositions) due to the introduction of control mechanisms external to the system and the ubiquitous coupling to the environment. The use of electromagnetic pulses as a control mechanism does not cause severe decoherence, which has motivated the proposal of all-optical quantum computation schemes [25, 26, 27].

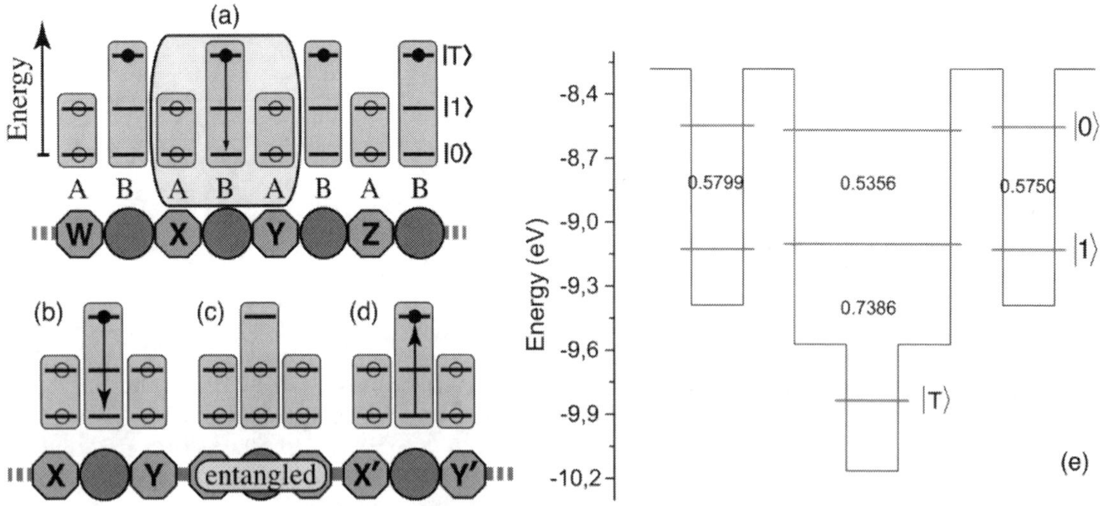

Figure 1. (a) Qubits (A species) coupled through a barrier (B species). (b) The qubits are denoted in the computational basis $\{|0\rangle, |1\rangle\}$, and the central system possesses a third accessible state $|T\rangle$. The barrier is "removed" and all the qubits interact so that they (c) get entangled. (d) The barrier is raised again, once the lateral qubits X' and Y' are decoupled from the central qubit. (e) Conduction-band profile and monoelectronic levels of the proposed molecular architecture, calculated by means of the extended Hückel method. Energies are given in eV.

2. Model

Various control schemes have been proposed to control decoherence, among which the spin-barrier one stands out [7, 26, 27], which employs two types of two-level systems in a linear arrangement: one type carries the information (qubits) and the other type acts as passive barriers of global control between qubits. In this work we propose the implementation of a model based on the following architecture [27]: Fig. 1(a) shows a linear arrangement of qubits (A species) and control units (B species) which possess a third additional level $|T\rangle$ of electronic energy well off resonance with the levels of the computational qubits ($\{|0\rangle, |1\rangle\}$). The linear arrangement is illustrated schematically in the lower part of Fig. 1(a), where the circles represent the barriers and the octagons the computational qubits (W, X, Y, Z).

Fig. 1(b) shows the basic block ABA of Fig. 1(a) at work. In it, the barrier is prepared in the state $|T\rangle$ by the excitation with an external laser, which precludes the interaction between the lateral qubits X and Y (Fig. 1(b)), since $|T\rangle$ is out of resonance with X and Y. To carry out a logical gate, the central system decays to the subspace $\{|0\rangle, |1\rangle\}$ (Fig. 1(c)), and the three systems become entangled under free evolution, which allows the quantum gates to be carried out (Fig. 1(c)). To turn off the interaction, the laser is applied again in such a way that the state $|T\rangle$ is populated again, restoring the barrier, as shown in Fig. 1(d), where a product state of the qubits has been recovered, now represented as $|X'\rangle \otimes |Y'\rangle$.

3. Molecular architecture based on π-conjugated multi-block copolymers

In this work, we implement the model illustrated above employing negatively doped π-conjugated multi-block copolymers [23], where the carriers of the physical information are electrons interacting through the Coulomb force. This architecture has the advantage of permitting a strong coupling between qubits, since it is molecular [27], interacting weakly with phonons, since it is quasi-linear [23], and being scalable. By adjusting the length of the wells and barriers, the nine-block architecture 8 PPP-7 PDA-8 PPP-3 PA-7 (CCH acene)-3 PA-8 PPP-7 PDA-8 PPP was devised, where PPP, PDA, PA and CCH acene refer to poly-(p-phenylene), poly-diacetylene, *trans*-polyacetylene and poly-acene substituted with ethynyl (CCH) groups.

According to the envelope molecular orbital theory [23], the conduction band of the nine-blocks can be visualised as the multiple well illustrated in Fig. 1(e), where the mono-electronic levels localised in each well and their energy differences are indicated. Such levels were determined employing the semiempirical extended Hückel method [28], which includes all the valence electrons and takes into account the overlap between all atomic basis functions.

4. Results

To visualise compactly the shape of the molecular orbitals in one dimension, a localisation parameter on the j-th atom is defined as [23]

$$L_j^{(i)} = \sum_n |c_{jn}^{(i)}|^2, \tag{1}$$

where n labels the atomic orbitals that enter the expression of the i-th Hückel molecular orbital in terms of a linear combination of atomic orbitals

$$|\psi_i\rangle = \sum_r c_r^i |\phi_r\rangle. \tag{2}$$

Since $c_{jn}^{(i)}$ is the contribution of the n-th basis AO centered on the j-th atom to the i-th MO, $L_j^{(i)}$ provides a measure of the electron density at that site of the molecule. Since the value of this parameter is a function of the position along the chain, specified by j, the set of $L_j^{(i)}$ reveals the spatial shape of the MO. In Figs. 2(a)-(h) the localisation parameters are shown for the MOs of the conduction band of the heterostructure.

As Fig. 2 shows, the MOs are localised within one of the wells, which justifies the assignment of each eigenstate to a certain spatial region of the molecule, as indicated in Fig. 1(e). When each of the three wells is doped with an electron, the states $|0\rangle$ and $|1\rangle$ of the lateral wells act as qubits, whereas the state of the central well permits activation and deactivation of the resonant coupling, and therefore the entanglement between them through stimulated optical transitions $|0\rangle \longleftrightarrow |T\rangle$. It must be noted that the ordering of the energy levels is the inverse of the original scheme of Fig. 1(a).

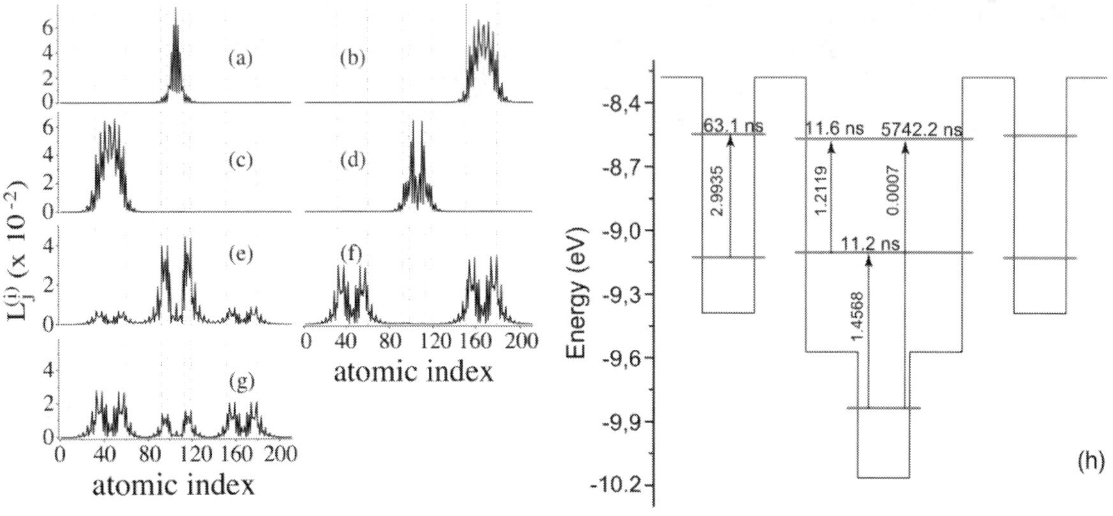

Figure 2. Behaviour of the localisation parameter for the MOs of the conduction band of the heterostructure. (a) LUMO, (b) LUMO+1, (c) LUMO+2, (d) LUMO+3, (e) LUMO+4, (f) LUMO+5, (g) LUMO+6. The dashed lines indicate the position of the barriers of each well. (h) Mean lifetimes (in ns) and oscillator strengths for the molecular architecture, calculated by means of the semiempirical method ZINDO/S [30].

From a calculation of the oscillator strengths for the various possible transitions, we obtain the Einstein coefficients for spontaneous emission, which, in turn, are related to the mean lifetimes of the excited states by $\gamma = \tau^{-1}$ [29]. These lifetimes provide information about the relaxation times of the system and give an indication about the plausibility of the architecture proposed. The calculated values of τ are reported in Fig. 2(h). Assuming a simple exponential decay of the populations of the form $e^{-\gamma t}$, we relate γ with the damping parameter K: $\gamma = K\Delta\varepsilon$, where $\Delta\varepsilon$ is the energy difference between the computational levels considered. For the designed heterostructure proposed here, K is of the order of 10^{-7}. Although there is some small degree of energy mismatch in the designed heterostructure, this can be minimized by refining the quantum-chemistry calculations, and the main source of decoherence would come from electron-phonon coupling effects.

5. Conclusions

The results obtained indicate that the proposed architecture is a plausible system for the physical implementation of global quantum computation schemes. It fulfills the requirement of scalability, and, in addition, the control of the dynamics can be carried out using pulses of the order of picoseconds, which would allow the efficient realisation of one and two qubits unitary transformations, the elementary building blocks for performing a conditional quantum dynamics. The initial estimations made for the heterostructure relaxation rates indicate that the system is not very susceptible to prompt environment-driven decoherence, which is a good indicator that the quantum coherent control proposed in this work is feasible.

Acknowledgments
We acknowledge financial support from Colciencias under contract 1106-45-221296, the scientific exchange program PROCOL (DAAD-Colciencias), and by the Excellence Initiative of the German Federal and State Governments.

References

[1] Dowling J P and Milburn G J 2003 *Phil. Trans. R. Soc. Lond.* A **361** 1655
[2] Nielsen M A and Chuang I L 2000 *Quantum Computation and Quantum Information* (Cambridge: Cambridge University Press)
[3] Shor P W 1997 *SIAM J. Comput.* **26** 1484
[4] Deutsch D and Jozsa R 1992 *Proc. R. Soc. London* A **439** 553
[5] Palma M, Suominen K-A and Ekert A K 1996 *Proc.: Math., Phys. Eng. Sci.* **452** 567
[6] Reina J H, Quiroga L and Johnson N F 2002 *Phys. Rev.* A **65** 032326
[7] Benjamin S C 2004 *New J. Phys.* **6** 61
[8] Cirac J I and Zoller P 1995 *Phys. Rev. Lett.* **74** 4091
[9] Briegel H-J *et al.* 2000 *J. Mod. Opt.* **47** 415
[10] Jones J A and Mosca M 1998 *J. Chem. Phys.* **109** 1648
[11] O'Brien J L 2007 *Science* **318** 1567
[12] Loss D and DiVincenzo D P 1998 *Phys. Rev.* A **57** 120
[13] Reina J H, Quiroga L and Johnson N F 2000 *Phys. Rev.* A **62** 012305
[14] Lovett B, Reina J H, Nazir A and Briggs A 2003 *Phys. Rev.* B **68** 205319
[15] Thorwart M and Hänggi P 2002 *Phys. Rev.* A **65** 012309
[16] Gershenfeld N and Chuang I L 1998 *Sci. Am.* **278** 66
[17] Tesch C M and Vivie-Riedle R 2002 *Phys. Rev. Lett.* **89** 157901
[18] Thorwart M, Eckel J, Reina J H and Weiss S (*Preprint* cond-mat.mes-hall/0808.2906)
[19] Reed M A 1993 *Sci. Am.* **268** 118
[20] Chen G *et al.* 2000 *Science* **289** 1906
[21] DiVincenzo D P 2005 *Science* **309** 2173
[22] Berezovsky J *et al.* 2008 *Science* **320** 349
[23] Mujica Martinez C A 2006 *Effective mass molecular orbital theory applied to quasi-one-dimensional polyacenes and their heterostructures*, undergraduate thesis, Universidad del Valle
[24] Yu P T and Cardona M 1999 *Fundamentals of Semiconductors: Physics and Material Properties* 2nd ed (Berlin: Springer-Verlag)
[25] Calarco T *et al.* 2003 *Phys. Rev.* A **68** 012310
[26] Benjamin S C and Bose S 2003 *Phys. Rev. Lett.* **90** 247901
[27] Benjamin S C , Lovett B W and Reina J H 2004 *Phys. Rev.* A **70** 060305R
[28] Hoffmann R 1963 *J. Chem. Phys.* **39** 1397
[29] McHale J L 1999 *Molecular Spectroscopy* (New York: Prentice Hall)
[30] Ridley J E and Zerner M C 1973 *Theo. Chem. Acta* **32** 111

Reactivity indexes for different geometries of palladium leads

S C Gómez-Carrillo[1] , J Ortega[2] and P G Bolcatto[1,3]

[1] Departamento de Física. Facultad de Ingeniería Química, Universidad Nacional de Litoral, Santiago del Estero 2829 S3000AOM Santa Fe Argentina

[2] Departamento de Física Teórica de la Materia Condensada. Universidad Autónoma de Madrid

[3] Facultad de Humanidades y Ciencias, Universidad Nacional del Litoral

E-mail: scgomez@fiq.unl.edu.ar

Abstract. Electronic transport through metallic break junctions or molecules is clearly dependent not only on the electronic structure of the central nanodevice connecting the leads, but also the shape and crystalline orientation of the contacts which can define the possible conduction channels. In this work we examine different geometries of contacts of palladium characterizing them through global and local reactivity indexes as electrophilicity, chemical hardness and Fukui functions. In molecules, these indicators are essentially defined by the energies of the frontier molecular orbitals and in solids they are related with the local and partial density of states. We use for this purpose an *ab-initio* based code (FIREBALL), applied to plane contacts with (001) fcc faces and also pyramidal tips grown following a (001) and (111) packaging. The results allow us to have an insight about the chemical features of this type of nanojunctions.

1. Introduction

Controllable metallic break junctions connected by molecules are good candidates to realize electronic transport at the nanoscale. Many of this molecular junctions are made with molecules with terminal functional groups as it is the case of thiols or amine [1; 2]. Conductance measurements for these systems show an ample range of values for transport properties, and this diversity of values is mainly attributed to different chemical arrangement between functional groups and metallic leads. Besides, the fabrication method of break junctions not always guarantees that the leads have the same crystalline structure [3]. Thus, it becomes interesting to have tools which allow to give quick insights about the feasibility of different geometrical configuration for leads to react chemically with radicals connecting them. In this work we progress in this sense by calculating reactivity indexes as electrophilicity, chemical hardness and softness and Fukui functions. From the chemical point of view, these indicators are essentially defined by the energies of the frontier molecular orbitals but in solids they are related with the local and partial density of states (DOS). We calculate DOS from an *ab-initio* based code (FIREBALL) using Density Functional Theory. The selected systems are contacts of palladium with different geometries. One of the attractive on this metal resides on it can be form nanowires at the last stage of the breaking [4] with transport measurements presenting a peak at $0.5G_0$ (being G_0 the quantum of conductance) whose interpretation is controversial. Some authors attributes this peak to a spontaneous polarized current [4; 5] and others to a effect of an hydrogen atmosphere [6]. The results allow us to have an insight about the chemical features of this type of nanojunctions.

2. Reactivity indexes

The density functional theory (DFT) is a description of many-body systems with the electronic density ρ as a functional parameter. ρ is related with the number of electrons, N, by $\int \rho(\mathbf{r})d\mathbf{r} = N$ and the value which minimizes the total energy depends on the external potential $v(\mathbf{r})$. Thus, the ground state energy of a system E can be expressed as a functional of N and v, $E[N, v]$ and then a differential change in the ground state energy is:

$$dE = \left(\frac{\partial E}{\partial N}\right)_v dN + \int \left[\frac{\delta E}{\delta v(r)}\right]_N dv(\mathbf{r})d\mathbf{r} = \mu dN + \int \rho(r)dv(\mathbf{r})d\mathbf{r}, \tag{1}$$

where it was defined the chemical potential as $\mu = \left(\frac{\partial E}{\partial N}\right)_v$, and $\rho(\mathbf{r}) = \left[\frac{\delta E}{\delta v(r)}\right]_N$ in a first order of perturbation [7]. From the chemical point of view, μ is identified as the negative of the electronegativity and in solids, for 0 K, is the Fermi energy, E_f.

Equivalently, a differential change in the chemical potential is:

$$d\mu = \left(\frac{\partial \mu}{\partial N}\right)_v dN + \int \left[\frac{\delta \mu}{\delta v(\mathbf{r})}\right]_N dv(\mathbf{r})d\mathbf{r} = 2\eta dN + \int f(r)dv(\mathbf{r})d\mathbf{r}, \tag{2}$$

where η and f are defined by

$$\eta = \frac{1}{2}\left(\frac{\partial \mu}{\partial N}\right)_v = \frac{1}{2}\left(\frac{\partial^2 E}{\partial N^2}\right)_v, \quad f(\mathbf{r}) = \left[\frac{\delta \mu}{\delta v(\mathbf{r})}\right]_N = \left[\frac{\partial \rho(\mathbf{r})}{\partial N}\right]_v. \tag{3}$$

η is the (chemical) hardness, which is basically a half of the difference between ionization and affinity levels in atoms and molecules, and it is a half of the band gap in insulators and semiconductors [8]. $f(\mathbf{r})$ is the so-called Fukui function. This is a local indicator which account for how much of the frontier orbitals are localized at $\mathbf{r}$ position. In this sense, Fukui function measures the reactivity towards a radical reagent and it is normalized as $\int f(\mathbf{r})d\mathbf{r} = 1$ [9].

With the chemical potential and the hardness we can define two more indexes which give us information about as the system could react chemically with an adsorbate or molecule. These are the softness S and the electrophilicity ω:

$$S = \frac{1}{2\eta} = \left(\frac{\partial N}{\partial \mu}\right)_v, \quad \omega = \frac{\mu^2}{2\eta}. \tag{4}$$

S and ω are global indexes and by using the Fukui function it is possible to define local softness and electrophilicity as $s(\mathbf{r}) = f(\mathbf{r})S$, and $\omega(\mathbf{r}) = f(\mathbf{r})\omega$ [7].

Finally, to complete the prescription of calculation we give the relationship between these quantities and the microscopic electronic properties of the material forming the contacts. If $g(E, \mathbf{r})$ and $g(E)$ are the local and total density of states respectively, it can be demonstrated that for 0 K [7; 9]:

$$s(\mathbf{r}) = g(\mathbf{r}, E_f), \quad S = g(E_f), \quad f(\mathbf{r}) = \frac{g(\mathbf{r}, E_f)}{g(E_f)}. \tag{5}$$

3. Results and discussion

For the calculation of the DOS we use an *ab-initio* DFT-based code (FIREBALL)[10; 11]. We use the Local Density Approximation for the exchance-correlation functional. The basis set is composed by numerically generated pseudo-atomic *fireball* wavefunctions and the core effective potentials were calculated following Ref. [12]. We have studied two pyramidal contacts grown following (001) and (111) packaging and a plane contact with a (001) fcc crystalline surface. Figure 1-3 show these geometries and the corresponding DOS. The energy scale was changed so that the Fermi level is fixed to 0 eV and the insets show a detail of DOS around E_f. The reactivity indexes are summarized in Tables 1-3.

In the (001) pyramidal contact, the atom with higher f function is the number 1, although atom 2 has a value quite close to atom 1 [Table (1)]. Atoms 3 and 4 are practically equivalent but they have lower f values. Thus, in this configuration it is expected that an adsorbate is located either near the apex atom or in a bridge place between atoms 1 and 2.

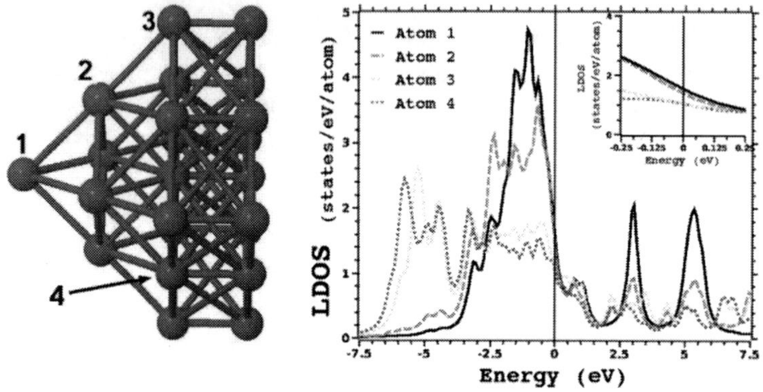

Figure 1. Pyramidal lead grown following (001) packaging and LDOS for labelled atoms.

Table 1. Reactivity indexes for (001) pyramidal tip. η, ω and $\omega(\mathbf{r})$ are given in eV, S in eV^{-1}, $s(\mathbf{r})$ in 1/eV/atom and $f(\mathbf{r})$ is dimensionless.

Global indexes				Local indexes		
η	S	ω		$f(\mathbf{r})$	$s(\mathbf{r})$	$\omega(\mathbf{r})$
0.44	1.13	52.98	Atom 1	1.38	1.57	73.36
			Atom 2	1.25	1.42	66.29
			Atom 3	0.94	1.06	49.71
			Atom 4	0.92	1.04	48.82

For the (111) tip the scenario changes a little from the (001) case. Here, we have again the atom 1 as the more reactive while atoms 2 and 3 have almost the same low value for Fukui function [Table (2)]. Atom 4 is the less favorable site for a bonding. Differences between atoms 3 and 4 can be understood from the fact that atom 3 form part of the edge of the pyramid and then it has a different coordination compared with the atom 4, located at the face of the tip.

In the (001) surface all the atoms are symmetrically equivalent and then we have only one set of local indexes whose values are clearly lower than contacts with pyramidal shape [Table (3)]. In order to do a relative comparison among different geometries, the local electrophilicity is the proper index to observe. This is because $\omega(\mathbf{r})$, in contrast with f, contains information

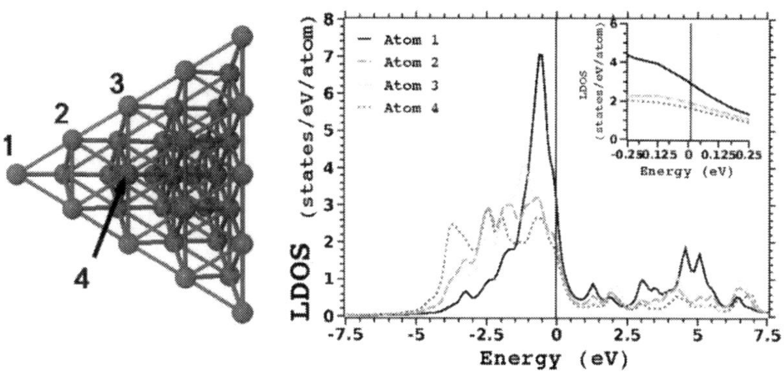

Figure 2. Pyramidal lead grown following (111) packaging and LDOS for labelled atoms.

Table 2. Reactivity indexes for (111) pyramidal tip. Same conventions that in Table 1.

Global indexes				Local indexes		
η	S	ω		$f(r)$	$s(r)$	$\omega(r)$
0.26	1.96	103.77	Atom 1	1.53	3.00	159.23
			Atom 2	0.98	1.92	101.66
			Atom 3	0.96	1.89	100.06
			Atom 4	0.84	1.64	86.98

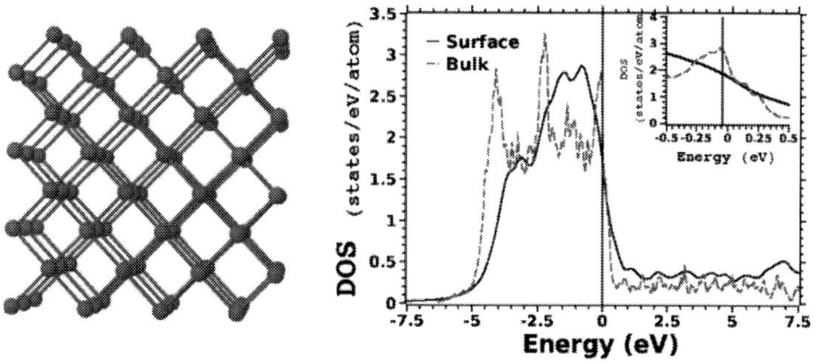

Figure 3. (001) superficial plane and DOS for (001) surface and bulk.

Table 3. Reactivity indexes for (001) plane. Same conventions that in Table 1.

Global indexes				Local indexes		
η	S	ω		$f(r)$	$s(r)$	$\omega(r)$
0.27	1.82	114.69	Surface	1.03	1.87	117.73

not only about the spatial distribution of the frontier states but also about the relative energetic ordering. Thus, Fukui functions are suitable to analyze reactivity in different places of a given system, while local electrophilicity becomes a better tool to compare chemical activity of regions of different systems. By comparing the maximal local electrophilicities for the three geometries we observe that the tendency to form chemical bond follow the ordering: (111) tip → (001) tip → (001) surface. On the other hand, for pyramidal contacts, it is predicted in both cases that the apex atom is the more convenient site to realize the contact, although the edge region can be an alternative site to do this. These result are in qualitatively agreement with previous ones for H adsorption on a (111) Pd tip and Au nanowires [13; 14] and for dissociation of O_2 and O_3 molecules on Al pyramidal contacts oriented in the (111) direction [15]. Nevertheless, here we have calculated indexes only for the contacts and then the results offer us a first order insight about the chemical reactivity in molecular junctions composites. More conclusive statements could be extracted if these indexes were calculated for each pair of molecule/lead combination. Moreover, following the ideas of chemical reactivity, it should be necessary the definition of (solid state) Fukui functions and local electrophilicity for electrophilic (acceptor) or nucleophilic (donor) attacks. One way to progress in this direction are suggested in Ref. [16].

4. Conclusions

Starting from *ab initio* based DFT calculation, we have evaluated reactivity indexes for different contacts of palladium. Local electrophilicity results indicate that (111) oriented tips are more reactive than (001) ones and these, more than the (001) surface. Besides, from Fukui function results, the preferred site for the chemical bonding is on the surroundings of an edge atom of the pyramid, being the apex atom the most convenient place for bonding. The results are in overall agreement with those obtained for similar atom/metal junctions.

Acknowledgments

SCGC and PGB acknowledge the financial support by CONICET. This work was supported by Grant nro 14248-128 by Fundación Antorchas.

References

[1]Reed M A, Zhou C, Muller C J, Burgin T P and Tour J M 1997 *Science* **278** 252
[2]Venkataraman L, Klare J E, Tam I W, Nuckolls C, Hybertsen M s and Steigerwald M L 2006 *Nano Lett.* **6** 458
[3]Rodrigues V, Fuhrer T and Ugarte D 2000 *Phys. Rev. Lett.* **85** 4124
[4]Rodrigues V, Bettini J, Silva P C and Ugarte D 2003 *Phys. Rev. Lett.* **91** 096801
[5]Romero M A, Gómez-Carrillo S C, Bolcatto P G and Goldberg E C *unpublished*
[6]Csonka S, Halbritter A, Mihály G, Shklyarevskii O I, Speller S and van Kempen H 2004 *Phys. Rev. Lett.* **93** 016802
[7]Parr R G and Yang W 1989 *Density-Funtional Theory of Atoms and Molecules* (New York: Oxford University Press)
[8]Parr R G and Pearson R G 1983 *J. Am. Chem. Soc.* **105** 7512
[9]Parr R G and Yang W 1984 *J. Am. Chem. Soc.* **106** 4049
[10]Sankey O F and Niklewski D J 1989 *Phys. Rev. B* **40** 3979
[11]Demkov A A, Ortega J, Sankey O F and Grumbach M P 1995 *Phys. Rev. B* **52** 1618
[12]Hamann D R 1989 *Phys. Rev. B* **40** 2980
[13]Pieczyrak B, González C, Jelínek P, Pérez R, Ortega J and Flores F 2008 in preparation
[14]Jelínek P, Pérez R, Ortega J and Flores F 2006 *Phys. Rev. Lett.* **96** 046803
[15]Wohlthat S, Pauly F, Viljas J K, Cuevas J C and Schon G 2007 *Phys. Rev. B* **76** 075413
[16]Dronskowski R 1992 *J. Am. Chem. Soc.* **114** 7230

Non-equilibrium entanglement in a driven many-body spin-boson model

Victor M Bastidas[1], **John H Reina**[1] **and Tobias Brandes**[2]
[1]Universidad del Valle, Departamento de Física, A. A. 25360, Cali, Colombia
[2]Institut für Theoretische Physik, Technische Universität Berlin, Hardenbergstr. 36, 10623 Berlin, Germany

E-mail: vicmabas@univalle.edu.co and j.reina-estupinan@physics.ox.ac.uk

Abstract. We study the entanglement dynamics in the externally-driven single-mode Dicke model in the thermodynamic limit, when the field is in resonance with the atoms. We compute the correlations in the atoms-field ground state by means of the density operator that represents the pure state of the universe and the reduced density operator for the atoms, which results from taking the partial trace over the field coordinates. As a measure of bipartite entanglement, we calculate the linear entropy, from which we analyze the entanglement dynamics. In particular, we found a strong relation between the stability of the dynamical parameters and the reported entanglement.

1. Introduction

Understanding the properties displayed by the entanglement of physical systems is one of the fundamental purposes of quantum information theory [1]. A quantum system composed of two or more entangled subsystems has the interesting property that although the state of the total system can be well defined, it is impossible to identify individual properties for each one of its parts. A subject that has attracted recent interest is linked to the relationship between entanglement and certain physical properties of many body quantum systems; this has taken special interest in relation to quantum phase transitions (QPTs) [2, 3, 4]. A QPT in a many body system strongly influences the behavior of the system near to the critical point, with the consequent appearance of long-range correlations in the ground state. Although several proposals exist, currently there is no complete theory of multipartite entanglement, and the common techniques are based on bipartite decompositions of the total system. This type of decomposition has made it possible to study the entanglement in the Dicke model in thermal equilibrium [5, 6, 7]. Here, we study the entanglement properties of a many-body system in the non-equilibrium regime, in which the total system has a unitary dynamics but is not an isolated system. For this purpose we study the single mode "externally-driven" Dicke Hamiltonian

$$\hat{H}(t) = \omega_0 J_z + \omega a^\dagger a + \frac{g(t)}{\sqrt{N}}(J_+ + J_-)(a^\dagger + a), \tag{1}$$

where $J_z = \sum_{i=1}^{N} J_z^{(i)}$, $J_\pm = \sum_{i=1}^{N} J_\pm^{(i)}$ are collective atomic operators, ω_0 is the level splitting of the atoms, ω is the frecuency of the bosonic mode and $g(t) = g + \Delta g \cos \Omega t$ is the time dependent atom-field coupling. We assume that Δg is a fraction of the static coupling g. An exact diagonalisation of the problem Eq. (1) has been previously carried out by means of the Holstein-Primakoff transformation for the case of a *static* atom-field coupling $g(t) \equiv g = constant$ [5, 6], but there is no known solution for the time dependent atom-field coupling case.

2. The reduced density operator

When studying a composed quantum system whose dynamic is unitary, it is interesting to study the dynamics of the parts of the system. In contrast to the dynamics of the whole system, the dynamics of the subsystems is not unitary. However, it is possible to make a description of the subsystem through the reduced density operator. The Schrödinger equation for the single mode "externally-driven" Dicke Hamiltonian has an exact solution when the field is in resonance with the atoms $\omega = \omega_0$. With the purpose of formulating the exact solutions, we introduce an abstract coordinate representation through the coordinates of the physical field x and the atoms coordinate y. In terms of these coordinates, the ground state of the universe (atoms+field) is given by the expression [8]

$$\Psi_{0_-,0_+}(x,y,t) = \exp(i\gamma_{0_-}(t))\exp(i\gamma_{0_+}(t))\Phi^-_{0_-}\left(\frac{x}{\sqrt{2}} - \frac{y}{\sqrt{2}}, t\right)\Phi^+_{0_+}\left(\frac{x}{\sqrt{2}} + \frac{y}{\sqrt{2}}, t\right), \quad (2)$$

where

$$\Phi^{\mp}_{0_{\mp}}(w,t) = \left(\frac{1}{2\pi|B^{\mp}(t)|^2}\right)^{1/4} \exp\left(i\frac{\dot{B}^{\mp}(t)}{2B^{\mp}(t)}w^2\right),$$

and the time dependent phase is given by

$$\gamma_{0_{\mp}}(t) = \int_0^t \langle \Phi^{\mp}_{0_{\mp}}, t| i\frac{\partial}{\partial t} - \hat{H}(t)|\Phi^{\mp}_{0_{\mp}}, t\rangle dt.$$

It is interesting to note that the dynamics of the system is influenced by the dynamics of the auxiliary parameters $B^{\mp}(t)$, which are solutions of the Mathieu equation [9]

$$\ddot{B}^{\mp}(t) + [(1 \mp 2g) \mp (2\Delta g)\cos\Omega t]B^{\mp}(t) = 0, \quad (3)$$

subject to the Wronskian condition

$$\dot{B}^{\mp}(t)(B^{\mp})^*(t) - B^{\mp}(t)(\dot{B}^{\mp})^*(t) = i.$$

We describe the dynamics of the total system in terms of the density operator, where the pure state of the universe is represented by the operator $\hat{\rho}_G(t) = |\Psi_{0_-,0_+}, t\rangle\langle\Psi_{0_-,0_+}, t|$. In the coordinate representation, the density matrix takes the form

$$\rho_G(x', y'; x, y, t) = \Psi^*_{0_-,0_+}(x,y,t)\Psi_{0_-,0_+}(x', y', t). \quad (4)$$

At this stage we develop a bipartite decomposition of the universe, which lets us study the reduced dynamics of the atoms through the reduced density matrix (RDM). In so doing, we calculate the partial trace over the physical field coordinate x,

$$\rho^{(R)}_G(y', y, t) = \int_{-\infty}^{+\infty} \Psi^*_{0_-,0_+}(x,y,t)\Psi_{0_-,0_+}(x,y',t)\, dx, \quad (5)$$

and as a result of this, the ground state of the universe $\Psi_{0_-,0_+}(x,y,t)$ given in Eq. (2) is Gaussian, which facilitates the integration of Eq. (5). After some algebraic calculations, we

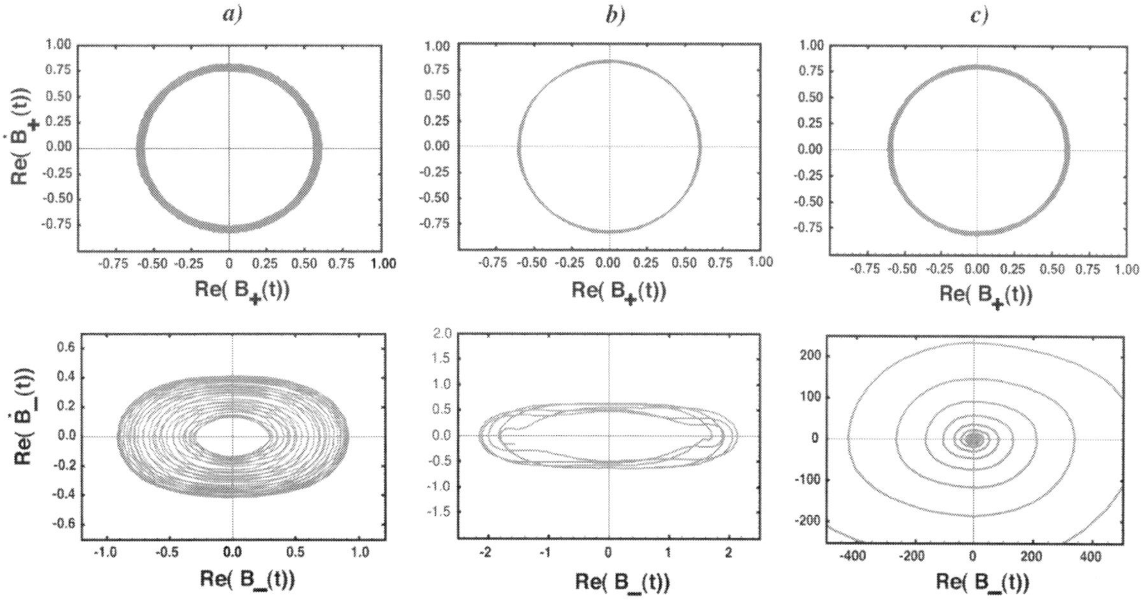

Figure 1. Phase space trajectories of the auxiliary dynamical parameters $B^{\pm}(t)$ in the case $\Delta g = 0.1g$, for a value of the static coupling in a) the stable region $g = 0.40$, b) the stable region $g = 0.46$, and c) the unstable region $g = 0.38$.

obtain the following result for the reduced density matrix

$$\rho_G^{(R)}(y', y, t) = \left(\frac{\Re e(\xi^-)\Re e(\xi^+)}{\pi(\Re e(\xi^-)c^2 + \Re e(\xi^+)s^2)}\right)^{1/2} \exp\left(\frac{c^2 s^2[y(\xi^- - \xi^+)^* + y'(\xi^- - \xi^+)]^2}{4(\Re e(\xi^-)c^2 + \Re e(\xi^+)s^2)}\right) \times \exp\left(-\frac{y^2}{2}[(\xi^-)^* s^2 + (\xi^+)^* c^2]\right) \exp\left(-\frac{(y')^2}{2}[(\xi^-)s^2 + (\xi^+)c^2]\right), \quad (6)$$

where $c = s = 1/\sqrt{2}$ y $\Re e(\xi^{\mp})$ is the real part of the function $\xi^{\mp}(t) = -i\frac{\dot{B}^{\mp}(t)}{B^{\mp}(t)}$; in particular, when $\Delta g = 0$ and $\omega = \omega_0 = 1$, this function is independent of time and is defined as $\xi^{\mp}(t) = \epsilon^{\mp} = \sqrt{(1 \mp 2g)}$. Hence, our model reproduces the results previously in the literature for the atom's reduced density matrix in the single mode Dicke model in thermal equilibrium [5, 6].

3. The linear entropy

We consider a pure state of a composed system AB (universe) represented by the density operator $\hat{\rho}_{AB}$. The linear entropy for the reduced density operator $\hat{\rho}_A = tr_B(\hat{\rho}_{AB})$ of the subsystem A is defined as $L_A = 1 - tr_A(\hat{\rho}_A^2)$. This gives a measure of purity of the reduced density operator $\hat{\rho}_A$ (o $\hat{\rho}_B$) of one part of the total system (in a bipartite decomposition of the universe); therefore, if the pure state of the universe is separable, the reduced density operator of one part of the system represents a pure state and as a result the linear entropy must be zero. Similarly, if the pure state of the universe is a maximally entangled state [1], the linear entropy is 1/2. In the context of the "externally driven" single mode Dicke model, it is possible to use the linear entropy as a measure of bipartite entanglement between the atoms and the field. The linear entropy for the time dependent reduced density operator of the atoms, Eq. (6), is thus given by

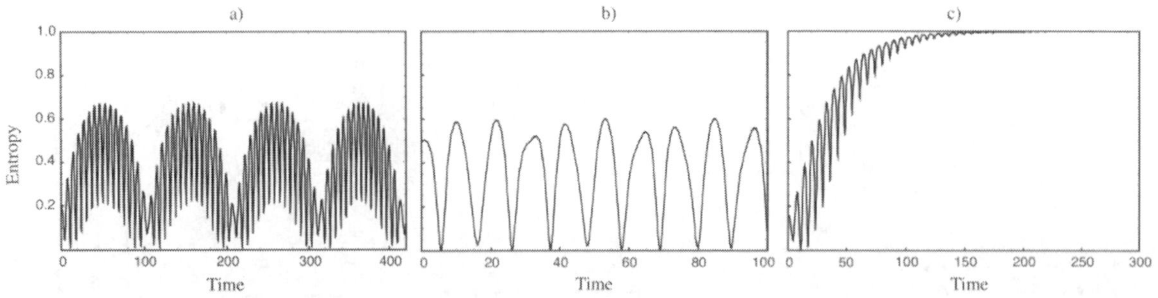

Figure 2. Linear entropy for $\Delta g = 0.1g$; a) $g = 0.40$, b) $g = 0.46$, and c) $g = 0.38$.

$L(t) = 1 - tr\left[\left(\hat{\rho}_G^{(R)}(t)\right)^2\right]$. With the aim of calculating the linear entropy explicitly, we can use the representation of the density matrix, Ec. (6)

$$L(t) = 1 - \int_{-\infty}^{+\infty}\int_{-\infty}^{+\infty} \rho_G^{(R)}(y', y, t)\rho_G^{(R)}(y, y', t)\; dy\; dy',$$

and after some algebra, we obtain the expression for the linear entropy

$$L(t, g) = 1 - \frac{\pi\Lambda^2}{\sqrt{4(\Re e(\alpha))^2 - \beta^2}}\,, \tag{7}$$

where the parameters α, β and Λ are defined as

$$\begin{aligned}\alpha &= \frac{(\xi^-)^* s^2 + (\xi^+)^* c^2}{2} - \frac{c^2 s^2[(\xi^- - \xi^+)^*]^2}{4(\Re e(\xi^-)c^2 + \Re e(\xi^+)s^2)},\; \beta = \frac{c^2 s^2(\xi^- - \xi^+)^*(\xi^- - \xi^+)}{2(\Re e(\xi^-)c^2 + \Re e(\xi^+)s^2)},\\ \Lambda &= \left(\frac{\Re e(\xi^-)\Re e(\xi^+)}{\pi(\Re e(\xi^-)c^2 + \Re e(\xi^+)s^2)}\right)^{1/2}.\end{aligned} \tag{8}$$

4. Results

Interestingly, the linear entropy exhibits a time dependence that is not determined by the global phase of the ground state wave function of the universe, Eq. (2), but by the auxiliary dynamical parameters $B^{\mp}(t)$. In the particular case $\Delta g = 0.1g$ and $\Omega = 1$, the stability zones of Eq. (3) are known [8], and the study of the corresponding solutions is based on the Floquet's theorem for second order differential equations with time periodic coefficients. For this, the solutions of Eq. (3) have the general form $B^{\pm}(t) = \exp(iF^{\pm}t)\phi^{\pm}(t)$, where $\phi^{\pm}(t+T) = \phi^{\pm}(t)$ and $F^{\pm}$ is the Floquet exponent, which depends on the shape of the driving. For driving functions for which $F^{\pm}$ is complex, the solution becomes unstable. In the stable regime, $F^{\pm}$ is a real number [9]. In Figs. 1(a), (b) we consider the phase space representation of the trajectories of the auxiliary dynamical parameters $B^{\pm}(t)$. For values of g that belong to a common stability zone ($g = 0.4$ and $g = 0.46$) [8] the solutions are bounded in the phase space; in order to establish a relationship between the stability and the entanglement dynamics, we consider the linear entropy in Figs. 2(a), and (b) for the values of the static coupling described above. Our choice of the initial conditions for Eq. (3) implies that the system starts in thermal equilibrium, but is described by a non-separable quantum state. Strikingly, the dynamics exhibits a behaviour whereby the system experiences a disentanglement process, before reaching a maximum value of entanglement, and then successive collapses and revivals.

Figure 1(c) shows another interesting behaviour: when the static coupling belongs to the instability zone ($g = 0.38$) of $B^-(t)$, the solution $B^+(t)$ is bounded while the solution B^- is unbounded in the phase space. For this value of the static coupling, the linear entropy oscillates, before it reaches the stationary state with linear entropy $L = 1$, as shown in Fig. 2(c). In [8] we discuss the relationship between the stability zones and the localization of the ground state wave function of the universe. In order to study this relation, we define the characteristic length $l^{\mp}(t) = \sqrt{2}|B\mp(t)|$, which is bounded in the stable zones and unbounded in the unstable zones. The numerical simulations of the ground state probability density show that for values of the static coupling that belong to the common stability zones the probability is localized in the abstract $x - y$ space and presents an oscillatory behaviour. In contrast, for values of the static coupling that belong to the unstable zones, the probability density has an oscillatory behavior. However, when the system evolves, the density is systematically stretched in a fixed direction, with a consequent dilation in the perpendicular direction.

5. Conclusions

We have obtained exact results for the atoms reduced density operator and the linear entropy for the "externally driven" Dicke model, in the thermodynamic limit. These results allows us to study the entanglement dynamics through the auxiliary dynamical parameters. In contrast, in the Dicke model in thermal equilibrium [5, 6], the atoms-field entanglement does not have dynamics because the temporal dependence of the ground state of the universe is given by a global phase of the wave function. In this work the auxiliary dynamical parameters and its stability properties determine the entanglement dynamics of the system and the behavior of the universe ground state wave packet.

Acknowledgments

We acknowledge financial support from Colciencias under contract 1106-45-221296, and the scientific exchange program PROCOL (DAAD-Colciencias).

References

[1] Nielsen M A and Chuang I L 2000 *Quantum Computation and Quantum Information* (Cambridge University Press)
[2] Legeza Ö and Sólimon J 2006 *Phys. Rev. Lett.* **96** 116401
[3] Dusuel S and Vidal J 2004 *Phys. Rev. Lett.* **93** 237204
[4] Wu L A, Sarandi M S and Lidar D A 2004 *Phys. Rev. Lett.* **93** 250404
[5] Lambert N, Emary C and Brandes T 2004 *Phys. Rev. Lett.* **92** 073602
[6] Lambert N, Emary C and Brandes T 2005 *Phys. Rev.* A **71** 053804
[7] Tolkunov D and Solenov D 2007 *Phys. Rev.* B **75** 024402
[8] Bastidas V M, Reina J H and Brandes T 2008 (unpublished)
[9] Abramowitz M and Stegun I A 1972 *Handbook of Mathematical Functions* (Dover)

Valence band states in Si-based p-type delta-doped field effect transistors

Martínez-Orozco J. C., Stoyan J. Vlaev.
Unidad Académica de Física. Universidad Autónoma de Zacatecas.
Calzada Solidaridad esquina con Paseo la Bufa S/N, C.P. 98060. Zacatecas, Zac. México.
E-mail: `jcmover@correo.unam.mx`.

Abstract. We present tight-binding calculations of the hole level structure of δ-doped Field Effect Transistor in a Si matrix within the first neighbors sp^3s^* semi-empirical tight-binding model including spin. We employ analytical expressions for Schottky barrier potential and the p-type δ-doped well based on a Thomas-Fermi approximation, we consider these potentials as external ones, so in the computations they are added to the diagonal terms of the tight-binding Hamiltonian, by this way we have the possibility to study the energy levels behavior as we vary the backbone parameters in the system: the two-dimensional impurity density (p_{2d}) of the p-type δ-doped well and the contact voltage (V_c). The aim of this calculation is to demonstrate that the tight-binding approximation is suitable for device characterization that permits us to propose optimal values for the input parameters involved in the device design.

1. Introduction

The field effect transistor with δ-doping (δ-FET) was proposed firstly by F. Schubert and K. Ploog [1, 2, 3]. They constructed a FET with a Schottky base (metal-semiconductor contact) grown by MBE. The δ-doped layer of n-type impurities in this device is located at a fixed distance from the metal-semiconductor contact and spreads the whole region from the source terminal to the drain terminal. The authors compare this transistor with the FET and HEMT (high electron mobility transistor), the latter having an inversion layer channel. The δ-FET is a system where a high carrier concentration of the two-dimensional electron gas can be achieved. As a consequence, high transversal conductance (g_m) results, the latter being inverse proportional to the transversal resistance, then a low dissipation is expected. Additionally, the δ-FET has a higher avalanche breaking potential than the other transistors. The reported mobility is higher than 3000 $cm^2\ V^{-1}\ s^{-1}$ at $T = 77$ K. The authors interpret the system as a quasi-two-dimensional electron cloud strongly localized in a biased triangular quantum well (V-type) with a constant electrostatic field ϕ_b/d where ϕ_b is the Schottky barrier height and d is the distance between the well and the barrier.

A δ-FET in Si is constructed by Wu *et al.* [4, 5] with a B δ-doped quantum well as conduction channel (δ-FET in Si of p-type). Hall measurements observe a hole mobility of 120 $cm^2\ V^{-1}\ s^{-1}$ at T=300 K and of 180 $cm^2\ V^{-1}\ s^{-1}$ at T=77 K. The authors estimate an impurity density of the carriers in the δ-doped layer of 1.8×10^{12} cm^{-2} approximately and demonstrate this δ-FET has an extrinsic transversal conductance g_m of 640 $\mu s/mm$ for a gate length of 5 μm.

A deep understanding of the physical phenomena in δ-FETs based on different host materials is needed to improve the optimization of the device parameters. The first step in this process has to be a realistic calculation of the electronic structure of these systems. Semi-empirical tight-binding considerations [6] seem to provide additional information to the results obtained by simple effective-mass models. In the present work we have conducted the first tight-binding calculation of the potential barrier influence on the electronic structure of V-shaped quantum well in p-type δ-doped Silicon FET.

2. Model

The main assumption of the model describing the valence band of the semiconductor crystal in p-type Si δ-doped FET is that the potential profile can be constructed within the depletion region approximation in the proximity of the metal-semiconductor contact and by the V-shaped potential profile obtained in the Thomas-Fermi approach [7] that describes correctly the properties of the isolated p-type δ-doped well, this fact is clearly explained in reference [8]. The model has been applied to study the differential capacitance profile of the device [9, 10], the electronic spectra of δ-doped quantum wells of n- and p- types in GaAs [11] and p-type B δ-doped Si quantum wells [12]. The potential model is:

$$V(z) = -\frac{2\pi e^2}{\varepsilon_r} P_d (z-l)^2 \theta(l-z) + \frac{\alpha^2}{(\alpha|z-d|+z_0)^4}\theta(l_p - z), \tag{1}$$

where $\alpha = \frac{2m_a^{3/2}}{(15\pi)}$ and $z_0 = \left(\frac{\alpha^3}{\pi p_{2D}}\right)^{1/5}$, with $m_a = \left[1 + (m_{lh}/m_{hh})^{3/2} + (m_{so}/m_{hh})^{3/2}\right]^{2/3}$. P_d being the background impurities density, ε_r is the electric permittivity constant of Silicon, d is the distance at which the p-type δ-doped well is positioned, p_{2D} is the two-dimensional impurities density of the well. And finally l is the screening distance for the electrical field given by:

$$l = \sqrt{\frac{\varepsilon_r V_c}{4\pi e^2 P_d}}, \tag{2}$$

where V_c is the contact voltage, that is an input parameter in our calculations.

To take into account an external constant electric field applied to a planar heterostructure in the growth direction in the tight-binding formalism, the value of the external potential is introduced to all diagonal elements of the Hamiltonian matrix in each atomic layer [13]:

$$TB_{ii}(n) = TB_{ii}(0) + neF, \tag{3}$$

In the equation 3 the discrete variable n is used, instead of the continuous variable z, to label the atomic layers in the growth direction. The zero of the external potential is in the atomic layer n=0, e is the electron charge, F is the magnitude of the constant electric field applied along to the growth direction, and TB_{ii} are the diagonal tight-binding parameters (i being the atomic orbitals index). This approximation is considered a reasonable one, at least as a first step, and will not change the nondiagonal tight-binding parameters. This equation shows a shift in the energetic positions of all atomic orbitals in a given atomic layer n with the potential drop neF of the field potential.

In the present work we consider the potential V(z) from Eq. 1 as an external potential applied to a finite region of the projected bulk in the growth direction. Similarly to Eq. 3, we have

$$TB_{ii}(n) = TB_{ii}(0) + V(n) \tag{4}$$

The potential V(n) is the potential V(z) from equation 1, written in discrete notation. The tight-binding (TB) calculations in the present paper are made in spin-dependent sp^3s^* basis (nearest neighbors) at the center of the two-dimensional Brillouin zone for the (001) growth direction of δ-doped Silicon. The Silicon tight-binding parameters were taken from the work [14]

We applied this parameterization to calculate the Stark shifts in quantum wells with different potential profiles [15, 16] and obtained very good results. A next step could be the incorporation of the charge neutrality in a self-consistent way, as in [17], to obtain quantitative dependence of the electronic properties on the contact potential.

The semi-empirical tight-binding model complements the EFA approximation giving information about the dependence of the delta-FET electronic structure on the crystal and band structure of the host material and on the orbital components. Our model can provide also a more detailed study of the total density of states taking into account the electron plane motion considering explicitly all three dimensions of the system. This fact could be a serious advantage in the calculations of many macroscopical properties of the δ-FET.

Results and discussion

We present the electronic spectrum description of the δ-FET in a Silicon host material with p-type δ-doping. The following values of the system parameters are used: the background impurities density $P_d = 1.0 \times 10^{18}$ cm^{-3}, the contact potential V_c was varied between -100 meV and -1200 meV, the two-dimensional impurities density is $p_{2D} = 3.5 \times 10^{12}$ cm^{-2}, the Silicon electric permittivity ε_r=11.7, the heavy hole effective mass m_{hh}=0.52, the light hole effective mass m_{lh}=0.16 and the split-off one is m_{so}=0.23. The distance between the Schottky barrier and the p-type δ-doped well us fixed to 300 Å.

We have considered a slab of 600 monolayers containing the whole system. This size is sufficiently large to avoid the truncation effect for the confined states. The small imaginary part added to the energy as well as the energy step in the Green function numerical calculations were chosen to be 1 meV. We have calculated the density of states to obtain the energies of the confined states and then for each energy we have calculated also the spatial distribution of the state over the slab to determine the symmetry and the type of the confined hole states. The finite slab models our semi-infinite system. One of the surfaces matches the vacuum and here we have a free surface without reconstruction. The second surface is artificial, this is because we consider a large slab. By this way we avoid the influence of the second free surface to the discrete energy spectrum. We have applied this model in [18] obtaining very good results.

In figure 1 we show the density of states (DOS) of p-type δ-doped Silicon FET for different contact potential values. The energy zero coincides with the flat band region of the valence band.The confining potential and the spatial distributions of the ground heavy and light hole states are presented in figure 2 when V_c=-500 meV. We can understand the electronic spectrum of our system in terms of the confining potential and interpret qualitatively the features of all spectra in terms of this potential, but only the realistic calculation based on the resolution of the Schrödinger equation in the framework of the model presented in the previous section permit us to obtain semi-quantitative results. Finally in figure 3 we show influence of the contact potential on the energies of the confined hole states for this system, the ground states are mainly affected for contact potentials higher than -800 meV and we obtain that the first heavy hole excited state escapes at contact potential value of -900 meV.

Tight-binding calculations of the electronic spectrum in p-type δ-doped quantum wells in Si are presented in [19]. The confining potential is treated as an external one and the diagonal elements of the tight-binding Hamiltonian are modified adding the corresponding terms as in the present work. The authors have used the TB parameters from [14], the same set of parameters we have used in our calculations. The comparison of the results obtained within the EFA and TB approximations given in [19] demonstrates that such kind of approach is suitable for this type of calculations. The agreement between the calculations and the experimental data reported in [19] is quite well.

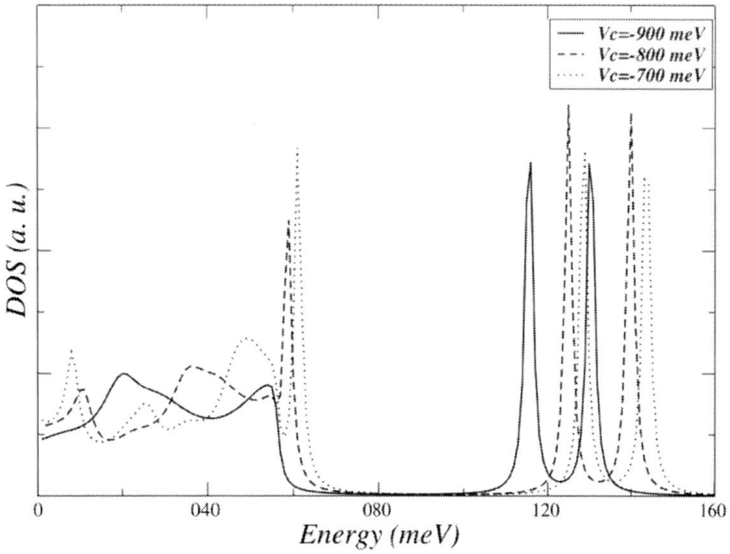

Figure 1. DOS for a p-type δ-doped Silicon FET when $p_{2D} = 3.5 \times 10^{12}$ cm^{-2}, and V_c=-700 meV (pointed line), V_c=-800 meV (dashed line) and V_c=-900 meV (solid line).

3. Conclusions

In summary, we present the first tight-binding calculation of the electronic structure of a p-type δ-doped silicon FET. The computation relies on analytical expressions for the Schottky barrier and δ-doped potentials obtained previously [10] within the depletion and Thomas-Fermi approximation, respectively. These potentials are considered as external ones in the tight-binding formalism and added to the diagonal terms of the tight-binding Hamiltonian. It is shown that the tight-binding formalisms is suitable for the device characterization because the computed energy level behavior reflects the presence of the Schottky barrier potential. The presented scheme permits also to consider the dependence of the electron properties of the δ-doped FET on the band structure, growth direction and orbital components.

Acknowledgments

J.C.M.O and S.J.V acknowledge to the National Science and Technology Council (CONACyT) from México, through the retention program, J.C.M.O also acknowledges to the Universidad Autónoma de Zacatecas (UAZ) and to the Consejo Zacatecano de Ciencia y Tecnología (COZCyT) for the financial support.

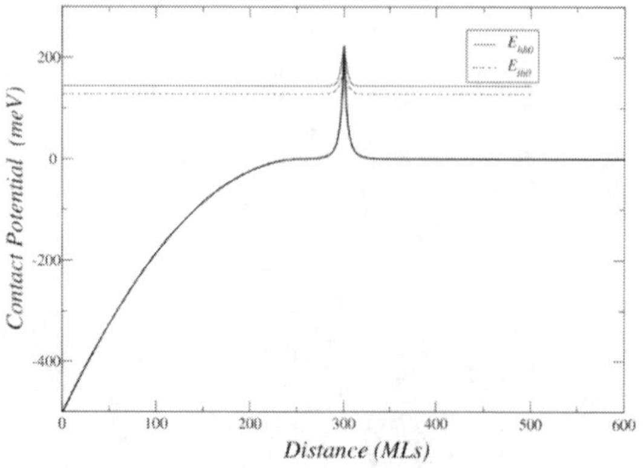

Figure 2. Potential profile and hole wave function for a p-type δ-doped Silicon FET with $p_{2D} = 3.5\times10^{12}$ cm^{-2} and V_c=-500 meV.

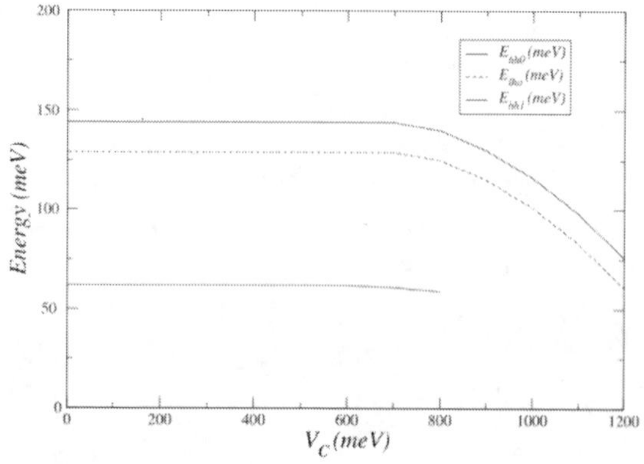

Figure 3. Energy levels versus the contact voltage for a p-type δ-doped silicon FET with $p_{2D} = 3.5 \times 10^{12}$ cm^{-2} and d=300 Å.

4. References

[1] E. F. Schubert and K. Ploog. Jpn. J. Appl. Phys. **24**, L608 (1985).
[2] E. F. Schubert, A. Fischer and Klaus Ploog, IEEE Transactions on electron devices, **33**, 625 (1986).
[3] K. Ploog, Physica Scripta **T19A**, 136 (1987).
[4] T. K. Carns, X. Zheng, K. L. Wang, S. L. Wu and S. J. Wang. Journal of Vacuum Science and technology: Microelectronic and Nanometer Structures. **12**, 1203 (1994).
[5] S. L. Wu, T. K. Carns, S. J. Wang, K. L. Wang, Appl. Phys. Lett. **63**, 1363 (1993).
[6] J.C. Martínez-Orozco, I. Rodríguez-Varga, S.J. Vlaev and L.M. Gaggero-Sager. Rev. Mex. Fis. S**53**, 128 (2007).
[7] L. Ioriatti, Phys. Rev. B. **41**, 8340 (1990).
[8] Rodríguez-Vargas I and Gaggero-Sager L M 2007 *Physica B* **389** 227.
[9] J. C. Martínez-Orozco, L. M. Gaggero-Sager and S. J. Vlaev. Materials Science and Engineering B. **B84**, 155-158 (2001).
[10] L. M. Gaggero-Sager and R. Pérez-Alvarez. J. Appl. Phys. **78**(7), 4566-4569 (1995).
[11] S. Vlaev, L. M.Gaggero-Sager. Physical Review B, **58**, 1142(1998).
[12] L. M. Gaggero-Sager, S. Vlaev, G. Monsivais. Computational Materials Science, **20**, 177 (2001).
[13] M. Graf and P. Vogl, Phys. Rev. B **51**, 4949 (1995).
[14] G. Klimeck, R. Chris Bowen, T. B. Boykin, C. Salazar-Lazaro, T. A. Cwik, A. Stoica. Superlattices and Microstructures, **27**, 77 (2000).
[15] S. J. Vlaev, A. M. Miteva, D. A. Contreras-Solorio and V. R. Velasco. Surface Science, **424** 331 (1999).
[16] S. J. Vlaev, A. M. Miteva, D. A. Contreras-Solorio and V. R. Velasco. Superlattices and Microstructures, **26**, 325 (1999).
[17] J. O. Sofo and C.A. Balseiro, Phys. Rev. B, **42**, 7292 (1990).
[18] S. J. Vlaev, M. R. Muro-Ortega, D. A. Contreras-Solorio and V. R. Velasco, Surface Science, **418**, 536 (1998).
[19] I. Rodríguez Vargas, J. Madrigal Melchor, S. J. Vlaev, Tight-binding study of the hole subband structure properties of p-type delta-doped quantum well structures in Si, J. Phys.: Condensed Matter, in press, 2009.

Hole states in diamond p-delta-doped field effect transistors

J. C. Martínez-Orozco[1], **I. Rodríguez-Vargas**[1], **M. E. Mora-Ramos**[2]
[1] Unidad Académica de Física, Universidad Autónoma de Zacatecas, Calzada Solidaridad Esquina con Paseo la Bufa S/N, CP 98060 Zacatecas, ZAC., Mexico
[2] Facultad de Ciencias, Universidad Autónoma del Estado de Morelos, Av. Universidad 1001, Col. Chamilpa, CP 62209 Cuernavaca, MOR., México

E-mail: `jcmover@correo.unam.mx`

Abstract. The p-delta-doping in diamond allows to create high density two-dimensional hole gases. This technique has already been applied in the design and fabrication of diamond-based field effect transistors. Consequently, the knowledge of the electronic structure is of significant importance to understand the transport properties of diamond p-delta-doped systems. In this work the hole subbands of diamond p-type delta-doped quantum wells are studied within the framework of a local-density Thomas-Fermi-based approach for the band bending profile. The calculation incorporates an independent three-hole-band scheme and considers the effects of the contact potential, the delta-channel to contact distance, and the ionized impurity density.

1. Introduction

Diamond has revealed as a very promising semiconductor material. As very attractive properties, CVD diamond presents unusually high electron and hole mobilities, thermal conductivity and breakdown electric field. This makes it specially suitable for applications in high temperature, high voltage, high frequency and high power electronics [1]. Delta doping techniques have been used to create two-dimensional conductive channels with a high density of charge carriers. Delta-doped quantum wells (DDQW) have associated two-dimensional electron and hole gases provided the type of doping. In the case of p-delta-doping in diamond, the typical situation is the use of boron as the acceptor dopant. Doping concentrations are above 10^{20} cm^{-3}, allowing two-dimensional sheet carrier densities exceeding 10^{13} cm^{-2} [2]. The properties of the two-dimensional hole gas (2DHG) in p-type DDQW have been theoretically investigated in previous works [3, 4, 5].

On the other hand, the interest in diamond p-type delta-doped systems, from the experimental point of view, has been focused in the design and fabrication of electronic devices, mainly Field Effect Transistors (FET). Among these it is the δ-channel JFET with a boron/nitrogen junction [6, 7, 8].

In this article we are going to present the results of a preliminary investigation about the main features of the hole level spectrum in a boron δ-doped diamond FET. To do such a study we use a simplified description of the heavy, light and split-off valence band profiles in the system. This approach uses the so-called depletion approximation to deal with the contact potential barrier and the adjacent uniformly doped region. The potential function associated with the valence

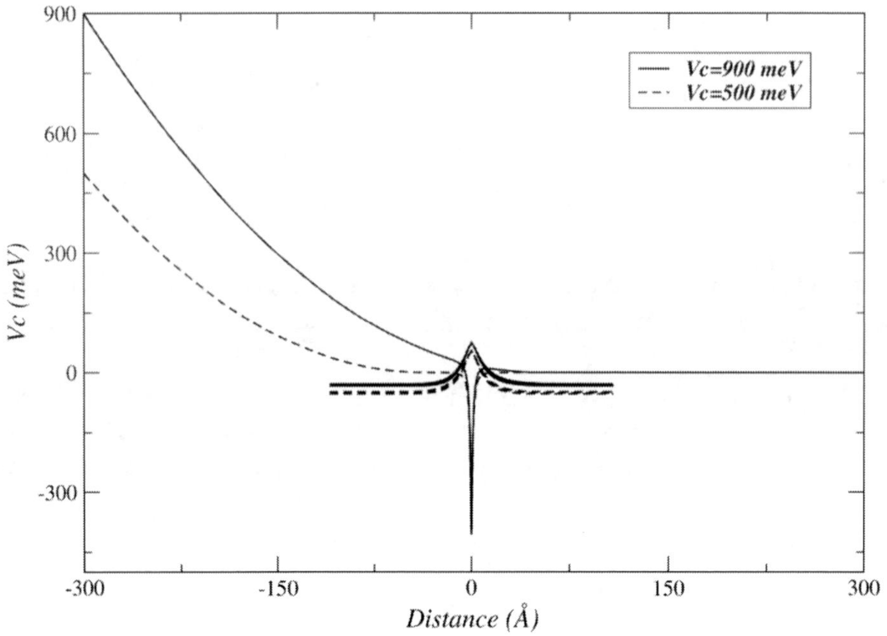

Figure 1. Potential profile and hole wave function for a p-type δ-doped FET in diamond. The two dimensional hole gas in this case is $p_{2D} = 5.0 \times 10^{14}$ cm^{-2}. In this figure the dashed lines correspond to the case of $V_c = 500$ meV and the solid one to $V_c = 900$ meV.

band warping is derived along the lines of the Thomas-Fermi theory [4]. In this approximation, Hartree contributions are explicitly included in the analytic expression for $V(z)$.

2. Model and results

The expression for the Thomas-Fermi potential associated to the delta-like ionized acceptor layer is [9]:

$$V_{TF}(z) - E_F = -\frac{\alpha^2}{(\alpha\,|z - l_d| + z_0)^4}. \tag{1}$$

Here it is assumed to be in the low temperature limit. l_d is the distance between the delta-channel and the contact barrier. Heavy, light and split-off hole bands are considered uncoupled. This constitutes a fairly good description of the hole dynamics at $\mathbf{k} = \mathbf{0}$. Atomic units are used throughout, referred to an effective Bohr radius $a_0^* = \varepsilon\hbar^2/(e^2 m_{hh})$, and to an effective Rydberg $R_y^* = e^2/(2\varepsilon a_0^*)$. The static dielectric constant of diamond is taken as $\varepsilon = 5.72$. In the above equation E_F is the Fermi level energy, and $\alpha = 2m_a^{3/2}/15\pi$, with $m_a = \left[1 + (m_{lh}/m_{hh})^{3/2} + (m_{so}/m_{hh})^{3/2}\right]^{2/3}$. m_j $(j = mh, lh, so)$ are the corresponding valence bands effective masses in units of the free electron mass, which are 4.27, 3.66 and 3.94 [10]. Additionally,

$$z_0 = \left(\frac{\alpha^3}{\pi p_{2D}}\right)^{1/5}, \tag{2}$$

where p_{2D} is the two-dimensional hole concentration associated to the plane of ionized impurities.

The potential function of equation (1) must combine with the valence band potential energy expression related with the Schottky barrier at the metallic contact. For this, we propose the following modification of the depletion approximation:

$$V_d(z) = 8\pi N_d (z - L_1)^2, \tag{3}$$

where N_d is the three-dimensional density of ionized acceptor impurities in the barrier region. The modified depletion length L_1 is given by:

$$L_1 = \sqrt{L^2 + \frac{1}{8\pi N_d}\frac{\alpha^2}{(\alpha l_d + z_0)^4}}, \tag{4}$$

where

$$L = \sqrt{\frac{V_c}{8\pi N_d}}, \tag{5}$$

is the so-called depletion length associated to the Schottky barrier of contact potential V_c. Finally, the potential function modeling the valence band profile in the system is $V(z) = V_{TF}(z) + V_d(z)$, which is written as,

$$V(z) = 8\pi N_d (z - L_1)^2 \theta(z - L_1) - \frac{\alpha^2}{(\alpha |z - l_d| + z_0)^4}\theta(z - l_d). \tag{6}$$

3. Results and Discussion

Table 1. Energy levels for a diamond δ-FET. The values of energy are given with respect to the well bottom. We also report -in the last column- the width of the depletion region.

V_c (meV)	E_{hh}	E_{lh}	E_{so}	l_p (Å)
100	352.56	357.64	355.25	117.0
200	352.56	357.64	355.25	165.0
300	352.56	357.64	355.25	203.0
400	352.56	357.64	355.25	234.0
500	352.56	357.64	355.25	260.0
600	352.57	357.66	355.26	279.0
700	353.69	358.84	356.41	289.0
800	360.37	365.48	363.08	293.0
900	372.61	377.67	375.29	295.0

The basic input parameters are: background impurities density of $N_d = 1.0 \times 10^{18}$ cm^{-3}, and contact voltage potential V_c from 100 to 900 meV. We locate the p-type δ-doped well 300 Å from the metal-semiconductor contact, and the two-dimensional hole concentration associated with this p-type δ-doped well are $p_{2D} = 5.0 \times 10^{14}$ and $p_{2D} = 7.0 \times 10^{14}$ cm^{-2}.

The first aspect to be analyzed has to do with the way the hole structure is affected as a function of the contact potential. To do this we firstly fix V_c to a very low value, in such a way that we reproduce the results reported for an isolated delta quantum well in ref. [5]. Then we start changing the value of the contact potential and find that for $p_{2D} = 5.0 \times 10^{14}$ cm^{-2}

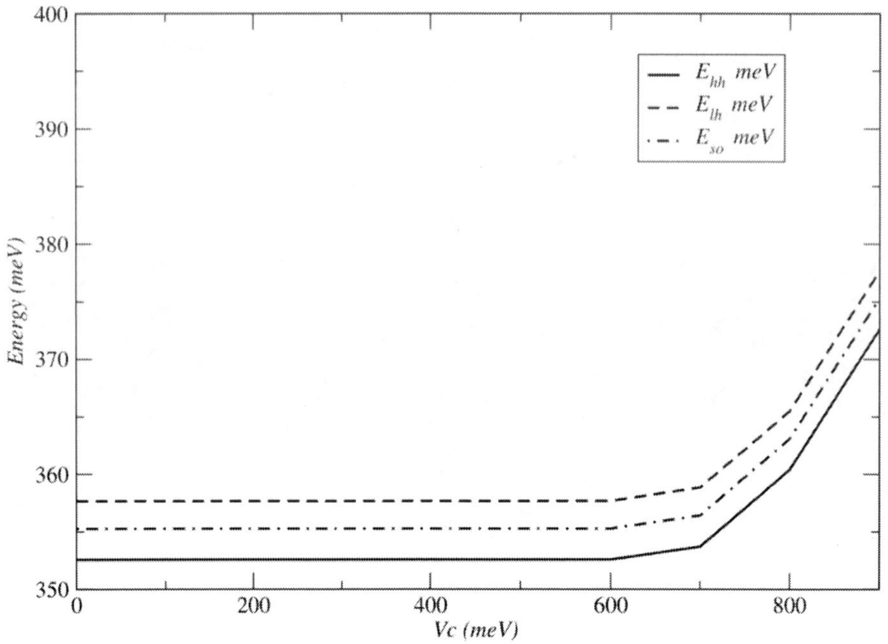

Figure 2. Energy levels as a function of contact potential V_c for an p-type δ-doped FET in diamond.

the well has a depth of 406 meV, and it binds the ground levels of the three hole bands. In figure 1 we present the wave functions calculated for the three ladder of levels. In this figure we present two values of the contact potential, 500 and 900 meV. In table 1 the calculated energies are presented for each of the effective masses m_j $(j = mh, lh, so)$. The energies are reported as functions of V_c. In addition, we give the width of the impurity depletion region, adjacent to the metal-semiconductor contact.

Finally, in figure 2 we show how the energy levels vary as a function of V_c. We find that such levels are practically not affected when $V_c < 600$ meV, and for greater values of this potential the hole level structure is strongly dependent on this parameter.

4. Conclusion

In this work we present the first calculation of the hole level structure in a p-type δ-doped FET in diamond. We find that only three ground states are bound within the well for the set of parameters considered. We also observed regions for which the contact potential significantly affects the electronic energy level structure in the device ($V_c > 600$ meV). Additionally, the proposed model allows us to modify the main parameters of the system and find the optimum values of them in order to achieve the desired features.

Acknowledgments

J.C.M.O acknowledges to the National Science and Technology Council (CONACyT) from México, through the retention program as well as to secretaria General of the Universidad Autónoma de Zacatecas (UAZ) and to the Consejo Zacatecano de Ciencia y Tecnología for the financial support. M.E.M.R. acknowledges support from Mexican CONACYT through Grant No. CB-80846/2007

References

[1] Wort C J H and Balmer R S 2008 *Materials Today* **11** 22.
[2] Aleksov A, Vescan A, Kunze M, Gluche P, Ebert W, Kohn E, Bergmaier A and Dollinger G 1999 *Diam. Relat. Mater.* **8** 941.
[3] Mora-Ramos M E 2002 *phys. stat. sol. (b)* **234** 481; 2003 *phys. stat. sol. (b)* **240** 665 (erratum).
[4] Mora-Ramos M E 2003 *Diam. Relat. Mater.* **12** 550.
[5] Rodríguez-Vargas I, Duque C A, Mora-Ramos M E and Velasco V R 2007 *phys. stat. sol. (c)* **4** 415.
[6] Aleksov A, Kubovic M, Kaeb N, Spitzberg U, Bergmaier A, Dollinger G, Bauer Th, Schreck M, Stritzker B and Kohn E 2003 *Diam. Relat. Mater.* **12** 391.
[7] Aleksov A, Denisenko A, Kunze M, Vescan A, Bergmaier A, Dollinger G, Ebert W and Kohn E 2003 *Semicond. Sci. Technol.* **18** S59.
[8] Denisenko A and Kohn E 2005 *Diam. Relat. Mat.* **14** 491.
[9] Ioriatti L 1990 *Phys. Rev. B* **41** 8340.
[10] Willatzen M, Cardona M and Christensen N E 1994 *Phys. Rev. B* **50** 18054.

Optical properties of alpha spodumene: orientation of its principal optical axes

S O Souza, A F Lima and M V Lalic[1]

Universidade Federal de Sergipe, Departamento de Fisica, P.O.Box 353, 49100-000 Sao Cristovao – SE, Brasil

E-mail: mlalic@fisica.ufs.br

Abstract. We studied the orientation of the three orthogonal principal optical axes of the alpha spodumene crystal. This orientation is determined relative to the crystallographic axes, and expressed as function of the incident radiation wavelength in ultraviolent region. The calculations were performed by density functional theory based, full potential augmented plane wave method.

1. Introduction

The spodumene ($LiAlSi_2O_6$) is a natural crystal of significant technological interest. It is a principal source of lithium and is widely used in glass and ceramic industry. Since considered as semi-precious gemstone, it is interesting for a jewel industry as well. Due to all these characteristics the spodumene has been subject of many investigations of physicists and mineralogists [1-5].

Recently we reported a brief theoretical study of electronic and optical properties of the pure spodumene [6]. There we discussed its electronic structure and several optical properties (absorption, refraction index and extinction coefficient) calculated along the three principal optical axes, for the case when the incident radiation energy ranges from 0 to 25 eV [6]. In the present paper we continue this study by focusing our attention to the orientation of the spodumene principal optical axes with respect to its crystallographic axes. This orientation is determined for any wavelength of the incident light ranging from zero to 35 eV.

2. Calculations

All the calculations were performed by the first-principles, full potential linear augmented plane wave (FP-LAPW) method [7], based on density functional theory (DFT) [8] and implemented into the WIEN2k computer code [9]. The technical details of calculations were the same as reported in our recent paper [6].

The α-spodumene adopts the monoclinic crystal structure with the space group number 15, with the two possible settings of the axes: *C2/c* or *B2/b*. In our calculations the unit cell of the spodumene was fully optimized, i.e. the lattice parameters and all the atomic positions were relaxed in order to reach the values that correspond to minimum of energy.

[1] Corresponding author

Optical response of the spodumene is determined by calculating its complex dielectric tensor ε. Imaginary part of this tensor is directly proportional to the optical absorption spectrum of the material. It is computed on the basis of the calculated electronic band structure, in the limit of linear optics, neglecting electron polarization effects and within the frame of random phase approximation [6]. The Im(ε) is computed for the photon energies ranging from zero to ħω = 35 eV, with a mesh of 55 k-points in the irreducible wedge of the first Brillouin zone.

3. Orientation of the principal optical axes of the spodumene

Due to monoclinic symmetry of the spodumene the dielectric tensor is not diagonal in the arbitrary Cartesian coordinate system {x,y,z}. However, it can be reduced to its diagonal form by transforming this coordinate system into another one $(\vec{n}_1,\ \vec{n}_2,\ \vec{n}_3)$, which is also orthogonal and usually called the principal axes system. This transformation is illustrated below,

$$\operatorname{Im}\varepsilon(\omega)=\begin{bmatrix}\varepsilon_{11}(\omega) & \varepsilon_{12}(\omega) & 0\\ \varepsilon_{21}(\omega) & \varepsilon_{22}(\omega) & 0\\ 0 & 0 & \varepsilon_{33}(\omega)\end{bmatrix} \Rightarrow \operatorname{Im}\varepsilon'(\omega)=\begin{bmatrix}\varepsilon'_{11}(\omega) & 0 & 0\\ 0 & \varepsilon'_{22}(\omega) & 0\\ 0 & 0 & \varepsilon'_{33}(\omega)\end{bmatrix} \qquad (1)$$

where $\varepsilon_{ij}(\omega)$ and $\varepsilon'_{ij}(\omega)$ stand for $\operatorname{Im}\varepsilon_{ij}(\omega)$ and $\operatorname{Im}\varepsilon'_{ij}(\omega)$ for the reason of shorter notation. The eigenvalues $\varepsilon'_{ij}(\omega)$ are expressed in terms of the calculated $\varepsilon_{ij}(\omega)$ components by formulas:

$$\left.\begin{aligned}\varepsilon'_{11}(\omega)&=\frac{1}{2}[(\varepsilon_{11}+\varepsilon_{22})+\sqrt{(\varepsilon_{11}-\varepsilon_{22})^2+4\varepsilon_{12}^2}\,]\\ \varepsilon'_{22}(\omega)&=\frac{1}{2}[(\varepsilon_{11}+\varepsilon_{22})-\sqrt{(\varepsilon_{11}-\varepsilon_{22})^2+4\varepsilon_{12}^2}\,]\\ \varepsilon'_{33}(\omega)&=\varepsilon_{33}\end{aligned}\right\} \qquad (2)$$

They are directly proportional to the intensity of the optical absorption of the material along its three principal optical axes, which are determined by the eigenvectors $\vec{n}_1$, $\vec{n}_2$ and $\vec{n}_3$ respectively:

$$\left.\begin{aligned}\vec{n}_1&=x(\omega)\vec{e}_x+\vec{e}_y\\ \vec{n}_2&=y(\omega)\vec{e}_x+\vec{e}_y\\ \vec{n}_3&=\vec{e}_z\end{aligned}\right\} \qquad (3)$$

$\vec{e}_x$, $\vec{e}_y$ and $\vec{e}_z$ denote the orts of the x, y and z directions of the original coordinate system, and:

$$x(\omega)=\frac{\varepsilon_{11}-\varepsilon_{22}-2\varepsilon_{12}+\sqrt{(\varepsilon_{11}-\varepsilon_{22})^2+4(\varepsilon_{12})^2}}{\varepsilon_{11}-\varepsilon_{22}+2\varepsilon_{12}-\sqrt{(\varepsilon_{11}-\varepsilon_{22})^2+4(\varepsilon_{12})^2}} \qquad (4)$$

$$y(\omega)=\frac{\varepsilon_{11}-\varepsilon_{22}-2\varepsilon_{12}-\sqrt{(\varepsilon_{11}-\varepsilon_{22})^2+4(\varepsilon_{12})^2}}{\varepsilon_{11}-\varepsilon_{22}+2\varepsilon_{12}+\sqrt{(\varepsilon_{11}-\varepsilon_{22})^2+4(\varepsilon_{12})^2}} \quad (5)$$

Our objective is to connect the principal axes directions with the directions of the monoclinic crystallographic axes *a*, *b* and *c*. The spodumene has the base centered monoclinic unit cell, and the WIEN2k program accepts the *B2/b* settings of its axes. In this setting, the three primitive vectors of the Bravais lattice have the following coordinates: $(0,\ b,\ 0)$, $(\frac{a}{2}\sin\gamma,\ \frac{a}{2}\cos\gamma,\ -\frac{c}{2})$ and $(\frac{a}{2}\sin\gamma,\ \frac{a}{2}\cos\gamma,\ +\frac{c}{2})$, expressed in the {x,y,z} coordinate system (figure 1).

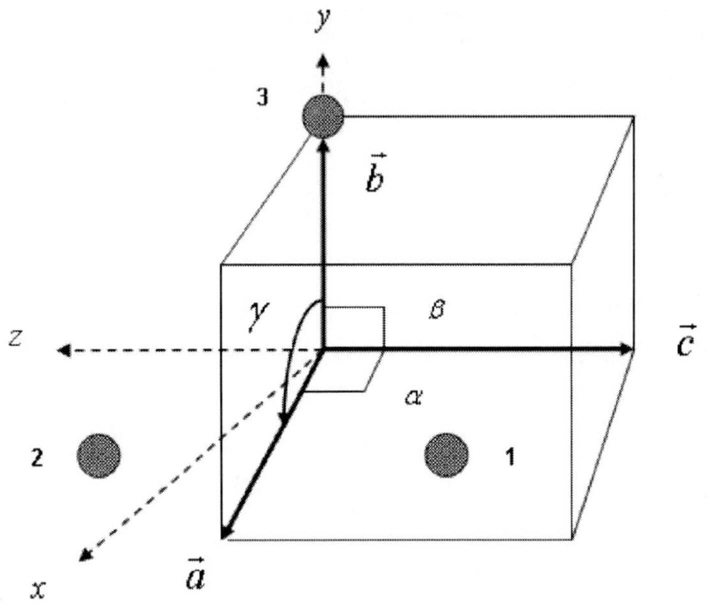

Figure 1: The monoclinic unit cell of the spodumene with the *B2/b* setting of its crystallographic axes. The angles α and β are the right angles, while the γ is 110,13°. The coordinate system in which the dielectric tensor is calculated is denoted as {x,y,z} and crystallographic axes as *a*, *b* and *c*. The x-axis lies in the *a-b* plane, inclined by the angle γ–90° with respect to the *a*-axis. The primitive vectors of the Bravais lattice connect the points 1, 2 and 3 of the lattice.

From the figure 1 it can be easily established a connection between the original coordinate system $(\vec{e}_x,\ \vec{e}_y,\ \vec{e}_z)$ and the system of crystallographic axes $(\vec{e}_a,\ \vec{e}_b,\ \vec{e}_c)$:

$$\left.\begin{aligned}\vec{e}_x&=\vec{e}_a/\sin\gamma-(ctg\,\gamma)\vec{e}_b\\ \vec{e}_y&=\vec{e}_b\\ \vec{e}_z&=-\vec{e}_c\end{aligned}\right\} \quad (6)$$

Combining this result with the formula (3) one can determine the orientation of the principal optical axes of the spodumene in terms of its crystallographic directions:

$$\left.\begin{aligned}\vec{n}_1 &= \frac{x(\omega)}{\sin\gamma}\vec{e}_a + [1 - x(\omega)ctg\,\gamma]\vec{e}_b \\ \vec{n}_2 &= \frac{y(\omega)}{\sin\gamma}\vec{e}_a + [1 - y(\omega)ctg\,\gamma]\vec{e}_b \\ \vec{n}_3 &= -\vec{e}_c\end{aligned}\right\} \quad (7)$$

The principal axis $\vec{n}_3$ is parallel to the crystallographic c-axis, while the other two principal axes ($\vec{n}_1$ and $\vec{n}_2$) lie in the plane defined by the other two crystallographic axes (a and b). It can be verified that the $\vec{n}_1$ and $\vec{n}_2$ are mutually orthogonal for arbitrary value of the frequency ω of incident radiation. Their orientation with respect to the axes a and b, however, depend on the radiation wavelength, i. e. $\vec{n}_1$ and $\vec{n}_2$ rotate around the c axis as a function of the incident photon wavelength.

Since direction of the $\vec{n}_3$ is fixed along the c-axis and the angle between the $\vec{n}_1$ and $\vec{n}_2$ is always 90°, the orientation of the principal axes is determined by just one angle, θ, between the axis a and the axis $\vec{n}_1$. Manipulating the formula (7) it is easy to verify that this angle depends on frequency as:

$$\cos\theta = \frac{x(\omega)\sin\gamma + \cos\gamma}{\sqrt{1 + x^2(\omega)}} \quad (8)$$

The dependency of θ with respect to the incident radiation wavelength is shown in the figure 2.

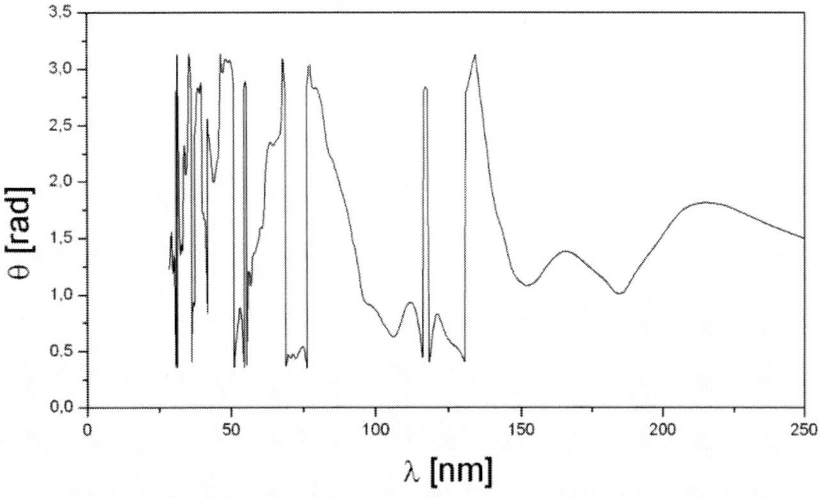

Figure 2: The angle θ between the principal axis $\vec{n}_1$ and crystallographic axis a of the spodumene as a function of the incident photon wavelength.

The figure 2 finally determines complete orientation of the principal optical axes of the pure spodumene, along which the optical absorption and other optical characteristics have been discussed in ref. [6]. It is seen that the values of the angle θ range from approximately 21° to the 180° and exhibit very sharp changes with respect to the slight variation of the wavelength. The reason for such abrupt changes lies in the fact that the off-diagonal dielectric tensor component $\varepsilon_{12}(\omega)$ (1) (not shown in this paper) is very small and frequently reverses its sign as function of wavelength. Each time when the sign reversal occurs, the denominator of $x(\omega)$ (4) becomes extremely small, also reversing its sign. Thus, the small variation of incident wavelength can cause extreme variation of $x(\omega)$: when ε_{12} is positive, the $x(\omega)$ is very large and positive, and when ε_{12} is negative, the $x(\omega)$ is very large and negative. The abrupt changes of $x(\omega)$ provoke the abrupt changes on the angle θ within the range which can be estimated on the basis of the formula (8). For $x(\omega)$ very large and positive, $\sqrt{1+x^2} \approx x$ and (8) simplifies at $\cos\vartheta \approx \sin\gamma$,which approximately determines the lower limit $\theta \approx 21^0$. On the other side, for $x(\omega)$ very large and negative, $\cos\vartheta \approx -\sin\gamma$ and $\theta \approx 160^0$.

4. Conclusions

The present work was focused on determination of the orientation of the principal optical axes of the pure α-spodumene. It is found that one principal axis ($\vec{n}_3$) is parallel to the crystallographic axis *c* and the other two ($\vec{n}_1$ and $\vec{n}_2$) rotate around the *c* in the plane defined by the crystallographic axes *a* and *b*, as functions of the incident light wavelength. The orientation of these axes is precisely determined for each wavelength in the ultraviolet range (up to 35 eV). All calculations were performed by the first-principle, DFT based, FP-LAPW method.

Acknowledgements

The authors gratefully acknowledge the CNPq and FAPITEC-SE (Brazilian funding agencies) for financial support.

References

[1] Claffy E V 1953 *American Mineralogist* **38** 919
[2] Fujii A T and Isotani S 1988 *An. Acad. Bras. Ci.* **60** 127
[3] Pommier C J S, Senton M B and Downs R T 2003 *Journal of Raman spectroscopy* **34** 769
[4] Souza S O, Ferraz G M and Watanabe S 2004 *Nucl. Instr. Meth. in Phys. Res. B* **218** 259
[5] Lipatov E I, Orlovskii V M, Tarasenko V F and Solomonov V I 2007 *J. Lumin.* **126** 817
[6] Lima A F, Souza S O and Lalic M V 2008 *Optical Materials* **30** 1048
[7] Andersen O K 1975 *Phys. Rev. B* **12** 3060; Singh D J 1994 *Plane Waves, Pseudopotentials and the LAPW method* (Dodrecht: Kluwer Academic)
[8] Hohenberg P and Kohn W 1964 *Phys. Rev.* **136** B864; Kohn W and Sham L J 1965 *Phys. Rev.* **140** A1133
[9] Blaha P, Schwarz K, Madsen G K H, Kvasnicka D and Luitz J 2001 ***WIEN2k***, *An Augmented Plane Wave + Local Orbitals Program for Calculating Crystal Properties* (Karlheinz Schwarz, Techn. Universität Wien, Austria)

Journal of Physics: Conference Series **167** (2009) 012067 doi:10.1088/1742-6596/167/1/012067

Impurity effect on spectrum of nanoring

W Gutiérrez[1], L F García, and I D Mikhailov
Escuela de Física, Universidad Industrial de Santander, Colombia A.A. 678
E-mail: willigun@gmail.com

Abstract. The effect of the donor position on the energy levels and far-infrared spectrum of a finite-barrier toroidal-shaped quantum ring in the presence of the external magnetic field applied along the axis is studied. It is shown that the displacement of the donor from the centre toward the ring produces a localization of the low lying states, quenching of the Aharonov–Bohm oscillations of the corresponding levels and a drastic modification of the far-infrared spectrum.

1. Introduction

In recent years there has been a great progress in the ability to confine charge carriers in low-dimensional semiconductor structures that have allowed constructing nanoscopic systems with ring-like geometry [1]. Quantum rings (QRs) with single or two captured electrons may be considered as the best candidates to display various pure quantum effects such as the Aharonov–Bohm (AB) oscillations under external magnetic field applied along the ring axis [1]. This fact has stimulated much interest in the theoretical and experimental study of the electronic and optical properties of QRs in the last two decades. The spectral properties related to the axial symmetry of the one- and two-electron single and vertically coupled QRs have been studied by various authors [2]. On the other hand, one could expect that these properties should be changed essentially in the presence of any disorder. One of the possibilities for the localization of the electron motion inside the QR is related to the presence of donors. Ground state binding energies and the density of impurity states of shallow donors randomly distributed inside a toroidal QR have been calculated recently by using the variational method [3]. Similar calculations for neutral and negatively charged donors have been realized by using the matrix diagonalization method [4] for a hard-well two-dimensional model and the adiabatic approximation [5] for narrow soft-edge-barrier three-dimensional QR. It has been shown that in the presence of the off-centre donor, that breaks the axial symmetry, the low-lying states become strongly localized in the region close to donor. One could expect that such localization should quench the AB oscillation of the levels corresponding to these states. Therefore it would be interesting to analyze the influence of the presence of the off-centre donor on the AB oscillation. To this end we above extend the method proposed in the paper [5] for calculating the low-lying energies in a narrow QR with arbitrary cross-section profile in order to investigate the effect of the electron localization due to the presence of a donor on the AB oscillations and the far infrared (FIR) spectra

2. Theoretical framework

We study a model of an embedded in $Al_{0.35}Ga_{0.65}As$ matrix toroidal semi-elliptically-shaped $In_{0.55}Al_{0.45}As$ QR, presented in figure 1. In the model the QR is a thin layer characterized in XY plane

[1] Corresponding author: e-mail willigun@gmail.com , Phone: +57-7-6443234

by the exterior radius R_{ext} and the interior radius R_{int}, along the z-axis by the greatest height d_0, and by the dependence of the heterostructure height, d on the distance ρ from the axis.

$$d(\rho)=d_0\sqrt{1-(\rho-R_a)^2/R_t^2}\,\vartheta(\rho-R_{int})\vartheta(R_{ext}-\rho);\ R_a=(R_{ext}+R_{int})/2;\ w=R_{ext}-R_{int} \quad (1)$$

In this paper we consider a case when the donor is located at the base plane with position vector whose cylindrical coordinates (ρ,φ,z) are $(\xi,0,0)$. The confinement potential due to the conduction band discontinuity in the junctions of heterostructure is given by the piecewise constant function $V(\rho,z)$ which is equal to zero inside the $In_{0.55}Al_{0.45}As$ layer and to V_0 outside of it.

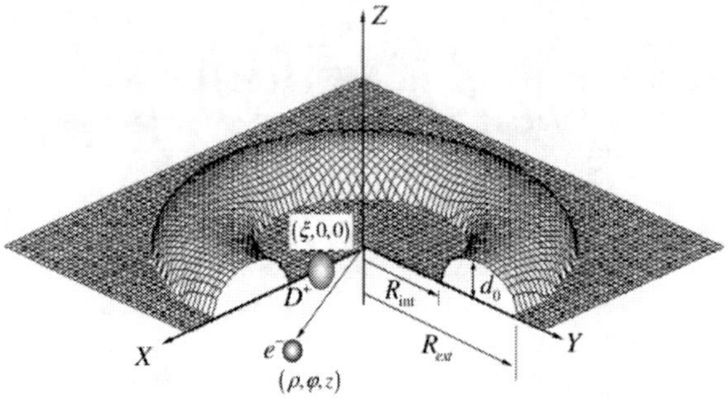

Figure 1 Scheme of a toroidal QR with semielliptically-shaped cross section

The following values of physical parameters [6] are used in our calculations: dielectric constant $\varepsilon=12.71$, the effective masses in the $In_{0.55}Al_{0.45}As$ material layer and in the $Al_{0.35}Ga_{0.65}As$ matrix material for the electron $m_{ew}=0.076\,m_0$ and $m_{eb}=0.097\,m_0$, respectively, the conduction bands offset in junctions is $V_0=358meV$. We scale all lengths in terms of the effective Bohr radius $a_0^*=\hbar^2\varepsilon/m^*e^2\approx 10.4nm$, all energies in the effective Rydberg $Ry^*=e^2/2a_0^*\varepsilon\approx 5.2meV$. Parameter $\gamma=e\hbar B/2m^*Ry^*$ is used as dimensionless units of the strengths of the applied magnetic field. The dimensionless Hamiltonian, that describes in the effective mass approximation the off-axis donor with $\tau=1$ and the electron with $\tau=0$, in cylindrical coordinates (ρ,φ,z) is given by:

$$H=-\frac{\partial^2}{\partial\rho^2}-\frac{1}{\rho}\frac{\partial}{\partial\rho}-\frac{1}{\rho^2}\frac{\partial^2}{\partial\varphi^2}-\frac{\partial^2}{\partial z^2}+V(\rho,\varphi)+\frac{\gamma^2\rho^2}{4}-i\gamma\frac{\partial}{\partial\varphi}-\frac{2\tau}{\sqrt{\rho^2+2\rho\xi\cos\varphi+\xi^2+z^2}} \quad (2)$$

The trial function commonly used in the framework of the adiabatic approximation in order to separate the fast motion in z-direction from the slow in-plane motion [7] in the Hamiltonian (2) is:

$$\psi=f_z(\rho,z)\phi(\rho,\varphi), \quad (3)$$

where $f_z(\rho,z)$ is the ground state wave function for the electron fast motion in the z-direction along the line distant at ρ from the axis, which is given by the solution for the one-dimensional rectangular quantum well with barrier height V_0 and width $d(\rho)$. The corresponding energy $E_z(\rho)$ can be found as a solution of the well-known transcendent equation. Following the procedure described in references [5, 7] one can derive the two-dimensional wave equation for the function $\phi(\rho,\varphi)$

$$-\frac{\partial^2\phi(\rho,\varphi)}{\partial\rho^2}-\frac{1}{\rho}\frac{\partial\phi(\rho,\varphi)}{\partial\rho}-\frac{1}{\rho^2}\frac{\partial^2\phi(\rho,\varphi)}{\partial\varphi^2}-i\gamma\frac{\partial\phi(\rho,\varphi)}{\partial\varphi}+V_{eff}(\rho,\varphi)\phi(\rho,\varphi)=E\phi(\rho,\varphi) \quad (4)$$

The effective potential in the equation (4) is given by:

$$V_{eff}(\rho,\varphi) = E_z(\rho) + \frac{\gamma^2\rho^2}{4} - \int_{-\infty}^{\infty} \frac{2f_z^2(\rho,z)dz}{\sqrt{\rho^2 + 2\rho\xi\cos\varphi + \xi^2 + z^2}} \quad (5)$$

For on-centre donor $(\xi = 0)$ the equation (4) presents a particular case of the two-dimensional central force problem for which the radial and angular motions can be separated and the equation can be solved exactly. For off-centre donors $(\xi \neq 0)$ such separation can be fulfilled the adiabatic approximation describe in details in the paper [5] only for very narrow rings. Once the wave function (3) and eigenvalues of the problem (4) are found one then can calculate the charge distribution, energies and the corresponding absorption coefficients in the dipole approximation by using the following relation [4]

$$\alpha(E) = cE\sum_f \left|\langle\psi_f|\vec{e}\cdot\vec{d}|\psi_0\rangle\right|^2 \delta(E - \Delta E_{f0}), \quad (6)$$

where $\vec{d}$ is the electric dipole operator, c is a constant factor, and ΔE_{f0} is the energy difference between the inicial state $|\psi_0\rangle$ and the final state $|\psi_f\rangle$.

3. Results

First, we analyze the effect of the cross-section size and its profile on the AB oscillations by varying the parameter R_{int} in the relation (1). In figure 2 we display the energies of the eight low-lying levels as functions of the magnetic field strength γ for the on-centre donor in QRs with the same heights $d_0 = 4nm$, exterior radii $R_{ext} = 20nm$ and three different interior radii R_{int}. The decrease of the interior radius from 18nm up to zero changes the shape of the QR from the ring-like up to almost lens-shaped geometry. It is seen from figure 2 that such change of the geometry leads to modification of the ground state energy oscillation which becomes less pronounced as R_{int} decreases.

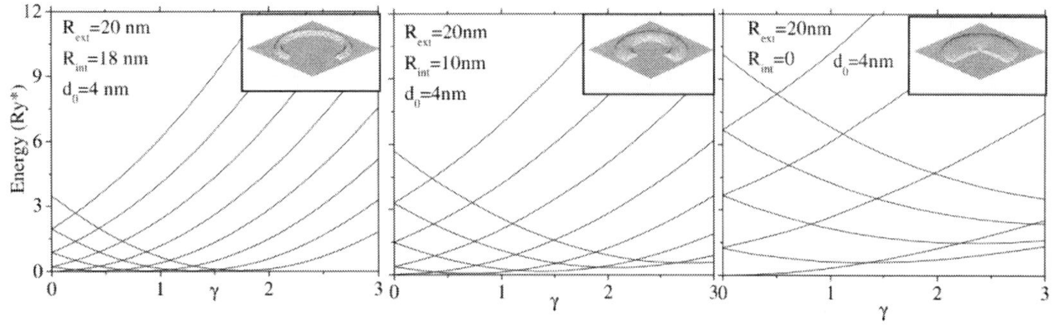

Figure 2. Low-lying energies of on-centre donor as functions of γ in QRs with different shapes

The results presented in figure 2 are exact, while the energies for off-centre donor can be found only approximately. Below, to analyze the effect of the donor position on the spectrum properties we show the calculation results for off-centre donors obtained for narrow rings $(R_t << R_a)$ in the adiabatic approximation. In Figure 3 we show off-centre donor energies as functions of the distance from the donor location to the axis for narrow ring $R_t = 2a_0{}^*$ of radius $R_a = 10a_0{}^*$ (a) without and (b) with magnetic field $(\gamma = 0.5)$. It is seen that energies of all levels initially decrease as the donor displaces from the axis up to the interior ring region when the parameter $\beta = \xi/R_a$ increases from 0 to 1. As the donor displaces further and the parameter β increases from 1 to infinity the energies becomes to grow and tend to the values corresponding to the rigid rotor. The dependencies of the energies on the donor position is much more pronounced for the ground and some low-lying excited states while for higher energy levels the dependence on the donor position is very weak. The donor displacement toward the rings leads to the strong localization related to the transformation of the electron rotational motion around the axis in the oscillation close to the point inside the ring the nearest to the donor location.

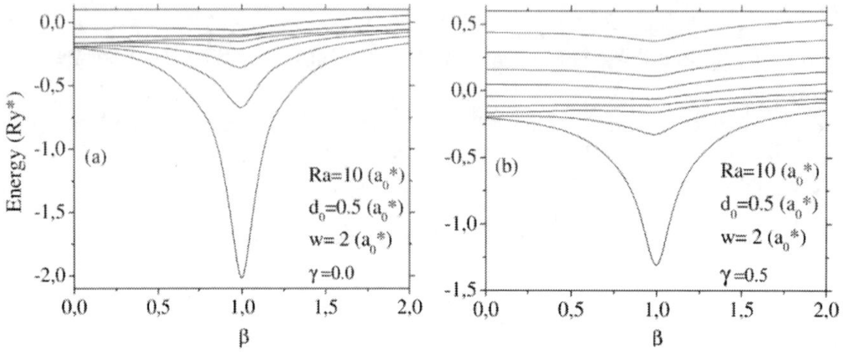

Figure 3. Electron energies as functions of the distance from the donor position to the ring axis

It should be noted that for donors located inside the ring the ground state energy tends to $-\infty$ similarly to the 1D hydrogen atom as the thickness of the ring tends to zero. Therefore the narrower the ring the lower are minima en curves of the energies dependencies on the donor position in figure 3. The localization of the electron motion due to the presence of the off-centre donor may be weakened by external magnetic field. One can see in figure 3(b) that the relatively weak external magnetic field $(\gamma = 0.5)$ change essentially electron energies making the major part of them less sensible to the donor position. Also the magnetic field splits energy levels presented in figure 3(a) in two, the lower for negative angular momentum and the upper for positive one.
The low-lying electron energies dependencies on the magnetic field strength in the presence of the off-centre donor are presented in figure 4.

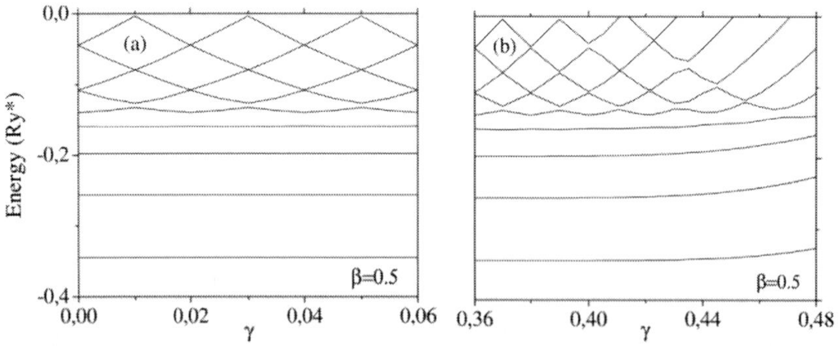

Figure 4 Low-lying electron energies as functions of magnetic field strength in the presence of the off-centre donor.

It is seen that the energies of the first 4 low-lying levels almost do not depend on the magnetic field strength while the energy dependencies of higher are similar to those for on-center donor. Such difference between the curves is due to different character of the motions corresponding to different energy levels in the presence of the off-center donor; in the first case the motion is oscillatory, while for the higher levels it is rotational.

In order to analyze the influence of donor position on the FIR spectra we have plotted the corresponding FIR spectra in figure 5 for on-centre (a) and off-centre donors located outside (b) and inside (c) the ring. As it shown in figure 5(a), there are discontinuous drop in the FIR spectra for on-centre donor with increasing magnetic field because of the oscillation of the ground state energy. Also only one peak appears in FIR in this case because of the electron motion in this case is a pure rotation. As the donor displaces from the axis toward the ring the pure rotational states are mixed and in the FIR appear more and more peaks, while the discontinuous drop related to Aharonov-Bohm oscillation becomes less and less pronounced. For donor location inside the ring in figure 5(c) peaks positions almost are not dependent on the external magnetic field because of a strong localization of the low-lying levels.

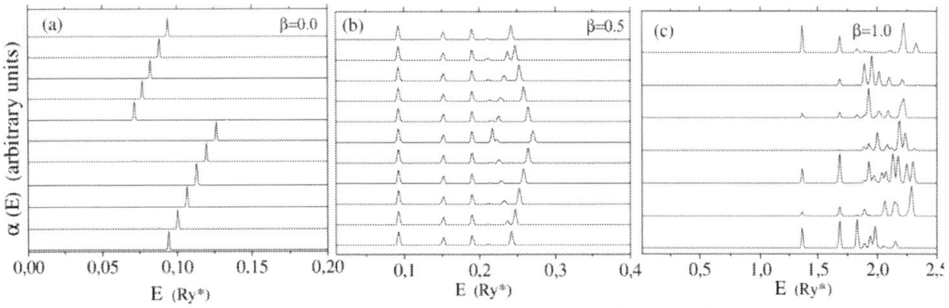

Figure 5. FIR spectra of right circularly polarized light as a function of E for (a) on-centre and (b) off-centre donors. (In each subplot, from botton to top, γ is increased from zero in steps of 0.002)

In conclusion, we present a simple method to analyze the energy spectrum of the donor located inside or close to narrow quantum ring with arbitrary cross-section profile in the presence of the external magnetic field applied along the axis. It is shown that with the donor approach to the ring the rotational motion of the electron inside the ring gradually transform in oscillation while the Aharonov-Bohm oscillations become less pronounced. As the donor is located inside the narrow ring no Aharonov-Bohm oscillations appear in the FIR spectrum for the first low-lying levels.

Acknowledgements

This work was financed by the Industrial University of Santander (UIS) through the Dirección General de Investigaciones (DIF Ciencias, Cod. Cod. 5124) and the Excellence Center of Novel Materials ECNM, under Contract No. 043-2005 and the Cod. No. 1102-05-16923 subscribed with COLCIENCIAS

References

[1] Lorke A, Luyken R J, Govorov A O, Kotthaus J P, García J M, and Petroff P M 2000 *Phys. Rev. Lett.* **84,** 2223
[2] Zhu J L, Dai Z, and Hu X. 2003 *Phys. Rev. B* **68**, 045324 ; Marin J H, García F, and Mikhailov I D. 2006 *Braz. J. Phys* **36**, No. 3B, 940
[3] Bruno-Alfonso A and Latgé A. 2000 *Phys. Rev.* B **61**, 15887
[4] Pan H and Zhu J L. 2003 *J. Phys: Condensed Matter* **15**, 7287
[5] Betancur J, Gutiérrez W, Mikhailov I D and Paredes H. 2006 *Braz. J. Phys* **36**, No. 3B, 882
[6] Pryor C., Pistol M.-E., and Samuelson L. 1997 *Phys. Rev.* B **56** 10404
[7] Marín J H, Mikhailov I. D, and Betancur F. J. 2006 *J. Phys.: Condens. Matter* **18,** 1005

Monte Carlo study of the double and super-exchange model with lattice distortion

J R Suárez[1], **E Vallejo**[1], **O Navarro**[1] **and M Avignon**[2]

[1]Instituto de Investigaciones en Materiales, Universidad Nacional Autónoma de México, Apartado Postal 70-360, 04510 México D. F., México.
[2]Institut Néel, Centre National de la Recherche Scientifique (CNRS) and Université Joseph Fourier, BP 166, 38042 Grenoble Cedex 9, France.

E-mail: jrsuarez@iim.unam.mx

Abstract. In this work a magneto-elastic phase transition was obtained in a linear chain due to the interplay between magnetism and lattice distortion in a double and super-exchange model. It is considered a linear chain consisting of localized classical spins interacting with itinerant electrons. Due to the double exchange interaction, localized spins tend to align ferromagnetically. This ferromagnetic tendency is expected to be frustrated by anti-ferromagnetic super-exchange interactions between neighbor localized spins. Additionally, lattice parameter is allowed to have small changes, which contributes harmonically to the energy of the system. Phase diagram is obtained as a function of the electron density and the super-exchange interaction using a Monte Carlo minimization. At low super-exchange interaction energy phase transition between electron-full ferromagnetic distorted and electron-empty anti-ferromagnetic undistorted phases occurs. In this case all electrons and lattice distortions were found within the ferromagnetic domain. For high super-exchange interaction energy, phase transition between two site distorted periodic arrangement of independent magnetic polarons ordered anti-ferromagnetically and the electron-empty anti-ferromagnetic undistorted phase was found. For this high interaction energy, Wigner crystallization, lattice distortion and charge distribution inside two-site polarons were obtained.

1. Introduction

Theoretical studies to explain the ferromagnetism of manganites are widely based on the so-called double exchange (DE) model introduced by Zener [1, 2]. The origin of the DE mechanism lies in the intra-atomic Hund's spin coupling J_H between itinerant and localized electrons [1, 2, 3, 4, 5]. The key point is that this coupling implies that the hopping depends on the configuration of the neighbor spins and explains how carriers improve their kinetic energy by forcing the localized spins to become ferromagnetically ordered. This ferromagnetic (F) tendency is expected to be frustrated by anti-ferromagnetic (AF) super-exchange (SE) interactions between localized spins $\overrightarrow{S}_i$ as first discussed by de Gennes [6] who conjectured the existence of canted states. Since then, it has become clear that microstructured spin configurations exist instead of macroscopic canted states resulting from such competition. Recently, an unifying picture in one dimension (1D) for classical local spins has shown the existence of two and three-sites ferromagnetic polarons separated by AF links in the whole range of electron density [7, 8]. As a result of the spin dependent hopping, carriers are localized in the ferromagnetic bonds, giving rise to bond ordered states for commensurate fillings. In turn, this will induce significant lattice distortions in systems in which the electrons interact with the lattice by affecting the hopping amplitude as in the Su-Schrieffer-Heeger (SSH) model [9, 10]. This point implies an important connection between the magnetic structure and lattice distortions. Recently, Vallejo et al. [11, 12, 13] have shown that three-leg ladders in the oxyborate system Fe_3BO_5 may provide evidence for the interplay between the magnetic structure and the observed structural and charge ordering transition, such that long and short bonds on the rungs alternate along the ladder axis [14]. In this Fe-ludwigite X-ray diffraction studies show contraction of the rungs [15].

The main goal of this work is to study the interplay between magnetic interactions and lattice distortion in one-dimensional systems. We consider the SSH model for the electron-lattice coupling together with the double and super-exchange model. As in [7] we will determine the phase diagram as a function of the band filling, the super-exchange interaction energy and we will also examine the effect of including coupling with the lattice.

We consider the exchange Hamiltonian to describe the localized and itinerant electrons [5, 16],

$$H = -\sum_i t_{i,i+1} \cos\left(\frac{\theta_{i,i+1}}{2}\right)(c_i^+ c_{i+1} + h.c.) + J\sum_i \vec{S}_i \cdot \vec{S}_{i+1}, \tag{1}$$

where $c_i^+ (c_i)$ are the fermions creation (annihilation) operators of conduction electrons at site i, $t_{i,i+1}$ is the nearest-neighbor (n.n) hopping parameter and $\vec{S}_i$ is the localized spin at site i. The first term represents the DE contribution favouring F ordering of local spins. The last term is a SE coupling between n.n localized spins $\vec{S}_i$, J being an AF interaction energy which stabilizes an AF phase for $x = 0$ and competes with DE for intermediate fillings. In this approach we consider the local spins as classical, $\vec{S}_i \to \infty$, a reasonable approximation in many cases in view of the similarity with the well known results [17, 18, 19, 20]. Itinerant electrons are either parallel or antiparallel to the local spins are thus *spinless*. Finally, $\theta_{i,i+1}$ $(0 \leq \theta_{i,i+1} \leq \pi)$ is the relative angle between the classical localized spins at sites i and $i+1$.

Now we introduce the effect of bond deformation on the itinerant electrons. In the SSH model, the complicated inter-atomic potential is represented using the electron-lattice coupling constant $g_t \simeq \partial t/\partial y$ $(g_t < 0)$ which describes the change of the hopping amplitude under a small change of the n.n bond length y,

$$H_{e-l} = g_t \sum_i y_{i,i+1}(c_i^+ c_{i+1} + h.c.) + \frac{K_t}{2}\sum_i y_{i,i+1}^2, \tag{2}$$

$y_{i,i+1}$ is the change of the $i, i+1$ lattice distance. It is important to mention that the original equilibrium lattice spacing a_0 (and so is t) in the absence of H_{e-l} results from the bonding produced by all other electrons in the system except the itinerant ones that we are considering in (1). The elastic constant K_t refers to this equilibrium lattice.

The electron-lattice part may be written in the following standard form, introducing the dimensionless parameters for the deformation $\delta_{i,i+1} = g_t y_{i,i+1}/t$ and the usual coupling constant $\lambda = 2g_t^2/\pi t K_t$:

$$H_{e-l} = -t\sum_i \delta_{i,i+1}(c_i^+ c_{i+1} + h.c.) + \frac{t}{\pi\lambda}\sum_i \delta_{i,i+1}^2. \tag{3}$$

The hopping term changes as $t_{i,i+1} = t\,(1 + \delta_{i,i+1})$ with $|\delta_{i,i+1}| \ll 1$. The complete Hamiltonian is given by

$$\begin{aligned} H &= -t\sum_i (1 + \delta_{i,i+1}) \cos\left(\frac{\theta_{i,i+1}}{2}\right)(c_i^+ c_{i+1} + h.c.) \\ &\quad + JS^2\sum_i \cos(\theta_{i,i+1}) + B\sum_i \delta_{i,i+1}^2, \end{aligned} \tag{4}$$

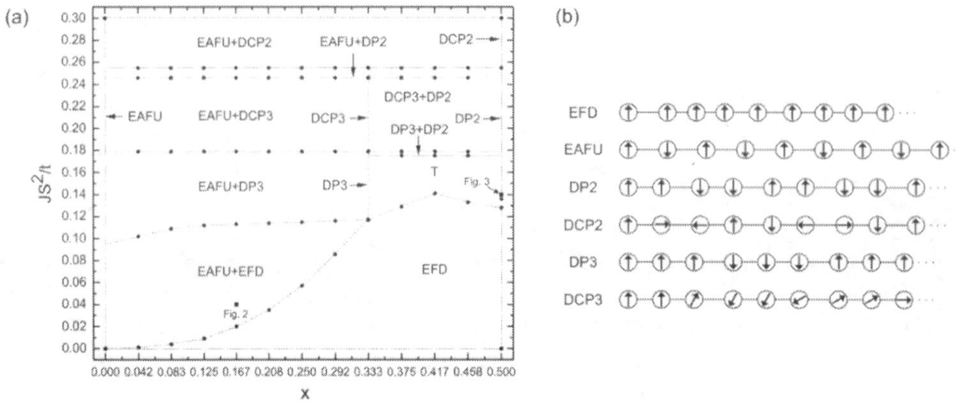

Figure 1. (a) Magneto-elastic phase diagram as a function of the SE interaction energy JS^2/t and the conduction electron density x, for a typical value of the elastic energy $B/t = 30$. A dotted line in this diagram represents a guide for the eyes. (b) Partial spin configuration snapshot of the different phases is also shown. The phases are described in the text.

where $B/t = 1/\pi\lambda$. Due to the dependence of the hopping amplitude on the spin configuration, the contribution of the itinerant electrons to the metallic bonding varies with the magnetic state and so will do the resulting lattice spacing.

2. Numerical results and discussion

The phase diagram for the model given by (4) is obtained as a function of the conduction electron density x ($0 \leq x \leq 0.5$ because of hole-electron symmetry) and the SE interaction energy J. $T = 0K$ and open boundary conditions on a linear chain of $N = 24$ sites were used. $N-1$ angles $\theta_{i,i+1}$ and $N-1$ values of $\delta_{i,i+1}$ had to be optimized. For this goal, we used an analytical optimization and a classical Monte Carlo method. The analytical solution was tested as a starting point in the Monte Carlo simulation. The whole magneto-elastic phase diagram obtained here is shown in figure 1(a). Partial spin configuration snapshot of the most important phases is shown in figure 1(b). For low super-exchange interaction energy, $JS^2/t \lesssim 0.11$, an electron-full ferromagnetic distorted (EFD) and an electron-empty anti-ferromagnetic undistorted (EAFU) phase transition was found. Analytical optimization implies angles $\theta_{i,i+1} = 0$ exactly for the EFD phase. For commensurate electronic fillings at $x = 1/3$ and at $x = 1/2$ displacements look like Peierls distortion. Because of the typical value of the elastic energy, uniform displacements ($\delta_U = \sin(\pi x)/(\pi B/t)$) and uniform charge distribution ($n_i = x$) are expected in the thermodynamic limit ($N \to \infty$). The $\delta_{i,i+1} = 0$ displacement and $\theta_{i,i+1} = \pi$ angle solutions were obtained for EAFU phase at $x = 0$. Phase separation (EAFU+EFD in figures 1(a) and 2) between EFD and EAFU phases, consist of one large electron-full F distorted polaron within an electron-empty AF undistorted background. Charge distributions n_i are also presented. Analytical optimization implies angles 0 and π exactly for the ferromagnetic and anti-ferromagnetic domains respectively. In the limit $B/t \to \infty$, the magnetic-only F-AF phase transition was previously reported [7]. Above EAFU+EFD phase separation, another phase separation (EAFU+DP3) between T-phase (for high conduction electron density) and EAFU phase (for $x = 0$) can be observed in figure 1(a). T-phase is a more general complex distorted phase found by the Monte Carlo method and can be polaronic like or not. At $x = 1/2$

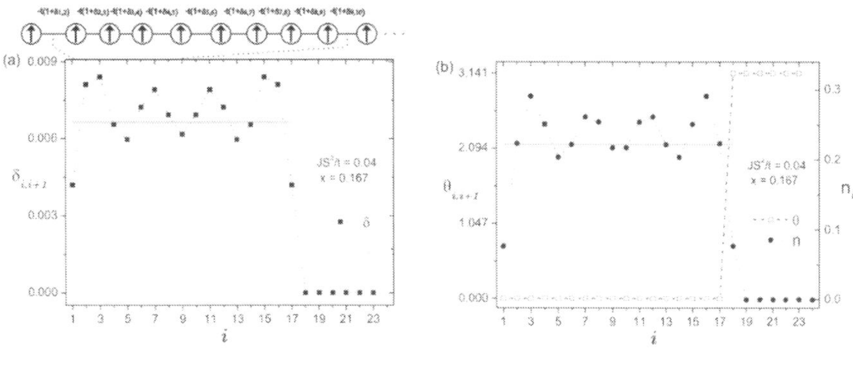

Figure 2. EAFU+EFD phase separation for 4 electrons ($x = 0.167$) and $JS^2/t = 0.04$ showing: (a) $\delta_{i,i+1}$ and (b) angles and charge distribution. Partial spin configuration snapshot is also shown.

and at $x = 1/3$ T-phase becomes basically DP2 (see figure 3) and DP3 phases respectively. DP2 and DP3 phases are two and three-site-distorted periodic arrangement of independent F polarons ordered anti-ferromagnetically. DP2 and DP3 are magneto-elastic Peierls phases. Displacements are $(2B/t)^{-1} \sim 0.0167$ and $(2\sqrt{2}B/t)^{-1} \sim 0.0118$ for DP2 and DP3 phases respectively. EAFU+DP3 phase separation is degenerate with phases where the polarons can be ordered or not, while keeping the number of F and AF bonds fixed; phases obtained within the "spin-induced Peierls instability" [21] belong to this class. The former degeneracy unifies ideas like phase separation and individual polarons and gives a natural response to the instability at the Fermi energy and to an infinite compressibility as well. It is worth to notice that in the magnetic-only case, a phase AF+P3 (similar to EAFU+DP3) was identified using $S = 3/2$ quantum spins [22]. P2 and P3 phases were also previously reported for classical [7, 21] and $S = 1/2$ quantum [17] local spins. Figure 1(a) shows EAFU+DP3 phase separation for a typical value of the SE interaction energy $JS^2/t = 0.14$. There is a single electron inside each three-site independent magnetic polaron, so a Wigner crystallization is formed. For DP3 and EAFU+DP3 phases, we find lattice distortion within each three-site magnetic polaron. For high SE interaction energy, DP3 and DP2 become DCP3 and DCP2 canted phases (distorted canted P3 and P2 phases respectively). In the absence of electron-lattice interaction (limit $B/t \to \infty$), the CP3 phase present a continuos angular degeneracy [7]. Now, this degeneracy is broken by the lattice distortion. Instead of the continuos degeneracy, only four set of angles inside each CP3 polaron were found [23]. For instance, DCP3 phase in figure 1(a) is a distorted canted phase with angles 0 and θ. Displacements decrease when $\theta \to \pi$. The expected DP2-DCP3, DCP3-EAFU and DP2-EAFU phase transitions are also obtained for values of the SE interaction energy above the T-EAFU phase transition (see figure 1(a)). Finally, the DCP2-EAFU phase transition occurs for high SE interaction energy $JS^2/t > 0.2542$. DCP2 phase becomes AFU (anti-ferromagnetic undistorted) in the limit $JS^2/t \to \infty$. In this phase each electron is trapped in a single site forming a Wigner crystallization. It is important to mention that the size chosen for the linear chain (N=24 sites), does not change the nature of the phases involved in the phase diagram.

In conclusion we have studied the rich phase diagram resulting from the interplay between magnetic interactions and lattice distortion within an exchange model in one-dimensional systems using large Hund's coupling and classical localized spins. Basically, microstuctured phases with small ferromagnetic polarons result from the double and super-exchange interaction. Our results for low SE interaction energy show phase separation between ferromagnetic and anti-ferromagnetic phases. In this case, the ferromagnetic domain contains all the electrons

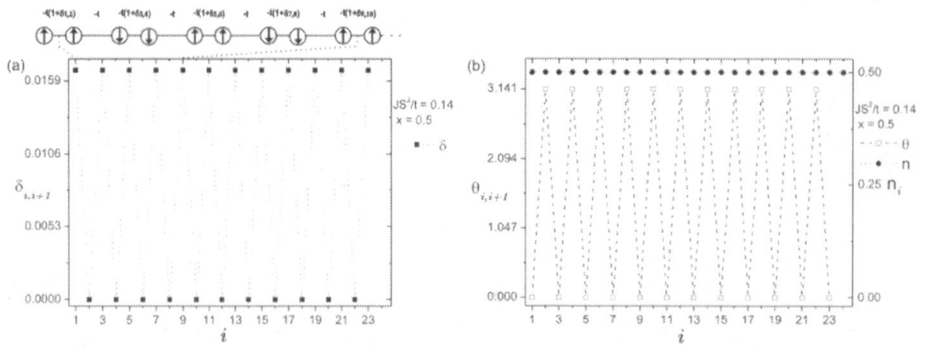

Figure 3. DP2 phase at $x = 0.5$ showing (a) $\delta_{i,i+1}$ and (b) angles and charge distribution. Partial spin configuration snapshot is also shown.

and a lattice contraction occurs only within this domain. For larger SE interaction energy, we found phase separations involving small two and three-site distorted polarons, in which a Wigner crystallization can be identified. The important magneto-elastic effect obtained here leads to local bond contractions that consequently change the lattice parameters which should be observable. We expect this effect to occur in low-dimensional systems in which magnetic ions coupled via a double-exchange type interaction are present, as for example molecular magnets, halogen bridged metal chains or charge transfer salts. Ladders in the ludwigite family seem to be good candidates also.

Acknowledgments

We want to acknowledge partial support from CONACyT Grant-57929 and PAPIIT-IN108907 from UNAM. E. V. acknowledge DGAPA-UNAM for financial support.

References

[1] Zener C 1951 *Phys. Rev.* **82** 403
[2] Zener C 1951 *Phys. Rev.* **81** 440
[3] Jonker G H and Van Santen J H 1950 *Physica* **16** 337
[4] Jonker G H and Van Santen J H 1950 *Physica* **16** 599
[5] Anderson P W and Hasegawa H 1955 *Phys. Rev.* **100** 675
[6] de Gennes P G 1960 *Phys. Rev.* **118** 141
[7] Vallejo E, López-Urías F, Navarro O and Avignon M 2008 *J. Magn. Magn. Mater.* **320** e425
[8] Vallejo E, López-Urías F, Navarro O and Avignon M 2009 *Solid State Commun.* **149** 126
[9] Su W P, Schrieffer J R and Heeger A J 1979 *Phys. Rev. Lett.* **42** 1698
[10] Su W P, Schrieffer J R and Heeger A J 1980 *Phys. Rev.* B **22** 2099
[11] Vallejo E and Avignon M 2006 *Phys. Rev. Lett.* **97** 217203
[12] Vallejo E and Avignon M 2007 *Rev. Mex. Fís.* S **53** (7) 1
[13] Vallejo E and Avignon M 2007 *J. Magn. Magn. Mater.* **310** 1130
[14] Mir M *et al.* 2001 *Phys. Rev. Lett.* **87** 147201
[15] Mir M, Janczak J and Mascarenhas Y P 1996 *J. Appl. Cryst.* **39** 1356
[16] Müller-Hartmann E and Dagotto E 1996 *Phys. Rev.* B **54** R6819
[17] Garcia D J *et al.* 2000 *Phys. Rev. Lett.* **85** 3720
[18] Garcia D J *et al.* 2002 *Phys. Rev.* B **65** 134444
[19] Yunoki S *et al.* 1998 *Phys. Rev. Lett.* **80** 845
[20] Dagotto E *et al.* 1998 *Phys. Rev.* B **58** 6414
[21] Koshibae W, Yamanaka M, Oshikawa M and Maekawa S 1999 *Phys. Rev. Lett.* **82** 2119
[22] Neuber D R *et al.* 2006 *Phys. Rev.* B **73** 014401
[23] Vallejo E 2008 *Microelectron. J.* **39** 1266

AUTHOR INDEX

AUTHOR INDEX

AUTHOR INDEX